ILSI Human Nutrition Reviews

Series Editor: Ian Macdonald

Zinc in
Human Biology

Edited by Colin F. Mills

With 19 Figures

Springer-Verlag Berlin Heidelberg GmbH

Colin F. Mills, MSc, PhD, CChem, FRSC, FRSE
Director of Postgraduate Studies, Rowett Research Institute,
Bucksburn, Aberdeen, AB2 9SB, Scotland, UK

Series Editor
Ian Macdonald, MD, DSc, FIBiol
Department of Physiology, Guy's Hospital Medical and Dental
School, St Thomas' Street, London, SE1 9RT, UK

ISBN 978-1-4471-3881-5

British Library Cataloguing in Publication Data
Mills, C.F. (Colin Frederick), *1926–*
   Zinc in human biology.
   1. Man. Health. Effects of zinc
   I. Title 613.2'8
ISBN 978-1-4471-3881-5   ISBN 978-1-4471-3879-2 (eBook)
   DOI 10.1007/978-1-4471-3879-2
Library of Congress Cataloging-in-Publication Data
Zinc in Human Biology.
   (ILSI human nutrition reviews)
   Bibliography: p.
   Includes index.
   1. Zinc in the body. 2. Zinc—Metabolism.
3. Zinc deficiency diseases. I. Mills, Colin F.
(Colin Frederick), 1926–. II. Series.
QP535.Z6Z57 1989   612'.01524   88–24817
ISBN 978-1-4471-3881-5

© Springer-Verlag Berlin Heidelberg 1989
Originally published by Springer-Verlag Berlin Heidelberg New York in 1989
Softcover reprint of the hardcover 1st edition 1989

The use of registered names, trademarks etc. in this publication does not imply,
even in the absence of a specific statement, that such names are exempt from the
relevant laws and regulations and therefore free for general use.

Product Liability: The publisher can give no guarantee for information about drug
dosage and application thereof contained in this book. In every individual case the
respective user must check its accuracy by consulting other pharmaceutical literature.

Filmset by Photo-Graphics, Honiton, Devon

2128/3916–543210 (Printed on acid-free paper)

Foreword

The present volume is one of a series concerned with topics considered to be of growing interest to those whose ultimate aim is the understanding of the nutrition of man. Volumes on *Sweetness, Calcium in Human Biology* and *Sucrose: Nutritional and Safety Aspects*, have already been published, and another, on *Dietary Starches and Sugars in Man: A Comparison*, is in preparation.

Written for workers in the nutritional and allied sciences rather than for the specialist, they aim to fill the gap between the textbook on the one hand and the many publications addressed to the expert on the other. The target readership spans medicine, nutrition and the biological sciences generally and includes those in the food, chemical and allied industries who need to take account of advances in these fields relevant to their products.

Funded by industry but with an independent status, the International Life Sciences Institute (ILSI) is a non-profit organization founded to deal objectively with the numerous health and safety issues that today concern industry internationally. ILSI sponsors scientific research, organizes conferences and publishes monographs relative to these problems.

London
March 1988

Ian Macdonald
Series Editor

Preface

This volume has been prepared at a time when interest in both the biological roles of zinc and its nutritional significance is growing rapidly.

The essentiality of zinc for animals has been beyond dispute for more than 50 years. The extent to which deficiencies or excesses of zinc influence human health and wellbeing has been the subject of more extensive controversy. Such controversy has two fundamental causes. No other element in the periodic table has had ascribed to it such a wide range of biochemical functions. Despite this, difficulty still remains in relating these known functions to diagnostically specific signs of zinc deficiency or excess in man. The second cause of controversy lies in our limited ability to interpret data on the dietary supply and availability of zinc and its relationship to health.

This is a time of intense debate on the influence of diet or health and on the wisdom or folly of "pill popping" with vitamin or trace element supplements to rectify real or imagined micronutrient deficiencies or toxicities arising from the diet or environment. As with many of the essential trace elements a mixture of justifiable and extravagant claims has been made regarding the value of zinc for health. However, behind this confusion lies a growing weight of evidence concerning the importance of zinc as a regulator of growth, development and of responses to injury and infection. This volume has addressed the problems of identifying or anticipating circumstances in which defects in the metabolism or supply of zinc can be of significance for the health of man.

Although each specialist contributor has been invited to address a specific aspect of these questions, they were also encouraged to review critically those studies of zinc in experimental animals which are likely to have human relevance. The resulting chapters paint a picture of a micronutrient with a multitude of vital functions but which, in many instances, has presented particular difficulties in the diagnosis of deficiency or excess in human populations.

No restraint has been placed upon authors to avoid controversial topics when discussing the biological importance of zinc. The emergence of such controversy is typified by that surrounding the behavioural implications of zinc deficiency (regarded as secondary manifestations of a general malaise by some but, by others, as specifically attributable to metabolic changes induced by lack of zinc). Perhaps the very existence of such controversies reflects a need for greater care in the selection and interpretation of critical experimental approaches in studies of such complex features of nutritional disorders.

Similar opportunities for improving the precision with which we approach typical nutritional problems emerged during editing of those sections of this volume which consider either requirements for zinc or the efficiency with which dietary zinc is utilized. Virtually all editors of nutritional texts which deal with these topics face one problem in common – a semantic fog, the origin of which is the technical difficulty of precise experimentation in fields in which net effects are governed by a wide range of biological variables. Contributors to this volume encountered these age-old problems. Your Editor was faced with two alternatives. One was to enforce arbitrary uniformity of terminology upon contributors. The other was to accept the typically wide variety of terms used to describe subtle and ostensibly different but frequently unmeasurable parameters associated with nutrient needs, nutrient flux or nutrient utilization. The second option, less prejudicial to the Editor's relationships with his Authors, was selected. Doing so should not be construed as accepting without question the wisdom of a technical nomenclature which encourages use of ill-defined terms. These often obscure the technical and interpretive limitations of many techniques for determining nutrient requirements and availability. The challenges presented by the terminology used in these areas of nutritional investigation are reminiscent of Lewis Carroll.

"When I use a word" Humpty Dumpty said "it means what I choose it to mean neither more nor less – *adjectives* you can do anything with". "The question is", said Alice "whether you can make words mean so many different things".

(Lewis Carroll, 1871, *Through the Looking Glass*)

Alternatives to the use of at least eight differing qualifying terms to define requirements are suggested in the final chapter of this book. Similar appeals also could have been included to discontinue usage of adjectives such as "apparent" and "true" to qualify measurements of absorption. The former confers an unwarrantable dignity upon measurements whose limitations are rarely evident until an attempt is made to use them as absolute criteria from which to predict the nutritional adequacy of diets.

Although such arguments and their influence on our views as to the nutritional significance of zinc for health will be fully revealed in the following chapters, they will be accompanied by abundant evidence of the importance of zinc in human biology. Together, such topics reveal not only the challenges presented by studies of the roles of this microelement but also the importance of providing new investigational techniques whereby its influence on health and development can be estimated more readily than at present. The extensive technical resources needed to undertake such studies may limit our opportunities and slow our progress. Despite this, it is clear that adequate appraisal of the significance of zinc is contingent upon detailed quantitative investigation of the many factors that influence its impact on growth, development and health.

Finally, I would like to express my gratitude to authors and the publishers for the patience and understanding that has made this volume possible.

Aberdeen   C. F. Mills
May 1988

Contents

Contributors

Dr. P. J. Aggett
Department of Child Health, Aberdeen University Medical
School, Aberdeen, AB9 2ZD, Scotland, UK

Dr. I. Bremner
Rowett Research Institute, Bucksburn, Aberdeen, AB2 9SB,
Scotland, UK

Professor G. E. Bunce
Department of Biochemistry and Nutrition, Virginia Polytechnic
Institute and State University, Blacksburg, Virginia 24061, USA

Dr. C. Castillo-Duran
Institute of Nutrition and Food Technology, University of Chile,
PO Box 15138, Santiago, Chile

Dr. J. K. Chesters
Rowett Research Institute, Bucksburn, Aberdeen, AB2 9SB,
Scotland, UK

M. S. Clegg
Department of Nutrition, University of California, Davis,
California 95616, USA

Professor R. J. Cousins
Food Science and Human Nutrition Department, Institute of
Food and Agricultural Sciences, University of Florida,
Gainesville, Florida 32611, USA

Dr. I. E. Dreosti,
Commonwealth Scientific and Industrial Research Organisation,
Division of Human Nutrition, Kintore Avenue, Adelaide, SA
5000, Australia

Dr. Sandra E. File
The School of Pharmacy, University of London, 29/39 Brunswick
Square, London, WC1N 1AX, England, UK

Dr. M. R. S. Fox
Division of Nutrition, Food and Drug Administration,
Washington, DC 20204, USA

Dr. Barbara E. Golden
Tropical Metabolism Research Unit, University of the West
Indies, Mona, Kingston 7, Jamaica, West Indies

Dr. M. H. N. Golden
Tropical Metabolism Research Unit, University of the West
Indies, Mona, Kingston 7, Jamaica, West Indies

Professor R. A. Good
Department of Pediatrics, All Children's Hospital, University of
South Florida, St. Petersburg, Florida 33701, USA

Professor K. M. Hambidge
University of Colorado Health Sciences Center, 4200 East Ninth
Avenue, Denver, Colorado 80262, USA

Dr. Lucille S. Hurley *
Departments of Nutrition and Internal Medicine, University of
California, Davis, California 95616, USA

Dr. M. J. Jackson
University Department of Medicine, Royal Liverpool Hospital,
PO Box 147, Liverpool, L69 3BX, England, UK

Dr. C. L. Keen
Departments of Nutrition and Internal Medicine, University of
California, Davis, California 95616, USA

Professor Janet C. King
Department of Nutritional Sciences, University of California,
Berkeley, California 94720, USA

Professor B. Lönnerdal
Departments of Nutrition and Internal Medicine, University of
California, Davis, California 95616, USA

Dr. P. M. May
School of Mathematical and Physical Sciences, Murdoch
University, Murdoch, Western Australia 6150, Australia

Dr. C. F. Mills
Rowett Research Institute, Bucksburn, Aberdeen, AB2 9SB,
Scotland, UK

Professor B. L. O'Dell
Department of Biochemistry, University of Missouri, Columbia,
Missouri 65211, USA

Dr. P. G. Reeves
Department of Biochemistry, University of Missouri, Columbia,
Missouri 65211, USA

M. Ruz
Department of Family Studies, University of Guelph, Guelph,
Ontario, Canada

Dr. A. Brittmarie Sandström
Research Department of Human Nutrition, Royal Veterinary and
Agricultural University, Rolighedsvej 25, DK-1958 Fredericksberg
C, Denmark

Dr. N. W. Solomons
Center for Studies of Sensory Impairment, Ageing and
Metabolism, Hospital de Ojos y Oidos, Diagonal 21 y 19 Calle,
Zona 11, Guatemala City, Guatemala

Dr. Judith R. Turnlund
USDA and Western Human Nutrition Research Center, Albany,
California 94710, USA

Professor R. J. P. Williams
Inorganic Chemistry Laboratory, Oxford University, South Parks
Road, Oxford, OX1 3QR, England, UK

Dr. R. L. Willson
Department of Biology and Biochemistry, Brunel University of
West London, Uxbridge, UB8 3PH, England, UK

Chapter 1

Physiology of Zinc: General Aspects

M.J. Jackson

Introduction

Zinc is one of the so-called trace elements present in the body, but in nature it is present in the Earth's crust to the extent of about 0.02% and is 23rd in the order of abundance of the elements. In this context it is far scarcer than many elements which are less familiar to biologists today (e.g. zirconium, vanadium or strontium). That zinc is essential for normal life in both humans and animals is now beyond doubt, having been first shown to be required for normal growth in rats and mice in the 1930s (Todd et al. 1934; Bertrand and Bhattacherjee 1935).

The following chapter concentrates on the unique and extensive role that zinc plays in the biochemistry of enzymes and other biological molecules, but it must be remembered that, by utilizing zinc in such a role, evolution has provided the body with the complex task of ensuring the delivery of the element to the correct site in the correct concentration, in order to perform the appropriate function. The body therefore has developed sophisticated mechanisms to remove zinc from dietary constituents (where little is likely to be found in the free form), to absorb this element (apparently by a specific mechanism) from a milieu which contains 30–40 other elements in an appropriate quantity, to prevent either depletion or excess, to transport it safely to all tissues of the body and to its site of action within these tissues, to maintain the appropriate quantity in tissues over an indefinite period of time despite loss from tissues and turnover of biological substances which bind zinc (proteins, lipids etc.) and finally to excrete appropriate amounts of zinc safely from the body.

In spite of the apparent complexity of these problems, overt zinc deficiency or excess is relatively rare in man and animals despite quite large variations in the intake of this element and of other substances which effect its metabolism (i.e. other cations and macromolecules). This in itself is excellent evidence that the body has developed efficient homeostatic mechanisms for the control of body zinc levels, as has now been demonstrated in a number of experimental studies (e.g. see McCance and Widdowson 1942; Cotzias et al. 1962; Miller 1969; Jackson et al. 1980; Jackson et al. 1984).

The use of zinc for the functions to be described in the following chapter has presumably evolved because of its suitability for these biochemical roles and because of its availability in the ecosystems of the lower organisms and early mammals from which man evolved. This implies that zinc must have been present in adequate quantities in the food chain for the roles placed upon it and that any contemporary claims for a widespread occurrence of nutritional zinc deficiency in man must imply a fundamental change in this situation.

The aim of this chapter is to provide an overview of the distribution and handling of zinc by the body and of the homeostatic systems which it has developed to maintain zinc levels, together with a consideration of the ways which these systems are affected by various physiological and pathological stimuli.

Distribution of Zinc within the Body

Zinc is present in all organs, tissues, fluids and secretions of the body. The average concentrations of zinc found in the major tissues of normal man are shown in Table 1.1, together with a calculation of the total amount of zinc present in each tissue. Calculations such as these are inherently inaccurate as they assume that compositional values obtained from biopsy samples or sections of tissues contain quantities of zinc which reflect the levels in other parts of the same tissue and in different individuals. In practice this is unlikely to be true and the values obtained only provide a general guide; nevertheless the

Table 1.1. Approximate zinc content of major organs and tissues in normal adult man (70 kg)

Tissue	Approximate zinc concentration (μg/g wet wt)	Total zinc content (g)	Proportion of total body zinc (%)
Skeletal muscle	51	1.53	57 (approx)
Bone	100	0.77	29
Skin	32	0.16	6
Liver	58	0.13	5
Brain	11	0.04	1.5
Kidneys	55	0.02	0.7
Heart	23	0.01	0.4
Hair	150	<0.01	0.1 (approx)
Blood plasma	1	<0.01	0.1 (approx)

Data derived from Spray and Widdowson (1950), Widdowson et al. (1951), Lehmann et al. (1971), Documenta Geigy (1975), Underwood (1977), Jackson et al. (1982).

data in Table 1.1 serve as an indicator of the relative distribution of zinc in different tissues of man.

Because of the large bulk of skeletal muscle this tissue contains the greatest portion of the body zinc although its actual concentration is not excessively large. Together with bone, these two tissues apparently account for more than 80% of the total body zinc. It is also apparent from the figures shown that zinc is a primarily intracellular ion with intracellular zinc contributing well over 95% of the total body zinc. It can therefore be assumed that at least some of the important sites of zinc function are intracellular and in studies of the possible roles of zinc in human pathology an assessment of intracellular zinc content is likely to be extremely important.

The concentration of zinc in extracellular fluids is relatively low (e.g. blood plasma contains only approximately 1 µg zinc/ml) compared to intracellular stores. If the total amount of blood plasma is taken to be 45 ml/kg body weight (Documenta Geigy 1975) then a 70-kg man will have approximately 3 l of blood plasma which will contain only about 3 mg of zinc or about 0.1% of the total body content. This has major implications for the assessment of total body zinc status since the vast majority of studies to examine zinc status have utilized the measurement of plasma zinc concentrations.

Since the total amount of zinc present in the major tissues is much larger than the total present in plasma, relatively small variations in the zinc content of tissues such as liver can have dramatic effects on plasma zinc. An increase of the zinc content of the liver by 1% (or 1.3 mg zinc in total) could deplete the plasma zinc by over 40% but if distributed evenly would increase the liver concentration by less than 1 µg/g wet weight, which would not be detectable by available analytical techniques. Relatively small redistributions in body zinc may therefore cause large variations in plasma zinc. Likewise, since all absorbed zinc must pass via the plasma to the tissues and is likely to be in the order of 5 mg per d (assuming a dietary intake of 15 mg per d and an absorption rate of 30%), it can be seen that if plasma zinc levels are to remain relatively constant there must be a rapid flux of zinc through the plasma (Chesters and Will 1981). Unless the plasma zinc is under close homeostatic control and is capable of being maintained from some intracellular store, a rapid elevation or depression of the plasma zinc will result from an increase or decrease in the dietary zinc intake. This situation has been shown to occur following zinc deprivation in animals where very rapid depression of the plasma zinc concentration is known to follow the introduction of a zinc-depleted diet (Leucke et al. 1968; Wilkins et al. 1972).

An analogy can be drawn between zinc and the major intracellular cation, potassium. It is widely known that potassium is a primarily intracellular ion and measurements are made of both plasma potassium as an index of extracellular potassium content and of whole body potassium (either by counting the naturally occurring ^{40}K or by isotope dilution techniques with radioactive potassium) as an index of intracellular stores (Lye and Winston 1979). However no such useful non-invasive means of assessing intracellular zinc exist for use in man. Measurements of the zinc of leucocytes or of other formed elements of blood (Jones et al. 1981) may provide access to this pool but it should be noted that again the total quantity of zinc present in these cells only represents a very small proportion of the total body zinc in man. The problem of assessing the physiological relevance of changes in the zinc status of man will be discussed later in the chapter by Dr. M.H.N. Golden (Chap. 20). Meanwhile, the fundamental limitations of the techniques we have available, currently, for assessment of zinc status should also be borne in mind.

Distribution of Zinc within Tissues

There is a considerable lack of information concerning the distribution of zinc within tissues and the nature of its intracellular binding. It appears that at least a small amount of zinc is found within all organelles of the cell but this is to be expected since zinc-containing enzymes are so widely distributed. In 1959 Bartholomew et al. demonstrated that injected ^{65}Zn was distributed within the nuclear, mitochondrial and supernatant fractions of mouse liver with the largest proportion in the supernatant. This finding has been confirmed by other workers (Alfonzo and Heaton 1974; Smeyers-Verbeke et al. 1977) and in addition, Bettger and O'Dell (1981) have suggested that there is a fraction of the intracellular zinc specifically bound to membranes which may be very important in terms of the onset of deficiency symptoms. Zinc thus appears to be ubiquitously distributed within cells although whether any particular subcellular fraction pool is more susceptible to the effects of zinc depletion remains to be fully explored.

The nature of the binding of zinc within cells is much less clearly understood. Currently available biochemical techniques invariably involve destruction of the cells prior to analysis of the zinc-binding substances and there is thus opportunity for zinc to redistribute during the analysis. Nevertheless it appears that the major proportion of the zinc within cells is protein-bound (e.g. for liver, see Cousins 1985; for muscle, see Jeffreys et al. 1982) and it is unclear whether any substantial amounts of free zinc or zinc bound to amino acids exist within cells. Dialysis experiments and experiments with isolated proteins (Bartholomew et al. 1959) indicate that different proteins have different affinities for zinc and it appears that zinc may have a controlling role in certain enzymes such as fructose-1,6-diphosphatase (Pedrosa et al. 1977) rather than be a firmly bound and integral part of the molecule. However if zinc can play such a controlling role within cells this implies that there is a "pool" from which the zinc is readily available or some binding substances which facilitate the activation of the enzyme by zinc. The nature of these pools or substances is presently undefined.

Factors Affecting Body Distribution of Zinc

Development and Growth

There are substantial differences in the relative sizes of tissues in neonates compared with adult man together with differences in the zinc content of certain tissues. In particular, the zinc content of the liver is much larger in the newborn infant than in adult man (Shaw 1979) and the relative size of the liver is much larger. Widdowson et al. (1972) have therefore calculated that the liver zinc contributes about 25% of the total body zinc in neonates, compared with the 6% calculated for adult man in Table 1.1. The purpose of this apparent accumulation is unknown. In part, it reflects the relative increase in size of the liver at birth, but an elevated concentration of zinc in neonatal liver has been described in various species and is at least partly due to an increase in the metallothionein-bound zinc in the liver (Bremner et al. 1977).

An increased proportion of body zinc is probably also present in the bones of newborn infants compared with adults. Shaw (1979) has calculated that this may represent up to 40% of the body zinc in term infants and has suggested that much of this may become available for soft tissue growth if zinc is scarce after birth since neonatal bones undergo extensive remodelling after birth. Since liver and bone contribute such a large proportion of body zinc at birth in man, it is apparent that the proportion contributed by skeletal muscle must be dramatically reduced in comparison with adults. Cassens and co-workers have demonstrated that highly oxidative (red) skeletal muscles contain considerably (three to five times) more zinc than fast glycolytic (white) muscles (Cassens et al. 1967). The major site of this extra zinc in oxidative muscles appears to be as an essential component of one of the isoenzymes of carbonic anhydrase (carbonic anhydrase III) which is found in large quantities in oxidative skeletal muscle (Jeffreys et al. 1982). Since foetal and newborn muscles are relatively underdeveloped and contain much less carbonic anhydrase III than adult muscle (Carter et al. 1979) it can be seen that muscle zinc must contribute a smaller proportion of total body zinc in neonates than in adults.

It is therefore apparent that during development there are considerable alterations in the distribution of zinc in the body. Some of this can be explained by the changes in the relative sizes of organs and tissues which occur or in the necessary development of the organs, i.e. in the increase of the oxidative capacity of the muscle. However the relatively large amounts of zinc in the liver at birth can only be partially explained in these terms. The liver is the major organ of zinc metabolism in the body (Cousins 1985) and is intimately involved in the mechanisms by which the body handles zinc. It is therefore tempting to speculate that the liver accumulation of zinc in utero acts as a buffer against the possibility of relative deficiency at birth or soon after but there are few experimental data in support of this. Other alternative explanations can be presented such as that the processes for "export" of protein-bound zinc from the liver are immature in utero and immediately after birth which could lead to a net accumulation of zinc because of the diminished ability of the liver to excrete the ion.

Hormones and Stress

The distribution of zinc between extracellular fluid pools and tissues appears to be sensitive to variations in hormonal balance and to various stress situations. Much of the experimental work in elucidating the metabolism of zinc by the liver has been undertaken by Cousins and co-workers and he has stated that "hormonal interactions on zinc metabolism suggest that changes in liver zinc metabolism are collectively mediated, at least partly by glucagon, insulin and glucocorticoids" (Cousins 1985). Since the liver is the major organ of the body for zinc metabolism, the actions of these hormones will have major effects on whole-body zinc metabolism.

Glucocorticoids stimulate zinc uptake by HeLa cells (Cox 1969), hepatic cells in culture (Failla and Cousins 1978a, b) and probably by the intact human liver (Henkin et al. 1984) and corticosteroids are known to decrease the serum or plasma zinc concentration (Flynn et al. 1973) as did infusions of adrenocorticotropic hormone (ACTH) in humans (Falchuk 1977). Glucagon also stimulated zinc uptake and/or exchange by rat liver parenchymal cells (Kuipers and Cousins 1984).

In addition to these well-described effects of hormones on zinc metabolism by the liver, significant links between zinc and growth hormone (Kirchgessner and Roth 1985) and zinc and various sex hormones have been proposed (Prasad 1982; Abbassi et al. 1980).

It has been known for a considerable number of years that infection or other stresses could depress plasma zinc concentration. Both Weinberg (1974) and Chvapil (1976) have suggested important physiological roles which the reduction in the plasma zinc content could play in protection of the body against stress. The mechanisms by which the effect occurs appear to be at least partially hormonally controlled. Falchuk (1977) suggested that the fall in plasma zinc was entirely due to a decrease in the non-α_2-macroglobulin-bound fraction and may be mediated by changes in ACTH and Wannemacher et al. (1975) suggested the defect was mediated by a low-molecular-weight polypeptide hormone (leucocyte endogenous mediator) which mediated the redistribution of zinc by intrahepatic sequestration. It now appears that this mediator is interleukin-1 and that it stimulates a redistribution of plasma zinc to liver in many "stress" situations. It appears to exert this effect via both glucagon and glucocorticoids and possibly by a direct action on the liver. The net effect appears to be the induction of synthesis of metallothionein and a sequestration of some of the circulating zinc in the plasma (see Cousins 1985, for review).

"Stress" can therefore lead to a redistribution of zinc from plasma to liver. However, the total amount of zinc involved is likely to to be very small since, as previously stated, a depression of the plasma zinc by greater than 40% can be achieved by removal of less than 2 mg of zinc from the plasma of a 70–kg man. However it may be important not to confuse the low plasma zinc caused by this redistribution with that due to zinc deficiency, since administration of zinc supplements was found to be lethal to some pigs in which a low plasma zinc had been induced by the stress of endotoxin infection (Chesters and Will 1981).

If the "stress" is sufficient to cause tissue catabolism it may also have other effects on zinc metabolism. Severe muscle catabolism leads to a substantial loss of zinc in the urine (Cuthbertson et al. 1972; Fell et al. 1973) and, in severe cases, in the faeces (Jackson and Edwards 1982). However it appears that this loss is entirely accounted for by the reduction in muscle tissue bulk (see Fig. 1.1) and will not lead to any depression of the "active" zinc pool in the body. Indeed, in cases of severe zinc depletion in experimental animals, catabolism of muscle tissue may release zinc for other uses (Pierce et al. 1985). A reduced dietary calcium supply also appears to facilitate the release of zinc bound within bone and markedly decreases the incidence of foetal malformations in zinc-deficient pregnant rats (Tao and Hurley 1975). Further evidence that the balance between catabolism and anabolism has a dramatic effect on zinc supply comes from animal studies in which the plasma zinc concentration was found to increase during dietary protein restriction and decrease during dietary

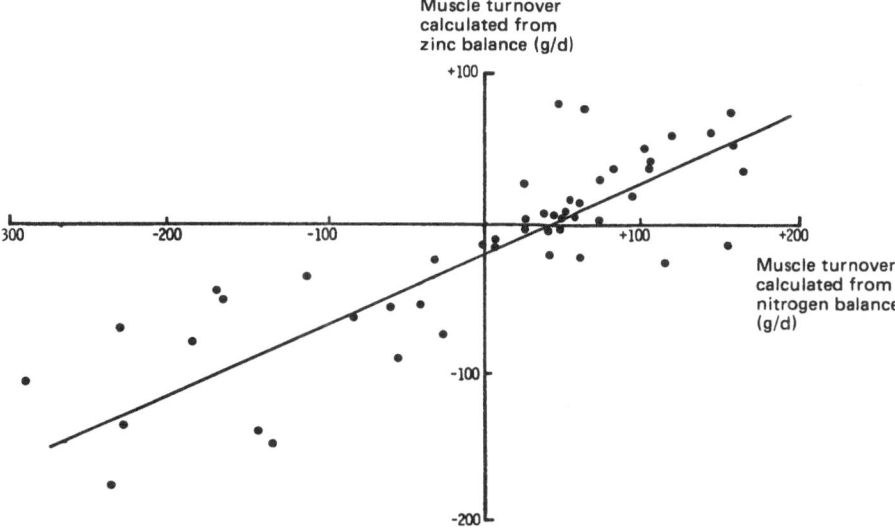

Fig. 1.1. A plot of muscle turnover calculated from zinc balance against the muscle turnover calculated from the nitrogen balances assuming that human muscle contains 51 μg zinc/g wet weight. Points represent individual balance periods for three patients during periods of acute muscle degeneration and regeneration. Reproduced with permission from Jackson and Edwards (1982).

protein repletion (Mills 1973). These data, together with evidence of a significant inverse relationship between food intake and the plasma zinc concentration (Chesters and Will 1973), suggest that zinc flux into tissues is enhanced by anabolic processes and zinc efflux into plasma from tissues is promoted by tissue catabolism (Mills 1981).

Inherited Disorders of Zinc Metabolism

There are few data available on any redistribution of zinc which occurs due to inherited disorders of zinc metabolism in man. Only two overt inherited disorders are known; acrodermatitis enteropathica (Danbolt and Closs 1942) which manifests itself as a severe dermatological disorder, apparently due to zinc deficiency (Moynahan 1974), and familial hyperzincaemia (Smith et al. 1976). In the latter condition no apparent ill-effects have been found but the affected family members have a plasma zinc level of up to six times normal. This is an extremely large increase in the plasma zinc and is equivalent to the concentration found in a patient who had died from an inadvertent intravenous injection of 2 g of zinc (Brooks et al. 1977). However, the lack of any toxicity symptoms in this inherited disorder may mean that the defect is confined to the plasma and that only the plasma concentration is elevated to such a significant extent. In this case the increase may only represent an increase of approximately 15 mg in the total body zinc content. Unfortunately no examinations of tissue levels of zinc appear to have been undertaken in subjects with this disorder so that estimates of the total increase in body burden cannot be made.

On the other hand the cause and extent of the defect in zinc metabolism which occurs in acrodermatitis enteropathica has been the subject of considerable study and speculation. It is now generally accepted that the pathogenic defect in this disease is an abnormality in zinc absorption from the gut (Lombeck et al. 1975; Atherton et al. 1979) leading to zinc deficiency (Moynahan 1974). It can be corrected by oral zinc supplements. In untreated patients with the disease the plasma zinc concentration is usually abnormally low (Michaelsson 1974; Moynahan 1974; Portnoy and Molokhia 1974; Nelder and Hambidge 1975), but this is not a universal finding (Garretts and Molokhia 1977). Abnormally low amounts of zinc in urine (Nelder and Hambidge 1975) and hair (Amador et al. 1975) have also been reported. There is therefore good evidence for a depletion of extracellular zinc and hair zinc in most patients with acrodermatitis enteropathica. While these low levels do not appear to be repleted during therapy with diodoquin [the drug used prior to zinc which effectively ameliorates the clinical symptoms (Strain et al 1975; Nelder and Hambidge 1975)], they are corrected by oral zinc therapy (Moynahan 1974; Nelder and Hambidge 1975; Walravens et al. 1978).

There have been few attempts to measure intracellular zinc levels in this disorder. Walravens et al. (1978) have shown that the red cell zinc content was either low or normal in three patients and that this failed to respond to short-term zinc therapy. They also described an abnormally low content of zinc in a skin biopsy sample from macroscopically normal skin in a breast-milk-fed patient during remission. However, Garretts and Molokhia (1977) described a patient in whom the skin, plasma and hair zinc contents were normal prior to treatment and yet he responded clinically to oral zinc supplements and relapsed clinically once zinc supplements were withdrawn. Measurements of muscle zinc content appear to have been undertaken in only one (treated) patient with this disorder; values were found to be within the normal range on three occasions despite large variations in the plasma zinc content (Brenton et al. 1981).

In summary, the evidence suggests that there is a depletion of body zinc in patients with acrodermatitis enteropathica but that the depletion is likely to be remarkably small in spite of the acute clinical manifestations of the disease. In our studies of one patient during commencement of zinc therapy, balance studies revealed that she retained approximately 120 mg of zinc in total which was less than 10% of her expected total body zinc (Jackson 1977). Thus it may be that the depletion of only a small specific "pool" of zinc (such as the membrane-bound zinc, as proposed by Bettger and O'Dell 1981, or the "exchangeable pool" of zinc described by Jackson et al. 1984) is sufficient to produce severe zinc deficiency symptoms (see also Chap. 17).

Whole-Body Zinc Homeostasis

It has been stated previously that the relatively constant levels of zinc in tissues and body fluids in situations where the composition of the diet varies remarkably suggests that there is an efficient mechanism for whole body homeostasis. Some of the earliest workers in human trace element research examined this in man (McCance and Widdowson 1942) by injecting large quantities of zinc intravenously and then following the amount of zinc excreted in both urine and faeces. They found that the majority of the injected ion was excreted in the faeces with little excreted in the urine thus demonstrating an effective homeostatic response to an increased body burden of zinc. Extensive studies of the process of zinc absorption in animals have subsequently been undertaken which invariably demonstrate an efficient homeostatic control of zinc absorption (e.g. see Becker and Hoekstra 1971 and Chap. 3, this volume) although the control of zinc excretion has been much less extensively studied.

It appears that homeostasis is maintained by manipulation of both gastrointestinal absorption and gastrointestinal excretion of the element. Little zinc is excreted in the urine and therefore changes in the urine zinc content can contribute little to the maintenance of whole-body homeostasis at normal dietary zinc intakes. However, it appears that the urinary zinc content responds to large changes in body status such that overt depletion of zinc causes a reduction in the urinary zinc content (Nelder and Hambidge 1975). An increase in the proportion of zinc absorbed from the diet is a well-described consequence of a diet of low zinc content in man (e.g. see Istfan et al. 1983) but decreases in the rate of excretion of zinc into the gastrointestinal tract are also known to occur in response to a reduction in dietary zinc (Baer and King 1984). Conversely an elevation of dietary zinc results in a reduced fractional absorption of zinc and an increase in the rate of gastrointestinal excretion of zinc (Jackson et al. 1984). In this situation the changes in gastrointestinal excretion of zinc appear to respond more rapidly to changes in dietary intake than do changes in gastrointestinal absorption although the magnitude of changes in the amount of zinc absorbed was considerably larger (see Fig. 1.2).

It is thus apparent that the human body is able to respond to relatively large variations in dietary zinc to maintain a relatively constant body content of zinc. However when the homeostatic mechanisms are insufficient to cope with a large reduction or excess of zinc in the diet it can be argued that a second line of defence comes into play. An excess intake of zinc seems to lead to an excessive loss of zinc via non-physiological routes. Thus the amount of zinc deposited in hair is very susceptible to an increase in zinc intake (Jackson 1980) and, conversely, following feeding of a zinc-deficient diet there is a selective loss from certain body fluids and tissues. Animal studies demonstrate that plasma zinc levels can fall rapidly after introduction of a low-zinc diet (Leucke et al. 1968; Wilkins et al. 1972) and that liver zinc levels reflect the recent dietary intake (Underwood 1977; Jackson et al. 1982). However, tissues which show symptoms of acute zinc deficiency (e.g. skin, hair and gut) show little initial change in their total zinc content (Jackson et al. 1982). It is therefore possible to argue that the rapid fall in plasma zinc content following initiation of a very low dietary zinc intake may in fact be a part of the homeostatic mechanism which the body has developed for preferential maintenance of zinc levels in tissues susceptible to zinc deficiency symptoms. In much the same way, the preferential deposition of excess zinc in hair is likely to be a response to prevent damage to other tissues caused by an excess of zinc.

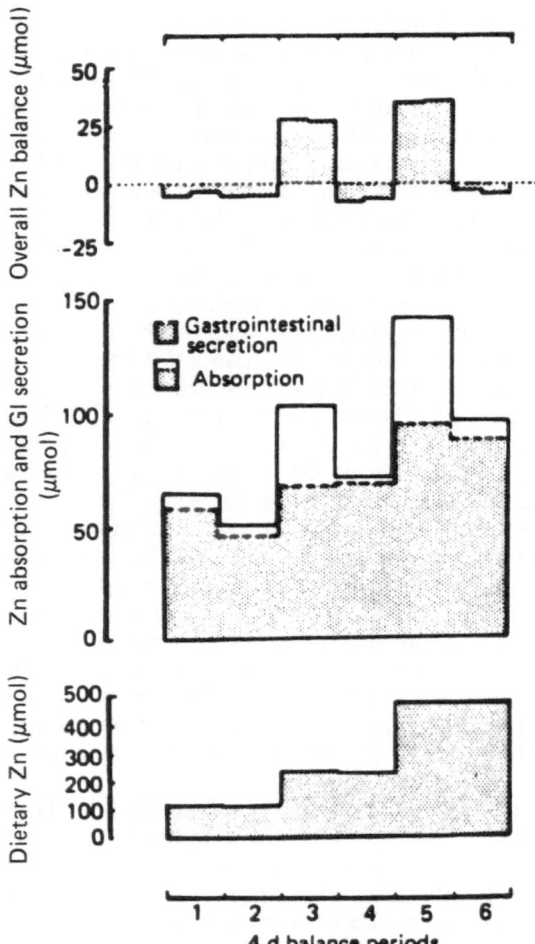

Fig. 1.2. Comparison of changes in gastrointestinal (GI) absorption and secretion of zinc measured by stable isotope techniques with the overall balance for zinc at different dietary intakes. Reproduced with permission from Jackson et al. (1984, p. 203).

Transport of Zinc within the Body

The systemic transport of zinc within the body will be covered in detail by other authors in this volume (e.g. Chap. 5, this volume), but it is important for an appreciation of the overall picture of the physiology of zinc to consider it briefly here. The plasma is the major route of zinc transport around the body and, within this fluid, zinc appears to be primarily bound to albumin with a small amount bound to the globulin fraction, probably α_2-macroglobulin (Prasad and Oberleas 1968; Parisi and Vallee 1970). A very small amount of the zinc in the plasma is also bound to amino acids.

Since the content of zinc in the plasma is in the order of 1 μg/ml and in most tissues ranges from 11–150 μg/g (Table 1.1) it is obvious that there must be an active or facilitated uptake of zinc by all tissues, but little is known of the nature of this uptake mechanism. A small amount of work has been undertaken on the mechanism by which zinc is transported into red blood cells which suggests zinc uptake is stimulated by bicarbonate ions and that an anion transport system may be involved (Kalfakou and Simons 1986), while it appears transport of zinc into hepatocytes is an energy-dependent process (Failla and Cousins 1978a).

Even less is known about the transport of zinc out of tissues into the plasma. It is evident that certain tissues (e.g. the pancreas or mucosal cells of the gastrointestinal tract) secrete zinc as part of their normal physiological function. However, there is also a turnover of zinc in other tissues which results in a loss of zinc from the cells via unknown mechanisms, although it appears that a constant renewal exchange process is necessary to maintain sufficient levels of cellular zinc (Cousins 1985).

During conditions of tissue breakdown, losses of zinc from tissues may be excessive. For example, during muscle catabolism there is a large loss of zinc from skeletal muscle which results in an increased urinary and faecal excretion of zinc (Jackson and Edwards 1982). This appears to be related directly to the amount of muscle mass which is lost since, in patients with acute muscle disorders, the zinc balance correlates well with the nitrogen balance during periods of overt loss or gain of muscle tissue (see Fig. 1.1). The grossly elevated urinary zinc which occurs during muscle catabolism may be related to the loss of peptides and amino acids from the muscle during breakdown which could increase the ultrafiltrable "pool" of zinc in plasma. Alternatively, the zinc could be associated with the myoglobin lost during the myoglobinuria which accompanies severe muscle breakdown.

Acknowledgements. The author would particularly like to thank Professor R. H. T. Edwards and Dr. D. A. Jones for their enthusiastic support and collaboration in his work on zinc metabolism and would like to acknowledge financial support from the Muscular Dystrophy Group of Great Britain, Nestlé Nutrition SA and F. Hoffman-La Roche and Co. He also wishes to acknowledge the inspiration and great help he received from the late Professor C. E. Dent at the beginning of his interest in zinc.

References

Abbassi AA, Prasad AS, Rabbani P, DuMouchelle E (1980) Experimental zinc deficiency in man. Effect on testicular function. J Lab Clin Med 96: 544–550

Alfonzo B, Heaton FW (1974) The subcellular distribution of copper, zinc and iron in liver and kidney. Changes during copper deficiency in the rat. Br J Nutr 32: 435–445

Amador M, Pena M, Garcia-Miranda A, Gonzalez A, Hermelo M (1975) Low hair zinc concentration in acrodermatitis enteropathica. Lancet I: 1379

Atherton DJ, Muller DPR, Aggett PG, Harries JT (1979) A defect in zinc uptake by jejunal biopsies in acrodermatitis enteropathica. Clin Sci 56: 505–507

Baer MT, King JC (1984) Tissue zinc levels and zinc excretion during experimental zinc depletion in young men. Am J Clin Nutr 39: 556–570

Bartholomew ME, Tupper R, Wormall A (1959) Incorporation of ^{65}Zn in the sub-cellular fractions of the liver and spontaneously occurring mammary tumours of mice after the injection of zinc-glycine containing ^{65}Zn. Biochem J 73: 256–261

Becker WM, Hoekstra WG (1971) The intestinal absorption of zinc. In: Skoryna SC, Waldron-Edward E (eds) The intestinal absorption of metal ions, trace elements and radio nucleides. Pergamon, New York, pp 229–256

Bertrand G, Bhattacherjee RC (1935) Recherches sur l'action combineé du zinc et des vitamines dès l'alimentation des animaux. Bull Soc Sci d'Hygiene Alimentaire et d'Alimentation Rationelle de l'Homme 23: 369–376

Bettger WJ, O'Dell BL (1981) A critical physiological role of zinc in the structure and function of bio-membranes. Life Sci 28: 1425–1438

Bremner I, Williams RB, Young BW (1977) Distribution of copper and zinc in the liver of the developing sheep foetus. Br J Nutr 38: 87–92

Brenton DP, Jackson MJ,. Young A (1981) Two pregnancies in a patient with acrodermatitis enteropathica treated with zinc sulphate. Lancet II: 500–502

Brooks A, Reid H, Glazer G (1977) Acute intravenous zinc poisoning. Br Med J I: 1390–1391

Carter N, Jeffrey S, Shiels A, Edwards Y, Tipler T, Hopkinson DA (1979) Characterisation of human carbonic anhydrase III from skeletal muscle. Biochem Genet 17: 837–852

Cassens RG, Hoekstra WG, Faltin EC, Briskey EJ (1967) Zinc content and subcellular distribution in red vs. white porcine skeletal muscle. Am J Physiol 212: 688–692

Chesters JK, Will M (1973) Some factors controlling food intake by zinc deficient rats. Br J Nutr 30: 555–586

Chesters JK, Will M (1981) Measurement of zinc flux through plasma in normal and endotoxin-stressed pigs and the effects of zinc supplementation during stress. Br J Nutr 46: 111–118

Chvapil M (1976) Effects of zinc on cells and biomembranes. Med Clin North Am 60: 799–812

Cotzias GC, Borg DC, Selleck B (1962) Specificity of zinc pathway through the body: turnover of ^{65}zinc in the mouse. Am J Physiol 202: 359–363

Cousins RJ (1985) Absorption, transport and hepatic metabolism of copper and zinc: special reference to metallothionein and ceruloplasmin. Physiol Rev 65: 238–309

Cox RP (1969) Hormonal induction of increased zinc uptake in mammalian cell cultures: requirement for RNA and protein synthesis. Science 165: 196–199

Cuthbertson DP, Fell GS, Smith CM, Tilstone WJ (1972) Metabolism after injury. 1. Effects of severity, nutrition and environmental temperature on protein, potassium, zinc and creatine. Br J Surg 59: 925–931

Danbolt N, Closs K (1942) Acrodermatitis enteropathica. Acta Derm Venereol 23: 127–129

Documenta Geigy Scientific Tables (1975) 6th edn, Geigy, Basle

Failla ML, Cousins RJ (1978a) Zinc uptake by isolated rat liver parenchymal cells. Biochim Biophys Acta 538: 435–444

Failla ML, Cousins RJ (1978b) Zinc accumulation and metabolism in primary cultures of rat liver cells: regulation by glucocorticoids. Biochim Biophys Acta 543: 293–304

Falchuk KH (1977) Effect of acute disease and ACTH on serum zinc proteins. N Engl J Med 296: 1129–1134

Fell GS, Fleck A, Cuthbertson DP et al. (1973) Urinary zinc levels as an indicator of muscle catabolism. Lancet I: 280–282

Flynn A, Pories WJ, Strain WH, Hill OA (1973) Zinc deficiency with altered adrenocortical function and its relation to delayed healing. Lancet I: 789–790

Garretts M, Molokhia M (1977) Acrodermatitis enteropathica without hypozincaemia. J Pediatr 91: 492–494

Henkin RI, Foster DM, Aamodt RL, Berman M (1984) Zinc metabolism in adrenal cortical insufficiency: effects of carbohydrate active steroids. Metabolism 33: 491–501

Istfan NW, Janghorbani M, Young VR (1983) Absorption of stable ^{70}Zn in healthy young men in relation to zinc intake. Am J Clin Nutr 38: 187–194

Jackson MJ (1977) Zinc and di-iodohydroxy quinoline therapy in acrodermatitis enteropathica. J Clin Pathol 30: 284–287

Jackson MJ (1980) Tissue zinc stores and whole body homeostatic mechanisms in man and the rat. PhD thesis, University of London

Jackson MJ, Edwards RHT (1982) Zinc excretion in patients with muscle disorders. Muscle Nerve 5: 661–663

Jackson MJ, Jones DA, Lilburn M (1980) Demonstration of zinc homeostasis in man. J Physiol 305: 53p–54p

Jackson MJ, Jones DA, Edwards RHT (1982) Tissue zinc levels as an index of body zinc status. Clin Physiol 2: 333–343

Jackson MJ, Jones DA, Edwards RHT, Swainbank IG, Coleman ML (1984) Zinc homeostasis in man: studies using a new stable isotope dilution technique. Br J Nutr 51: 199–208

Jeffreys D, Edwards YH, Jackson MJ, Jeffrey S, Carter ND (1982) Zinc and carbonic anhydrase III distribution in mammalian muscle. Comp Biochem Physiol 73B: 971–975

Jones RB, Keeling PWN, Hilton PJ, Thompson RPH (1981) The relationship between leukocyte and muscle zinc in health and disease. Clin Sci 60: 237–239

Kalfakou V, Simons TJB (1986) The mechanism of zinc uptake into human red blood cells. J Physiol 381: 75P

Kirchgessner M, Roth HP (1985) Influence of zinc depletion and zinc status on serum growth hormone levels in rats. Biol Trace Element Res 7: 263–268

Kuipers PJ, Cousins RJ (1984) Zinc accumulation in rat liver parenchymal cells in primary culture and response to glucagon and dexamethasone. Fed Proc 43: 1403

Lehmann BH, Hansen JDL, Warren PJ (1971) The distribution of copper, zinc and manganese in various regions of the brain and in other tissues of children with protein-calorie malnutrition. Br J Nutr 26: 197–202

Leucke RW, Olman ME, Baltzer BV (1968) Zinc deficiency in the rat: effect on serum and intestinal alkaline phosphatase activities. J Nutr 94: 344–350

Lombeck I, Schnippering HG, Ritzl F, Feinendegen LE, Bremer HJ (1975) Absorption of zinc in acrodermatitis enteropathica. Lancet I: 855

Lye M, Winston B (1979) Whole body potassium and total exchangeable potassium in elderly patients with cardiac failure. Br Heart J 42: 568–572

McCance RA, Widdowson EM (1942) The absorption and excretion of zinc. Biochem J 36: 692–696

Michaelsson G (1974) Zinc therapy in acrodermatitis enteropathica. Acta Derm Venereol 54: 377–381

Miller WJ (1969) Absorption, tissue distribution, endogenous excretion and homeostatic control of zinc in ruminants. Am J Clin Nutr 22: 1323–1331

Mills CF (1973) Trace element nutrition. In: Jonxis JHP, Visser HKA (eds) Therapeutic aspects of nutrition. Stenfert Kroes, Leiden, pp 13–27

Mills CF (1981) Some outstanding problems in the detection of trace element deficiency diseases. Philos Trans R Soc London [Biol] 294: 199–213

Moynahan EJ (1974) Acrodermatitis enteropathica: a lethal inherited human zinc deficiency disorder. Lancet II: 399–400

Nelder KH, Hambidge KM (1975) Zinc therapy of acrodermatitis enteropathica. N Engl J Med 292: 879–882

Parisi AF, Vallee BL (1970) Isolation of a zinc α_2-macroglobulin from human serum. Biochemistry 9: 2421–2426

Pedrosa FO, Pontremoli S, Horecker BL (1977) Binding of Zn^{2+} to rat liver fructose-1, 6-bisphosphatase and its effect on the catalytic properties. Proc Natl Acad Sci USA 74: 2742–2745

Pierce P, Jackson MJ, Tomkins A, Millward DJ (1985) Zinc is highly conserved in the severely zinc deficient rat. Proc Nutr Soc 44: 78A

Portnoy B, Molokhia M (1974) Zinc in acrodermatitis enteropathica. Lancet I: 663–664

Prasad AS (1982) Clinical and biochemical spectrum of zinc deficiency in human subjects. In: Prasad AS (ed) Clinical, biochemical and nutritional aspects of trace elements. Alan R. Liss, New York, pp 3–62

Prasad AS, Oberleas D (1968) Zinc in human serum: evidence for an amino acid bound fraction. J Lab Clin Med 72: 1006

Shaw JCL (1979) Trace elements in the foetus and young infant. Am J Dis Child 13: 1260–1268

Smeyers-Verbeke J, May C, Drochmans P, Massart DL (1977) The determination of Cu, Zn and Mn in subcellular rat liver fractions. Ann Biochem 83: 746–753

Smith JC Jr, Zeller JA, Brown E, Dandong SC (1976) Elevated plasma zinc: a heritable anomaly. Science 193: 496–498

Spray CM, Widdowson EM (1950) The effect of growth and development on the composition of mammals. Br J Nutr 4: 332–353

Strain WH, Hirsh FS, Michel B (1975) Increased copper/zinc ratios in acrodermatitis enteropathica. Lancet I: 1196–1197

Tao SH, Hurley LS (1975) Effect of dietary calcium deficiency during pregnancy on zinc utilisation. J Nutr 105: 220–225

Todd WJ, Elvehjem CA, Hart EB (1934) Zinc in the nutrition of the rat. Am J Physiol 107: 146–156

Underwood EJ (1977) Trace elements in human and animal nutrition, 4th edn. Academic Press, London

Walravens PA, Hambidge KM, Nelder KH et al. (1978) Zinc metabolism in acrodermatitis enteropathica. J Pediatr 93: 71–73

Wannemacher RW, Pekarek RS, Klainer AS et al. (1975) Detection of a leukocytic endogenous mediator-like mediator of serum amino acid and zinc depression during various infectious illnesses. Infect Immun 11: 873–875

Weinberg ED (1974) Iron and susceptibility to infectious disease. Science 184: 952–956

Widdowson EM, McCance RA, Spray CM (1951) The chemical composition of the human body. Clin Sci 10: 113–125

Widdowson EM, Chan H, Harrison GE et al. (1972) Accumulation of Cu, Zn, Mn, Cr and Co in the human liver before birth. Biol Neonate 20: 360–367

Wilkins PJ, Grey PC, Dreosti IE (1972) Plasma zinc as an indicator of zinc status in rats. Br J Nutr 26: 113–120

Chapter 2

An Introduction to the Biochemistry of Zinc

R.J.P. Williams

Introduction

The Value of Metal Ions in Biology

There are certain advantageous properties of metal ions both in their binding strengths and their rates of exchange of ligands which make them a potential source of chemically and hence biologically interesting selective properties. Zinc is the case in point here. Its most obvious distinction is its highly concentrated charge, Zn^{2+}. It is also a small ion, radius 0.65 Å. In itself, however, an electrostatic charge of two and small size gives quite modest binding even to anions such as carboxylate and phosphate and even where several anions occur together. This is due to competition from water of hydration. Furthermore the electrostatic binding by zinc is a property shared almost equally with Mg^{2+} and, to a lesser degree, Ca^{2+} as well as with other metal cations such as Cu^{2+} and Ni^{2+}, but it is not strong amongst organic ions. Thus although this part of its chemistry is special to metal ions it is not peculiar to zinc.

The second characteristic of metal ions is their high affinity for electrons as shown by the energy required to get from the gas atom M to M^{2+}, i.e. their ionization potentials. The ionization potential is a rough guide to Lewis acid strength of a cation. Here zinc differs markedly from magnesium and calcium as well as from organic cations since its electron affinity is much higher, but while it is therefore a strong Lewis acid it remains similar to copper and nickel in all properties described so far (see Fig. 2.1). All these last three ions can then bind strongly to donors such as thiolates and amines, which is a property of strong Lewis acids, and is not shown by magnesium or calcium. Zinc does differ from copper and nickel in yet other respects however. Zinc does not show variable valence (oxidation state change) and could then be preferred in certain circumstances for this reason alone since redox activity introduces the

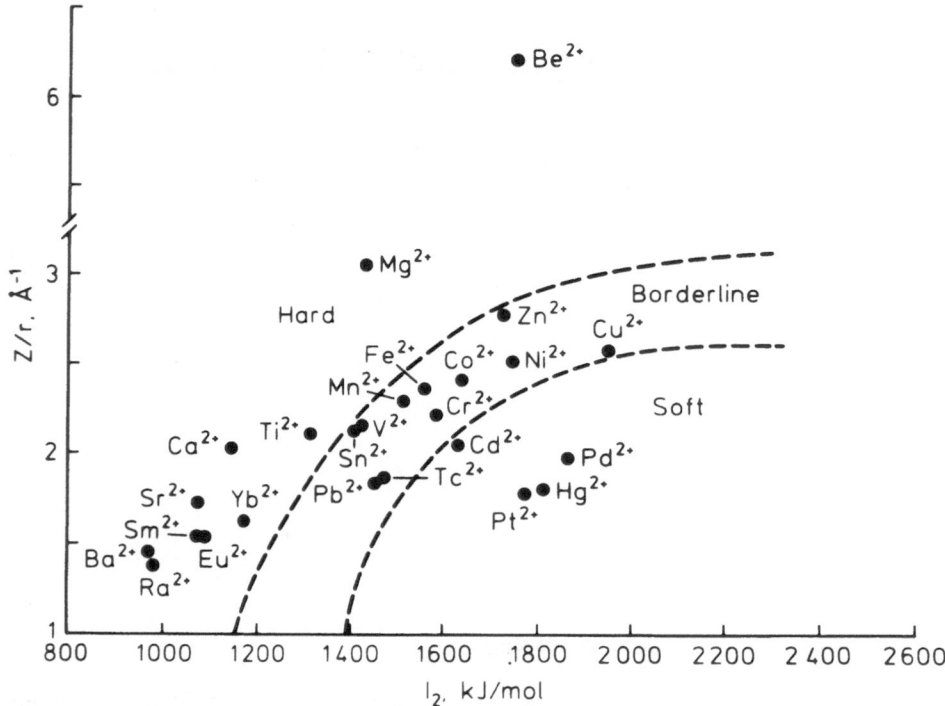

Fig. 2.1. A plot of electron affinity of divalent cations, M^{2+}, against their charge : radius ratio. The "a"-class (hard) cations such as Mg^{2+} lie towards the top left while the "b"-class lie towards the bottom right. The trend towards "b"-class is easily seen in the Irving–Williams series $Mn < Fe < Co < Ni < Cu > Zn$. The intermediate position of zinc is apparent.

risk of free radical reactions. However, zinc is not quite unique. Though it is highly selective as a cation, it differs from cadmium only somewhat in much of its chemistry and thus cadmium remains a competitive poison at certain zinc sites. However, Cd^{2+} is the same size as Ca^{2+} and, just as calcium does not resemble magnesium very closely, so zinc differs from cadmium. Zinc then carries a special set of thermodynamic properties (Table 2.1), but these are further enhanced in rates of reactions as follows.

Reactions require re-organization of atoms. Fast reactions will be more easily carried out by metal ions which, while binding well, both take up and release molecules from their coordination spheres rapidly. We can write K, binding constant, as a ratio of rate constants

$$K = \frac{k_{on}}{k_{off}}$$

The values for both k_{on} and k_{off} for zinc usually greatly exceed those of nickel, for example, while K is roughly the same. Moreover, the re-arrangements of groups on the surface of zinc without release, fluctional reactions, are usually fast unlike those of nickel. The differences arise since in a field of ligands the

Table 2.1. Why zinc is valuable in biology

---

1. Availability (> nickel or proton)
2. Strongly retained (> manganese, iron(II) or magnesium)
3. Fast ligand exchange (> nickel or magnesium)
4. No redox reaction possible (contrast with copper, iron and manganese)
5. Flexible coordination geometry (> nickel or magnesium)
6. Good Lewis acid (only copper better)
7. Supplies hard base, hydroxide [only copper(II), iron(III) and manganese(III) better]

---

zinc ion is not as effectively polarized, its ligand field stabilization energy is lower. As a result zinc, but not nickel, passes readily from one symmetry of its surrounds to another without exchange and exchanges ligands quickly. Such factors allow for better catalytic activity in that they allow rapid atom flow including substrate binding, intermediate to intermediate changes, and product release in enzymes. If we use I_{o2}, the ionization potential to the doubly charged state, multiplied by k_{on} (i.e. water release from the coordination sphere) as a guide to catalysis zinc looks like a better catalyst than almost any other divalent cation readily available to biology except copper which is objectionable on redox grounds. It may even be that if we examine more highly charged cations, such as Al^{3+}, Fe^{3+} and Co^{3+}, zinc (Zn^{2+}) would still turn out to be the best catalyst in neutral aqueous media in which the higher valent ions are hydrolysed and their functional strength is lost by binding to hydroxide ions.

Metal ions need not be used in catalysis alone. The role of calcium in triggering muscle activity is well-known. We know that such triggering, like catalysis, requires good binding with selective recognition but also reasonably rapid on and off reactions since there is often a need to stimulate a biological system quickly but to relax it quickly too. Amongst the divalent cations zinc, like calcium, has the required properties for a trigger or control ion since, as described above, it binds well but exchanges rapidly. Naturally it is used in different circumstances from those in which calcium is used as the calcium ion binds to O-donor ligands only but zinc binds largely to S- and N-donors in biological systems. The similarities (and differences) between zinc and calcium lead to peculiar parallel activities in biology where both metal ions are of extreme importance in control and cross-linking.

Zinc in Biology

Before making particular statements, a few general remarks are worth making about the three roles of zinc, as an acid catalyst, a control ion and a structural ion. The spatial distribution of these roles in cellular systems is worth careful note. At the risk of over-generalization I make the following statements:

1. Zinc is the most common metal ion in the cytoplasm of cells after groups IA and IIA metal ions. Iron is much more common in membranes, manganese in vesicles or organelles and copper is largely outside the

cytoplasm and outside cells; nickel and cobalt and all other metal ions are
not common.

2. Zinc digestive enzymes are outside cells or in vesicles.
3. Zinc is rarely intimately associated with a membrane.
4. The structural role of zinc extends from filamentous (keratin-like structures)
    to the organization of chromosomes, largely inside cells.

The peculiarities of the chemistry of zinc, summarized in Table 2.1, are then
put to particular use in particular parts of biological space (see Fig. 2.2).

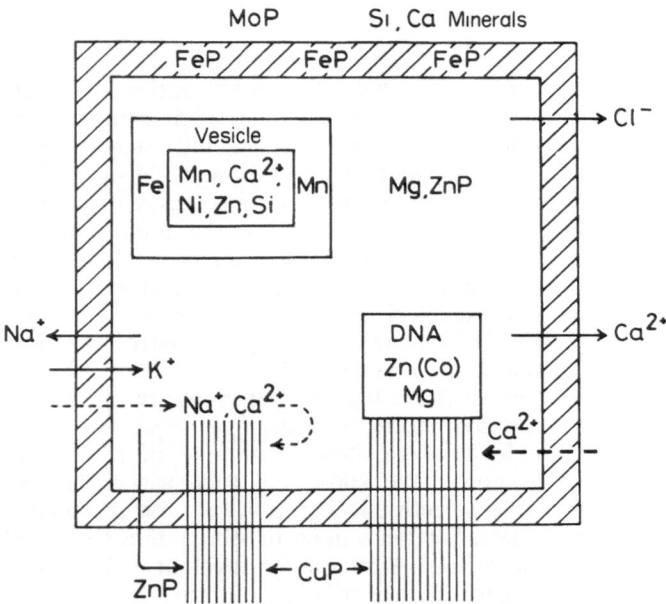

Fig. 2.2.   The distribution of some metal ions in biological space is shown for a eukaryotic cell.
Zinc has many roles but is especially prominent relative to other elements in the cytoplasm, in
the nucleus and in vesicular storage. Of the other elements shown, copper is prevalent outside
cells and iron in membranes and in organelles. The combined use of zinc and copper in the
degradation of filamentous structures is indicated.

Availability of Metal Ions

Biological systems are energy consuming. Energy economy is therefore
important and this implies that biology will use available elements rather than
those that have similar properties but are much less available. Zinc is relatively
available even in sea water (see Table 2.2) and is amongst the most available
of the trace elements. It is more available than copper, nickel, iron, cobalt or
cadmium. Considering its potential value as described above it is not surprising

Table 2.2. Logarithmic concentration of elements in sea water (mol/dm)

Na	−0.3	K	− 2.0	Fe	− 9.0
Mg	−1.3	Ca	− 2.0	Co	−10.7
Al	−7.7	Sc	−11.0	Ni	− 8.1
Si	−4.0	Ti	− 8.0	Cu	− 8.4
P	−5.7	V	− 7.5	Zn	− 8.1
S	−1.6	Cr	− 8.4	Mo	− 7.0
Cl	−0.3	Mn	− 9.3	Sr	− 4.0

Data from Bruland (1983).

that it is relatively common in biological systems, about as common as iron, and that here and there it is quite highly concentrated. The relative concentration of free zinc ions within biological systems varies from $\leqslant 10^{-9}$ M in the cytoplasm of many cells to $\geqslant 10^{-3}$ M in some vesicles. (Calcium has a very similar localized concentration in vesicles.) Clearly a mechanism must be available for moving the element into selected compartments by pumping. These pumps like all the other activities associated with zinc and described above require selective binding. I turn next to the types of chemical neighbours with which zinc is associated in biology and I start with proteins since a great deal is now known about zinc proteins from crystallographic and chemical studies.

Types of Protein Associated with Zinc

Elsewhere we have described two very different types of protein, the extended β-pleated sheet and sets of α-helices, in relation to their relative mobility and therefore their functional potential. Such characteristics of some typical metal proteins are indicated in Table 2.3. We observe in the table that the "trigger" proteins, e.g. those which bind calcium, are almost totally helical and mobile and that this is also true of the *allosteric* oxygen-carrier proteins which use iron or copper. On the other hand the electron transfer proteins which contain copper, the blue proteins, are based on immobile β-sheets. Cytochrome *c*,

Table 2.3. Major fold characteristics of some metal-proteins

Helical	β-Sheet	Mixed[a]	Random
Haemocyanin, Cu	Azurin, Cu	Phospholipase, Ca	Metallothionein[b], Cu,
Calmodulin, Ca	Carbonic anhydrase,	Carboxypeptidase, Zn	Zn
Haemoglobin, Fe	Zn	Alcohol	Chromogranin A, Ca
Insulin, Zn	Superoxide dismutase,	dehydrogenase, Zn	σ-Factor IIIA, Zn
S–100, Zn, Ca	Zn, Cu		

[a] Folds only on binding metal ions.
[b] The active site is always closely associated with the β-sheet.

another electron transfer protein, although it is helical is cross-linked three times and is also of low internal mobility. Such electron transfer proteins do not require gross mobility which in fact would be disadvantageous. Turning our attention to enzymes it is usually true that active sites are held in proteins close to or on β-sheet-like structures and we may suppose that the relative rigidity of this structure helps to control specificity of binding and the directionality of attack by acid/base or redox centres. The stability of the fold also prevents exchange of metal ions from such a protein. We now look at zinc proteins in this context.

Crystal structures are available for several enzymes which require zinc at the active site (Table 2.3). It is immediately clear that they too are based on extensive β-sheets and we presume that it is the stability of the fold which generates the peculiar and particular active site geometries of the metal ion. The imposition of a special geometry at the metal by the strength of the fold we have described as an entatic (strained) condition of the metal geometry (Fig. 2.3). We return to this feature later since the metal binding strains the protein too. An enzyme in which the zinc is not at the active site is superoxide dismutase, another β-sheet protein, where copper is the active metal ion. We note that here both the binding of the zinc into a β-sheet fold and the β-sheet itself now stabilize the unusual entatic, rhombic geometry around the copper ion.

Turning to other kinds of proteins to which zinc is bound and not at the active site there are crystallographic data for a few (Fig. 2.3). In some proteins zinc acts as a cross-link between thiolate groups, usually four, or perhaps a combination of thiolates and histidines to make a total of four ligands. This is a double or knotted cross-link and is found in intracellular proteins, e.g. alcohol dehydrogenase. The cross-link is in part a replacement for the –S–S– bridges of extracellular proteins since inside a cell most –S–S– bridges are unstable to reduction and calcium cross-links are unstable since in cells there is very little calcium. In the cross-links in alcohol dehydrogenase and aspartate transcarbamylase zinc has a geometry which is now that of a conventional tetrahedron.

So far all the zinc binding we have noticed is strong, with binding constants $\geqslant 10^9$. In insulin complexes where the zinc and insulin are bound in a vesicle, the complexes are weak and the concentration of free zinc is high. The zinc is cross-linked to this *helical* protein in two different but conventional ways through histidines of the protein. In the two forms the helices are slightly differently disposed relative to one another. There is a remote link here to

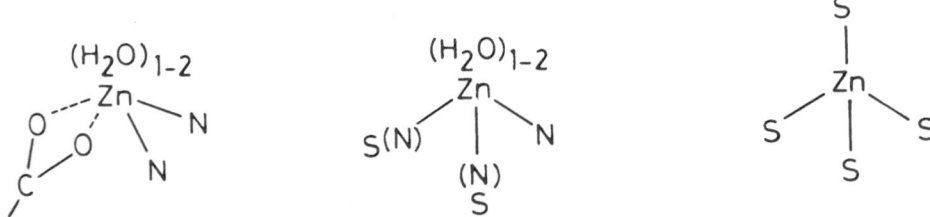

Fig. 2.3. The geometries at the active sites of some enzymes. The geometries are usually irregular, to the left carboxypeptidases, centre carbonic anhydrase and alcohol dehydrogenase, in contrast with zinc sites which are just structural, shown on the right.

calcium and zinc binding to calmodulins (see S-100 below) where the metal ions bind to and modulate the properties of a helical protein. In insulin the zinc site geometries are closely related to an octahedron. The possible reason for the presence of zinc in insulin vesicles will be analysed later though we do not know why insulin should have a mobile surface but a cross-linked interior. Notice particularly however that the conventional zinc binding geometry, α-helical structure and non-enzymic function appear to occur together.

Metallothionein is, in the absence of zinc, a random coil and only on binding zinc, cadmium or copper does it form a series of turns so as to hold the metal ions in two types of cluster formed by regular tetrahedral, thiolate donors. The probable reason for having such clusters of metal ions in this protein will be discussed later. The structure of metallothionein bound by metals was not found to be the same when examined by nuclear magnetic resonance (NMR) and by x-ray diffraction but the essential core structure is fixed. There are two clusters of M_4 and M_3 respectively. It should be noticed that while it is reasonable to assume that all the other proteins described so far are very highly zinc selective, this is not true of metallothionein which is in effect a scavenger of heavy metal ions and, remembering the introduction, notice that three of the metals scavenged are cadmium, zinc and copper. The protein clearly adapts itself to the metal ion here and no question of an entatic state arises. The value of such a common scavenger will be examined again especially in the discussion of the interactive role of zinc and copper in biology.

While the structures of other non-enzyme proteins which bind zinc well are not known there are strong predictions as to their character. Two of these whose metabolic significance will emerge later in this volume are the S-100 so-called "brain" protein and the σ-factor, transcriptase IIIA. Of these it is very likely that S-100 is helical and is a trigger protein using imidazole groups to bind the zinc. It is also triggered by calcium. The second protein is a transcription factor and is likely to bind zinc through thiolates and imidazoles in a tetrahedral geometry. This is a control factor for DNA inspection and control and there is likely to be ready exchange of the zinc. It also binds from 6 to 10 zinc atoms but not in clusters and we must seek a reason for this. Undoubtedly the conformation in the absence of zinc is mobile.

In passing it is worthwhile comparing some other small metalloproteins (Table 2.2). As stated above, the simple electron transfer proteins are invariably β-sheets or are cross-linked and this restriction of mobility is correct for it would be contrary to the value of those centres if they were permitted to lose metal ions or if the metal ions had to pass over large activation energy barriers (relaxations) while changing oxidation state. However, when electron transfer is coupled to proton transfer or triggering then the electron transfer proteins are helical and not cross-linked, e.g. cytochromes b and a. The examples of helical enzymes on the other hand are very restricted to a few special cases. Notice particularly enzymes for reactions which go by preferred order, e.g. citrate synthetase and cytochrome P-450 and enzymes which carry out extremely easy reactions such as H_2O_2 activation at haem. In the case of preferred order reactions adjustment of the enzyme is required as substrates are bound and, if the discussion above is correct, only helical proteins can manage this effectively. The reactions of H_2O_2 at haem are such that it is not just catalysis that is required but a very hidden catalytic site to avoid release of harmful intermediates. There is no need to activate metals for such reactions.

When some general rule is being established it is always wise to place stress upon interesting and possibly indicative exceptions. Zinc in glyoxalase is bound by histidines (and perhaps carboxylates) and it appears to be 6-coordinate.

The Number of Binding Ligands to Zinc

The proteins listed in Table 2.3 nearly all bind zinc through imidazole and cysteine. In extreme cases four such groups are used when the zinc acts as a cross-link and is blocked from interaction with any small molecule substrate. In the enzymes, zinc can have as few as two imidazoles when it is also bound to a carboxylate, carboxypeptidase, or it can have three such donors. The fourth (and fifth) positions are then occupied by water molecules which become part of the catalytic act or are readily replaced allowing access of substrate to the zinc. The zinc site is then organized in several ways differing with respect to (a) type of ligands, (b) angular distribution of ligands around the zinc, (c) bond distances, (d) leaving groups and (e) the next nearest and so-on surrounds of the zinc. The last term includes the whole of the active site groove in which the zinc sits. We should be able to rationalize all the peculiarities of these different parts of the zinc sites in terms of the reactions which they carry out. Carbonic anhydrase is given as an example below.

Zinc as a Catalytic Group

A zinc ion is a good electron acceptor, or "Lewis acid" (see the Introduction and Fig. 2.1), in inorganic and organic chemistry. As such it polarizes groups to which it binds and either increases the attacking power of the bound group or increases the probability of attack on the bound group. An example is bound water which becomes an effective bound OH^- at some pH, usually above 9 for model complexes:

$$Zn^{2+}(H_2O) \rightarrow Zn^{2+}(OH^-)..H^+$$

This means that zinc as an acid can generate a very effective localized hydroxide ion (base) which can attack other molecules. Such an action is proposed for the function of zinc in carbonic anhydrase where the OH^- attacks CO_2:

$$O{=}C{=}O + OH^- \rightarrow HCO_3^-$$

Note that water is a substrate in the reaction here. In carbonic anhydrase the zinc ligands, all neutral imidazole, are designed to make the $Zn(H_2O)$ as acid as possible so that the ionization $H_2O \rightarrow OH^-$ occurs at pH=7. In addition the movement of the substrate atoms along the path of intermediates is aided by the rapid switching of coordination geometry of the zinc which reduces barriers between steps. Furthermore the groove which leads in the enzyme from the surface to this active site functions as a proton conductor due to the incorporation of remote imidazoles along the protein groove. It is the combination of the geometry, bond angle and bond length, and the ligand groups, the site and its mobility, its immediate surrounds, effectively its solvent, of the zinc complex together with the more remote proton conduction groups which makes the enzyme so powerful.

An example of a second use of zinc is in a direct zinc ester or amide substrate complex where the zinc acts as an attacking acid on this substrate while making water attack it:

$$Zn^{2+} \leftarrow O{=}C\underset{OEt}{\overset{R}{\big<}} + OH^- \rightarrow Zn^{2+} \leftarrow {}^-O{-}C\underset{OH}{\overset{R}{\big<}}{-}OEt \rightarrow Zn^{2+}$$

$$+ \ EtOH \ + \ RCOOH$$

No clearcut example of this reaction path can be given but it is probable that zinc in alcohol dehydrogenase activates the alcohol by polarizing it in a similar way and the above path could be that of carboxypeptidase. Note again how firmly the zinc is bound in these enzymes.

Why are there Zinc Enzymes?

The need for zinc in catalytic sites of enzymes may not be thought to be obvious since many enzymes which hydrolyse esters or amides, the major substrates of zinc enzymes, do not require metals. As stated above, zinc has a highly localized charge and electron affinity and it is therefore a very effective attacking group in a sense quite different from that of any organic centre. Looking at the substrates it attacks (in enzymes) we see firstly that small molecules such as CO_2 and C_2H_5OH are one target and that another major target is amino- or carboxyl-terminal peptide links. A clear possibility therefore is that enzymes with a metal ion in them can be used to attack in a chemically non-selective but a physically constrained site. Enzymes without such powerful attacking centres as zinc can then attack only large substrates (with very high selectivity) by using many points of attachment to the substrate. The active site groups are here relatively poor attacking groups per se and attack in these non-metal enzymes is based on very well-regulated orientation and strain in the substrate generated by the precision of binding. Zinc enzymes would then be reserved (a) for attack on small substrates which cannot be handled by organic groups in a multi-point attachment and (b) for less specific but very rapid attack when they could work even with relatively weak binding of substrate. Lack of selectivity which would then of necessity go with powerful attacking groups is of obvious advantage in digestion. Speed (also at the price of selectivity) could be advantageous in hormone release and destruction or fertilization for example (see below), and we notice that zinc is particularly associated with peptide hormones and their destruction. (Once again there are close connections with calcium chemistry since calcium is another metal ion intimately linked with hormone action.)

Why have Zinc as a Cross-link?

Zinc is often found in an internal cross-link of intracellular proteins where it would seem to perform a function similar to that of –S–S bridges of extracellular proteins. (Disulphide bridges are not very stable in the reducing atmosphere of cells.) There are three striking differences between the two types of bridge. Zinc can cross-link between thiolates, imidazoles and perhaps between carboxylates though the last is a role more usually found to be performed by calcium (or magnesium) and again so far has mostly been seen outside cells. The zinc bridging can also be between a mixture of imidazole and thiolate donors (see Fig. 2.3). A second difference is that the zinc cross-link is usually multi-dentate, the zinc atom being coordinated to some four donors (Fig. 2.3). A further novel feature of the zinc cross-link is that it can be made readily reversible. In this way the zinc cross-link could become a functional switch control over a protein unlike the covalent –S–S bridge which can only be broken by chemical reduction. Two forms, zinc cross-linked (A) and zinc-free (B) are then in relatively fast exchange:

$$\text{(strongly folded)}A(Zn^{2+}) \rightleftarrows B \text{ (weakly folded)} + Zn^{2+}$$

In some cases, e.g. transcriptase factor IIIA, the $A(Zn^{2+})$ form has a new binding function and regulates DNA directly. In other cases the B form is functional in that the B form of the protein releases a hormone (zinc-free insulin) or allows a pro-enzyme to be converted to an enzyme. (It goes without saying that calcium could be substituted verbally for zinc in almost every sentence of this section.)

From the above, we then see the advantage not just of the acidity of zinc in binding but also of the fluctional character of its complexes in allowing re-orientation of helical (in particular) parts of proteins.

Use of Zinc in Biology Space and Time

Zinc and Filamentous Structures

It is difficult today to know where all the zinc in cellular systems is located. The stress of biological studies on chemical change and therefore on enzymes has left the structural and mechanical parts of cellular systems very much in the background. However cells are highly internally structured in dynamic forms. From DNA outwards to ribosomes and to membranes there are vast numbers of filamentous constructions in cells and additional ones outside cells. The role of zinc in these structures is as yet unknown. Methods are needed for the full analysis of the zinc associated with DNA and RNA. At the present time, only the function in some polymerases is known. (Note that very fast synthesis just like very fast degradation requires very good catalytic sites.) Here we can only mention such polymers as keratins (see Table 2.4).

Table 2.4. Zinc and filamentous structures

Filamentous structure	Role of zinc
Collagens	Zinc collagenases
Proteoglycans	Zinc stromelysin
Denatured collagens	Zinc gelatinase
(Chromosomal proteins)	Zinc structural role?
Keratins	Zinc cross-links

Regulatory Role of Zinc

Zinc is probably a major regulatory ion in the metabolism of cells. Evidence for this comes from gross biological considerations.

As is the case for regulatory calcium, the regulatory role of zinc can be made to depend on a high power of the free zinc ion concentration by the binding of several metal ions to one protein. This type of cooperativity is familiar in haemoglobins and calmodulins, and possibly in insulin, but more obviously too in the transcription σ-factor IIIA where as many as ten zinc atoms may occur. The equilibrium reaction must now be written in a series of steps up to n, where n is maximum cooperativity:

$$A(Zn^{2+})n \rightleftarrows B + nZn^{2+}$$

Alternatively each zinc complex $A(Zn^{2+})$, $A(Zn^{2+})_2$, $A(Zn^{2+})_3$, and so on could have a separate role to play so that the gearing of reactions in a cell could be regulated by zinc concentration in a smooth fashion. It is the author's opinion that these control roles of zinc in which the zinc bound states can be active or inhibitory will be found in many more cases than we know about at present, especially associated with fertilization and growth. Of particular interest are filamentous structures in cells, proteins such as S–100 and the internal chromosome where metal ions are found associated with core proteins.

The high dependence on Zn^{2+} concentration can also be of value in homeostasis (Fig. 2.4). Here we note the clusters of metallothionein again. As described elsewhere, the most sensitive homeostatic mechanism is based on a high power, n, of the free metal ion concentration, here $[Zn^{2+}]^n$. The homeostasis of zinc is managed in all probability by metallothionein. However we have to note that the same protein seemingly maintains the homeostasis of copper (and cadmium). Let us suppose that the random coil protein, metallothionein, folds according to the metals to which it binds. There can then be two sets of receptors, one each for copper and zinc metallothioneins as well as one for the apoprotein on DNA. But this means that the metabolisms of the two metals have become inextricably linked. Why might this be of value? I return to this point below.

Zinc Controls over Synthesis

The excitement over the discovery of zinc in transcriptase factor IIIA and its possible involvement in the core proteins of chromosomes must not detract

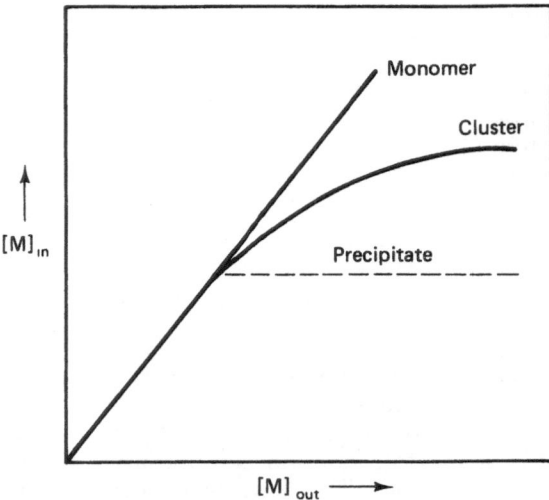

Fig. 2.4. The buffering effect (homeostasis) of a simple one-to-one complex ML, of a cluster complex M_nL and of a precipitate. The effect of change of outside concentration on the inside buffered concentration is shown.

from the importance of the discovery of zinc in reverse transcriptase and in RNA polymerase (Table 2.5). Once again notice that these enzymes have high rates but may not be very discriminating with regard to the nucleotide bases. Zinc is most valuable where speed is most important but its aggression as a Lewis acid implies that care must be taken in the selection of targets or else a general attack must be anticipated on peptide, phospho-diester or polysaccharide bonds.

Summary of the Roles of Intracellular Zinc

Inspection of the zinc activities (a) in transcription factor, (b) in the core proteins of DNA, (c) in RNA synthetase and reverse transcriptase and (d) in many cytoplasmic enzymes leads one to consider the possibility that the internal homeostasis of zinc by metallothionein is finely tuned so that zinc is one of the dominant cytoplasmic growth hormones. Rapid cell division will then

Table 2.5.   Zinc enzymes in synthesis

Synthetic step	Zinc enzyme
RNA	RNA polymerase
DNA	Reverse transcriptase
Viral synthesis	Terminal dNT transferase
Transfer RNA	t-RNA synthetase

depend upon the presence of stored zinc and its transfer. Note the concentrations of zinc in the male reproductive tract. This is not more than an hypothesis but the importance of zinc, which equals iron in concentration in living systems, could be missed easily because it is so hard to follow by analysis. We shall see the all-embracing role of this metal next when we turn to extracellular activities. Perhaps in early evolution (and at present in prokaryotes) iron and manganese played the role of communicating with DNA but in the eukaryotes this role has been taken over by zinc in the cytoplasm.

Export of Zinc Proteins

Extracellular enzymes can be generated in two ways: first by direct export of protein across the outer membrane when, if they are zinc enzymes, they find and fold around zinc in the extracellular fluids; second by synthesis in vesicles. Here the protein is secreted into a space in which zinc is also secreted. Both modes are common in zinc biochemistry and we should ask why both are used.

It is particularly noticeable that zinc proteases are stored in vesicles in such systems as the pancreas (digestive carboxy- and aminopeptidases) and in venoms (Table 2.6). The proteases associated with hormone and transmitter release are frequently zinc dependent too. Thus, in nutritional studies, zinc deficiency can show up in many ways which reflect the variety of modes of zinc proteases, e.g. collagen, transmitters, hormones, digestion. It must not be forgotten that zinc occurs too in other hydrolytic enzymes, e.g. phospholipases, nucleases and saccharases, all in digestion.

Zinc Concentrations and Pumps

Zinc is relatively very common in cellular and extracellular systems, including vesicular compartments. The actual concentrations of free zinc are not known but they must vary from around 10^{-9} M to 10^{-10} M in the cytoplasm (to account for the binding to cytoplasmic enzymes) to much higher concentrations

Table 2.6. Zinc in degradation processes

Degradative process	Role of zinc
Pancreatic juice action	Carboxypeptidase Zn enzyme
Venom haemorrhagic action	Zn proteases
Extracellular digestion (yeast)	Zn aminopeptidase
Extracellular digestion (bacteria)	Thermolysin
Breakdown of DNA and RNA	3'Nucleotidase

in some vesicles, say 10^{-4} M (to account for the binding to such proteins as insulin). To maintain the very high zinc concentrations in special compartments it is necessary to postulate energy-driven pumps for zinc, although these have not been found. The pump may be of the same kind as that for calcium which transports a simple ion or similar to the iron pumps which pump either small chelates such as citrate complexes or even large proteins such as transferrin. There are thought to be zinc carrier proteins but the role of these and of zinc metallothionein in zinc transport across membranes is unknown.

Zinc and Hormonal (Trigger) Activities

A feature of the zinc concentration is its localization in particular organs. Parts of the male reproductive systems of both plants and animals are well-known as being zones of high zinc concentration but in addition there are areas of the brain, especially the hippocampus, which also have high zinc concentrations. Why?

There is a requirement for the activity of many triggers and hormones that both their release and their removal must be very rapid. Consider the usual reaction sequence for a peptide hormone:

Stored precursor $\rightarrow$ Hormone (transmitter) $\rightarrow$ Local hydrolysis

The familiar parallel is with stored acetylcholine or adrenaline and *local* rapid degradation after action by esterases and amine oxidases (copper proteins). The rapidity of catalysis by E_1 and E_2 (Figs. 2.2 and 2.5) and their localization are equally important. Zinc enzymes are at their best in such situations and the known distribution of zinc in certain parts of the brain may well be linked to this function. Certainly the endopeptidase 24.11 and its close relatives are associated directly with the destruction of enkephalins and other peptide hormones (Table 2.7).

The role of zinc associated with transcriptase factor IIIA has become more extensively recognized. Two recent reviews by Klug and Rhodes (1987) and by Berg (1986) do much to update the work recorded above and lend it much extra emphasis. Of particular interest is the relationship with steroid hormones

Table 2.7. Zinc proteins related to hormonal action

Hormone	Mode of association with hormones
Insulin	Zinc associated with hormone storage
Angiotensin	Zinc in antiotensin-converting enzyme
Enkephalin	Zinc in the Enzyme enkephalinase
Several peptides	General hydrolysis by endopeptidase 24.11
Neurotensin	Degradation by zinc-dependent enzyme
Other peptides	Amino- and Carboxypeptidase (Zn) at synapses

in that the receptor in the cell for these hormones is very probably TFIIIA-like. Here is a suggestive connection between intracellular organic hormones, sterols and a regulatory metal ion, zinc, which parallels that between extracellular hormones and calcium ions (see Williams 1987). The possibility arises of a multiplicatory relationship between the organic compounds and several inorganic trigger zinc ions. Particularly fascinating is the known link of the steroid hormones and zinc with developmental features of sexual activity involved in puberty and fertilization. There is also the likelihood that other hormones which work at the DNA rather than the membrane level work through zinc-TFIIIA factors. An example is the thyroid hormone, again related to growth. In this respect the possible network of communication from EGF and other peptide hormones to zinc and DNA is worthy of study. Note that the zinc control would work with a *long* time constant, say > 1 h, like a steroid hormone (Giedroc et al. 1986).

Joint Role for Zinc and Copper Enzymes?

Cells generally have to be mobile and do not maintain one position in organisms indefinitely. The need to grow and multiply must be made consistent with the need for organization. This means that cells must be able to be confined, e.g. by a strong external network of proteins such as collagens or by chitins, but they must be able to break out from this confinement under different (growth) conditions. This break-out development demands the hydrolysis of the extracellular constraining matrix. Zinc enzymes, collagenases and other proteases are involved in this second activity. This contrasts with the synthesis of the stable collagen matrix which is controlled in its final stages by external cross-linking due to oxidative copper enzymes (see Fig. 2.2). We begin to see that a proper balance between zinc and copper is an essential ingredient of the alternating demands of stability and growth of the soft structures of an organism. Is the link through metallothioneins which bind to both metal ions but in different ways?

Perhaps we are beginning to see the ways in which the primitive digestive, and probably not very specific, zinc enzymes of prokaryotes have not only kept this function in higher organisms but have become adapted to handle the network of molecular messages (hormones and triggers) and filamentous communications and connections (collagens and chitins) between soft structures. We are left to wonder if there are similar requirements inside cells or if in fact that chemistry is left to calcium. Calcium and zinc are two very fast-acting acids with no redox properties.

Zinc in Cross-linking of Extracellular Soft Matrices

There is growing evidence that zinc is used to cross-link phenolic materials which strengthen the teeth of insects and lower animals. These teeth or mandibles are made from chitins. The presence of high zinc concentrations in the teeth has not been definitely associated with phenolate side-chains but this remains highly probable. The movement of the zinc down hollow central tubes of the teeth towards the cutting edges is thought to occur by the transfer of vesicles with high zinc concentrations.

Zinc in Solid-State Devices

Zinc occurs in some simple crystalline materials within biology. The most obvious are the deposits of zinc phosphates in what may be storage vesicles of such animals as snails. The real purpose of the vesicles is unknown though they could act as depositories of unwanted excess zinc until it is rejected by exocytosis. Another certainly useful zinc deposit is crystalline zinc cysteinate in the tapetum of night-seeing animals. It seems likely that these crystalline deposits were also laid down in vesicles. The crystals act as reflectors giving added sensitivity to the eye.

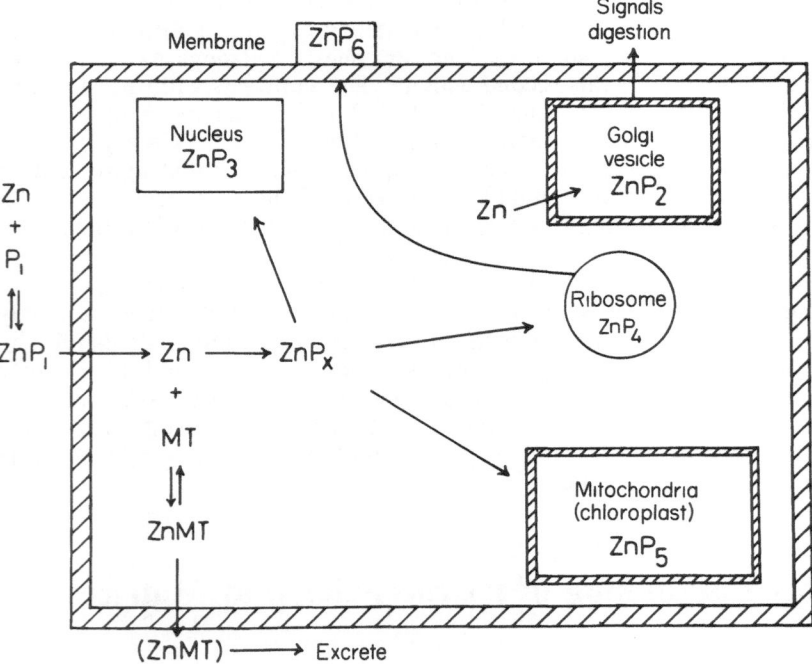

Fig. 2.5. An expanded version of Fig. 2.2 showing the full involvement of zinc in many parts of biological space.

Conclusion

This chapter demonstrates the diversity of uses of zinc which are summarized in Fig. 2.5. It is not quite true to say that zinc has as many uses as sulphur but the time has come to look at the biology of zinc in a similar way to the way in which we look at that of sulphur (see Tables 2.1–2.7). This may stop the search for too many specific pathological features of zinc excess or zinc deficiency states, since such states may well generate a general impairment of many functions.

The references accompanying this chapter, which is an introduction to the principles behind the use of zinc in biochemistry, are deliberately few. The inorganic chemistry of zinc is well described in modern textbooks (see below). The bio-inorganic chemistry has been described in detail in three review books by experts on particular topics (also see below). The biology of zinc follows in this book. The author of this chapter saw it as his purpose to pull all these lines together so as to provide a guide through chemistry and biochemistry, the main substance of the book. Naturally this has had to involve some speculative comment but the deep involvement of zinc in biology is gradually becoming apparent.

References

Berg J (1986) Potential metal-binding domains in nucleic acid binding proteins. Science 252: 485–493

Bertini I, Luchinat C, Maret W, Zeppezaner M (eds) (1986) Zinc enzymes. Progress in biochemistry and biophysics, vol 1. Berkhauser, Basle

Bruland KW (1983) Trace elements in sea water. In: Riley JP, Chester R (eds) Chemical oceanography, vol 8. Academic Press, London, pp 84–96

Giedroc DP, Keating KM, Martin CT, Williams KR, Coleman JE (1986) Zinc metalloproteins involved in replication and transcription. J Inorg Biochem 28: 155–169

Klug A, Rhodes D (1987) Zinc fingers. Trends Biochem Sci 12: 464–469

Siegel H (ed) (1983) Zinc and its role in biology and nutrition; metal ions in biological systems, vol 15. Dekker, New York

Spiro TG (ed) (1983) Zinc enzymes; metal ions in biology, vol 5. Wiley, New York

Williams RJP (1987) The functions of structure and dynamics in proteins, peptides and metal ion complexes. Carlsberg Res Comm 52: 1–30

Chapter 3

Intestinal Absorption of Zinc

B. Lönnerdal

Introduction

It may appear somewhat surprising that, in spite of all recent concern about zinc nutrition in humans and research on zinc metabolism in man and experimental animals, we know so little about the mechanism of zinc transport across the small intestine. This step in zinc metabolism is likely to be a main regulatory mechanism of zinc homeostasis and therefore increased knowledge about the molecular events at the cell membrane and inside the enterocyte will be crucial for interpreting the control of zinc influx and efflux. It is less surprising though if one compares our knowledge about mechanisms of zinc absorption with what is known about absorption of other trace elements such as iron. Even though the knowledge about homeostatic regulation of iron absorption and identification of the extra- and intracellular carriers of iron was obtained long before the era of intense research on zinc metabolism, we still do not know how the intestinal cell regulates iron absorption. With our increased availability of more sensitive methods in molecular biology and knowledge of cell biology it should, however, be reasonable to expect that the absorptive mechanisms can be elucidated in the near future. At present, it is necessary to identify what is known and what remains to be learned about the mechanisms of zinc absorption.

Postulated Mechanisms of Zinc Absorption – "Zinc Carriers"

The zinc ion in the intestinal lumen is likely to be coordinated to some ligand with affinity towards this cation. Digestion of ingested foods will lead to the presence of amino acids, organic acids, phosphates and other cations capable of associating with zinc (see Chap. 4). Such complexes of zinc are then presented to the surface of the mucosal cell. Thus it is uncertain whether "free" zinc ions, such as those presented when $ZnCl_2$ is used in in vitro experiments, occur under normal conditions in vivo. Therefore, although $ZnCl_2$ may serve as a convenient standard to which other zinc complexes can be compared, the "enhancing" or "inhibitory" effect of various ligands should be judged with caution.

Role of the Glycocalyx

When approaching the enterocytes lining the intestinal mucosa, intraluminal zinc complexes would encounter the glycocalyx first, a mucous layer of polysaccharides covering the brush border membrane. Few studies have assessed the role of the glycocalyx in zinc absorption, possibly because of the fragile and labile nature of this layer. Two possible roles can be postulated for the glycocalyx: (a) zinc is released from its complexes in the lumen because of a higher affinity of the element for some surface-exposed groups of the glycocalyx or (b) zinc complexes adhere to the glycocalyx and are transported across this mucous layer intact. Quarterman (1985) has performed studies of the binding of iron and zinc to the glycocalyx isolated from rat intestine. It was found that fasting increased the concentration of glycocalyx in the intestine and that this increase was accompanied by increased adsorption and transfer of iron and zinc. Thus, a role of the glycocalyx in mineral absorption could be inferred, although the precise mechanism is unknown.

Role of the Brush Border Membrane

Events at the surface of the mucosal cells involved in absorption have been investigated using brush border membrane vesicle preparations. This approach has been used for several minerals and trace elements and can provide important information about carrier molecules. Criticisms can be raised with regard to the validity of using such vesicles, however, and it is crucial that many control experiments are carried out to establish that the right side of the membrane is facing out to the medium, that osmolarity, pH and the endogenous concentration of minerals and the element to be studied are within a physiological range, etc. Menard and Cousins (1983) used brush border membrane vesicles (BBMVs) from rat intestine and found saturable zinc uptake at 0.2 mM extravesicular zinc while zinc uptake was non-saturable at 1.0 mM. Initial rate of uptake was linear up to 0.3 mM extravesicular zinc. Since some binding of zinc occurred at zero intravesicular space, it was indicated that membrane binding may occur during transport. Sodium ions and ATP did not affect zinc uptake. Kinetic studies suggested a K_m of 0.38 mM and a J_{max} of 5.4 nmol/min/mg protein. These authors concluded that at lower zinc concentrations (0.2 mM) uptake is saturable and occurs by a carrier-mediated process that does not appear to be energy dependent. At higher zinc concentrations, uptake is linear, indicating a passive diffusion process. It was cautioned, however, that a negative effect of ATP observed at low zinc concentration could be due to either binding of zinc to ATP, making less zinc available, or to the fact that maximum uptake by the saturable process already had occurred and that uptake with ATP only represented the non-saturable (passive diffusion) component occurring at higher zinc concentrations. Further studies appear necessary to resolve the question about the possible energy dependency of the carrier-mediated transport. It appears evident, however, that part of the uptake is due to binding of zinc to the mucosal surface.

Zinc deficiency did not appear to affect the affinity (K_m) for zinc but the uptake capacity (J_{max}) for the carrier-mediated transport system was considerably higher (12.0 nmol/min/mg protein). This would be consistent with the higher zinc absorption observed during zinc deficiency (see below). Finally, the authors were able to show a similarity between data obtained in vivo for K_m (Davies 1980) and J_{max} (Smith and Cousins 1980), supporting the validity of the BBMV system used.

Recently, Hoadley and Cousins (1987) solubilized brush border membranes and studied ^{65}Zn binding to solubilized membrane proteins. Most of the ^{65}Zn binding was localized to membrane proteins with molecular weights $> 100\ 000$. No induction of a membrane carrier protein was observed when rats were fed a zinc-restricted diet although J_{max} was found to be increased threefold for the saturable component of zinc uptake. Seal and Heaton (1987) also studied zinc-binding proteins in rat mucosa and found two main components with molecular weights (MW) of 45 000 and 6500. These authors speculated that zinc first binds to the 6500 MW component passively and then is transferred actively to the 45 000 MW component. There is, however, some question regarding these molecular weights since the 45 000 MW component could not be distinguished from the void volume of the column, indicating suboptimal chromatography conditions. It has been cautioned that sample pretreatment and column techniques will dramatically affect the results when separating mucosal zinc-binding proteins (Lönnerdal et al. 1980a).

Role of Intracellular Ligands

The potential function of various intracellular zinc-binding ligands in controlling zinc absorption has been a subject of much controversy (Cousins 1985). A key role for metallothionein (MT) in this process was suggested by Richards and Cousins (1975). By this concept, mucosal MT is reflecting dietary zinc status, and the amount of MT present in the mucosa is negatively correlated to zinc absorption, thereby exerting homeostatic control of zinc metabolism. Although some evidence for this was originally obtained, other reports have been in conflict with this concept. Starcher et al. (1980) found increased zinc absorption in mice that had been previously injected intraperitoneally with zinc to induce MT. Olafson (1983) also studied zinc absorption in mice and its relationship to MT in zinc-deficient and zinc-replete animals but failed to support Cousins' findings. Similarly, Flanagan et al. (1983) reported that induction of MT by dietary zinc was only transitory and that stress-induced synthesis of MT caused by zinc deficiency was responsible for the effects observed.

Steel and Cousins (1985) showed in a vascularly and luminally per-fused system that newly absorbed zinc became bound to a pattern of intra-

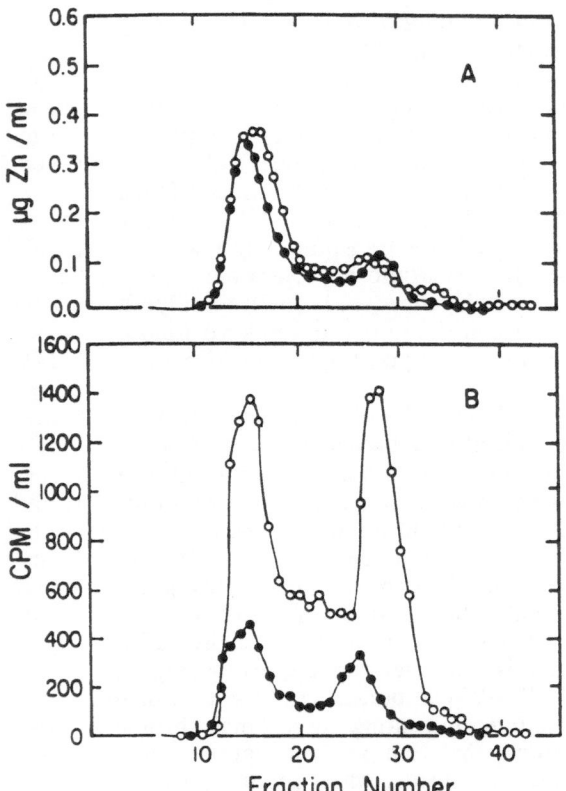

Fig. 3.1. Gel filtration chromatography on Sephadex G–75 of mucosal cytosol from intestines of rats fed either a zinc-adequate (●———●) or zinc deficient (○———○) diet prior to infusions. **a** Cytosolic zinc distribution prior to perfusion. **b** Cytosolic ^{65}Zn distribution at end of perfusion period (60 min) where luminal zinc concentration was 8 μM. (From Steel and Cousins 1985)

cellular ligands that differed from that of rats prior to the experiment. Both high-molecular-weight proteins and metallothionein were found to bind newly absorbed zinc in a 60 : 40 ratio for zinc-adequate and zinc-deficient rats, but zinc-deficient rats retained considerably larger quantities of ^{65}Zn in both peaks than control rats. Prior to the experiment, only 20% of zinc was bound to metallothionein and there was no difference between zinc-deficient and control rats (Fig. 3.1). The levels of zinc used were not considered high enough to induce mucosal synthesis of metallothionein. It appears therefore that preformed metallothionein provides a labile pool of zinc during high rates of transmucosal flux.

Recent studies by Coppen and Davies (1987), while not in conflict with Cousins' original hypothesis, offer no confirmatory evidence. These authors found a linear inverse relationship between zinc absorption from a test meal and zinc in intestinal MT when the zinc level in the diet was between 5 and 80 mg/kg but no relationship at higher dietary zinc levels. Thus, when zinc bound to intestinal MT doubled, zinc absorption was unaffected. The authors suggested that MT may regulate zinc excretion when dietary zinc levels exceed the requirement for growth. Therefore, at levels of 20 mg zinc/kg diet and higher, an increased turnover and loss of body zinc may be a consequence of intestinal MT levels. However, as pointed out by the authors, further studies in which both zinc and MT are labelled are required to investigate the processes regulating the intracellular trafficking of zinc and MT.

Role of the Basolateral Membrane

This step in zinc absorption has received relatively little attention. Kowarski et al. (1974) reported that 2,4-dinitrophenol, which inhibits oxidative phosphorylation, causes a reduction in zinc absorption by segments of rat jejunal mucosa. Since the studies by Menard and Cousins (1983) suggested that brush border transport of zinc is not energy dependent, these authors proposed that zinc transport at the basolateral membrane may be the step that requires energy. This would be consistent with our knowledge of calcium transport, which is known to occur by active transport at the basolateral membrane (Hildman et al. 1979). Evidently, more research is needed on this component of zinc transport.

Integrity of the Mucosal Barrier

The increased zinc transport occurring during zinc deficiency feasibly could be due to a structural alteration of the intestinal mucosa which affects the influx/efflux of nutrients. Ghishan (1984) found that the net transport of water and sodium from the small and large intestines of zinc-deficient rats was lower than in pair-fed controls. Net absorption of glucose and potassium was not influenced by such dietary treatments, however, in their in vivo perfusion system. It is possible that the decreased sodium transport is due to altered membrane permeability as suggested from studies from Bettger et al. (1981). The lower water and sodium transport could possibly explain the diarrhoea observed in zinc deficiency. Somewhat conflicting results were presented by Southon et al. (1984) who found that the maximum transport rate of a hexose (galactose, 3-O-methylglucose) measured in everted jejunal rings of zinc-deficient rats was significantly higher than in pair-fed controls. It was also found that the mucosal morphology was changed in the zinc-deficient rats in that villi from the proximal jejunum were shorter and narrower across the base, but present in greater numbers per serosal area than in control rats. It was also found that mucosal enzymes such as alkaline phosphatase and disaccharidase had lower activities in zinc-deficient rats than in controls and that the alterations were not due to changes in intestinal microflora (Southon et al. 1986). With regard to the ultrastructure of the mucosa, Moran and Lewis (1985) could not find any abnormalities in tight junction permeability or any damaged membranes as assessed by a lanthanum distribution study using electron microscopy. Rather, these authors suggested that effects on nutrient transport in zinc deficiency are due to "impaired transepithelial transport processes where carrier proteins are involved."

Factors Regulating Zinc Absorption

Absorption of zinc is most likely regulated by several control mechanisms. Although there have been many studies on "zinc bioavailability", largely relating dietary components to zinc retention as measured by body zinc or by radiozinc, few have attempted to define the quantitative significance of each control mechanism. The inherent capacity of each segment of the small intestine to absorb zinc and to be regulated, the effect of long-term zinc intake and thus zinc status versus the zinc content of a particular meal, age or whether a subject is pregnant or lactating, the use of liquid diet and concomitant water flux versus solid diets, as well as the body's ability to excrete zinc into the small intestine and re-utilize it, are all factors which need to be considered. Although most recommendations on dietary zinc intake are based on the requirements of healthy individuals, the ability of patients with chronic diseases to absorb and utilize zinc also needs to be considered.

Site of Zinc Absorption in the Small Intestine

Although we can now rule out both the stomach and caecum/colon as sites for zinc absorption (Methfessel and Spencer 1973a; Sandström et al. 1986), the quantitative importance of the different segments of the small intestine has not yet been accurately assessed. Early studies indicated that absorption of ^{65}Zn in the rat was higher from duodenum than from other segments when an in vivo ligated sac model was used (Methfessel and Spencer 1973a). Studies by Antonson et al. (1979), however, suggest that zinc absorption is much higher in the ileum (approximately 60%) than in the duodenum (19%) and jejunum (20%). These authors emphasized that the ligated sac method does not allow normal peristalsis, nor does it account for variations in intestinal contents. On the other hand, it could be argued that Antonson et al. completely removed the intestinal contents by perfusion and followed the absorption from a water solution, a situation not likely to occur under physiological conditions. Antonson et al. also emphasized the problem of disruption of vascular supply and possible hypoxic injury to transport mechanisms. The studies of both Methfessel and Spencer (1973a) using solid diets and Jackson et al. (1981) using water solutions in the rat indicate, however, a significant involvement of the upper small intestine in zinc absorption since the maximum rate of uptake of a test dose into plasma was achieved very rapidly (< 30 min). It may be that the ileum has the largest potential capacity for zinc absorption. However, the duodenum, with a relatively lower capacity to absorb zinc, has the first opportunity to absorb the element from digesta and thus much less is left for absorption by the ileum. This possibility was not taken into account by Antonson et al.

Zinc Transport Across the Intestinal Cell

Two mechanisms for control of zinc homeostasis were suggested in 1962 by Cotzias et al. They suggested that higher absorption at lower body zinc load and increased excretion at high zinc intake both contribute to zinc homeostasis.

To investigate the role of the mucosal cell in the control of zinc absorption and homeostasis, Jackson et al. (1981) performed a series of experiments in which several current hypotheses were refuted or supported. By giving ^{65}Zn by gavage to rats and following the distribution of radioisotope with time it was found that control and zinc-depleted rats retained the same amount of ^{65}Zn in the intestinal wall even though more ^{65}Zn was found in the carcass of the zinc-depleted rats. The proportion of radioisotope in the intestinal wall declined with time, but the rate of decline was similar for both groups of rats. About 60% of the dose was absorbed from a water solution of ^{65}Zn (with 0.25 µmol zinc) in control rats, while about 90% was absorbed in zinc-depleted rats. These experiments clearly showed homeostatic control of zinc absorption in rats. Whether this is true in humans is not yet clear, but it is interesting to note that zinc in water solution is also absorbed to about 60% in humans (Sandström et al. 1985).

The experiments of Jackson et al. (1981) also refuted the earlier hypothesis of Richards and Cousins (1975) which suggested that control of absorption was exerted by induction of metallothionein, a zinc-binding protein, which, when synthesized inside the intestinal cell, could sequester zinc and make it unavailable for transfer to the circulating plasma. Since the intestinal wall retained the same proportion of ^{65}Zn in both control and zinc-deficient animals, this could not have been the case. When high levels of zinc were injected intraperitoneally, zinc absorption decreased after 6–24 h, possibly by induction of metallothionein. Subsequent studies by Cousins (1985) have also shown that relatively large quantities of zinc are needed to induce metallothionein and that this is not likely to occur under physiological levels.

Another physiologically significant observation by Jackson et al. was the lack of evidence of a mucosal "efflux" of zinc; thus the decreasing amounts of ^{65}Zn inside the intestinal wall with time were completely accounted for by a corresponding increase in carcass ^{65}Zn. Thus, zinc newly absorbed from the lumen into the mucosal cell does not appear to be re-excreted from the cell into the lumen. This is in agreement with results of studies by Davies (1980), who found that all ^{65}Zn in ligated rat duodenal loops was transferred to the carcass. Over a longer time perspective, however, it is evident that zinc will leave the body via the small intestinal mucosa, although whether this is by desquamation of cells or actual secretion is uncertain. Methfessel and Spencer (1973b) showed that ^{65}Zn injected intraperitoneally appeared in segments of the small intestine within 30 min after injection and reached peak values after 3 h (Fig. 3.2). On a per weight basis, duodenal, jejunal and ileal segments contained a similar percentage of ^{65}Zn, while the stomach and caecum/colon contained little ^{65}Zn. That the injected ^{65}Zn actually left the body was shown by a faecal excretion of 18% at 24 h after injection. Some evidence for this zinc being secreted into the intestinal lumen was obtained by Davies and Nightingale (1975). With regard to control of zinc uptake, Jackson et al. found high plasma zinc levels during the first 6 h after intraperitoneal injection but no reduction in zinc absorption, thereby refuting the hypothesis of Evans et al. (1975) which suggested that saturation of serum zinc carriers (albumin) would lead to lower release of zinc at the basolateral membrane.

The observation that the percentage of ^{65}Zn bound to the intestinal cell is constant and independent of the total amount of zinc absorbed at any time

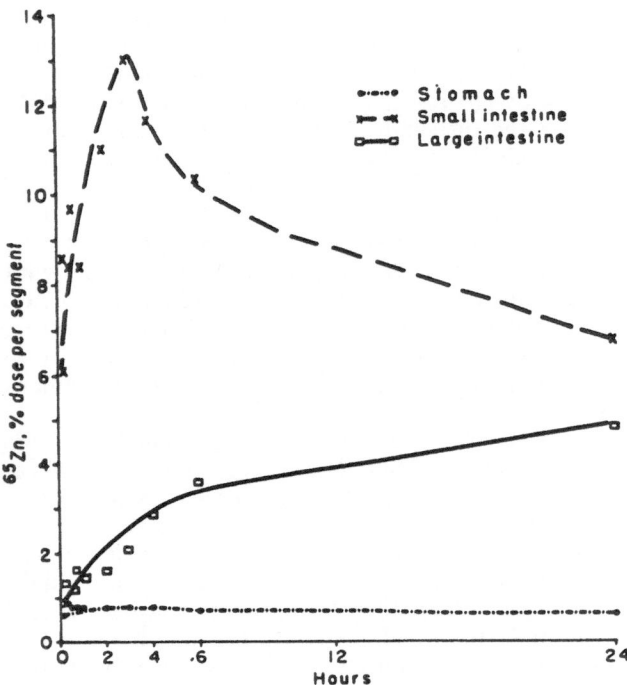

Fig. 3.2. Total ^{65}Zn content of stomach, small intestine and large intestine in rats given a single intravenous injection of ^{65}ZnCl$_2$. (From Methfessel and Spencer 1973b)

point led Jackson et al. to suggest that there are two mechanisms for zinc absorption: one mechanism which is obligatory and involves binding of zinc to the mucosa and a second one by which zinc is rapidly transferred into the body without being retained in the mucosa for any substantial time. The latter would be located in the upper small intestine as all absorption by this process is complete within 1 h after administration of ^{65}Zn. An alternate explanation would be a change in number or turnover of mucosal binding sites; this was argued against by the similarity of the rate of decrease of mucosal zinc binding in control and zinc-deficient animals.

Davies (1980) described two phases of zinc absorption by using the ligated rat duodenal loop technique. These two phases, one slow and one rapid, are likely to represent two distinct processes of zinc absorption. Jackson et al. tried to define the "triggering point" for the rapid, high affinity process and found it to be active in rats fed a diet containing 0.15 μmol/g (10 mg/kg) zinc and partially active when they were fed a diet with 0.24 μmol/g (16 mg/kg). The characteristics of the slow process of zinc absorption are consistent with a carrier-mediated transport with Michaelis–Menten kinetics. Saturation would occur between 1 and 5 μmol of zinc/g (60–300 mg/kg) in the diet. In rats given control levels of zinc, less than 1 μmol of zinc would be likely to occur in the lumen at any time; therefore, the absorption would follow the linear part of the curve and any increase in zinc intake would be followed by a proportional increase in absorbed zinc. In order to maintain homeostasis, increased gastrointestinal secretion of zinc must occur. The increase in luminal zinc occurring after intraperitoneal zinc injection supports this idea. Weigand and Kirchgessner (1978) have also suggested that under normal conditions, zinc homeostasis is controlled by gastrointestinal excretion. A limited study on zinc homeostasis in a human subject (Jackson et al. 1984) supports the idea that smaller fluctuations in zinc intake are followed by a change in gastrointestinal secretion of zinc, but that this short-term mechanism is of limited capacity. Larger fluctuations in zinc intake appear to cause a regulation in the absorptive process, but although this regulation has a larger capacity to adapt, it is much slower to respond (1–4 d).

Jackson et al. (1981) also refuted the contention that intraluminal ligands control zinc absorption. By exchanging duodenal washings from zinc-deficient rats and control rats, no changes in zinc absorption were observed. Thus, intraluminal factors from zinc-depleted rats do not stimulate zinc absorption, nor do these factors from control rats inhibit zinc absorption. This finding also supports the hypothesis of a second high-affinity mechanism for zinc absorption and the authors suggested an inducible zinc-binding ligand in the mucosal cell of the proximal small intestine.

Recently, Steel and Cousins (1985) investigated the kinetics of zinc transport across the small intestine by perfusing zinc-adequate and zinc-deficient rats both vascularly and luminally. The authors took several precautions to ensure that each perfusate contained appropriate endogenous components (such as serum albumin and glutamine) and that no absorption of a non-absorbable marker occurred. It was found that the zinc absorption rate in the zinc-deficient rat was much higher than in the control rat (Fig. 3.3). In the last 10 min of the perfusion, steady-state transfer was reached and distinct differences were found for control and zinc-depleted rats (Fig. 3.4). When comparing their results with those from a previous study using brush border membranes (Menard and Cousins 1983), in which similar K_m but different J_{max} were found in control and zinc-deficient rats, they suggested that the saturable membrane transport mechanism (but not affinity) increases when zinc supply decreases. Therefore, at low luminal zinc levels, the mediated uptake accounts for a larger percentage of zinc uptake than when zinc levels are higher (Fig. 3.5). The authors argued that differences in experimental design and the level of dietary zinc may explain why Jackson et al. (1981) concluded that greater transfer of zinc into the body accounted for higher zinc absorption in the zinc-deficient rat. Saturation of the mediated component of zinc absorption was reached at about 100 µM luminal zinc when absorption was measured directly by Steel and Cousins, which is in close agreement with that calculated by Bonewitz et al. (1983) from their recirculation perfusion system. With regard to vascular-to-lumen flux of zinc, Steel and Cousins found no difference between zinc-deficient and control rats and also that the vascular concentration of zinc did not significantly affect the results when it was within a physiological range. They suggested that transport from the serosal to the mucosal side of the small intestine is relatively low under normal circumstances. Their data indicate that brush border membrane uptake is the rate-limiting step in zinc absorption and that although vascular zinc enters the mucosal cell, most of this is transferred back to the circulation.

In order to further delineate the mechanisms controlling the steps of zinc uptake from the intestinal lumen to the vascular system, Hoadley et al. (1987)

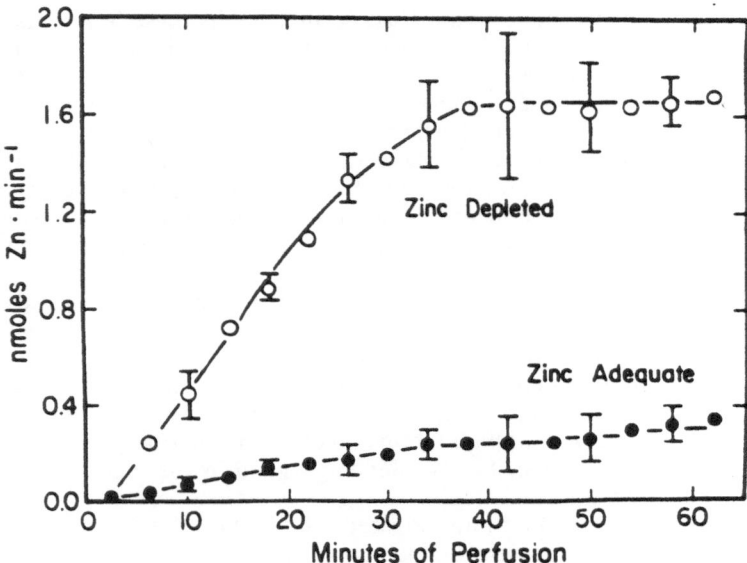

Fig. 3.3. Time course curve for increase in rate of zinc absorption (transfer to portal vascular perfusate) by rat intestine where luminal and vascular supplies were perfused simultaneously. Before perfusion studies, rats had been fed either zinc-adequate or zinc-deficient diets (zinc-depleted). Luminal and vascular perfusates contained 26 and 8.7 μM zinc, respectively. To measure absorption rates, 2-min fractions of vascular effluent were collected, and ^{65}Zn content was determined and corrected for specific activity of luminal perfusate. Each *curve* is mean absorption rate at each time derived from perfusion of four separate intestines. Only mean of every other fraction is shown. Mean ± SD is shown for every 4th fraction starting at 10 min of perfusion. (From Steel and Cousins 1985)

studied luminal-to-mucosal and mucosal-to-vascular transfer of zinc in vascularly perfused rats. They found that mucosal uptake of zinc was significantly higher in zinc-deficient than in zinc-adequate rats when luminal zinc concentration was 50 μM or less. Similar to their previous study, K_m was not different while J_{max} was threefold higher in zinc-deficient rats as compared with controls. The diffusion constant (K_d) for the first-order non-saturable component was the same in the two dietary groups. Thus, the difference in zinc uptake at low luminal zinc concentration (5 μM) was accounted for by a relatively larger involvement of the saturable mechanism in zinc-deficient rats (73%) than in zinc-adequate rats (43%). Increasing mucosal zinc led to an increase both in luminal (secreted) and vascular (absorbed) zinc while decreasing the percentage of zinc retained by the mucosal tissue. However, a major portion of the zinc (75%–90%) was retained by the mucosa and this was independent of zinc status. After "loading" the mucosa with zinc at varying concentrations in the luminal perfusate for 30 min, the fate of the mucosal zinc was studied by perfusing the lumen with a zinc-free medium. Zinc-depleted rats were found to secrete considerably less zinc into the lumen than zinc-adequate rats and this was largely independent of mucosal zinc. Transfer of zinc from the mucosa to the vascular perfusate increased with increasing mucosal zinc and zinc-deficient rats tended to absorb more zinc than control rats.

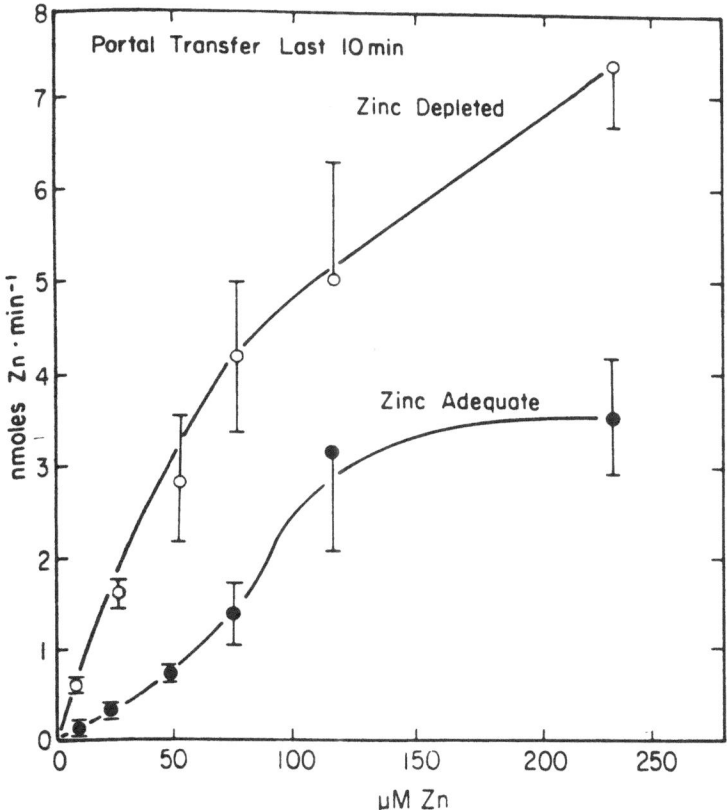

Fig. 3.4. Rate of zinc absorption (transfer to vascular perfusate) plotted against luminal zinc concentration measured during last 10 min of perfusion period. Intestines were from rats fed either zinc-adequate or zinc-deficient (zinc-depleted) diets. Means ± SD of at least four intestines from rats of each dietary group perfused at each luminal zinc concentration are shown. (From Steel and Cousins 1985)

The authors concluded that intestinal zinc transport is homeostatically regulated by both absorption and secretion, although the bulk of zinc uptake is not associated with the short-term absorption process nor with homeostatically regulated mechanisms. They suggested that the binding capacity for zinc within the mucosa is limited and when mucosal zinc increases, a larger proportion of loosely attached zinc will cause a greater degree of dissociation.

In summary, it appears evident that mucosal uptake of zinc occurs via two processes; one non-saturable process which is not affected by dietary zinc intake and a saturable process which is stimulated by zinc depletion, possibly via induction of a carrier-mediated mechanism (Hoadley et al. 1987). Since no evidence was found for saturability of zinc absorption beyond the initial uptake, the brush border membrane is suggested to provide the primary homeostatic control of zinc absorption. When dietary zinc intake is low, lower secretion of zinc from the mucosa to the lumen and higher absorption from the mucosa to the vascular compartment contribute to the higher absorption of zinc observed

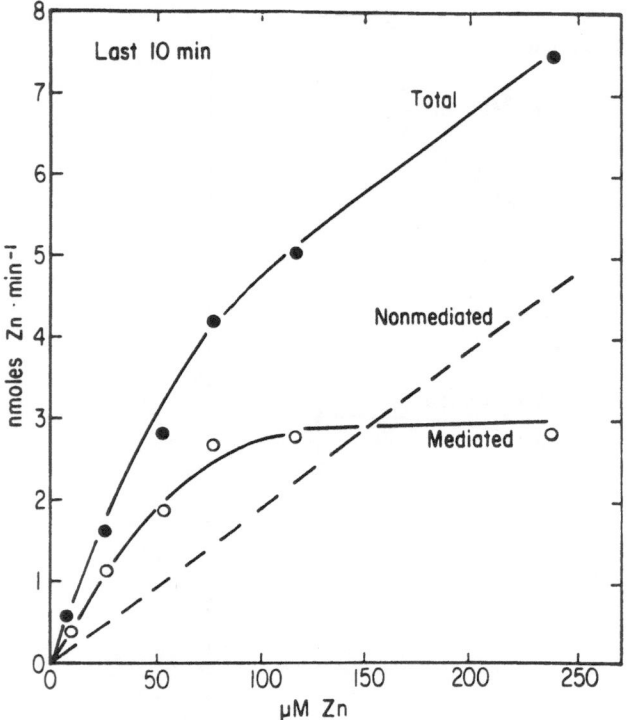

Fig. 3.5. Total absorption (vascular transfer) rate for intestines from zinc-depleted rats during last 10 min of perfusion period (shown in Fig. 3.4) when separated into non-mediated and mediated components. Total transfer was primarily derived from mediated transfer at lower luminal zinc concentrations. (From Steel and Cousins 1985)

under this condition (Fig. 3.6). This will result in a significantly shorter half-life of zinc in zinc-deficient (13 min) than in zinc-adequate animals (24 min).

Level of Dietary Zinc/Zinc Status

The relationship between previous intake of dietary zinc as well as level of zinc in the meal and zinc absorption was recently studied by Coppen and Davies (1987). By feeding rats diets with 5, 10, 20, 40, 80 or 160 mg zinc/kg for 2 weeks and giving a test meal with ^{65}Zn (added to the 20 mg zinc/kg diet), the authors found that the ^{65}Zn absorbed varied inversely with dietary zinc intakes of between 5 and 40 mg zinc/kg. At higher dietary zinc levels, however, efficiency of absorption was not affected by dietary zinc. Turnover and body loss of zinc were particularly slow at the lowest levels of dietary zinc (5–10 mg zinc/kg), but increased proportionally with increasing dietary zinc above those levels. This study showed that absorption of zinc can be regulated up to a certain level of dietary zinc, but that excretion of zinc mainly regulates zinc homeostasis in the body above this level of zinc.

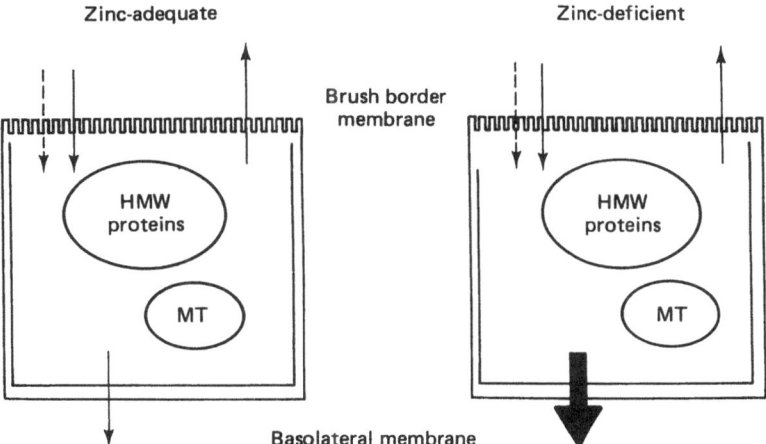

Fig. 3.6. Schematic representation of zinc absorption in zinc-adequate and zinc-deficient rats. (-----, saturable, mediated transport; ———, non-saturable transport, ← ←, boldness of arrows and ligands indicates magnitude.)

Varying the dose of dietary zinc in rats maintained on the same level of zinc resulted in a relationship between meal zinc and zinc absorption that appeared curvilinear. At levels of dietary zinc between 40 and 160 mg/kg this relationship was linear, indicating that at lower levels of dietary zinc, absorption is achieved by a saturable process but that above a certain level, the capacity of this saturable process is exceeded (Coppen and Davies 1987). It is interesting to note that in agreement with the results of Coppen and Davies, the estimate for saturation of the putative high-affinity carrier for zinc given by Jackson et al. was 60–300 mg/kg, i.e. the efficiency of absorption would be lower at higher levels of zinc.

Liquid Diets as Compared with Solid Diets

Although it is known that zinc is absorbed to a high extent when given as a salt in water solution, this area has received little research interest. This could, however, be of importance, especially as it relates to hospital patients fed liquid enteral diets. It has been shown that zinc absorption from human and cow's milk, although different, appears to be higher than from most solid diets in human adults (Sandström et al. 1983). The reasons for this could be several: the relatively low caloric density, leading to more efficient digestion; the presence of a ligand such as citrate that may facilitate uptake of zinc; or increased water flux across the intestinal mucosa. The relative importance of these factors has not been adequately studied. It is interesting to note, however, that Ghishan (1984) found higher sodium and water transport in zinc-deficient animals as compared to controls. Citrate has been shown to increase jejunal water and sodium transport in humans at a concentration similar to that found in milk (5 mM) (Rolston et al. 1986). Therefore, it is possible that citrate either is co-transported with zinc into the mucosal cell or that zinc chelated to citrate is presented to the mucosa in an easily accessible form, is bound and released from citrate to a membrane carrier and citrate is transported separate from zinc into the cell. Higher concentrations of citrate (50 mM) did not have this stimulatory effect and it is therefore possible that when citrate was added to milk, which is already high in citrate, no increase in zinc absorption was found (Sandström et al. 1983). It is evident that further studies are needed in the area of fluid and electrolyte transport and zinc absorption.

Physiological State

The recommended dietary allowance for zinc is highest for infants and pregnant and lactating women; this takes into account the increased demands on zinc supply by the foetus and milk production. There is, however, also a possibility that the efficiency of zinc absorption by the small intestine is increased during these physiological states.

Zinc Absorption During Pregnancy and Lactation

Davies and Williams (1977) investigated zinc absorption in pregnant and lactating rats using isolated duodenal loops in situ. They found that zinc absorption expressed as zinc absorbed per loop increased progressively after day 12 of pregnancy and continued to rise during lactation. However, expressed on a dry matter basis, absorption increased during pregnancy but there was no further rise during lactation. To investigate whether the increase in zinc absorption was due to a general increase in nutrient absorption, they studied the absorption of lysine in the same animals. Lysine uptake was not increased in pregnant rats as compared to non-pregnant rats; however, during lactation there was a progressive increase in zinc absorption. This increase in lysine uptake was not found when absorption was expressed on a dry weight basis. The authors concluded that retention of zinc during the first part of pregnancy was a reflection of maternal growth only, since whole-body zinc retention increased but absorption by the duodenal loops was the same as in non-pregnant animals. During the latter part of pregnancy, zinc absorption is increased specifically, probably to meet the increased need for zinc of the foetus, since maternal zinc retention after subtracting foetal zinc was unchanged. The increase in absorption of zinc during lactation was accompanied by an increase in lysine uptake and these increases were not found when expressed on a dry weight basis. This suggests that mucosal hypertrophy is responsible for the general increase in absorption that was found. Thus, there are increases in the absorptive capacity of the small intestine during pregnancy and lactation, but the mechanisms behind these increases are different.

Studies on zinc absorption during pregnancy and lactation in humans are very limited. Swanson et al. (1983) used a stable isotope of zinc and found no significant difference in zinc absorption between pregnant and non-pregnant women. It is possible, however, that only a small difference in zinc absorption during pregnancy in the human may be sufficient for adequate zinc accrual to cover the need of the foetus, while the rat with many foetuses and a short pregnancy must require much faster zinc accrual.

Development of Zinc Transport Mechanisms with Age

It is well-known that the newborn animal absorbs zinc to a higher degree than the older animal (Weigand and Kirchgessner 1979). Ghishan and Sobo (1983) used an intestinal in vivo perfusion technique in the rat at three different ages: suckling (2 weeks old), weanling (3 weeks old) and adolescent (6 weeks old). These investigators found a curvilinear relationship between absorption and luminal zinc in the proximal small intestine and a linear relationship at the distal part of the small intestine. In weanling and adolescent rats, "curvilinear" absorption occurred in both segments of the intestine. The saturable process followed Michaelis–Menten kinetics with K_m increasing with age (35, 62 and 99 μM for the proximal part at 2, 3 and 6 weeks, respectively) as well as J_{max} (2–3 μmol/g dry weight/h). The data were interpreted as evidence for a transport system with higher affinity for zinc at a young age. Kennedy and Lönnerdal (1987), using brush border membrane vesicles from 1–, 7–, 14– and 21–day–old rat pups, found saturation at zinc concentrations of 400–500 μM, but no difference in K_m or J_{max} for the ages investigated. Total zinc uptake (expressed as μg Zn/mg total membrane protein), however, increased with age and all values were higher than those found for adult rats. The saturable process and the decreased binding of zinc to the membranes with increasing zinc concentration suggest that one or more luminal membrane components are involved in the process.

Endogenous Ligands Binding Zinc

Several studies have established that zinc homeostasis is in part regulated by endogenous zinc excretion either via desquamation of cells or in intestinal juices. Consequently, zinc will be presented to the intestine bound to various endogenous ligands. Several such ligands have been reported with varying biochemical evidence. What is more important, however, is whether these ligands secreted via pancreatic juice, bile or mucosal "backflux" are required in zinc uptake processes.

An involvement of pancreatic secretions in zinc absorption was suggested by Evans et al. (1975). Their studies indicated lower absorption of zinc after ligation of the pancreatic duct in rats as compared with controls. In vitro studies implicated a low-molecular-weight compound purified from dog pancreas as responsible for this effect. The compound was successively characterized as a peptide, N,N,N'-trimethyl-1,2-ethane diamine or picolinic acid (Evans and Johnson 1980). These studies suffered from methodological difficulties and the nature of the "low-molecular-weight ligand" is still dubious; e.g. picolinic acid could not be detected in pancreatic fluid (Rebello et al. 1982). It should be recognized that the separation and isolation of compounds from pancreatic juice is difficult. The presence of several proteolytic enzymes capable of hydrolysing proteins and peptides in the fluid and high concentrations of anions such as bicarbonate and citrate are confounding factors. It is well-known, however, that carboxypeptidases A and B have zinc as a co-factor (Vallee and Neurath 1954). When rat pancreatic fluid is kept on ice and chromatographed rapidly to avoid autolysis, carboxypeptidase was found to bind the vast majority of zinc present in the fluid (Reinstein et al. 1987). It is possible that the observation of lower zinc absorption when the pancreatic duct was ligated is due to the absence of proteolytic enzymes. This would then limit protein digestion and release of absorbable zinc. In bile, zinc is associated with low-molecular-weight ligands (Lönnerdal et al. 1980b).

The effect of the pancreas and bile fluid on zinc transport in the rat was studied by Lyall et al. (1981). Ligation of the common bile duct or subtotal pancreatectomy failed to affect zinc uptake in vivo as compared to sham-operated controls. In addition, when collected bile was added to 11-day-old rat pup intestine in vitro, an age at which only a passive transport mechanism was supposed to be operational, no difference in zinc uptake was observed. In contrast, Schricker and Forbes (1978) found a negative effect on zinc absorption of ligating the common bile duct and Antonson et al. (1979) found increased zinc uptake by duodenal segments when bile and pancreatic fluid were excluded. It is difficult to evaluate these experimental data, but it is likely that the in vivo results are the better estimate.

Data from Methfessel and Spencer (1973b) also argue against a significant involvement of pancreatic fluid and bile in zinc excretion. They found that excretion of ^{65}Zn injected intraperitoneally was similar in all segments of the small intestine and argued that if pancreatic fluid and bile were quantitatively important, initial zinc excretion into ligated segments of the intestine should have been much higher in the duodenum. A critical role for bile and pancreatic fluid in normal zinc absorption was ruled out by Vanderhoof et al. (1983) who found normal serum and hair zinc in a group of patients who had undergone complete bilio-pancreatic bypass 12–56 months earlier. Since these patients neither took therapeutic pancreatic enzymes nor increased their zinc intake, it was concluded unlikely that these fluids are of quantitative significance in zinc homeostasis in humans.

It is possible, however, that the pancreas is more important for zinc homeostasis in other species. Berger and Schneeman (1986) suggested that pancreatic enzymes have an important role in endogenous zinc secretion in the rat. The authors demonstrated higher output of zinc and carboxypeptidase in bile-pancreatic fluid from cannulated rats when fed a soya protein isolate as compared to casein. It should be recognized, however, that the significant difference in output of zinc does not necessarily have any quantitative importance in zinc homeostasis. The rats consumed a non-purified diet which probably contained at least 100 mg zinc/kg which would result in a dietary zinc intake of at least 2000 μg/d. Weigand and Kirchgessner (1980) have shown that true absorption of zinc at this intake is about 400–500 μg/d and that the endogenous losses are about 200–300 μg/d. In the study by Berger and Schneeman (1986) the difference in zinc output was 1.4 μg/h after stimulation by diet infusion and, assuming that rats are not constantly stimulated, the difference in daily output would be significantly less than 20 μg of zinc. Therefore, output of pancreatic fluid and bile is not likely to have a major role in zinc homeostasis. This would agree with studies by Stake et al. (1974) who demonstrated that zinc in pancreatic juice contributes only a minor part of the endogenous zinc losses in cows.

Prostaglandins were formerly believed to be zinc-binding ligands; however, stoichiometrically this is not a likely situation. There does appear to be an effect of prostaglandin metabolites on zinc absorption and also of zinc status and metabolism on prostaglandin metabolism (Song and Adham 1985). The effect is more likely to be a physiological, systemic consequence than a chelation effect. Since both the synthetic glucocorticoid dexamethasone and adrenalectomy are known to affect zinc absorption kinetics (Bonewitz et al. 1983) differently, the implication is that some hormonal control of the absorption process in the intestine is likely.

Pathological Conditions

The possibility of impaired intestinal zinc absorption in some disease states, further complicating the condition of the patient, is an area of concern. Logically, zinc absorption could be expected to be affected by conditions directly affecting the gut, such as coeliac disease, Crohn's disease and small bowel resection. However, there is also concern about other chronic debilitating diseases such as cirrhosis of the liver and diabetes.

Alcoholic Cirrhosis

Low liver zinc concentrations accompanied by low plasma zinc and high urinary loss of zinc in patients with alcoholic cirrhosis was first described by Vallee et al. (1957). The possibility of fibrous connective tissue in the livers of these patients causing a "dilution" of liver zinc was ruled out by Kiilerich et al. (1980) who expressed liver zinc in relation to the amount of hepatocytes and still found significantly lower liver zinc in these patients. The question whether the low liver zinc values were caused by malabsorption was addressed by Milman et al. (1983). These investigators used ^{65}Zn and whole-body counting and found higher zinc absorption in patients with compensated alcoholic cirrhosis (69%) as compared with healthy subjects (42%). A longer half-life of ^{65}Zn in these patients suggested a state of zinc depletion. However, Mills et al. (1983) also found increased zinc absorption in compensated alcoholic cirrhosis patients, but no evidence of depleted zinc stores.

Valberg et al. (1985) also studied zinc absorption in cirrhosis patients using a dual isotope technique. They found lower than normal zinc absorption in patients with alcoholic cirrhosis but normal zinc absorption in patients with non-alcoholic cirrhosis. However, the group of patients with alcoholic cirrhosis had complications such as ascites and encephalopathy; another group without complications had normal zinc absorption. Since zinc status in non-alcoholic cirrhosis, as assessed by leucocyte and liver zinc, is normal, these authors argued that chronic liver damage itself does not impair intestinal absorption of zinc. Instead, they suggested that the low zinc absorption found in decompensated alcoholic cirrhosis, which contributes to zinc deficiency, is due to an effect of alcohol on the zinc absorption process in the small intestine.

Crohn's Disease

Similar to alcoholic cirrhosis, Crohn's disease is usually accompanied by low plasma zinc values. Solomons et al. (1977) found that 25% of the patients studied had low plasma zinc and Fleming et al. (1981) reported an incidence of 46% in their patients. The latter study also revealed lower concentrations of serum albumin, in spite of an adequate dietary intake of zinc and other nutrients. Disease activity was found to be correlated to urinary zinc excretion but not to serum zinc, suggesting that the severity of the disease may affect zinc metabolism. Sturniolo et al. (1980) used ^{69m}Zn and followed the plasma appearance curve to assess zinc absorption. Patients with Crohn's disease showed a considerably lower degree of zinc uptake than healthy controls. Valberg et al. (1986) measured zinc absorption using ^{65}Zn in patients with Crohn's disease and found lower than normal (43%) zinc absorption in these patients as compared with controls (54%). Low zinc absorption was correlated to poor nutritional status, but not to disease activity or zinc status. In patients with mild-to-moderate forms of the disease and minimal nutritional impairment, zinc absorption was found to be normal. Thus, only patients with poor nutritional status and moderate or severe disease activity exhibited low zinc absorption, but poor zinc status was rarely noted. The authors suggested that low dietary intake of protein may have had a negative effect on zinc absorption and that direct effects on mucosal villi may not be the cause, as other nutrients appear to be absorbed to the same extent as in healthy individuals. It is possible that increased mucosal growth in parts of the intestine other than that directly affected by the disease may compensate for a lack of absorptive capacity. Urban and Campbell (1984) studied zinc absorption in rats after extensive small bowel resection and found significant mucosal growth, unchanged zinc transport per gram mucosa and increased zinc entry flux at the luminal side of the mucosa. Since both entry and exit fluxes at the basal surface of the mucosa increased, it was suggested that the mucosal zinc pool has a rapid turnover but the pool size is not changed.

Coeliac Disease

Impaired zinc status in patients with coeliac sprue, as assessed by low plasma zinc concentration and reduced taste acuity, has been described by Solomons et al. (1976). Although loss of zinc caused by binding to fat lost in the stool or by the protein-losing enteropathy accompanying the disease may in part explain a zinc deficiency state, the authors suggested that excessive zinc sequestration in the small intestine itself due to enhanced mucosal turnover may also be a causative factor. The primary site(s) of zinc absorption appear to be the duodenum and jejunum, which also are the sites of the most pronounced mucosal lesions in the disease. Coeliac disease in paediatric patients has also been shown to cause low plasma zinc values (Naveh et al. 1983; Guerrieri et al. 1986). Chronic diarrhoea in this age group can also lead to low plasma zinc, and the extent and duration of the diarrhoea may affect the precipitation of a zinc deficiency state (Guerrieri et al. 1986). There are few studies directly assessing zinc absorption in coeliac disease.

Diabetes

Most research on zinc absorption in diabetes has been performed in the streptozotocin-diabetic (STZ) rat and the relevance for diabetic humans still needs to be assessed. Craft and Failla (1983) found higher absorption of zinc and copper, but not of iron, in STZ rats as compared to controls. This higher absorption was concomitant with hyperphagia, hyperplasia of the small intestine and an increased glucagon/insulin ratio. All of these factors may stimulate zinc absorption and therefore need to be evaluated separately. The finding of similar quantities of zinc in the duodenal loop of STZ and control rats made the authors suggest that the homeostatic regulation of zinc absorption is impaired in the STZ rat. This is supported by Ghishan and Greene (1983) who found lower loss of endogenous zinc in the STZ rat than in controls. They, in turn, suggested that the lower endogenous loss of zinc was caused by a homeostatic response to the increased urinary loss of zinc occurring in the STZ rat. Studies in humans do not show any increase in zinc absorption; rather, data by Kinlaw et al. (1983) show that serum zinc uptake is lower in type II diabetes mellitus patients. These authors found a correlation between urinary zinc loss and glucose infusions and suggested that impaired intestinal absorption and increased urinary loss may produce zinc deficiency in these patients.

Acrodermatitis Enteropathica

The inborn error of zinc metabolism, acrodermatitis enteropathica (AE), was first described by Moynahan (1974). This disorder, which usually first appears in infants at the time of weaning, was earlier lethal but can now be treated with oral zinc supplements. The fact that the drug diodoquin can have a beneficial effect on the disease has been ascribed to its chelating capacity, which thereby enhances zinc absorption. This suggested an impairment in zinc absorption in these patients. Lombeck et al. (1975) assessed zinc absorption in AE patients by using ^{65}Zn and whole-body counting. The three patients investigated had significantly lower zinc absorption (16%–42%) than healthy subjects (58%–77%). Ultrastructural changes in Paneth cells of the small intestine were also observed and the authors suggested that the lesion in the disease may be defective intestinal absorption. Weismann et al. (1979) also measured zinc absorption in AE patients and found lower values than in controls. Interestingly, they found a considerably more marked defect in zinc absorption in two children with AE than in adult patients, suggesting a more dramatic response in the more rapidly growing individual. A defect in zinc uptake by jejunal biopsy samples from AE patients was subsequently described by Atherton et al. (1979). The authors therefore suggested that the primary abnormality responsible for the zinc deficiency of AE is a defective uptake of zinc by the enterocyte.

Conclusions

By dissecting the process of zinc absorption in the intestine into its integral components, i.e. the approach of the zinc ion (most likely coordinated to a ligand) to the glycocalyx and the unstirred water layer, the transfer of the zinc ion first across the brush border membrane and later over the basolateral membrane and finally, the fate of zinc within the enterocyte, it becomes evident that much further knowledge is needed about each of these steps. We now know that the net flux of zinc across these barriers is affected by zinc status, physiological state, disease etc., but little is known about the mechanisms controlling this flux. Although endogenous signals such as hormones and prostaglandins have been shown to exert some control, their quantitative importance is uncertain and they have rarely been studied by an integrative approach, i.e. by looking at several of the steps involved in zinc absorption both from a short-term perspective in vitro and/or in vivo and from a long-term perspective. These avenues of research should be a priority in zinc metabolism research for the near future in order to provide us with an enhanced knowledge of how the body can homeostatically regulate zinc absorption.

References

Antonson DL, Barak AJ, Vanderhoof JA (1979) Determination of the site of zinc absorption in rat small intestine. J Nutr 109: 142–147

Atherton DJ, Muller DPR, Aggett PJ, Harries JT (1979) A defect in zinc uptake by jejunal biopsies in acrodermatitis enteropathica. Clin Sci 56: 505–507

Berger J, Schneeman BO (1986) Stimulation of bile-pancreatic zinc, protein and carboxypeptidase secretion in response to various proteins in the rat. J Nutr 116: 265–272

Bettger WJ, Savage JE, O'Dell BL (1981) Extra-cellular zinc concentration and water metabolism in chicks. J Nutr 111: 1013–1019

Bonewitz RF Jr, Foullees EC, O'Flaherty EJ, Hertzberg VS (1983) Kinetics of zinc absorption by the rat jejunum: effects of adrenalectomy and dexamethasone. Am J Physiol 244: G314–320

Coppen DE, Davies NT (1987) Studies on the effects of dietary zinc dose on ^{65}Zn absorption in vivo and on the effects of Zn status on ^{65}Zn absorption and body loss in young rats. Br J Nutr 57: 35–44

Cotzias GC, Borg DC, Selleck B (1962) Specificity of zinc pathway through the body: turnover of Zn^{65} in the mouse. Am J Physiol 202: 359–363

Cousins RJ (1985) Absorption, transport, and hepatic metabolism of copper and zinc: special reference to metallothionein and ceruloplasmin. Physiol Rev 65: 238–309

Craft NE, Failla ML (1983) Zinc, iron, and copper absorption in the streptozotocin-diabetic rat. Am J Physiol 244: E122–E128

Davies NT (1980) Studies on the absorption of zinc by rat intestine. Br J Nutr 43: 189–203

Davies NT, Nightingale R (1975) The effects of phytate on intestinal absorption and secretion of zinc, and whole-body retention of zinc, copper, iron, and manganese in rats. Br J Nutr 34: 243–258

Davies NT, Williams RB (1977) The effect of pregnancy and lactation on the absorption of zinc and lysine by the rat duodenum in situ. Br J Nutr 38: 417–423.

Evans GW, Johnson PE (1980) Characterization and quantitation of a zinc binding ligand in human milk. Pediatr Res 14: 876–880

Evans GW, Grace CI, Votava HJ (1975) A proposed mechanism for zinc absorption in the rat. Am J Physiol 228: 501–505

Flanagan PR, Haist J, Valberg LS (1983) Zinc absorption, intraluminal zinc and intestinal metallothionein levels in zinc-deficient and zinc-repleted rodents. J Nutr 113: 962–972

Fleming CR, Huizenga KA, McCall JT, Gildea J, Dennis R (1981) Zinc nutrition in Crohn's disease. Dig Dis Sci 26: 865–870

Ghishan FK (1984) Transport of electrolytes, water and glucose in zinc deficiency. J Pediatr Gastroenterol Nutr 3: 608–612

Ghishan FK, Greene HL (1983) Intestinal transport of zinc in the diabetic rat. Life Sci 32: 1735–1741

Ghishan FK, Sobo G (1983) Intestinal maturation: in vivo zinc transport. Pediatr Res 17: 148–151

Guerrieri A, Catassi C, Pasquini E, Coppa GV, Benetti E, Giorgi PL (1986) Plasma zinc levels in children with chronic diarrhoea. Eur J Pediatr 145: 563–564

Hildman B, Schmidt A, Murer H (1979) Ca^{2+} transport in baso-lateral plasma membranes isolated from rat small intestinal epithelial cells. Pflugers Arch 382: R23–R34

Hoadley JE, Cousins RJ (1987) Cellular mechanisms for intestinal zinc absorption. Fed Proc 46: 598

Hoadley JE, Leinart AS, Cousins RJ (1987) Kinetic analysis of zinc uptake and serosal transfer by vascularly perfused rat intestine. Am J Physiol 252: G825–G831

Jackson MJ, Jones DA, Edwards RHT (1981) Zinc absorption in the rat. Br J Nutr 46: 15–27

Jackson MJ, Jones DA, Edwards RHT, Swainbank IG, Coleman ML (1984) Zinc homeostasis in man: studies using a new stable isotope-dilution technique. Br J Nutr 51: 199–208

Kennedy ML, Lönnerdal B (1987) Maturational differences in zinc uptake by brush border membrane vesicles (BBMV) from rat small intestine. In: Hurley LS, Keen CL, Lönnerdal B, Rucker RB (eds) Trace element metabolism in man and animals – TEMA 6. Plenum Press, New York (in press)

Kiilerich S, Dietrickson O, Loud FB et al. (1980) Zinc depletion in alcoholic liver diseases. Scand J Gastroenterol 15: 363–367

Kinlaw WB, Levine AS, Morley JE, Silvis SE, McClain CJ (1983) Abnormal zinc metabolism in type II diabetes mellitus. Am J Med 75: 273–277

Kowarski S, Blair-Stanek CS, Schachter D (1974) Active transport of zinc and identification of zinc-binding protein in rat jejunal mucosa. Am J Physiol 226: 401–407

Lombeck I, Schnippering HG, Kasperek K et al. (1975) Akrodermatitis enteropathica – eine Zinkstoffwechselstörung mit Zinkmalabsorption. Z Kinderheilk 120: 181–189

Lönnerdal B, Keen CL, Sloan MV, Hurley LS (1980a) Molecular localization of zinc in rat milk and neonatal intestine. J Nutr 110: 2414–2419

Lönnerdal B, Schneeman BO, Keen CL, Hurley LS (1980b) Molecular distribution of zinc in biliary and pancreatic secretions. Biol Trace Element Res 2: 149–158

Lyall VS, Majundar S, Prasad R, Nath R, Mahmood R (1981) Transport of zinc in rat intestine in vitro during growth and development. Indian J Biochem Biophys 18: 430–437

Menard MP, Cousins RJ (1983) Zinc transport by brush border membrane vesicles from rat intestine. J Nutr 113: 1434–1442

Methfessel AH, Spencer H (1973a) Zinc metabolism in the rat. I. Intestinal absorption of zinc. J Appl Physiol 34: 58–62

Methfessel AH, Spencer H (1973b) Zinc metabolism in the rat. II. Secretion of zinc into intestine. J Appl Physiol 34: 63–67

Mills PR, Fell GS, Bessent RG, Nelson LM, Russell RI (1983) A study of zinc metabolism in alcoholic cirrhosis. Clin Sci 64: 527–535

Milman N, Hvid-Jacobsen K, Hegnhøj J, Sølvsten Sørensen S (1983) Zinc absorption in patients with compensated alcoholic cirrhosis. Scand J Gastroenterol 18: 871–875

Moran JR, Lewis JC (1985) The effects of severe zinc deficiency on intestinal permeability: an ultrastructural study. Pediatr Res 19: 968–973

Moynahan EJ (1974) Acrodermatitis enteropathica: a lethal inherited human zinc-deficiency disorder. Lancet II: 399–400

Naveh Y, Lightman A, Zinder O (1983) A prospective study of serum zinc concentration in children with celiac disease. J Pediatr 102: 734–736

Olafson RW (1983) Intestinal metallothionein: effect of parenteral and enteral zinc exposure on tissue levels of mice on controlled zinc diets. J Nutr 113: 268–275

Quarterman J (1985) The role of intestinal mucus on metal absorption. In: Mills CF, Bremner I, Chesters JK (eds) Trace elements in man and animals – TEMA 5. Commonwealth Agricultural Bureaux, Farnham Royal, UK, pp 400–401

Rebello T, Lönnerdal B, Hurley LS (1982) Picolinic acid in milk, pancreatic juice, and intestine: inadequate for role in zinc absorption. Am J Clin Nutr 35: 1–5

Reinstein NH, Lönnerdal B, Keen CL, Schneeman BO, Hurley LS (1987) The effect of varying dietary zinc levels on the concentration and localization of zinc in rat bile-pancreatic fluid. J Nutr 117: 1060–1066

Richards MP, Cousins RJ (1975) Mammalian zinc homeostasis: requirements for RNA and metallothionein synthesis. Biochem Biophys Res Commun 64: 1215–1223

Rolston DDK, Moriarty KJ, Farthing MJG, Kelly MJ, Clark ML, Dawson AM (1986) Acetate and citrate stimulate water and sodium absorption in the human jejunum. Digestion 34: 101–104

Sandström B, Cederblad Å, Lònnerdal B (1983) Zinc absorption from human milk, cow's milk and infant formulas. Am J Dis Child 137: 726–729

Sandström B, Davidsson L, Cederblad Å, Lönnerdal B (1985) Oral iron, dietary ligands and zinc absorption. J Nutr 115: 411–414

Sandström B, Cederblad Å, Kivistö B, Stenquist B, Andersson H (1986) Retention of zinc and calcium from the human colon. Am J Clin Nutr 44: 501–504

Schneeman BO, Lönnerdal B, Keen CL, Hurley LS (1983) Zinc and copper in rat bile and pancreatic fluid: effect of surgery. J Nutr 113: 1165–1168

Schricker BR, Forbes RM (1978) Studies on the chemical nature of a low molecular weight zinc binding ligand in rat intestine. Nutr Rep Int 18: 159–166

Seal CJ, Heaton FW (1987) Zinc transfer among proteins in rat duodenum mucosa. Ann Nutr Metab 31: 55–60

Smith KT, Cousins RJ (1980) Quantitative aspects of zinc absorption by isolated vascularly perfused rat intestine. J Nutr 110: 316–323

Solomons NW, Rosenberg IH, Sandstead HH (1976) Zinc nutrition in celiac sprue. Am J Clin Nutr 29: 371–375

Solomons NW, Rosenberg IH, Sandstead HH, Vo-Khactu KP (1977) Zinc deficiency in Crohn's disease. Digestion 16: 87–95

Song MK, Adham NF (1985) Relationship between zinc and prostaglandin metabolisms in plasma and small intestine of rats. Am J Clin Nutr 41: 1201–1209

Southon S, Gee JM, Johnson IT (1984) Hexose transport and mucosal morphology in the small intestine of the zinc-deficient rat. Br J Nutr 52: 371–380

Southon S, Gee JM, Bayliss CE, Wyatt GM, Horn N, Johnson IT (1986) Intestinal microflora, morphology and enzyme activity in zinc-deficient and Zn-supplemented rats. Br J Nutr 55: 603–611

Stake PE, Miller WJ, Blackmon DM, Gentry RP, Neathery MW (1974) Role of pancreas in endogenous zinc excretion in the bovine. J Nutr 104: 1279–1284

Starcher BC, Glauber JG, Madaras JG (1980) Zinc absorption and its relationship to intestinal metallothionein. J Nutr 110: 1391–1397

Steel L, Cousins RJ (1985) Kinetics of zinc absorption by luminally and vascularly perfused rat intestine. Am J Physiol 248: G46–53

Sturniolo GC, Molokhia MM, Shields R, Turnberg LA (1980) Zinc absorption in Crohn's disease. Gut 21: 387–391

Swanson CA, Turnlund JR, King JC (1983) Effect of dietary zinc sources and pregnancy on zinc utilization in adult women fed controlled diets. J Nutr 113: 2557–2567

Urban E, Campbell ME (1984) In vivo zinc transport by rat small intestine after extensive small bowel resection. Am J Physiol 247: G88–94

Valberg LS, Flanagan PR, Ghent CN, Chamberlain MJ (1985) Zinc absorption and leukocyte zinc in alcoholic and nonalcoholic cirrhosis. Dig Dis Sci 30: 329–333

Valberg LS, Flanagan PR, Kertesz A, Bondy DC (1986) Zinc absorption in inflammatory bowel disease. Dig Dis Sci 31: 724–731

Vallee BL, Neurath H (1954) Carboxypeptidase, a zinc metalloprotein. J Am Chem Soc 76: 5006–5007

Vallee BL, Wacker WEC, Bartholomay AF, Hoch FL (1957) Zinc metabolism in hepatic dysfunction. II. Correlation of metabolic patterns with biochemical findings. N Engl J Med 257: 1055–1065

Vanderhoof JA, Scopinaro N, Tuma DJ, Gianetta E, Civalleri D, Antonson DL (1983) Hair and plasma zinc levels following exclusion of biliopancreatic secretions from functioning gastrointestinal tract in humans. Dig Dis Sci 28: 300–305

Weigand E, Kirchgessner M (1978) Homeostatic adjustments in zinc digestion to widely varying dietary zinc intakes. Nutr Metab 22: 101–112

Weigand E, Kirchgessner M (1979) Change in apparent and true absorption and retention of dietary zinc with age in rats. Biol Trace Element Res 1: 347–358

Weigand E, Kirchgessner M (1980) Total true efficiency of zinc utilization: determination and homeostatic dependence upon the zinc supply status in young rats. J Nutr 110: 469–480

Weismann K, Hoe S, Knudsen L. Sølvsten Sørensen S (1979) [65]Zinc absorption in patients suffering from acrodermatitis enteropathica and in normal adults assessed by whole-body counting technique. Br J Dermatol 101: 573–579

Chapter 4

Promoters and Antagonists of Zinc Absorption

Brittmarie Sandström and B. Lönnerdal

The absorption of zinc can be assumed to be determined to a large extent by the chemical environment at its site of absorption. Thus, absorptive efficiency will be influenced by the solubility of the zinc compounds present, by the presence of ligands of low molecular weight and by the competition between zinc and other minerals for carriers or uptake sites. To understand the mechanisms of zinc absorption, it is important to identify the factors promoting or inhibiting zinc uptake. Such knowledge can be used to improve zinc absorption from diets low in zinc and to develop special diets such as infant and enteral formulas intended for use at times when sensitivity to a suboptimal zinc supply may be increased. Efforts to eliminate strong antagonists of zinc absorption during food preparation and processing can also be made once these factors have been identified.

Methodological Considerations

Human Studies

Metabolic Balances

In contrast to macronutrients such as protein, fat and carbohydrates, only a fraction of most trace elements, including zinc, is absorbed. A number of physiological factors like gastric acid production, rate of gastric emptying and intestinal transit time can influence the degree of absorption. However, due to the methodological difficulty of studying these phenomena in man, knowledge of their relative importance is limited.

The way zinc is metabolized complicates absorption studies in man. Endogenous losses of zinc are large and often of a magnitude similar to the amount of zinc absorbed from the diet (Turnlund et al. 1986). Intestinal and faecal zinc is consequently a mixture of non-absorbed zinc from the diet and of the zinc from intestinal fluids and sloughed intestinal cells that has not been re-absorbed. Furthermore, intestinal transit time in the human is relatively long, especially through the colon.

It has been shown by Turnlund et al. (1982) that in achieving a complete collection of 1 day's zinc intakè, the faecal samples can also contain non-absorbed fractions of zinc from the previous 12 to 30 days' meals as well as fractions of endogenous zinc excreted over the same period of time. The conventional chemical balance technique, by which absorption is calculated as the difference between intake and faecal excretion, therefore requires long periods on a constant diet (Schwartz et al. 1986) to achieve equilibration and is not a sensitive method for the identification of individual factors that influence zinc absorption. Furthermore, with this technique only the so-called "apparent" absorption can be measured as there is no way of defining the extent to which absorption of orally ingested zinc or the endogenous secretion of zinc have been affected by dietary changes.

Isotope Techniques

With the use of zinc isotopes, stable or radioactive, many of the problems with the conventional balance technique can be overcome. Specific feeds or single meals can be labelled and the absorption of zinc determined with a high degree of accuracy. The large body pool of zinc means that the daily endogenous intestinal losses of zinc, although significant in relation to the amount of zinc absorbed, represent normally less than 1% per day of the body pool and thus do not invalidate the isotope measurements. Using a γ-emitting radioisotope of zinc, ^{65}Zn, absorption can be estimated from measurements of whole-body retention. With a sensitive counter the activities and the radiation doses can be kept low (Arvidsson et al. 1978).

Native zinc is a mixture of five stable isotopes. Techniques to measure zinc absorption with the use of zinc enriched with those naturally occurring in the smallest amounts, ^{67}Zn, ^{68}Zn and ^{70}Zn, have been developed (King et al. 1978; Janghorbani et al. 1982; Turnlund et al. 1982; Turnlund et al. 1986). With these techniques, absorption has to be determined from faecal monitoring as there is no tissue easily accessible in man in which a sufficient degree of enrichment can be achieved. Chemical separation of the sample is necessary to achieve sufficient precision for the determinations of zinc isotope ratios by sophisticated analytical techniques like thermal ionization mass spectroscopy. Another limitation with stable zinc isotope ratio techniques is that relatively large amounts of zinc have to be added to the test diet thus limiting their use mainly to formula diets or zinc-enriched foods. However, to identify factors that influence zinc absorption and for studies with infants and pregnant women, the stable isotope technique is valuable.

Extrinsic Versus Intrinsic Labelling. The isotopes used to measure absorption can be added to the prepared food or diet ("extrinsic" labelling) or can be biologically incorporated into edible tissues during growth of a plant or animal by addition to their nutrients or administration by injection ("intrinsic" labelling). When a diet is extrinsically labelled, isotope exchange is assumed to take place between the added label and the native element in the food or diet and absorption of the label is assumed to be directly related to the absorption from native zinc in the diet.

The validity of the extrinsic labelling technique has been studied in man only for a small number of foods. Flanagan et al. (1985) compared the absorption of zinc from turkey meat extrinsically and intrinsically labelled and found no difference. Using stable isotopes, Janghorbani et al. (1982) found a consistently lower absorption from an extrinsic label than from intrinsically labelled chicken. This is contrary to what could be expected as an incomplete exchange would mean that a smaller amount of the total zinc content of the meal was labelled which in turn would result in a higher fractional absorption (Sandström and Cederblad 1980; Istfan et al. 1983; Wada et al. 1985). In studies in rats by Ketelsen et al. (1984) no difference in the percentage of ^{65}Zn activity retained from intrinsically and extrinsically labelled meals containing soya flour or neutralized soya concentrate was observed. However, studies of other soya products indicate that processing techniques can impair isotope exchange. As only a limited number of foods can be intrinsically labelled further validation of the extrinsic labelling technique in man is important.

Plasma Zinc Response

The methodological difficulties and expense of isotope techniques have led to development of indirect ways to evaluate zinc absorption. By analogy with some methods of studying drug absorption, the plasma zinc response, i.e. the increase in plasma zinc after a zinc load, has been used as a method to identify dietary and physiological factors affecting zinc absorption (Pécoud et al. 1975; Andersson et al. 1976; Schelling et al. 1976; Solomons et al. 1979a, b, c; Solomons and Jacob 1981; Meadows et al. 1983; Solomons et al. 1983). One of the major limitations of this method is that the dose of zinc needed to produce detectable increases of plasma zinc concentration is much higher than the normal dietary level of zinc.

After non-pharmacological intakes of zinc the plasma turnover rate of absorbed zinc is very rapid (Molokhia et al. 1980). Since the plasma zinc level at any moment reflects the net balance of absorption and clearance, factors that influence rates of intestinal uptake and cellular uptake of zinc could influence the plasma response without necessarily reflecting the degree of absorption. At intakes of zinc more typical of those from diets, the plasma zinc concentration after a meal is normally unchanged or decreases without any relation to dose or the actual absorption from the meal (Sandström et al. 1980). Valberg et al. (1985) compared the plasma zinc response with radio-zinc absorption after an oral test dose of 25 mg zinc as zinc chloride or as zinc-enriched turkey meat. The radio-zinc absorption was identical from the two test meals whereas a significantly smaller increase in plasma zinc was observed after intake of the turkey meal. A correlation was found between the plasma zinc curve and the radio-zinc absorption in individual subjects after each meal. The precision in predicting absorption from increments in plasma zinc response was, however, unsatisfactorily low. Unless very standardized conditions are used, results from plasma zinc tests could lead to misinterpretation of the significance of factors affecting zinc absorption. Theoretically this method could be used to study the effect of addition to a meal of a component that was known not to affect gastric emptying, intestinal transit time or rate of plasma zinc clearance. Nevertheless, the limitation still applies that the high oral intakes of zinc needed to produce a plasma response may completely saturate intestinal and cellular binding sites for zinc and thus lead to errors in interpretation.

Animal Models

Because of the inherent difficulty in studying nutrient absorption in humans, our knowledge of factors affecting zinc absorption is mainly based on animal studies. Besides the limitations of animal models per se, both the effect of conditions preceding the test and the test levels of zinc and other interactive dietary variables used in such studies could influence the outcome and human relevance of such modelling. In many studies on experimental animals, stock diets are fed for long periods prior to the test period. These diets are usually formulated to provide at least a minimum supply of most nutrients and often contain quite "excessive" levels, i.e. levels several times higher than the expected nutrient requirements. Similarly, in purified diets offered to rats the control level of zinc is often as high as 100 mg/kg diet – considerably higher than the likely requirement of about 10 mg zinc/kg diet (Rogers et al. 1985).

Such excessive dietary intakes of zinc are unlikely to occur in humans, for whom most diets appear to provide zinc in amounts close to recommended dietary allowances (RDA) (see Chap. 21). Very high levels of dietary zinc can induce metallothionein synthesis in the mucosal cell (Cousins 1985), and as a consequence, the regulatory mechanisms for zinc absorption within the cell may differ from those operating during periods of normal zinc supply. The use of zinc-deficient diets in such experimental models could also induce systemic changes in the intestine that may influence the outcome of zinc absorption studies. Effects of zinc deficiency on the synthesis or metabolism of fatty acids, phospholipids and prostaglandins (Cunnane et al. 1981) may have implications for mucosal function and gastric acid secretion that, while possibly influencing zinc absorption, are not typical of subjects with normal zinc status.

In contrast to the constant levels of nutrients in the purified diets used in most experimental model studies, relationships between nutrients in the human diet can vary substantially among different meals. It is therefore essential that when factors affecting zinc absorption are tested in animal models, the levels of not only zinc and the tested factor, but also of other nutrients and putative interfering factors are realistic and reflect the range of normal human intakes.

Intestinal Uptake Models

A variety of experimental model systems have been developed to study, specifically, the transfer of zinc across the intestinal mucosa. The systems closest to the intact animal are those in which a segment of the small intestine is tied off or a loop of intestine is externalized. The intestine is then perfused with a solution of zinc and by measurements of changes in zinc concentration, transfer rates and kinetics of zinc transport can be determined. Potential drawbacks of this method include the anaesthesia used which may effect metabolism in general, trauma caused by the surgery and changes in vascularization of the intestine. Another approach is to use everted gut sacs; intestinal segments are surgically removed, everted and perfused. The above problems may be accentuated with this model system, as it obviously involves removal of blood vessels and an even more radical change in the microenvironment. A third system uses brush border membrane vesicles (BBMV) prepared by physically removing the mucosa and selectively precipitating the BBMVs. Although problems caused by vesicle surface reorientation, the "inside-out" phenomenon, can now be avoided, it is evident that this method is even further from the in vivo situation.

The advantage of the above approaches is of course that many conditions can be closely controlled and, thus, effects of pH, ionic strength, concentrations of ions, ligands, competitors and stimulatory factors can be investigated in detail and their influence on kinetics and binding constants can be determined. Results from such studies can then be valuable in the interpretation of data obtained from intact animals.

Radioisotope Studies in Animals

With the use of radiotracers, the zinc absorption process can be monitored in animals in a way not possible in human subjects. Information can be gained about the metabolic fate of the absorbed zinc; tissues can be homogenized and the molecular localization of the element can be determined. In addition, more pronounced deficiencies and excess of zinc as well as of other nutrients can be studied, in particular during vulnerable periods such as pregnancy, lactation and infancy. Although some of this information may provide insight into how zinc may be absorbed in humans, it should be recognized that most of these studies have been done in rodents whose zinc requirements and metabolism can differ significantly from those of humans. The applicability of the animal models used varies both with the species and the age of the animal. For example, we have shown that using intubation of rat pups and labelled formulas followed by measurements of the liver uptake of radio-zinc yields data which correspond well to the differences in absorption measured by whole-body retention in human adults (Sandström et al. 1983a). In contrast, data from weanling or adult rats showed considerably lower correlation to results from human adults.

Interacting Minerals and Trace Elements

The theoretical framework for the interaction between minerals and trace elements was set by the studies by Hill and Matrone (1970). These investigators postulated that the configuration and coordination number for a cation would be determinants for the mechanism of transport and that elements with similar physicochemical properties therefore may compete for common pathways.

Their subsequent studies in experimental animals confirmed their hypotheses and many others have since supported this concept. The zinc ion with its 10 outer electrons in d orbital forms a tetrahedral sp^3 chelate configuration and prefers a coordination number of four. These parameters are identical to those for Cu^+ and Cd^{2+} and therefore these elements would be expected to interact with zinc. Such biological interactions between zinc and copper or cadmium have been well documented. An element such as calcium with its filled 3p orbital, forming an octahedral d^2sp^3 chelate configuration, therefore preferring a coordination number of six, would not be expected to interact with zinc. In support of this, calcium and zinc do not appear to interact (see below).

Another interaction, less readily explicable on the basis of common structural features of the ions involved, is that between zinc and iron (Solomons 1986). In this case, their mutual affinity for a biological carrier such as transferrin, capable of binding many different elements, explains the interaction (see Chaps. 1 and 3).

Calcium

Of the cations in the human diet, calcium is present in the highest concentrations. Although calcium is thought to have its own closely regulated uptake mechanism, animal studies have shown that high calcium levels can impair the intestinal absorption of zinc (Heth and Hoekstra 1965). However, metabolic balance studies in human subjects offered calcium intakes of 200, 800 and 2000 mg/d (the two higher levels achieved by addition of calcium gluconate) did not show any effect of calcium level on zinc absorption or balance (Spencer et al. 1984). Neither 500 mg of elemental calcium as calcium carbonate or hydroxyapatite added to a standard test meal containing 3.62 mg of zinc (Dawson-Hughes et al. 1986) nor addition of calcium (as calcium chloride) to a cows' milk infant formula to simulate the levels of calcium found in cows' milk (Lönnerdal et al. 1984) altered the absorption of ^{65}Zn. Thus, there are no indications that calcium per se impairs zinc absorption at levels found in human diets.

Phosphorus

Increased dietary phosphorus has been found to decrease zinc absorption in rats fed semipurified diets (Heth et al. 1966). In metabolic balance studies in man no effects on the apparent absorption of zinc were observed after increasing dietary phosphorus content from 800 to 2000 mg/d by addition of sodium glycerophosphate (Spencer et al. 1984) or from 1000 to 2500 mg/d by addition of potassium monobasic phosphate (Greger and Snedeker 1980). It must be emphasized however that the phosphorus content of many mixed diets for humans is related to their content of protein and/or of unrefined cereals with a high phytic acid content and, as will be indicated, both such components have been found to affect zinc absorption.

Iron

Pharmacological doses of iron are often given to groups of subjects with high zinc requirements such as pregnant women and teenagers. Furthermore diets like infant formulas are often enriched with iron. A negative effect of iron on zinc absorption could therefore have serious implications for zinc supply to vulnerable groups. Observations of differences in plasma zinc levels of formula-fed infants have been attributed to the wide iron/zinc ratios of the formulas used (Craig et al. 1984). In experimental animals, iron has been shown to interfere with zinc absorption (Momcilovic and Kello 1979). A reduced plasma zinc uptake after an oral dose of 25 mg zinc has been observed after administration of elemental iron while haem iron did not have that effect (Solomons and Jacob 1981). Due to the large levels of zinc used in the latter study it is difficult to evaluate its relevance to situations involving a more normal dietary zinc intake. Using body retention of ^{65}Zn as an index of absorption, Valberg et al. (1984) found that both inorganic iron and haem iron inhibited zinc absorption from 6-mg doses of zinc as zinc chloride. Inorganic iron had no effect, however, on ^{65}Zn absorption from an extrinsically labelled turkey test meal. Similar differences in the effects of iron on zinc absorption depending upon whether the zinc is supplied in aqueous solution or from a meal have been observed by Sandström et al. (1985). The absorption of zinc from infant formulas by adults was not influenced by differences in the iron content of the two formulas (Lönnerdal et al. 1984).

Solomons has recently reviewed studies of the interaction between iron and zinc and suggests that the total amount of ionic species determines the effect on absorption of zinc (Solomons 1986). He claims that when the total amount of ions given as a single oral dose in solution is greater than 25 mg, a measurable effect on human zinc nutriture can be expected. The different effect of iron on zinc absorption from a solution and from a diet exemplifies the complex interactions between promoters and antagonists of absorption. Of importance for evaluation of the effect of iron on zinc absorption is also the fact that from the same amount of zinc a much higher fraction is absorbed from an aqueous solution than in the presence of food (Sandström et al. 1985). Thus in the studies of Valberg et al. (1984), zinc absorption from an aqueous zinc and iron solution was, even at an iron/zinc molar ratio of 10, higher than from a turkey meal without added iron. It was also shown that addition of a zinc-binding ligand, histidine, diminished the negative effect on zinc absorption of high iron concentrations (Sandström et al. 1985). The practical implication of these results could be that iron/zinc interactions can reduce the efficacy of a zinc supplement while normal levels of iron enrichment of food do not seem to significantly impair zinc supply. However, most studies of zinc absorption have been performed in healthy non-pregnant adults and the effect of iron could be different in groups with higher zinc or iron needs.

Tin

The levels of tin ordinarily found in the human diet are low and unlikely to interfere with zinc absorption. Food in unlacquered cans can, however, contain appreciable amounts of tin. Metabolic balance studies in humans have shown that supplements of 50 mg of tin increase faecal zinc excretion (Johnson et al. 1982) and radio-zinc studies have confirmed that tin reduces zinc absorption both from a test solution and from a meal (Valberg et al. 1984). In contrast, the plasma zinc response test failed to reveal any effect of tin on zinc absorption (Solomons et al. 1983).

Copper

Large quantities of ingested zinc can interfere with copper absorption and as copper is necessary for iron metabolism, anaemia may result (Patterson et al. 1985). Studies in rats using isolated, vascularly perfused rat intestines and dietary concentrations of 5, 30 and 180 mg/kg zinc and 1, 6 and 36 mg/kg copper, have, however, not shown mutually interactive effects on their absorption (Oestreicher and Cousins 1985). Using ^{65}Zn and whole-body counting in human subjects, Valberg et al. (1984) found no effect of 5 mg copper on the absorption of 0.5 mg zinc in water. In human diets the relation between zinc and copper is almost always a higher zinc than copper content and copper is in practice no serious inhibitor of zinc absorption.

Organic Substances Modifying Absorption

Proteins

The observation that the disorder of zinc metabolism, acrodermatitis enteropathica, could be treated with human milk while cows' milk was less efficacious (Moynahan 1974), led to the hypothesis that zinc in cows' milk is less available than in human milk. This was later confirmed by radionuclide studies with adults (Sandström et al. 1983b) and with rat pups (Sandström et al. 1983a) and is consistent with data from studies using the plasma zinc response to assess zinc absorption in human subjects (Casey et al. 1981). A large number of investigators have tried to find the explanation for the difference in the efficacy with which zinc can be utilized from the milk of different species.

Many observations suggest that the protein content and composition of the milk diets to a large extent influence the intestinal handling of zinc. In undigested cows' milk, zinc is found predominantly bound to the high-molecular-weight protein, casein (Cousins and Smith 1980; Harzer and Kauer 1982; Blakeborough et al. 1983) and it was suggested early on that zinc bound to casein may be inaccessible to the young infant (Lönnerdal et al. 1980). From in vitro dialysis experiments a zinc-binding capacity of 8.4 μg zinc/mg bovine casein has been observed at slightly alkaline pH (Harzer and Kauer 1982), and this exceeds the concentration of zinc normally present in cows' milk. No zinc remained bound to casein at pH 2. This study also indicated that zinc is complexed to the negatively charged phosphate groups of casein. Human milk contains a lower concentration of casein which is also less phosphorylated than bovine casein. Casein separated from human milk by ultracentrifugation was found to contain 14% of the total zinc content; serum albumin bound 28% and the remaining zinc was equally distributed between fat and low-molecular-weight ligands (Lönnerdal et al. 1982).

Eckhert (1985) separated a low-molecular-weight protein fraction (MW 12 500) from fat-free human milk by gel filtration chromatography and used the plasma zinc response to test its bioavailability. Addition of 30 mg of this isolated protein fraction to 148 ml cows' milk containing 5 g protein and loaded with 25 mg zinc resulted in a higher elevation of plasma zinc compared with zinc-loaded cows' milk alone. The number of subjects tested was not reported and no comparison with addition of other proteins was made. When dialysed against distilled, deionized water, only one out of 30 protein molecules contained zinc, indicating a very low affinity for zinc or a heterogeneous fraction unlikely to influence zinc absorption. Thus the observed effect could not have been zinc-specific. In contrast to the studies by Lönnerdal et al. (1982) and other similar studies, Blakeborough et al. (1983) recovered all zinc in human milk associated with high-molecular-weight proteins and suggested lactoferrin to be of importance for zinc absorption. However, according to the studies by Lönnerdal et al. (1982, 1985) lactoferrin in human milk does not bind native zinc or added ^{65}Zn, although purified lactoferrin in vitro can bind zinc when present in excess.

In human adults, dietary proteins are digested to 80%–95% in the small intestine where the protein-bound zinc is probably rendered available for absorption. Studies in piglets (Blakeborough et al. 1986) indicated that zinc in cows' milk and cows' milk-based infant formula was relatively poorly digested whereas zinc from human milk was found in a soluble form in the intestinal digesta. The actual degree of protein digestion or zinc absorption was, however, not measured; some of the proteins in the digesta seemed to be of endogenous origin and coprophagy could not be excluded, according to the comments by the authors.

A much lower absorption of zinc from a soya-based infant formula compared with human milk or cows' milk formula has been observed in adults (Sandström et al. 1983b) and growth studies of children recovering from protein–energy malnutrition also imply a poor availability of zinc from soya-based formulas (Golden and Golden 1981). This low availability could be due to the phytic acid usually associated with such soya protein fractions but could also be attributed to properties of the soya protein itself. Zinc in soya formula is associated with the large-molecular-weight proteins and the solubility of zinc bound to this soya protein fraction at physiological pH is low (Sandström et al. 1983a). Although a lower small intestinal protein digestion of a 25% soya protein diet compared with a meat diet has been observed in ileostomates (Sandström et al. 1986), this had no significant effect on apparent zinc absorption. When all meat protein was replaced by soya protein, a lower fractional absorption than expected from the dietary zinc content was observed in studies using the extrinsic labelling radionuclide technique to measure zinc absorption from a composite meal (Sandström et al. 1987a). In view of the high affinity of zinc for undigested protein, even a minor impairment in protein digestion could affect zinc absorption.

Istfan et al. (1983) measured zinc absorption, with the aid of the stable isotope ^{70}Zn, from a formula diet with either egg protein or a soya protein concentrate as the sole source of protein and with 8.5 or 12.8 mg of added zinc as zinc chloride. No significant difference in zinc absorption between the diets was observed. The large variations observed in zinc absorption within and also between individuals on a constant formula diet illustrated the inherent methodological problems with this technique. Similar estimates of zinc absorption were also obtained when soya isolate was compared with a skim milk formula (Solomons et al. 1982).

At present it is not possible to assess the relative role of dietary proteins as such or of other factors associated with the protein or the food source which influence the extent of zinc absorption. Both the type and the dietary supply of protein appear to affect the degree of zinc absorption. Significantly lower faecal zinc losses were observed in adult males fed a high protein diet (150 g/d), achieved by intake of a bread enriched with purified protein sources, as compared with an intake of 50 g protein/d and similar zinc intakes (Greger and Snedeker 1980). Urinary zinc excretion was however increased on the higher protein intake and the apparent retention of zinc was not statistically different. Greger and Snedeker (1980) also observed that supplementing the low protein diet with sulphur-containing amino acids increased urinary zinc excretion. While this could indicate a higher absorption of zinc, it could also be due to other changes in zinc metabolism induced by the changed amino acid intake (Snedeker and Greger 1981). Colin et al. (1983) found no difference in faecal zinc excretion on an intake of 50 g or 100 g protein/d. To further evaluate the influence of dietary protein, Snedeker and Greger (1983) studied the effects of a variety of dietary protein levels with and without supplements of histidine and cysteine on zinc absorption. Apparent absorption of zinc was higher when rats were fed a 45% lactalbumin diet or a 15% lactalbumin diet supplemented with histidine and cysteine than when fed the unsupplemented low-protein lactalbumin diet. An improvement in zinc absorption was also observed when cysteine and methionine were added to a soya protein diet; absorption was similar to that from a lactalbumin-containing diet (Greger and Mulvaney 1985). Wapnir et al. (1983) using an in vivo procedure found that ligand/zinc ratios equal to or less than 3 : 1 for certain amino acids and dipeptides were optimal while an excess of ligand reduced absorption by ileal segments of adult rats.

Vitamins

Folic Acid

The enzyme responsible for intestinal hydrolysis of dietary folates to their monoglutamate forms is probably zinc-dependent (Chandler et al. 1986). In zinc-depleted human subjects the increase in serum folate after an oral dose of pteroylheptaglutamate was reduced while pteroylheptaglutamate absorption appeared to be unchanged (Tamura et al. 1978). It has also been suggested that folic acid could impair the absorption of zinc. Greater faecal zinc losses and reduced urinary zinc excretion were observed in four men receiving 400 μg of pteroylheptaglutamate acid compared with four other subjects on a low-folate diet (Milne et al. 1984). This effect was observed at low zinc intakes, 3.5 and 7.5 mg/d, while no difference was observed for high-zinc diets. The observation has not yet been confirmed in other studies. Milne et al. (1984) suggested the formation of insoluble complexes between zinc and folic acid as the explanation for the indications of a lower zinc absorption. Ghishan et al. (1986) observed from both in vivo and in vitro studies a mutual inhibitory effect of folic acid and zinc on intestinal transport. Charcoal-binding studies indicated that complexes between zinc and folic acid were formed at pH 2. However, when pH was raised to 6.0 these complexes dissociated indicating that mechanisms other than formation of insoluble complexes have to be investigated.

Ascorbic Acid

Ascorbic acid is a potent promoter of non-haem iron absorption (Hallberg et al. 1986). Despite this, and the antagonistic effect of iron on zinc absorption, no effect of ascorbic acid on zinc absorption has been shown either by plasma zinc response studies with ascorbic acid intakes ranging from 0.5–2.0 g and zinc levels of 25 mg (Solomons et al. 1979c) or by radio-zinc absorption studies at more typical dietary levels of zinc (Sandström and Cederblad 1987).

Carbohydrates

Carbohydrates constitute the major component of most human diets. An enhancing effect of glucose on zinc absorption has been suggested by Steinhardt and Adibi (1984). Lactose is believed to improve calcium absorption (Allen 1982). However, when the carbohydrate sources in a soya formula (dextrins, maltose, glucose and starch) were substituted with lactose no difference in zinc absorption was observed (Lönnerdal et al. 1984).

A glucose polymer derived from corn starch, used as a low-osmolality caloric source in infant formula and enteral nutrition products, has been suggested to improve zinc balance (Bei et al. 1986). Using a triple-lumen technique in human subjects Bei et al. (1986) observed an increased net zinc absorption over the jejunal segment studied when 4 mM glucose polymer solution was perfused. To what extent this also corresponds to an overall higher absorption of zinc or merely a change in its rate of absorption remains to be shown with other techniques.

Fibre

The term dietary fibre covers a number of different non-starch polysaccharides and often, depending on analytical methods and definition, also includes lignins and uronic acids. Furthermore, fibre-rich foods often contain higher levels of minerals as well as of potential zinc absorption antagonists like phytic acid than do their low-fibre counterparts. The results from human studies, in which pure fibre fractions have been used, indicate that fibre per se is relatively inert as regards any effect on zinc availability. Turnlund et al. (1984) added α-cellulose in amounts providing 0.5 g/kg body weight to a basal low-fibre diet and found no effect on zinc absorption in young men. In studies using the radionuclide technique, the addition of beet pulp fibre to a bread- or a meat-based meal was also without effect on zinc absorption (Sandström et al. 1987b) as was 15 g of pectin added to a low-fibre diet in metabolic balance studies of ileostomy subjects (Sandberg et al. 1983). Any effect of pectins on zinc absorption could depend on the degree of methoxylation and thus the zinc-binding capacity of the pectin. Low-methoxylated apple pectin in amounts corresponding to 2.5% of the diet diminished the absorption and retention of zinc in pigs (Bagheri and Guéguen 1985) while no effect was observed for high-methoxylated pectin [similar to that used in the study of human subjects by Sandberg et al. (1983)]

Phytic Acid

Phytic acid (myo-inositol hexaphosphate), the major form of phosphorus in cereal grains and legumes, was early suggested as a potential antagonist of zinc absorption (O'Dell and Savage 1960). Addition of phytic acid to a casein diet reduced growth in chicks to about the same extent as that obtained with a soya-protein diet. The impairment of zinc utilization by phytic acid has been verified in a number of animal studies and in different species (Oberleas et al. 1962; Nwokolo and Bragg 1977; Davies and Olpin 1979; Morris and Ellis 1980a, b; Forbes et al. 1984).

The structure and chemistry of phytic acid appear to explain the mechanisms for this antagonistic effect (for review see Cheryan 1980). At pH values encountered in foods, phytic acid will be strongly negatively charged and will have a strong potential to bind positively charged molecules. In vitro studies have shown that next to copper, zinc forms the most stable complexes with phytic acid at pH 7.4 (Vohra et al. 1965).

Single-meal studies using radionuclide and stable isotope techniques have confirmed a negative effect of phytic acid on zinc absorption in man. Addition of 2.34 g of phytic acid to a liquid formula diet reduced the absorption of zinc from 33.8% to 17.5% (Turnlund et al. 1984). The level of phytic acid used in these studies was similar to those levels found in whole-grain cereal-based diets typical of those geographical areas where zinc deficiency signs in man have been observed (Halsted et al. 1972). A decrease in zinc absorption has also been observed when phytic acid was added to white bread to the same level as found in wholemeal bread (Nävert et al. 1985) and when it was added to cows' milk formula to produce phytic acid concentrations similar to those observed for soya-based formula (Lönnerdal et al. 1984).

Foods with a high phytic acid content often have a high content of dietary fibre as well as of zinc and other minerals. The presence of the enzyme phytase in unprocessed whole-grain cereals can be exploited during the leavening of bread or other fermentation processes to reduce their phytic acid content before consumption. Meals differing in phytic acid content but with an otherwise similar nutrient content can consequently be studied. In Fig. 4.1, data from several studies of zinc absorption from cereal-based meals in man are given (Sandström et al. 1980; Nävert et al. 1985; Sandström 1987a). For these meals with a zinc content of 30–60 μmol (2–4 mg), a low fractional absorption of zinc was observed above 400–500 μmol of phytic acid.

The molar ratio between phytic acid and zinc has been suggested as an index of zinc availability (Oberleas and Harland 1981). In individual food items for human consumption this ratio can be as high as 30–40 while in most mixed meals the ratio is less than 10. Davies and Olpin (1979), working with rats, showed by adjusting the phytate/zinc molar ratio through a range of 0 : 1 to

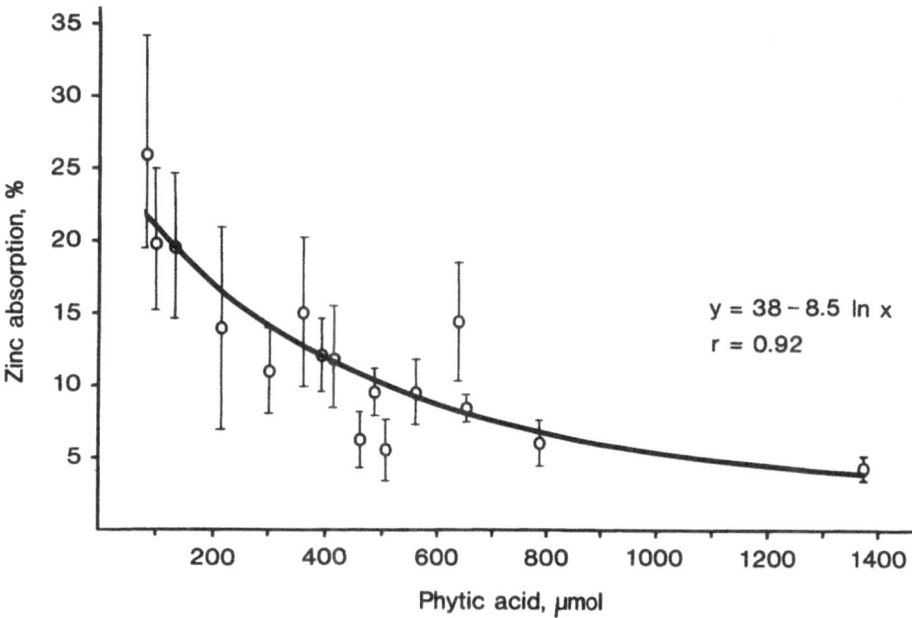

Fig. 4.1. Zinc absorption ($x \pm$ SD) from cereal-based meals in relation to phytic acid content. Each meal consisted of 60 g cereal, either as white bread and 10–30 g of wheat bran, or as whole-grain cereals (wheat, rye, oats, barley or triticale). Milk (200 g) was included in each meal (Sandström et al. 1980; Nävert et al. 1985; Sandström et al, unpublished observations). Reproduced from Sandström (1987b).

40 : 1 that ratios > 15 : 1 impaired rates of weight gain. Lo et al. (1981) found that a phytate/zinc ratio equal to or greater than 12.5 : 1 reduced [65]Zn absorption from a soya-based diet in rats. For the cereal-based diets presented in Fig. 4.1 low absorption figures were consistently observed above a ratio of approximately 5. Results from animal studies (Morris and Ellis 1980a, b) and from studies of zinc absorption from single meals in humans indicate that the molar ratio can, at best, be used to compare similar diets in which other nutrients, especially the protein and zinc content, are kept constant. In soya-protein-containing meals, the molar ratio was relatable to zinc absorption in bread meals with a low animal protein content and a relatively high calcium content but not in meals with a higher protein content (Fig. 4.2) (Sandström et al. 1987a). The possibility that diets high in calcium may potentiate the effects of phytic acid as a zinc antagonist is considered in more detail on p. 71.

During food processing, inositol polyphosphates can be dephosphorylated to lower inositol phosphates (DeBoland et al. 1975; Sandberg and Ahderinne 1986). It has been observed recently in studies with rat pups that inositol pentaphosphate has a similar partial inhibitory effect on zinc absorption as the fully substituted inositol (hexa) phosphate while lower phosphates (tetra- and tri-) do not reduce zinc absorption (Lönnerdal et al. 1987). The increasing use of processed food warrants the wider application of analytical techniques for determination of these differing forms of "phytic acid" and further studies of their effects on zinc absorption in man.

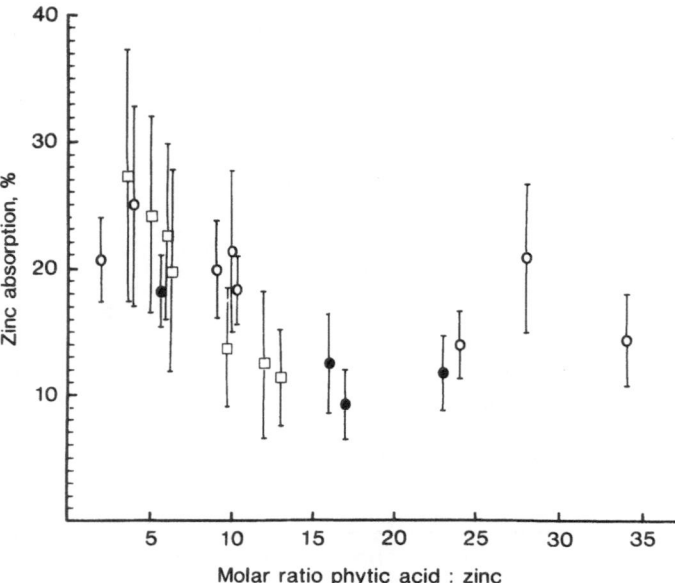

Fig. 4.2. Percentage zinc absorption ($x \pm$ SD) from soya-protein-containing meals based on bread (□), "meat sauce" with 30% or 100% soya protein (○) or "meat sauce" and milk (●) in relation to the molar ratio of phytic acid:zinc of the meals . Reproduced from J Nutr (Sandström et al. 1987a).

Zinc Chelators

Much attention has been given to the role that zinc chelators may play in zinc absorption. The unique beneficial properties of human milk in the treatment of acrodermatitis enteropathica have been suggested as being related to its content of specific low-molecular-weight ligands. A significant proportion of the zinc content of human milk is associated with citrate (Lönnerdal et al. 1980). However, increasing citrate levels of human milk or infant formula did not increase the fractional absorption of zinc in adults (Lönnerdal et al. 1984), possibly because the diets tested already contained significant quantities of citrate.

Picolinic acid is a metabolite of tryptophan which is present in milk and which has chemical similarities to the hydroxyquinolines used successfully to treat acrodermatitis enteropathica patients before it was realized that this disorder was attributable to a defect in zinc metabolism (Dillaha et al. 1953). Thus it has been proposed also that picolinic acid may be the variable constituent responsible for marked differences in the efficiency with which zinc from the milk of differing species can be utilized (Evans and Johnson 1980). However, high levels of picolinic acid enhanced both the absorption and excretion of zinc in rats (Seal and Heaton 1985) resulting in net negative retention, while no effect was observed at levels found in human milk (Schwarz et al. 1983). Picolinate did not improve zinc absorption in calves with

Adema disease, a zinc deficiency syndrome with similarities to acrodermatitis enteropathica in humans (Flagstad 1981). Studies using rat brush border membrane vesicles (Menard and Cousins 1983) suggest that neither citrate nor picolinic acid enhance zinc transport. Neither do in vitro experiments with isolated vascularly perfused rat intestine support the view that citrate or picolinate ligands promote more efficient absorption than that of zinc as its chloride (Oestreicher and Cousins 1982).

These observations do not exclude a role for low-molecular-weight ligands in promoting zinc absorption from human milk since many low-molecular-weight complexes of zinc in aqueous solution are probably absorbed readily regardless of their chemical form. Rather more probable is that zinc-chelating ligands such as citrate act by restricting the association of zinc with high-molecular-weight ligands which have a negative effect on zinc absorption.

Oxalic acid is known to decrease calcium availability in monogastric animals apparently by precipitation of calcium oxalates in the intestinal tract (Allen 1982). The relatively poor solubility of zinc oxalate and the possibilities of co-precipitation of zinc with calcium oxalates by occlusion in the precipitate, suggested that oxalic acid could impair zinc utilization. When this hypothesis was tested in zinc-deficient rats by Welch et al. (1977) no effect on zinc absorption from intrinsically labelled spinach leaves was observed. The effect of oxalic acid on zinc absorption in man has not been studied. Kelsay and Prather (1983) added spinach, with a high oxalic acid content, to a low-fibre diet and to a diet containing fruits and vegetables and determined mineral balances in 12 men. A lower apparent absorption of zinc observed on the combined high-oxalate, high-fibre, diet was suggested to result from formation of a fibre–mineral–oxalate complex.

Nutrient Interactions

Calcium–Phytic Acid Interaction

Early studies of the growth response of pigs to different diets imply an interaction between calcium, zinc and phytic acid. The inhibitory effect of a phytic-acid-containing diet on growth was much more pronounced at a high calcium level and high levels of zinc were needed to overcome this effect (O'Dell 1969). Oberleas et al. (1966) showed that the in vitro precipitation of insoluble zinc complexes at physiological pH is much more pronounced at high calcium concentrations. The potentiating effect of calcium on the antagonistic effect of phytate on zinc availability has also been confirmed in other experimental animal studies (Likuski and Forbes 1965; Davies and Olpin 1979; Morris and Ellis 1980a). Davies et al. (1985) suggested that the (calcium × phytate/zinc) ratio, expressed in mols/kg dry weight, would be a better index of available zinc than the ratio of phytic acid to zinc. Rat growth was inhibited at a ratio exceeding approximately 3. Calculations from available human data indicated that a ratio exceeding 0.5 may adversely affect zinc absorption. The predictive value of this ratio has not been systematically studied in man. One serious limitation of this ratio is that, in the human diet, the concentrations of these substances are not independent variables. Thus a high calcium or phytic acid intake is usually associated with a higher zinc intake. Furthermore other group II cations like magnesium could have a similar potentiating effect to calcium.

Dietary levels of calcium used in animal experiments are frequently much higher than those in human diets. Addition of milk and cheese, the calcium sources of quantitative importance in most human diets, to a phytic-acid-containing bread, improved the absorption of zinc (Sandström et al. 1980). However, the content of many other nutrients was also increased by the addition of milk and it was concluded that the increase in protein content of the meal was counterbalancing the negative effect of phytic acid. When calcium chloride was added to a phytic-acid-containing soya isolate formula, an improvement of zinc absorption was also observed (Lönnerdal et al. 1984). Recent studies in rats receiving diets with different levels of phytic acid and calcium showed that at the lowest level of calcium used, which was close to the level reached by heavy milk-drinkers, calcium had no effect on zinc availability (Davies et al., personal communication). These results indicate that at the levels of calcium and phytic acid found in most human diets, the calcium–phytate–zinc interaction observed in animal studies has limited relevance with regard to zinc absorption in humans. However, the potential risk exists that this interaction could restrict zinc utilization if calcium supplements are taken with phytic-acid-rich foods.

Phytic Acid–Protein Interactions

Both zinc and phytic acid have a high affinity for protein. Formation of stable protein–phytic acid–zinc complexes has been suggested as an explanation of observed differences in zinc absorption with different methods for processing of soya protein (Erdman et al. 1980). Forbes and Erdman Jr (1983) suggested that these complexes could be thermodynamically stable and resistant to proteolytic digestion in the gastrointestinal tract. The observation that complete recovery of phytic acid from a soya protein diet in small intestinal digesta is associated with a concomitant reduction in protein digestibility (Sandström et al. 1986) could support this suggestion. If these concepts are correct, the phytic acid of soya products would only affect absorption of the intrinsic zinc of the product and not the dietary zinc from other sources. This could explain the relative insensitivity of zinc absorption to phytate intake observed by Solomons et al. (1982) and Istfan et al. (1983) in studies of soya-protein diets using the stable isotope technique in which the majority of the zinc content was as added zinc.

A protein–phytic acid interaction not involving molecular inclusion of zinc could also explain the improvement in zinc absorption observed when animal protein was added to wholemeal bread with a high phytic acid content (Fig. 4.3) (Sandström et al. 1980). An alternative explanation could be that peptides and amino acid liberated during digestion of protein facilitated zinc absorption or had a greater affinity for zinc than phytic acid or a protein–phytic acid complex.

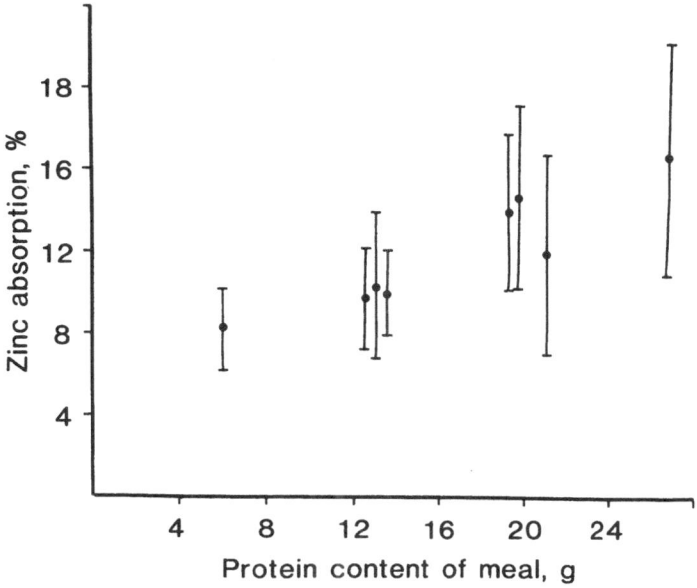

Fig. 4.3. The effect of an increasing amount of protein on zinc absorption ($x \pm$ SD) from wholemeal bread. The zinc contents of the meals were adjusted to 3.2–3.5 mg by addition of $ZnCl_2$ (data from Sandström et al. 1980).

Synthetic Chelating Agents and Drugs

Strong zinc chelators like EDTA have been found to improve zinc utilization in animals (Kratzer et al. 1959; Vohra and Kratzer 1964; Oberleas et al. 1966). In man, the effect on zinc absorption of the iron-fortifying agent, NaFeEDTA, in aqueous solution has been studied by the plasma zinc response method (Solomons et al. 1979b) from which there was tentative evidence that this supplement may reduce absorption of zinc. Many drugs possess chelating properties and could interfere with absorption and utilization of trace elements. Studies with rats have revealed that many drugs commonly used in man have significant effects on zinc absorption (Weismann 1986).

Concluding Remarks

In experimental models, effects on zinc absorption have been observed for strong chelators, for substances forming insoluble complexes with zinc and for high levels of other cations competing with zinc for binding sites and absorption mechanisms. The nutritional significance of many of these substances when occurring at levels found in human diets is, however, uncertain. Of the factors so far found to influence intestinal uptake of zinc, the significance of protein source, of protein digestibility and of the products of protein digestion upon the efficiency of zinc absorption and utilization require further research. Another important research area is to identify the form(s) and the dietary conditions under which phytic acid can exert negative effects on zinc absorption. Such studies may well reveal situations in which it is desirable to remove phytate(s) during food processing.

Finally, it is abundantly clear that failure to define the dietary content of constituents already known to modify the efficacy of zinc absorption is a common cause of difficulty in attempts to reconcile the frequently conflicting data arising from studies of the influence of diet on zinc utilization by human subjects. These difficulties will remain until the inorganic constituents of experimental diets are described more fully and accompanied by data on dietary contents of "fibre", protein and of those forms of the phytic acids which may restrict zinc absorption. Progress towards a fuller understanding of the nature and mode of action of dietary determinants of zinc absorption is contingent upon the adequate definition of such variables in future investigations.

References

Allen LHA (1982) Calcium bioavailability and absorption: a review. Am J Clin Nutr 35: 783–808

Andersson K-E, Bratt L, Dencker H, Lanner E (1976) Some aspects of the intestinal absorption of zinc in man. Eur J Clin Pharmacol 9: 423–428

Arvidsson B, Björn-Rasmussen E, Cederblad Å, Sandström B (1978) A radionuclide technique for studies of zinc absorption in man. Int J Nucl Med Biol 5: 104–109

Bagheri S, Guéguen L (1985) Effect of wheat bran and pectin on the absorption and retention of phosphorus, calcium, magnesium and zinc by the growing pig. Reprod Nutr Dév 25: 705–716

Bei L, Wood RJ, Rosenberg IH (1986) Glucose polymer increases jejunal calcium, magnesium, and zinc absorption in humans. Am J Clin Nutr 44: 244–247

Blakeborough P, Salter DN, Gurr MI (1983) Zinc binding in cow's milk and human milk. Biochem J 209: 505–512

Blakeborough P, Gurr MI, Salter DN (1986) Digestion of the zinc in human milk, cow's milk and a commercial babyfood: some implications for human infant nutrition. Br J Nutr 55: 209–217

Casey CE, Walravens PA, Hambidge KM (1981) Availability of zinc: loading test with human milk, cow's milk and infant formulas. Pediatrics 68: 394–396

Chandler CJ, Wang TY, Halsted CH (1986) Pteroylpolyglutamate hydrolase from human duodenal brush borders: purification and characterization. J Biol Chem 261: 928–933

Cheryan M (1980) Phytic acid interactions in food systems. CRC Crit Rev Food Sci Nutr 13: 297–335

Colin MA, Taper LJ, Ritchey SJ (1983) Effect of dietary zinc and protein levels on the utilization of zinc and copper by adult females. J Nutr 113: 1480–1488

Cousins RJ (1985) Absorption, transport and hepatic metabolism of copper and zinc: special reference to metallothionein and ceruloplasmin. Physiol Rev 65: 238–309

Cousins RJ, Smith KT (1980) Zinc-binding properties of bovine and human milk in vitro: influence of changes in zinc content. Am J Clin Nutr 33: 1083–1087

Craig WJ, Balbach L, Harris S, Vyhmeister N (1984) Plasma zinc and copper levels of infants fed different milk formulas. J Am Coll Nutr 3: 183–186

Cunnane SC, Huang Y-S, Horrobin DF, Davignon J (1981) Role of zinc in linoleic acid desaturation and prostaglandin synthesis. Prog Lipid Res 20: 157–160

Davies NT, Olpin SE (1979) Studies on the phytate: zinc molar contents in diets as a determinant of Zn availability to young rats. Br J Nutr 41: 591–603

Davies NT, Carswell AJP, Mills CF (1985) The effect of variation in dietary calcium intake on the phytate–zinc interaction in rats. In: Mills CF, Bremner I, Chesters JK (eds) Trace elements in man and animal. Commonwealth Agricultural Bureaux, Slough, pp 440–442

Dawson-Hughes B, Seligson FH, Hughes VA (1986) Effects of calcium carbonate and hydroxyapatite on zinc and iron retention in postmenopausal women. Am J Clin Nutr 44: 83–88

DeBoland AR, Garner GB, O'Dell BL (1975) Identification and properties of "phytate" in cereal grains and oilseed products. J Agric Food Chem 23: 1186–1189

Dillaha CJ, Lorinc AL, Aavik OR (1953) Acrodermatitis enteropathica. Review of the literature and report of a case successfully treated with diodoquin. J Am Med Ass 152: 509–512

Eckhert CD (1985) Isolation of a protein from human milk that enhances zinc absorption in humans. Biochem Biophys Res Commun 130: 264–269

Erdman JW, Weingartner KE, Mustakas GC, Schmutz RD, Parker HM, Forbes RM (1980) Zinc and magnesium bioavailability from acid-precipitated and neutralized soybean protein products. J Food Sci 45: 1193–1199

Evans GW, Johnson PE (1980) Characterization and quantitation of a zinc-binding ligand from milk. Pediatr Res 14: 876–880

Flagstad T (1981) Zinc absorption in cattle with a dietary picolinic acid supplement. J Nutr 111: 1996–1999

Flanagan PR, Cluett J, Chamberlain MJ, Valberg LS (1985) Dual-isotope method for determination of human zinc absorption: the use of a test meal of turkey meat. J Nutr 115: 111–122

Forbes RM, Erdman JW Jr (1983) Bioavailability of trace mineral elements. Ann Rev Nutr 3: 213–231

Forbes RM, Parker HM, Erdman JW Jr (1984) Effects of dietary phytate, calcium and magnesium levels on zinc bioavailability to rats. J Nutr 114: 1421–1425

Ghishan FK, Said HM, Wilson PC, Murell JE, Greene HL (1986) Intestinal transport of zinc and folic acid: a mutual inhibitory effect. Am J Clin Nutr 43: 258–262

Golden BE, Golden MHN (1981) Plasma zinc, rate of weight gain and energy cost of tissue deposition in children recovering from severe malnutrition on cow's milk or soya protein based diet. Am J Clin Nutr 34: 892–899

Greger JL, Mulvaney J (1985) Absorption and tissue distribution of zinc, iron and copper by rats fed diets containing lactalbumin, soy and supplemental sulfur-containing amino acids. J Nutr 115: 200–210

Greger JL, Snedeker SM (1980) Effect of dietary protein and phosphorus levels on the utilization of zinc, copper and manganese by adult males. J Nutr 110: 2243–2253

Hallberg L, Brune M, Rossander L (1986) Effect of ascorbic acid on iron absorption from different types of meals. Studies with ascorbic acid-rich foods and synthetic ascorbic acid given in different amounts and with different meals. Hum Nutr Appl Nutr 40A: 97–113

Halsted JA, Ronaghy HA, Abadi P et al. (1972) Zinc deficiency in man. Am J Med 53: 277–284

Harzer G, Kauer H (1982) Binding of zinc to casein. Am J Clin Nutr 35: 981–987

Heth DA, Hoekstra WG (1965) Zinc-65 absorption and turnover in rats. J Nutr 85: 367–374

Heth DA, Becker WM, Hoekstra WG (1966) Effect of calcium, phosphorus and zinc on zinc-65 absorption and turnover in rats fed semipurified diets. J Nutr 88: 331–337

Hill CH, Matrone G (1970) Chemical parameters in the study of in vivo and in vitro interaction of transition elements. Fed Proc 29: 1474–1488

Istfan NW, Janghorbani M, Young VR (1983) Absorption of stable ^{70}Zn in healthy young men in relation to zinc intake. Am J Clin Nutr 38: 187–194

Janghorbani M, Istfan NW, Pagounes JO, Steinke FH, Young VR (1982) Absorption of dietary zinc in man: comparison of intrinsic and extrinsic labels using a triple stable isotope method. Am J Clin Nutr 36: 537–545

Johnson MA, Baier MJ, Greger JL (1982) Effects of dietary tin on zinc, copper, iron, manganese metabolism of adult males. Am J Clin Nutr 35: 1332–1338

Kelsay JL, Prather ES (1983) Mineral balances of human subjects consuming spinach in a low-fiber diet and in a diet containing fruits and vegetables. Am J Clin Nutr 38: 12–19

Ketelsen SM, Stuart MA, Weaver CM, Forbes RM, Erdman JW Jr (1984) Bioavailability of zinc to rats from defatted soy flour, acid-precipitated soy concentrate and neutralized soy concentrate as determined by intrinsic and extrinsic labeling techniques. J Nutr 114: 536–542

King JC, Raynolds WL, Margen S (1978) Absorption of stable isotopes of iron, copper and zinc during oral contraceptive use. Am J Clin Nutr 31: 1198–1203

Kratzer FH, Allred JB, Davis PH, Marshall BJ, Vohra P (1959) The effect of autoclaving soybean protein and the addition of ethylenediaminetetracetic acid on the biological availability of dietary zinc for turkey poults. J Nutr 68: 313–322

Likuski HJA, Forbes RM (1965) Mineral utilization in the rat. IV. Effect of calcium and phytic acid on the utilization of dietary zinc. J Nutr 85: 230–234

Lo GS, Settle SL, Steinke FH, Hopkins DT (1981) Effect of phytate: zinc molar ratio and isolated soybean protein on zinc bioavailability. J Nutr 111: 2223–2235

Lònnerdal B, Stanislowski AG, Hurley LS (1980) Isolation of a low molecular weight zinc binding ligand from human milk. J Inorg Biochem 12: 71–78

Lönnerdal B, Hoffman B, Hurley LS (1982) Zinc and copper binding proteins in human milk. Am J Clin Nutr 36: 1170–1176

Lönnerdal B, Cederblad Å, Davidsson L, Sandström B (1984) The effect of individual components of soy formula and cow's milk formula on zinc bioavailability. Am J Clin Nutr 40: 1064–1070

Lönnerdal B, Keen CL, Hurley LS (1985) Manganese binding proteins in human and cow's milk. Am J Clin Nutr 41: 550–559

Lönnerdal B, Kunz C, Sandberg A-S, Sandström B (1987) Inhibitory effects of various inositol phosphates on zinc and calcium absorption. Fed Proc 46: 599

Meadows NJ, Grainger SL, Ruse W, Keeling PWN, Thompson RPH (1983) Oral iron and the bioavailability of zinc. Br Med J 287: 1013–1014

Menard MP, Cousins RJ (1983) Zinc transport by brush border membrane vesicles from rat intestine. J Nutr 113: 1434–1442

Milne DB, Canfield WK, Mahalko JR, Sandstead HH (1984) Effect of oral folic acid supplements on zinc, copper, and iron absorption and excretion. Am J Clin Nutr 39: 535–539

Molokhia M, Sturniolo G, Shields R, Turnberg LA (1980) A simple method for measuring zinc absorption in man using a short-lived isotope (^{69m}Zn). Am J Clin Nutr 33: 881–886

Momcilovic B, Kello D (1979) Fortification of milk with zinc and iron. Nutr Rep Int 20: 429–436

Morris ER, Ellis R (1980a) Effect of dietary phytate: zinc molar ratio on growth and bone response of rats fed semipurified diets. J Nutr 110: 1037–1045

Morris ER, Ellis R (1980b) Bioavailability to rats of iron and zinc in wheat bran: response to low-phytate bran and effect of the phytate-zinc molar ratio. J Nutr 110: 2000–2010

Moynahan EJ (1974) Acrodermatitis enteropathica: a lethal inherited human zinc-deficiency disorder. Lancet II: 399–400

Nävert B, Sandström B, Cederblad Å (1985) Reduction of the phytate content of bran by leavening in bread and its effect on absorption of zinc in man. Br J Nutr 53: 47–53

Nwokolo EN, Bragg DB (1977) Influence of phytic acid and crude fibre on the availability of minerals from four protein supplements in growing chicks. Can J Anim Sci 57: 475–477

Oberleas D, Harland BF (1981) Phytate content of foods: effect on dietary zinc availability. J Am Diet Assoc 79: 433–466

Oberleas D, Muhrer ME, O'Dell BL (1962) Some effects of phytic acid on zinc availability and parakeratosis in swine. J Anim Sci 21: 57–61

Oberleas D, Muhrer ME, O'Dell BL (1966) Dietary metal-complexing agents and zinc availability in the rat. J Nutr 90: 56–62

O'Dell BL (1969) Effect of dietary components upon zinc availability. Am J Clin Nutr 22: 1315–1322

O'Dell BL, Savage JE (1960) Effect of phytate on zinc availability. Proc Soc Exp Biol Med 103: 304–309

Oestreicher P, Cousins RJ (1982) Influence of intraluminal constituents on zinc absorption by isolated, vascularly perfused rat intestine. J Nutr 112: 1978–1982

Oestreicher P, Cousins RJ (1985) Copper and zinc absorption in the rat: mechanism of mutual antagonism. J Nutr 115: 159–166

Patterson WP, Winkelmann M, Perry MC (1985) Zinc-induced copper deficiency: megamineral sideroblastic anemia. Ann Intern Med 103: 385–386

Pécoud A, Donzel P, Schelling JL (1975) Effect of foodstuffs on the absorption of zinc sulfate. Clin Pharmacol Ther 17: 469–474

Rogers IM, Keen CL, Hurley LS (1985) Zinc deficiency in pregnant Long Evans hooded rats: teratogenicity and tissue trace elements. Teratology 31: 89–100

Sandberg A-S, Ahderinne R, Andersson H, Hallgren B, Hultèn L (1983) The effect of citrus pectin on the absorption of nutrients in the small intestine. Hum Nutr Clin Nutr 37C: 171–183

Sandberg A-S, Ahderinne R (1986) HPLC method for determination of inositol tri-, tetra-, penta-, and hexaphosphates in foods and intestinal contents. J Food Sci 51: 547–550

Sandström B (1987a) Cereals as a source of minerals in human nutrition. In: Morton I (ed) The First European Federation of Food Science. Technology congress – Cereals in a European context. Ellis Horwood, London, pp 241–248

Sandström B (1987b) Zinc and dietary fibre. Scand J Gastroenterol 22 (Suppl 129): 80–84

Sandström B, Cederblad Å (1980) Zinc absorption from composite meals. II. Influence of the main protein source. Am J Clin Nutr 33: 1778–1783

Sandström B, Cederblad Å (1987) Effect of ascorbic acid on the absorption of zinc and calcium in man. Int J Vitam Nutr Res 57: 87–90

Sandström B, Arvidsson B, Cederblad Å, Björn-Rasmussen E (1980) Zinc absorption from composite meals. I. The significance of wheat extraction rate, zinc, calcium and protein content in meals based on bread. Am J Clin Nutr 33: 739–745

Sandström B, Keen CL, Lönnerdal B (1983a) An experimental model for studies of zinc availability from milk and infant formulas using extrinsic labelling. Am J Clin Nutr 38: 420–428

Sandström B, Cederblad Å, Lönnerdal B (1983b) Zinc absorption from human milk, cow's milk and infant formulas. Am J Dis Child 137: 726–729

Sandström B, Davidsson L, Cederblad Å, Lönnerdal B (1985) Oral iron, dietary ligands and zinc absorption. J Nutr 115: 411–414

Sandström B, Andersson H, Kivistö B, Sandberg A-S (1986) Apparent small intestinal absorption of nitrogen and minerals from soy and meat protein-based diets. A study on ileostomy subjects. J Nutr 116: 2209–2218

Sandström B, Kivistö B, Cederblad Å (1987a) Absorption of zinc from soy protein meals in humans. J Nutr 117: 321–327

Sandström B, Davidsson L, Kivistö B, Hasselblad C (1987b) The effects of vegetables and beet fibre on the absorption of zinc in humans from a composite meal. Br J Nutr 58: 49–57

Schelling JL, Muller-Hess S, Thonney F (1976) Effect of food on zinc absorption. Lancet II: 968–970

Schwartz R, Apgar BJ, Wien EM (1986) Apparent absorption and retention of Ca, Cu, Mg, Mn, and Zn from a diet containing bran. Am J Clin Nutr 43: 444–455

Schwarz FJ, Kirchgessner M, Roth H-P (1983) Zum Einfluss von Pikolinsäure und Zitronensäure auf die Intestinale Zink-Absorption in vitro und in vivo. Res Exp Med 182: 39–48

Seal CJ, Heaton FW (1985) Effect of dietary picolinic acid on the metabolism of exogenous and endogenous zinc in the rat. J Nutr 115: 986–993

Snedeker SM, Greger JL (1981) Effect of dietary protein, sulfur amino acids, and phosphorus on human trace element metabolism. Nutr Rep Int 23: 853–864

Snedeker SM, Greger JL (1983) Metabolism of zinc, copper and iron as affected by dietary protein, cysteine and histidine. J Nutr 113: 644–652

Solomons NW (1986) Competitive interaction of iron and zinc in the diet: consequences for human nutrition. J Nutr 116: 927–935

Solomons NW, Jacob RA (1981) Studies on the bioavailability of zinc in humans: effects of heme and nonheme iron on the absorption of zinc. Am J Clin Nutr 34: 475–482

Solomons NW, Jacob RA, Pineda O, Viteri F (1979a) Studies on the bioavailability of zinc in man. J Lab Clin Med 94: 335–343

Solomons NW, Jacob RA, Pineda O, Viteri FE (1979b) Studies on the bioavailability of zinc in man. Effects of the Guatemalan rural diet and of the iron-fortifying agent NaFeEDTA. J Nutr 109: 1519–1528

Solomons NW, Jacob RA, Pineda O, Viteri FE (1979c) Studies on the bioavailability of zinc in man. II. Effects of ascorbic acid on zinc absorption. Am J Clin Nutr 32: 2495–2499

Solomons NW, Janghorbani M, Ting BTG et al. (1982) Bioavailability of zinc from a diet based on isolated soy protein: application in young men of the stable isotope tracer, ^{70}Zn. J Nutr 112: 1809–1821

Solomons NW, Marchini JS, Duarte-Favaro R-M, Vannuchi H, Dutra de Oliveira JE (1983) Studies on the bioavailability of zinc in humans: intestinal interaction of tin and zinc. Am J Clin Nutr 37: 566–571

Spencer H, Kramer L, Norris C, Osis D (1984) Effect of calcium and phosphorus on zinc metabolism in man. Am J Clin Nutr 40: 1213–1218

Steinhardt HJ, Adibi SA (1984) Interaction between transport of zinc and other solutes in human intestine. Am J Physiol 247: G176–182

Tamura T, Shane B, Baer MT, King JC, Margen S, Stokstad ELR (1978) Absorption of mono- and polyglutamyl folates in zinc-depleted man. Am J Clin Nutr 31: 1984–1987

Turnlund JR, Michel MC, Keyes WR, King JC, Margen S (1982) Use of enriched stable isotopes to determine zinc and iron absorption in elderly men. Am J Clin Nutr 35: 1033–1040

Turnlund JR, King JC, Keyes WR, Gong B, Michel MC (1984) A stable isotope study of zinc absorption in young men: effects of phytate and α-cellulose. Am J Clin Nutr 40: 1071–1077

Turnlund JR, Durkin N, Costa F, Margen S (1986) Stable isotope studies of zinc absorption and retention in young and elderly men. J Nutr 116: 1239–1247

Valberg LS, Flanagan PR, Chamberlain MJ (1984) Effects of iron, tin, and copper on zinc absorption in humans. Am J Clin Nutr 40: 536–541

Valberg LS, Flanagan PR, Brennan J, Chamberlain MJ (1985) Does the oral zinc tolerance test measure zinc absorption? Am J Clin Nutr 41: 37–42

Vohra P, Kratzer FH (1964) Influence of various chelating agents on the availability of zinc. J Nutr 82: 249–256

Vohra P, Gray GA, Kratzer FH (1965) Phytic acid–metal complexes. Proc Soc Exp Biol Med 120: 447–449

Wada L, Turnlund JR, King JC (1985) Zinc utilization in young men fed adequate and low zinc intakes. J Nutr 115: 1345–1354

Wapnir RA, Khani DE, Bayne MA, Lifshitz F (1983) Absorption of zinc by the rat ileum: effects of histidine and other low-molecular-weight ligands. J Nutr 113: 1346–1354

Welch RM, House WA, Van Campen D (1977) Effects of oxalic acid on availability of zinc from spinach leaves and zinc sulfate to rats. J Nutr 107: 929–933

Weismann K (1986) Chelating drugs and zinc. Dan Med Bull 33: 208–211

Chapter 5

Systemic Transport of Zinc

R.J. Cousins

Zinc transport through the systemic circulation can be viewed from many different perspectives. In the past, attention has been paid to the zinc concentration and zinc-binding proteins. More recently, the dynamic nature of zinc-binding constituents of the plasma as a metabolic compartment has been defined. Indeed, evidence suggests that hormonal regulation of cellular processes dramatically influences zinc in the systemic circulation.

Zinc in the Systemic Circulation

Distribution Among Blood Components

Zinc in whole blood represents $< 0.5\%$ of total body zinc. Plasma zinc comprises only about 10%–20% (depending upon species) of that in whole blood. Zinc associated with the erythron accounts for at least 75% of that found in blood (reviewed by Underwood 1977; Cousins 1985). The two isoenzymes of carbonic anhydrase and superoxide dismutase account for 87% and 5% of the total erythrocyte zinc content, respectively (Ohno et al. 1985). Leucocytes contain more zinc than erythrocytes, but exact estimates vary (Dennes et al. 1962). Mononuclear and polymorphonuclear cells from human blood average about 6 mg zinc/10^6 cells, whereas erythrocytes contain much less, about 1 mg zinc/10^6 cells (Milne et al. 1985). Improvements in cell isolation techniques place greater emphasis on more recent measurements of the zinc content of blood cells.

Zinc uptake and efflux by erythrocytes have been reported using cells from various sources and different culture conditions (Dennes et al. 1962; Kruckeberg and Brewer 1978). Non-specificity of binding was not considered however, in calculating uptake values. This may be important since erythrocyte ghosts bind zinc in vitro (Schmetterer 1978). Leucocytes appear to take up zinc but do not exchange it with the extracellular space (Dennes et al. 1962). Uptake of zinc by human lymphocytes has been demonstrated, both in the presence (Phillips et al. 1977) and absence of transferrin (Montgomery et al. 1977). It has been suggested that the zinc content of macrophages influences their phagocytic activity (Chvapil et al. 1982).

A small portion of plasma zinc is ultrafilterable. The exact percentage is not clear, but methods using membrane ultrafiltration suggest it is between 0.2% (Bloxam et al. 1984) and 1.0% (Whitehouse et al. 1982) of total plasma zinc. Older data place the value at between 2% and 5%. Prasad and Oberleas (1970) reported that histidine, glutamine, threonine, cystine and lysine account for the major portion of ultrafilterable zinc. From computed values, Hallman et al. (1971) proposed that cysteine and histidine would account for over half of zinc bound to amino acids in human plasma. Ultrafilterable zinc could act as a zinc donor for a high affinity, cellular zinc transport system or as an intermediate between protein-bound zinc and cell uptake mechanisms. Interaction between protein-bound zinc and that bound to amino acids may occur (Henkin 1974). Some readily absorbable zinc chelates may enter the circulation intact and contribute to the ultrafilterable pool (Suso and Edwards 1972).

A plethora of data has demonstrated that albumin is the principal zinc-binding protein in plasma. This represents a loosely bound, exchangeable zinc pool (Prasad and Oberleas 1970; Boyett and Sullivan 1973; Giroux et al. 1976; Chesters and Will 1981; Foote and Delves 1984). Transferrin binds zinc less firmly than albumin (Charlwood 1979). Two-dimensional immunoelectrophoresis and autoradiography revealed that about 12 proteins in human serum bind zinc in vitro (Scott and Bradwell 1983). The order of ^{65}Zn binding was albumin $\ll$ α_2-macroglobulin $<$ transferrin with only minor binding by other proteins.

Albumin-bound zinc represents the metabolically active pool of plasma as will be discussed below. Normally, the concentration of zinc associated with albumin is maintained within narrow limits. This relationship is curious since the abundance of albumin is such that on a molar basis the albumin : zinc ratio is 30 : 1. In humans, congenital hyperzincaemia has been attributed to plasma albumin with an abnormally high binding affinity for zinc (Failla et al. 1982). Of particular interest is the mutant rat strain that does not produce albumin. These analbuminaemic animals do not exhibit depressed plasma zinc levels (Suzuki et al. 1986).

Albumin appears to function as the carrier for zinc in portal plasma. Substitution of purified albumin for serum containing an equivalent amount of albumin substantially increased zinc absorption by the vascularly perfused rat intestine (Smith et al. 1979). This suggests that the plasma albumin concentration may exercise some control over zinc absorption from the small intestine. The finding could have direct relevance to the reduction in zinc absorption that accompanies diseases that produce hypoalbuminaemia. The rate of zinc transport across the basolateral membrane to albumin may help regulate the upper limit of the normal plasma concentration. It has been suggested that transferrin has a zinc transport function in portal plasma (Evans and Winter 1975). However, considerable evidence argues against this hypothesis. All zinc absorbed by the vascularly perfused rat intestine is quantitatively transferred to albumin (Smith et al. 1979). Iron-saturated or dialysed serum does not affect absorption in that system. These experiments utilized bicarbonate-containing vascular perfusate, which may favour binding of zinc to transferrin rather than to albumin (Harris 1983). Furthermore, albumin has a greater relative affinity for zinc than does transferrin which favours binding to albumin at the concentration of these proteins in plasma. The association constant for the zinc–albumin complex is 10^6 (Charlwood 1979; Chesters and Will 1981). Endocytosis of transferrin may contribute to a fraction of zinc uptake by some cells.

α_2-Macroglobulin comprises a tightly bound pool of plasma zinc. The zinc-α_2-macroglobulin association constant is $> 10^{10}$. Estimates of the proportion of plasma zinc bound to this protein range from 20% to 40% (Parisi and Vallee 1970; Boyett and Sullivan 1973; Giroux 1975). The metabolic fate and physiological significance of zinc circulating with α_2-macroglobulin is not clear. As an acute phase protein that is induced during inflammation, α_2-macroglobulin's zinc-binding capacity may be related to host defence functions. A histidine-rich glycoprotein in plasma with 58 000 molecular weight may bind small amounts of zinc with high affinity (Morgan 1981). This protein could elute with albumin during size-exclusion chromatography. The biological significance of this putative transport protein is questionable because binding affinity is high, renal filtration is possible and quantitatively its influence would be minimal.

Dietary Influences on Plasma Zinc

Exchange of zinc between the plasma and blood cells appears to be minimal. Data on this aspect of zinc metabolism are limited, however. Zinc uptake by erythrocytes is influenced by many factors and therefore is not an accumulate reflection of the dietary zinc supply (Chesters and Will 1978). In contrast, the plasma zinc concentration is decreased when dietary intake is insufficient (Shanklin et al. 1968). In rodents, there is a precipitous reduction in concentration within hours after a low-zinc diet is fed (Dreosti et al. 1968). This effect is not influenced by age. The reduction is usually to about 50% of the normal concentration and lasts for the duration of dietary zinc restriction. These observations have been made countless times and depressed plasma zinc concentrations have become a recognized standard of status assessment in uncomplicated zinc deficiency in experiments with laboratory animals maintained under carefully regulated environmental and dietary conditions. Furthermore, plasma zinc has been reported to exhibit circadian rhythmicity (Hurley et al. 1982). Serum generally has higher zinc concentrations than plasma (Smith et al. 1985).

Reduced plasma zinc concentrations correspond to a reduction in metallothionein-bound zinc in liver and intestine (Richards and Cousins 1976). Recently, Sato et al. (1984) demonstrated that reduced plasma zinc was correlated with reduced levels of metallothionein-I in both plasma and liver. In contrast to the plasma zinc depression, metallothionein-I reached undetectable levels. This suggests metallothionein in plasma may serve as a particularly sensitive indicator of dietary zinc status.

Experimental zinc deficiency in human subjects produced a reduction in plasma zinc concentration (Prasad et al. 1978). The decline was not as pronounced as observed in rodent studies, but the level of intake used was 2.7 mg/d, which, considered in relation to probable requirements, is not as severe a restriction as used in animal studies. Nevertheless, after 6 months at that intake, plasma zinc was reduced by 50%. When female subjects were fed a much lower amount of zinc (0.2 mg/d), a comparable reduction in plasma zinc was observed in 5 weeks (Hess et al. 1977). Acute deprivation of dietary zinc can produce reductions in serum zinc that are comparable to those found in animals, however (Gordon et al. 1982). A diet composed of egg white and EDTA-extracted soya protein providing 0.5 mg zinc/3000 cal produced declines in plasma zinc levels in less than 5 d. Circadian rhythmicity was observed (Lifschitz and Henkin 1971). This demonstrates that plasma zinc concentrations are closely associated with dietary zinc intake in both animals and humans. Long-term parenteral nutrition without adequate amounts of zinc will depress plasma zinc levels (Fleming et al. 1976; Tucker et al. 1976).

Abnormally high plasma zinc concentrations have been reported. In rats fed previously a diet low in zinc, repletion with normal or above normal intakes of zinc produces a dose-dependent increase in serum zinc concentration (Wilkins et al. 1972; Richards and Cousins 1976). Zinc repletion in this manner stimulates metallothionein synthesis in both liver and intestine (McCormick et al. 1981; Menard et al. 1981). Presumably, the increase in plasma zinc reflects absorption of zinc from the small intestine. A similar dose-dependent increase in serum zinc following oral doses of zinc has been observed in humans (Pecoud et al. 1975). This response to an oral zinc load was used as the basis for a zinc tolerance test to assess zinc status (Sullivan et al. 1979). Increase in plasma zinc may reflect relative bioavailability of zinc in food mixtures (Casey et al. 1981) and mineral interactions (Solomons and Jacob 1981), but it may not provide a reliable index of zinc status in humans (Fickel et al. 1986). Kinetic analysis has shown that oral zinc loading produces rapid equilibration with intracellular zinc pools (Babcock et al. 1982), which suggests an interactive relationship between the dietary supply and endogenous tissue zinc.

Diseases Associated with Low Plasma Zinc

Since Vallee (1959) focused attention on the relationship between zinc metabolism, function and disease, a plethora of data has supported the concept of such a relationship. A majority of diseases so related involve disorders of the kidney, liver or pancreas. More recent clinical data have related depressed plasma zinc to other diseases including cirrhosis (Schechter et al. 1976; Hartoma et al. 1977), foetal alcohol syndrome (Assadi and Ziar 1986), alcoholic seizures (Bogden and Troiano 1978), alcoholic pancreatitis (Williams et al. 1979), intestinal bypass surgery (Atkinson et al. 1978), severe malnutrition (Canfield et al. 1980; Golden and Golden 1981), biliopancreatic bypass (Vanderhoof et al. 1983), arthritis (Svenson et al. 1985), renal insufficiency (Lindeman et al. 1978; Zumkley et al. 1984), acne vulgares (Rebello et al. 1986), uraemia (Mahajan et al. 1979) and acrodermatitis enteropathica (Neldner and Hambidge 1975).

The metabolic basis for the depressed plasma zinc levels observed in the studies mentioned above and in numerous others is generally obscure. Hypozincaemia of protein–energy malnutrition and hepatic cirrhosis may be caused by hypoalbuminaemia (Schechter et al. 1976; Canfield et al. 1980). However, low plasma zinc associated with malnutrition has been observed without low plasma albumin (Golden and Golden 1981). Rat hepatocytes cultured in medium containing albumin exhibit different zinc uptake kinetics than cells cultured in albumin-free medium (Pattison and Cousins 1986a, b). Therefore, hepatic uptake produced in vivo during hypoalbuminaemia could produce atypical cellular redistribution and efflux to the plasma. Kidney damage and concomitant enhanced glomerular filtration of proteins could also produce a reduction in plasma zinc levels. This reduction could also occur through a shift of plasma zinc to the ultrafilterable pool. For example, infants with foetal alcohol syndrome exhibit hypozincaemia and hyperzincuria without appreciable changes in kidney function (Assadi and Ziar 1986), suggesting the ultrafilterable zinc pool is partly the source of the increased urinary zinc output observed in this disease. Congenital and acquired zinc malabsorption and pancreatic insufficiency can lead to reduced plasma zinc levels (reviewed in Solomons and Cousins 1984).

Hormonal Control of Plasma Zinc

An abundance of data from animal studies have shown that plasma zinc concentrations are hormonally regulated. Despite some circadian rhythmicity, the concentrations are held remarkably within limits. The regulatory signals that may be involved are complex, but they appear to represent those hormones that control key pathways of intermediary metabolism. This is in contrast to the unique control of systemic calcium. Some hormonal control of systemic zinc may be direct, involving concerted redirection of the metabolic flux of zinc, whereas indirect involvement could reflect changes in tissues which provide substrates for overall metabolism, e.g. muscle. Furthermore, biochemical parameters which characterize the acute phase of host defence include changes in plasma zinc concentrations.

There are considerable clinical data to suggest that stress, in a variety of forms, sufficiently alters the kinetics of zinc metabolism to produce depressions in the plasma zinc content. The effect is not specific, but may occur with any acute disease (Falchuk 1977). A mechanism involving ACTH release and reduction of zinc in an albumin-like fraction of plasma was proposed. The reduction in plasma zinc following trauma usually does not correlate with a reduction in plasma albumin (Moser et al. 1985), but concomitant depressions have been reported (Hallbook and Hedelin 1978). During trauma, depressed serum zinc accompanies hyperzincuria (McClain et al. 1986). In animal models, stress associated with exercise and inflammation (Oh et al. 1978) or bacterial infection (Sobocinski et al. 1978) was shown to produce hypozincaemia. Depressed serum zinc was observed in animals with tumours (Fenton and Burke 1983; Ujjani et al. 1986) and burn injury plus infection (Powanda et al. 1980), but not during restraint stress (Hildalgo et al. 1986). Conversely, in both humans (Henry and Elmes 1975) and rats (Richards and Cousins 1976), food restriction increases plasma zinc concentrations. In humans, plasma zinc increases observed during exercise were found to be related to zinc nutriture (Lukaski et al. 1984).

The hormonal basis for the changes in circulating plasma zinc is not defined, but there is considerable evidence that glucocorticoids, epinephrine, glucagon and interleukin-1 are important factors. Glucocorticoids administered to intact or adrenalectomized rats depress serum zinc levels within hours after administration (Etzel et al. 1979; Brady 1981; Cousins et al. 1986). Cyclic AMP, a mediator of glucagon and epinephrine action, similarly decreases plasma zinc. Cumulative effects of combined hormone treatments (glucagon or epinephrine) with glucocorticoids are observed only in adrenalectomized rats (Cousins et al. 1986). In humans, glucocorticoid therapy is associated with a reduction in plasma zinc (Flynn et al. 1971; Ellul-Micallef et al. 1976) which may be concomitant with plasma albumin decreases (Weismann and Hoyer 1986). Glucocorticoid-induced hypozincaemia was shown kinetically to involve a reduction in zinc in the plasma compartment with concomitant increases in cellular compartments (Henkin et al. 1984). The endocrine regulation of plasma zinc has been reviewed in detail earlier (Cousins 1985).

Leucocytic mediators when administered to rats produce hypozincaemia (Kampschmidt and Upchurch 1970; Pekarek and Beisel 1971). While the nomenclature, characterization and broad physiological actions of the mediator(s) involved are in an evolutionary phase, interleukin-1 (IL-1) will be considered as the leucocytic hormone that produces these effects (Dinarello 1984). Bacterial endotoxin produces similar effects on plasma zinc which are mediated through IL-1. In addition to hypozincaemia, both endotoxin and interleukin-1 produce hypercupraemia via increased plasma ceruloplasmin levels, a well-defined acute phase response (reviewed in Cousins 1985). Interleukin-1 and/or endotoxin has been reported to depress plasma zinc in hamsters (Etzel et al. 1982), chicks (Sas and Bremner 1979; Klasing 1984), streptozotocin-diabetic rats (Failla et al. 1983), pigs (Chesters and Will 1981), goats (Van Miert et al. 1984), turkey embryos (Klasing et al. 1987), rabbits (Goldblum et al. 1987) and rats (Sobocinski and Canterbury 1982; DiSilvestro and Cousins 1984a,b). Of particular interest is the demonstration that IL-1, stimulated in human subjects by vigorous aerobic exercise, produced plasma zinc depression when the crude human IL-1 was administered to rats (Cannon and Kluger 1983). This observation and others showing that IL-1 is a principal mediator of the "acute phase" response including fever and acute phase protein synthesis (Dinarello 1984) strongly implicates the depression in plasma zinc as part of host defence processes. Similarly, the decrease in plasma zinc during human pregnancy is considered physiological rather than nutritional (Breskin et al. 1983; Swanson and King 1983). IL-1 is produced by human placental monocytes (Flynn 1984), which could account for pregnancy-related hypozinc-aemia. IL-1 stimulates a PGE_2-induced activation of lysosomes which produces muscle protein breakdown (Baracos et al. 1983). This could increase urinary zinc output and eventually decrease plasma zinc levels even though muscle is a major reserve of zinc.

A unifying mechanism to explain the plasma zinc-depressive effects of hormones, catecholamines, endotoxin and perhaps other agents is difficult with existing information. Regulation of cellular uptake/retention or excretion could produce the effect. Hypozincaemia associated with administration of glucocorticoids, epinephrine, glucagon, dibutyryl-cAMP (Cousins et al. 1986), endotoxin or IL-1 (DiSilvestro and Cousins 1984a) is inhibited if actinomycin D, an inhibitor of mRNA synthesis, is given to rats prior to these agents. The latter has many effects, but it has been shown unequivocally to suppress gene transcription. Similarly, increased metallothionein gene expression follows administration of these hormones and mediators and this is blocked by actinomycin D. The close relationship between hormonally regulated hepatic metallothionein synthesis and serum zinc levels has been described (reviewed in Cousins 1985). As shown in Fig. 5.1, there is a close inverse relationship between these variables. These data drawn from Cousins et al. (1986) show that hormonal induction of hepatic metallothionein (μg/g), x, and its suppression by actinomycin D, is inversely related to the serum zinc concentration (μg/ml), y, as follows: $y = 2.48e^{-0.01x}$. These data suggest that, at least on a short temporal basis, increased hepatic metallothionein induction and hepatic zinc uptake exchange accounts for hypozincaemia. Recent ^{65}Zn kinetic studies have shown that ^{65}Zn transfer from the plasma component to the liver is directly proportional to binding of ^{65}Zn to metallothionein following increased expression of the gene by cAMP (Dunn and Cousins 1987). This conclusion agrees with data where the endotoxin- and IL-1-produced depression of serum iron was related to granulocyte-released lactoferrin, but the serum zinc depression was not mediated by granulocytes (Goldblum et al. 1987). Macrophages produce metallothionein after glucocorticoid stimulation (Patierno et al. 1983). This increases their stability to endotoxin stress. A similar uptake during macrophage proliferation could influence plasma zinc levels.

Cellular Uptake from the Systemic Circulation

The major tissues that appear to contribute to the regulation of systemic zinc are illustrated in Fig. 5.2.

Hepatic Uptake

The majority of zinc presented to hepatocytes under physiological conditions is bound to albumin. While it has not been established that cellular zinc is derived from this systemic source, the evidence suggests that albumin provides

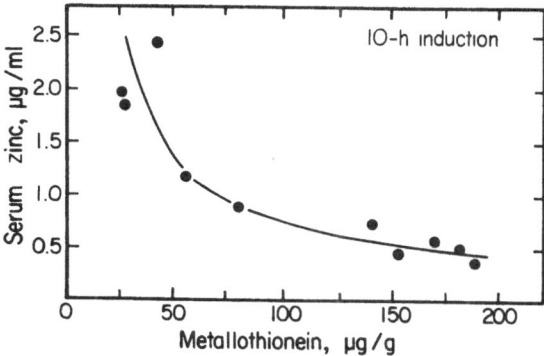

Fig. 5.1. Relationship between the serum zinc concentration and liver metallothionein after induction by hormones. The data are derived from Cousins et al. (1986).

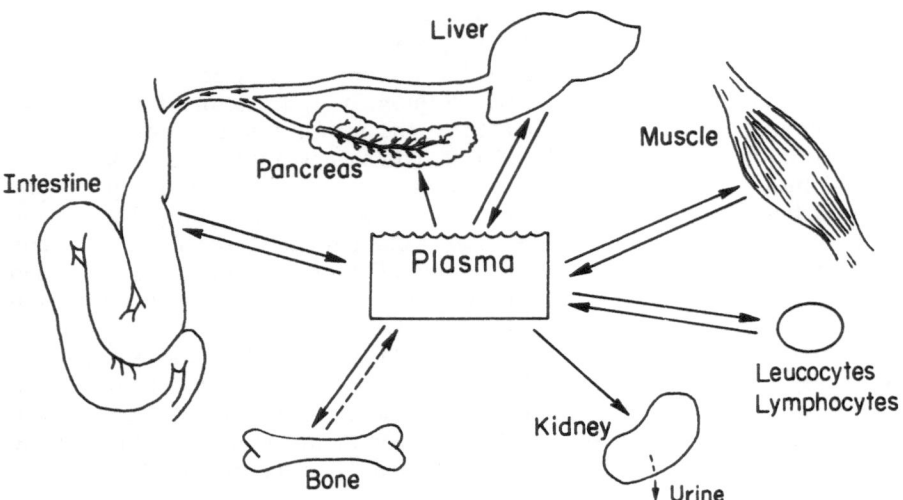

Fig. 5.2. The major tissues contributing to the regulation of systemic zinc.

this transport function. Specifically, zinc is transported from the intestine bound to albumin (Smith et al. 1979), hepatic uptake occurs rapidly after absorption (McCormick et al. 1981; Babcock et al. 1982) and hepatocytes take up zinc from albumin-containing medium in vitro (Failla and Cousins 1978a; Pattison and Cousins 1986a).

Marked changes in the dietary zinc supply produce only minimal changes in zinc content of the liver demonstrating that uptake and efflux are closely regulated. Early experiments with fish liver slices showed zinc uptake followed first-order kinetics, suggesting a passive mechanism (Saltman and Boroughs 1960). Recent experiments with rat liver parenchymal cells maintained in primary monolayer culture demonstrated energy dependence and saturable kinetics (Failla and Cousins 1978b). Uptake from the medium occurred in two phases. The slow phase represents exchange with cellular zinc with both saturable and linear components. Saturation is indicative of an associative step in transport before Zn^{2+} exchange which is rate limiting (Pattison and Cousins 1986a). Uptake with albumin-containing medium is a single saturable process with a half maximum exchange rate (K_m) at 9.5 μM and maximum exchange rate of 9.9 pmol Zn^{2+}/min/mg protein. The K_m of exchange is remarkably similar to the normal concentration of albumin-bound plasma zinc. Hepatocytes exchange zinc with a turnover time of 15 h ($t_{\frac{1}{2}}$ = 15 h). Although the bulk of zinc in hepatocytes is exchangeable with the extracellular compartment (plasma), they maintain a finite set of binding sites that are saturated under physiological conditions (Pattison and Cousins 1986b). New zinc-binding sites must be generated to produce major kinetic changes with the plasma zinc pool. Release of zinc from glucocorticoid-stimulated hepatocytes is less than from unstimulated cells, suggesting that stress-induced hypozincaemia is a manifestation of altered zinc efflux rather than only a change in uptake rate. Without albumin in the medium a linear component to slow zinc uptake/exchange is observed. This suggests that uptake can occur by a non-saturable process of significant magnitude within the physiological range of the plasma zinc concentration. The biological significance of this phase of uptake is not clear, but it could be a factor when ultrafilterable zinc accounts for a significant portion of total plasma zinc.

A variety of experimental approaches with both animals and humans indicate that uptake of radioactive zinc by the liver is extremely rapid, whether zinc is consumed orally (McCormick et al. 1981; Babcock et al. 1982) or intravenously (Jain et al. 1981; Henkin et al. 1984; Dunn and Cousins 1987). Treatments that induce hepatic metallothionein, e.g. glucocorticoids, Bt_2cAMP and endotoxin, appear to favour a flux of zinc in the direction of uptake/exchange (Chesters and Will 1981; Henkin et al. 1984; Dunn and Cousins 1987). A pharmacokinetic model of ^{65}Zn metabolism has demonstrated that the first-order mass transfer coefficient and linear binding constant are highest for the liver (Jain et al. 1981). This further suggests that control of plasma zinc levels involves hepatic mechanisms.

Kidney Uptake

Kinetic data have shown that the first-order mass transfer coefficient and binding constants for zinc place the kidney in line after the liver as a site of metabolic significance (Jain et al. 1981). Therefore, a major renal influence on the plasma zinc compartment would be expected.

Renal insufficiency and nephrotic syndrome result in reduced serum zinc concentrations. This is correlated to increased urinary zinc bound to ligands of smaller molecular weight than albumin (Lindeman et al. 1978). Plasma proteins were not changed in these conditions (Lindeman et al. 1978; Mahajan et al. 1979). In both rats (Freeman and Taylor 1977) and humans (Van Rij et al. 1979), infusion of amino acids into the systemic circulation increased urinary zinc excretion. Histidine and cysteine are the two plasma amino acids that have the greatest effect on renal filtration (Yunice et al. 1978). A shift of circulating zinc, from the protein-bound to ultrafilterable pool, could result in plasma zinc depression. This hypothesis is based in part on the observation that plasma proteins are not altered despite significant hypozincaemia.

Hormonal regulation of renal zinc excretion has been demonstrated. Insulin inhibits excretion of zinc by the perfused dog kidney when it is added to the vascular perfusate (Vander et al. 1983). Similarly, acute infusion of glucagon increases zinc excretion in the same model (Victery et al. 1982). In streptozotocin-induced diabetic rats, an insulin-reversible increase in urinary excretion of zinc, copper and iron has been shown (Lau and Failla 1984). This effect is accounted for by the drastic (18-fold) increase in daily urine output and a greater percentage of ultrafilterable zinc in plasma. Hyperphagia associated with this diabetic model could also contribute to urinary zinc loss since plasma levels do not change during this catabolic condition. The influence of renal metallothionein on renal handling of zinc has not been defined, but endotoxin and IL-1 administration increases metallothionein mRNA levels in kidney (DiSilvestro and Cousins 1984a). Furthermore, under practical feeding conditions, these mRNA levels are increased in kidney and intestine (Blalock et al. 1987), suggesting that these organs may influence zinc metabolism through metallothionein-related shifts in cellular uptake.

Brain Uptake

There is relatively little information available about the transport of zinc between the systemic circulation and the brain. The diffusion constant and linear binding constant are lower than for most other tissues (Jain et al. 1981). In man, the zinc content of brain is about 10 μg/g, but this increases with various types of tumours (Kokoglu et al. 1983). At least three zinc-binding proteins have been identified in brain. These have apparent molecular weights of 15 000, 25 000 and 210 000 (Itoh et al. 1983). The lowest molecular-weight protein could be a metallothionein as it is inducible under appropriate conditions (Ebadi et al. 1984). How these proteins may relate to zinc transport across the blood–brain barrier and influence uptake by brain cells is not known.

Placental Uptake

Placental transport of adequate amounts of zinc is necessary for optimum foetal development. The zinc content of the amniotic fluid appears to be a good index of the zinc supply needed for development (Gardiner et al. 1982). Human amniotic fluid contains somewhat higher amounts of albumin-bound zinc than does plasma (72%) and none is bound to α_2-macroglobulin. Analyses of fluid removed by amniocentesis reveal that no changes of zinc content occur with high-risk pregnancies (Rosick et al. 1983; Laitinen et al. 1984). Zinc concentrations in maternal plasma are less than normal, but in plasma from cord blood, they are significantly higher (Yamashita et al. 1985). Induction of foetal hepatic metallothionein could be related in part to this higher circulating plasma level.

Zinc transport across the placenta is a function of the number and size of foetuses present, probably because of the greater surface area involved (Feaster et al. 1955). Placenta-to-foetus transfer is rapid. During the last trimester of pregnancy in rabbits, intravenously administered ^{65}Zn is rapidly taken up by the conceptus with about half found in the foetuses within 3 h (Terry et al. 1960). This proportion increases during gestation. Most of the zinc that is newly acquired by the foetus is taken up by the liver. Retrograde placental zinc transport from foetal to maternal pools was also demonstrated.

Foetal alcohol syndrome may develop through the co-teratogenic effects of zinc deficiency and alcohol consumption (Keppen et al. 1985). Ghishan et al. (1982) proposed that in rats, ethanol consumption produced a 30%–40% reduction in placental zinc transport. This effect was not overcome with supplemental zinc (Ghishan and Greene 1983). Therefore, the hypozincaemia and hyperzincuria observed in infants with this syndrome (Assadi and Ziar 1986) could be manifestations of alcohol-related placental transport dysfunction.

Pancreatic Uptake

The pancreas exhibits a diffusion constant and a linear binding constant, derived from ^{65}Zn kinetic studies, that are comparable to those found for kidney and intestine (Jain et al. 1981). Furthermore, zinc deficiency is associated with decreased pancreatic secretory response (Sullivan et al. 1974). Uptake of zinc by isolated pancreatic islets follows saturation kinetics with K_m and V_{max} of 1.5 µM and 11.1 fmol/islet/min, respectively (Ludvigsen et al. 1979). A linear relationship between initial uptake rates was observed at higher zinc concentrations of the incubation medium. The involvement of the pancreas as a zinc excretory organ and the role of zinc in pancreatic synthesis of metalloenzymes and insulin production, suggest that the first-order nature of pancreatic zinc uptake at physiological plasma zinc concentrations may have physiological significance.

References

Assadi FK, Ziar M (1986) Zinc status of infants with fetal alcohol syndrome. Pediatr Res 20: 551–554

Atkinson RL, Dahms WT, Bray GA, Jacob R, Sandstead HH (1978) Plasma zinc and copper in obesity and after intestinal bypass. Ann Intern Med 89: 491–493

Babcock AK, Henkin RI, Aamodt RL, Foster DM, Berman M (1982) Effects of oral zinc loading on zinc metabolism in humans. II. In vivo kinetics. Metabolism 31: 335–347

Baracos V, Rodemann HP, Dinarello CA, Goldbert AL (1983) Stimulation of muscle protein degradation and prostaglandin E_2 release by leukocytic pyrogen (Interleukin-1). N Engl J Med 308: 553–558

Blalock TL, Dunn MA, Cousins RJ (1988) Metallothionein gene expression in rats: tissue-specific regulation by dietary copper and zinc. J. Nutr 118: 222–228

Bloxam DL, Tan JCY, Parkinson CE (1984) Non-protein bound zinc concentration in human plasma and amniotic fluid measured by ultrafiltration. Clin Chim Acta 144: 81–93

Bogden JD, Troiano RA (1978) Plasma calcium, copper, magnesium, and zinc concentrations in patients with the alcohol withdrawal syndrome. Clin Chem 24: 1553–1556

Boyett JD, Sullivan JF (1973) Distribution of protein-bound zinc in normal and cirrhotic serum. Metabolism 19: 148–157

Brady FO (1981) Synthesis of rat hepatic zinc thionein in response to the stress of sham operation. Life Sci 28: 1647–1654

Breskin MW, Worthington-Roberts BS, Knopp RH et al. (1983) First trimester serum zinc concentrations in human pregnancy. Am J Clin Nutr 38: 943–953

Canfield WK, Menge R, Walravens PA, Hambidge KM (1980) Plasma zinc values in children recovering from protein-energy malnutrition. J Pediatr 97: 87–89

Cannon JG, Kluger MJ (1983) Endogenous pyrogen activity in human plasma after exercise. Science 220: 617–619

Casey CE, Walravens PA, Hambidge KM (1981) Availability of zinc: loading tests with human milk, cow's milk, and infant formulas. Pediatrics 68: 394–396

Charlwood PA (1979) The relative affinity of transferrin and albumin for zinc. Biochim Biophys Acta 581: 260–265

Chesters JK, Will M (1978) The assessment of zinc status of an animal from the uptake of ^{65}Zn by the cells of whole blood in vitro. Br J Nutr 38: 297–306

Chesters JK, Will M (1981) Zinc transport proteins in plasma. Br J Nutr 46: 111–118

Chvapil ML, Stankova L, Weldy P et al. (1982) The role of zinc in the function of some inflammatory cells: clinical, biochemical and nutritional aspects of trace elements. Curr Top Nutr Dis 6: 1–577

Cousins RJ (1985) Absorption, transport, and hepatic metabolism of copper and zinc: special reference to metallothionein and ceruloplasmin. Physiol Rev 65: 238–309

Cousins RJ, Dunn MA, Leinart AS, Yedinak KC, DiSilvestro RA (1986) Coordinate regulation of zinc metabolism and metallothionein gene expression in rats. Am J Physiol 251: E688–E694

Dennes E, Tupper R, Wormall A (1962) Studies on zinc in blood. Transport of zinc and incorporation of zinc in leucocytes. Biochem J 82: 466–476

Dinarello CA (1984) Interleukin-1 and the pathogenesis of the acute-phase response. N Engl J Med 311: 1413–1418

DiSilvestro RA, Cousins RJ (1984a) Mediation of endotoxin-induced changes in zinc metabolism in rats. Am J Physiol 247: E436–E441

DiSilvestro RA, Cousins RJ (1984b) Glucocorticoid independent mediation of interleukin-1 induced changes in serum zinc and liver metallothionein levels. Life Sci 35: 2113–2118

Dreosti IE, Tao SH, Hurley LS (1968) Plasma zinc and leukocyte changes in weanling and pregnant rats during zinc deficiency. Proc Soc Exp Biol Med 128: 169–174

Dunn MA, Cousins RJ (1987) Regulation of metallothionein gene transcription and kinetics of zinc metabolism by dibutyryl-cAMP. Fed Proc 46: 1643 (abs)

Ebadi M, Wilt S, Ramaley R, Swanson S, Mebus C (1984) The role of zinc and zinc-binding proteins in regulation of glutamic acid decarboxylase in brain. Chem Biol Aspects A: 255–275

Ellul-Micallef R, Galdes A, Frenech FF (1976) Serum zinc levels in corticosteroid-treated asthmatic patients. Postgrad Med J 52: 148–150

Etzel KR, Shapiro SG, Cousins RJ (1979) Regulation of liver metallothionein and plasma zinc by the glucocorticoid dexamethasone. Biochem Biophys Res Commun 89: 1120–1126

Etzel KR, Swerdel MR, Swerdel JN, Cousins RJ (1982) Endotoxin-induced changes in copper and zinc metabolism in the syrian hamster. J Nutr 112: 2363–2373

Evans GW, Winter TW (1975) Zinc transport by transferrin in rat portal blood plasma. Biochem Res Commun 66: 1218–1224

Failla ML, Cousins RJ (1978a) Zinc uptake by isolated rat liver parenchymal cells. Biochim Biophys Acta 538: 435–444

Failla ML, Cousins RJ (1978b) Zinc accumulation and metabolism in primary cultures of rat liver cells: regulation by glucocorticoids. Biochim Biophys Acta 543: 293–304

Failla ML, Van de Veerdonk M, Morgand WT, Smith JC (1982) Characterization of zinc-binding proteins of plasma in familial hyperzincemia. J Lab Clin Med 100: 943–952

Failla ML, Craft ME, Weinberg GA (1983) Depressed response of plasma iron and zinc to endotoxin and LEM in STZ-diabetic rat. Proc Soc Exp Biol Med 172: 445–448

Falchuk KH (1977) Effect of acute disease and ACTH on serum zinc proteins. N Engl J Med 296: 1129–1134

Feaster JP, Hansard SL, McCall JT, Davis GK (1955) Absorption, deposition and placental transfer of zinc in the rat. Am J Physiol 181: 287–290

Fenton MR, Burke JP (1983) Changes in serum, liver, and tumor zinc levels during plasmacytoma growth in BALB/c mice. Proc Soc Exp Biol Med 173: 390–397

Fickel JJ, Freeland-Graves JH, Roby MJ (1986) Zinc tolerance tests in zinc deficient and zinc supplemented diets. Am J Clin Nutr 43: 47–58

Fleming CR, Smith LM, Hodges RE (1976) Essential fatty acid deficiency in adults receiving total parenteral nutrition. Am J Clin Nutr 29: 976–983

Flynn A (1984) Stimulation of interleukin-1 production from placental monocytes. Lymphokine Res 3: 1–5

Flynn A, Pories WJ, Strain WH, Hill OA, Fratianne RB (1971) Rapid serum-zinc depletion associated with corticosteroid therapy. Lancet II: 1169–1172

Foote JW, Delves HT (1984) Albumin bound and α_2-macroglobulin bound zinc concentrations in the sera of healthy adults. J Clin Pathol 37: 1050–1054

Freeman RM, Taylor PR (1977) Influence of histidine administration on zinc metabolism in the rat. Am J Clin Nutr 30: 523–527

Gardiner PE, Rosick E, Rosick U, Bratter P, Kynast G (1982) The application of gel filtration, immunonephelometry and electrothermal atomic absorption spectrometry to the study of the distribution of copper-, iron- and zinc-bound constituents in human amniotic fluid. Clin Chim Acta 120: 103–117

Ghishan FK, Greene HL (1983) Fetal alcohol syndrome: failure of zinc supplementation to reverse the effect of ethanol on placental transport of zinc. Pediatr Res 17: 529–531

Ghishan FK, Patwardhan R, Greene HL (1982) Fetal alcohol syndrome: inhibition of placental zinc transport as a potential mechanism for fetal growth retardation in the rat. J Lab Clin Med 100: 45–52

Giroux EL (1975) Determination of zinc distribution between albumin and α_2-macroglobulin in human serum. Biochem Med 12: 258–266

Giroux EL, Durieux M, Schechter PJ (1976) A study of zinc distribution in human serum. Bioinorg Chem 5: 211–218

Goldblum SE, Cohen DA, Jay M, McClain CJ (1987) Interleukin 1-induced depression of iron and zinc: role of granulocytes and lactoferrin. Am J Physiol 252: E27–E32

Golden BE, Golden MHN (1981) Plasma zinc, rate of weight gain, and the energy cost of tissue deposition in children recovering from severe malnutrition on a cow's milk or soya protein based diet. Am J Clin Nutr 34: 892–899

Gordon PR, Woodruff CW, Anderson HL, O'Dell BL (1982) Effect of acute zinc deprivation on plasma zinc and platelet aggregation in adult males. Am J Clin Nutr 35: 113–119

Hallbook T, Hedelin H (1978) Changes in serum zinc and copper induced by operative trauma and effects of pre- and postoperative zinc infusion. Acta Chir Scand 144: 423–426

Hallman PS, Perrin DD, Watt AE (1971) The computed distribution of copper (II) and zinc (II) ions among seventeen amino acids present in human blood plasma. Biochem J 121: 549–555

Harris WR (1983) Thermodynamic binding constants of the zinc–human serum transferrin complex. Biochemistry 22: 3920–3926

Hartoma TR, Sotaniemi EA, Pelkonen O, Ahlqvist J (1977) Serum zinc and serum copper and indices of drug metabolism in alcoholics. Eur J Clin Pharmacol 12: 147–151

Henkin RI (1974) Metal–albumin–amino acid interactions: chemical and physiological relationships. In: Friedman M (ed) Protein–metal interactions. Plenum, New York, pp 299–328

Henkin RI, Foster DM, Aamodt RL, Berman M (1984) Zinc metabolism in adrenal cortical insufficiency: effects of carbohydrate-active steroids. Metabolism 33: 491–502

Henry RW, Elmes ME (1975) Plasma zinc in acute starvation. Br Med J 4: 625–626

Hess FM, King JC, Margen S (1977) Zinc excretion in young women on low zinc intakes and oral contraceptive agents. J Nutr 107: 1610–1620

Hildalgo J, Armario A, Flos R, Garvey JS (1986) Restraint stress induced changes in rat liver and serum metallothionein and in Zn metabolism. Experientia 42: 1006–1010

Hurley LS, Gordon P, Keen CL, Merkhofer L (1982) Circadian variation in rat plasma zinc and rapid effect of dietary zinc deficiency. Proc Soc Exp Biol Med 170: 48–52

Itoh M, Ebadi M, Swanson S (1983) The presence of zinc-binding proteins in brain. J Neurochem 41: 823–829

Jain RK, Gerlowski LE, Weissbrod JM, Wang J, Pierson RN (1981) Kinetics of uptake, distribution, and excretion of zinc in rats. Ann Biomed Eng 9: 347–361

Kampschmidt RF, Upchurch HF (1970) The effect of endogenous pyrogen on the plasma zinc concentration of the rat. Proc Soc Exp Biol Med 134: 1150–1152

Keppen LD, Pysher T, Rennert OM (1985) Zinc deficiency acts as a co-teratogen with alcohol in fetal alcohol syndrome. Pediatr Res 19: 944–947

Klasing KC (1984) Effect of inflammatory agents and interleukin-1 on iron and zinc metabolism. Am J Physiol 247: R901–R904

Klasing KC, Richards MP, Darcey SE, Laurin DE (1987) Presence of acute phase changes in zinc, iron, and copper metabolism in turkey embryos. Proc Soc Exp Biol Med 184: 7–13

Kokoglu E, Guner G, Guner A (1983) Tumoural and normal brain tissue levels of zinc in man. IRCS Med Sci 11: 848

Kruckeberg WC, Brewer GJ (1978) The mechanism and control of human erythrocyte zinc uptake. Med Biol 56: 5–10

Laitinen R, Simes ASI, Vuori E, Salmela SS (1984) Amniotic fluid zinc in risk pregnancies. Biol Trace Element Res 6: 415–421

Lau AL, Failla ML (1984) Urinary excretion of zinc, copper and iron in the streptozotocin-diabetic rat. J Nutr 114: 224–233

Lifschitz MD, Henkin RI (1971) Circadian variation in copper and zinc in man. J Appl Physiol 31: 88–92

Lindeman RD, Baxter DJ, Yunice AA, Kraikitpanitch S (1978) Serum concentrations and urinary excretions of zinc in cirrhosis, nephrotic syndrome and renal insufficiency. Am J Med Sci 275: 17–31

Ludvigsen C, McDaniel M, Lacy PE (1979) The mechanism of zinc uptake in isolated islets of Langerhans. Diabetes 28: 570–576

Lukaski HC, Bolonchuk WW, Klevay LM, Milne DB, Sandstead HH (1984) Changes in plasma zinc content after exercise in men fed a low-zinc diet. Am J Physiol 247: E88–E93

McClain CJ, Twyman DL, Ott LG et al. (1986) Serum and urine zinc response in head-injured patients. J Neurosurg 64: 224–230

McCormick CC, Menard PM, Cousins PJ (1981) Induction of hepatic metallothionein by feeding zinc to rats of depleted zinc status. Am J Physiol 240: E414–E421

Mahajan SK, Prasad AS, Rabbani P, Briggs WA, McDonald FD (1979) Zinc metabolism in uremia. J Lab Clin Med 94: 693–698

Menard PM, McCormick CC, Cousins RJ (1981) Regulation of intestinal metallothionein biosynthesis in rats by dietary zinc. J Nutr 111: 1353–1361

Milne DB, Ralston NVC, Wallwork JC (1985) Zinc content of cellular components of blood: methods for cell separation and analysis evaluated. Clin Chem 31: 65–69

Montgomery DW, Dannenfelser SW, Chvapil M, Zukoski CF (1977) Characteristics of ^{65}Zinc uptake in vitro by human lymphocytes. Fed Proc 36: 4784 (abs)

Morgan WT (1981) Interactions of the histidine-rich glycoprotein of serum with metals. Biochemistry 20: 1054–1061

Moser PB, Borel J, Majerus T, Anderson RA (1985) Serum zinc and urinary zinc excretion of trauma patients. Nutr Res 5: 253–261

Neldner KH, Hambidge KM (1975) Zinc therapy of acrodermatitis enteropathica. N Engl J Med 292: 879–882

Oh SH, Deagan JT, Whanger PD, Weswig PH (1978) Biological function of metallothionein. V. Its induction in rats by various stresses. Am J Physiol 234: E282–E285

Ohno H, Doi R, Yamamura K et al. (1985) A study of zinc distribution in erythrocytes of normal humans. Blut 50: 113–116

Parisi AF, Vallee BL (1970) Isolation of a zinc α_2-macroglobulin from human serum. Biochemistry 9: 2421–2426

Patierno SR, Costa M, Lewis VM, Peavy DL (1983) Inhibition of LPS toxicity for macrophages by metallothionein-inducing agents. J Immunol 130: 1924–1929

Pattison SE, Cousins RJ (1986a) Kinetics of zinc uptake and exchange by primary cultures of rat hepatocytes. Am J Physiol 250: E677–E685

Pattison SE, Cousins RJ (1986b) Zinc uptake and metabolism by hepatocytes. Fed Proc 45: 2805–2809

Pecoud A, Donzel P, Schelling JL (1975) Effect of foodstuffs on the absorption of zinc sulfate. Clin Pharmacol Ther 17: 469–474

Pekarek RS, Beisel WR (1971) Characterization of the endogenous mediator(s) of serum zinc and iron depression during infection and other stresses. Proc Soc Exp Biol Med 138: 728–732

Phillips JL, Tuley JA, Bowman RP (1977) Zinc uptake in normal and leukemic lymphocytes: effect of poly-L-ornithine. J Natl Cancer Inst 58: 1229–1231

Powanda MC, Villarreal Y, Rodriguez, E, Braxton G, Kennedy CR (1980) Redistribution of zinc within burned and burned infected rats. Proc Soc Exp Biol Med 163: 296–301

Prasad AS, Oberleas D (1970) Binding of zinc to amino acids and serum proteins in vitro. J Lab Clin Med 76: 416–425

Prasad AS, Rabbani P, Abbasii A, Bowersox E, Fox MRS (1978) Experimental zinc deficiency in humans. Ann Intern Med 89: 483–490

Rebello T, Atherton DJ, Holden C (1986) The effect of oral zinc administration on sebum free fatty acids in acne vulgaris. Acta Derm Venereol 66: 305–310

Richards MP, Cousins RJ (1976) Metallothionein and its relationship to the metabolism of dietary zinc in rats. J Nutr 106: 1591–1599

Rosick E, Rosick U, Bratter P, Kynast G (1983) Trace element levels of amniotic fluid at term in normal and high risk pregnancies. Trace Elem Ann Chem Med Biol 2: 463–482

Saltman P, Boroughs H (1960) The accumulation of zinc by fish liver slices. Arch Biochem Biophys 86: 169–174

Sas B, Bremner I (1979) Effect of acute stress on the absorption and distribution of zinc and on Zn-metallothionein production in the liver of the chick. J Inorg Biochem 11: 67–69

Sato M, Mehra RK, Bremner I (1984) Measurement of plasma metallothionein-I in the assessment of the zinc status of zinc-deficient and stressed rats. J Nutr 114: 1683–1689

Schechter PJ, Giroux EL, Schlienger JL, Hoenig V, Sjoerdsma A (1976) Distribution of serum zinc between blumin and α_2-macroglobulin in patients with decompensated hepatic cirrhosis. Eur J Clin Invest 6: 147–150

Schmetterer G (1978) ATP dependent uptake of zinc by human erythrocyte ghosts. Z Naturforsch 33: 210–215

Scott BJ, Bradwell AR (1983) Identification of the serum binding proteins for iron, zinc, cadmium, nickel, and calcium. Clin Chem 29: 629–633

Shanklin SH, Miller ER, Ullrey DE, Hoefer JA, Luecke RW (1968) Zinc requirement of baby pigs on casein diets. J Nutr 96: 101–108

Smith JC, Holbrook JT, Danford DE (1985) Analysis and evaluation of zinc and copper in human plasma and serum. J Am Coll Nutr 4: 627–638

Smith KT, Failla ML, Cousins RJ (1979) Identification of albumin as the plasma carrier for zinc absorption by perfused rat intestine. Biochem J 184: 627–633

Sobocinski PZ, Canterbury WJ (1982) Hepatic metallothionein induction in inflammation. Ann NY Acad Sci 210: 354–367

Sobocinski PZ, Canterbury WJ, Mapes CA, Dinterman RE (1978) Involvement of hepatic metallothioneins in hypozincemia associated with bacterial infection. Am J Physiol 234: E399–E406

Solomons NW, Cousins RJ (1984) Zinc. In: Rosenberg IH, Solomons NW (eds) Absorption and malabsorption of mineral nutrients. Alan R. Liss, New York, pp 125–197

Solomons NW, Jacob RA (1981) Studies on the bioavailability of zinc in humans; effects of heme and nonheme iron on the absorption of zinc. Am J Clin Nutr 34: 475–482

Sullivan JF, Burch RE, Quigley HJ, Magee DF (1974) Zinc deficiency and decreased pancreatic secretory response. Am J Physiol 227: 105–108

Sullivan JF, Jetton MM, Burch RE (1979) A zinc tolerance test. J Lab Clin Med 93: 485–492

Suso FA, Edwards HM (1972) Binding of EDTA histidine and acetylsalicylic acid to zinc-protein complex in intestinal content, intestinal mucosa and blood plasma. Nature 236: 230–232

Suzuki KT, Ohta K, Sunaga H, Sugihira N (1986) Transport and distribution of copper injected into an albumin-deficient (analbuminemic) rat. Comp Biochem Physiol 84C: 29–34

Svenson KLG, Hallgren R, Johansson E, Lindh U (1985) Reduced zinc in peripheral blood cells from patients with inflammatory connective tissue diseases. Inflammation 9: 189–197

Swanson CA, King JC (1983) Reduced serum zinc concentration during pregnancy. Obstet Gynecol 62: 313–318

Terry CW, Terry BE, Davies J (1960) Transfer of zinc[65] across the placenta and fetal membranes of the rabbit. Am J Physiol 198: 303–308

Tucker SB, Schroeter AL, Brown PW, McCall JT (1976) Acquired zinc deficiency; cutaneous manifestations typical of acrodermatitis enteropathica. J Am Med Ass 235: 2399–2402

Ujjani B, Krakower G, Bachowski G, Krezoski S, Shaw CF, Petering DH (1986) Host zinc metabolism and the Ehrlich ascites tumour. Biochem J 233: 99–105

Underwood EJ (1977) Trace elements in human and animal nutrition. Academic Press, New York, pp 196–242

Vallee BL (1959) Biochemistry, physiology and pathology of zinc. Physiol Rev 39: 443–490

Vander AJ, Victery W, Germain C, Holloway D (1983) Insulin is a physiological inhibitor of urinary zinc excretion in anesthetized dogs. Am J Physiol 244: E536–E540

Vanderhoof JA, Scopinaro N, Tuma DJ, Gianetta E, Civalleri D, Antonson DL (1983) Hair and plasma zinc levels following exclusion of biliopancreatic secretions from functioning gastrointestinal tract in humans. Dig Dis Sci 28: 300–305

Van Miert ASJPAM, Van Duin CTM, Verheigden JHM, Schotman AJH, Nieuwenhuis J (1984) Fever and changes in plasma zinc and iron concentrations in the goat: the role of leukocytic pyrogen. J Comp Pathol 94: 543–557

Van Rij AM, Godfrey PJ, McKenzie JM (1979) Amino acid infusions and urinary zinc excretion. J Surg Res 26: 293–299

Victery W, Thomas D, Schoeps P, Vander AJ (1982) Lead increases urinary zinc excretion in rats. Biol Trace Elem Res 4: 211–219

Weismann K, Hoyer H (1986) Serum zinc levels during oral glucocorticoid therapy. J Invest Dermatol 86: 715–716

Whitehouse RC, Prasad AS, Rabbani PI et al. (1982) Zinc in plasma, neutrophils, lymphocytes, and erythrocytes as determined by flameless atomic absorption spectrophotometry. Clin Chem 28: 475–480

Wilkins PJ, Grey PC, Dreosti IE (1972) Plasma zinc as an indicator of zinc status in rats. Br J Nutr 26: 113–120

Williams RB, Russell RM, Dutta SK, Giovetti AC (1979) Alcoholic pancreatitis: patients at high risk of acute zinc deficiency. Am J Med 66: 889–893

Yamashita K, Ohno H, Doi R et al. (1985) Distribution of zinc and copper in maternal and cord blood at delivery. Biol Neonate 48: 362–365

Yunice AA, King RW, Kraikitpanitch S, Haygood CC, Lindeman RD (1978) Urinary zinc excretion following infusions of zinc sulfate, cysteine, histidine, or glycine. Am J Physiol 235: F40–F45

Zumkley H, Zidek W, Bertram HP (1984) Zinc substitution in renal insufficiency. Trace Elem Med 1: 43–46

Chapter 6

Systemic Interactions of Zinc

I. Bremner and P.M. May

Introduction

The behaviour of zinc in biological systems is affected by a wide range of nutritional and physiological factors. Interactions with dietary constituents which influence the absorption of zinc have been described in preceding chapters. However the systemic metabolism of zinc can also be affected by interactions with other metals or with organic metal-binding species.

In common with most trace elements, zinc readily forms complexes with a wide range of electron donors, such as the amino, carboxylate and thiolate groups in amino acids and proteins. Because of the high stability constants of these complexes, concentrations of free zinc ions in physiological fluids and tissues are always extremely low; consequently the properties of zinc in biological systems are determined largely by the nature of the complexes that are formed. For example, effects of diet on the intestinal absorption of zinc can often be attributed to a reaction between the metal and complexing agents present in the intestinal lumen and mucosa. Similarly, antagonistic effects of other metals on zinc metabolism can arise from competition for binding sites on proteins or other ligands that regulate the uptake and distribution of zinc.

In this chapter an account is given of how the systemic metabolism of zinc is influenced by the interactions between the metal and a variety of compounds which vary in molecular size, charge and lipophilicity. The effects of other trace minerals and of vitamins and certain xenobiotics on zinc metabolism are also considered.

Binding of Zinc to Low-Molecular-Weight Ligands

Experimental studies of metal ion interactions with low-molecular-weight ligands in vivo are made extraordinarily difficult by the very low concentrations which are involved and by the labile nature of most such associations (May et al. 1977). For this reason, there have been few direct experimental attempts to explore at a molecular level the systemic interactions of zinc with dietary constituents or with drugs; most reports have confined themselves to overall physiological effects. Accordingly, our present knowledge about the chemical binding which may, or may not, take place between zinc and low-molecular-weight agents in various biological fluids has had to be inferred largely from computer simulations of the equilibria which are thought to dominate the low-molecular-weight fraction of the metal ion.

Whilst such simulations have been, almost surprisingly, successful in rationalizing much of the experimental data on the behaviour of various chelating agents (May and Williams 1977; Jones et al. 1984; McGrath et al. 1986), the rather limited scope of the present models makes it important not to over-stretch the conclusions which can be drawn from them. Recent work clearly illustrates how the results of the simulations need to be continuously improved, particularly in respect of (a) upgrading the database (Berthon et al. 1986) and (b) extending the calculations, which now encompass some kinetic effects, most notably the changing concentration of administered agents (Jones and May 1987).

Naturally Occurring Ligands

In spite of the above reservations, some interesting insights into the in vivo interactions of zinc (II) in the low-molecular-weight fraction can be inferred from the computer models. The calculated distribution of zinc (II) in blood plasma is shown in Table 6.1. Binding is clearly dominated by cysteinate (Berthon et al. 1978), with histidine acting as the other important coordinating partner. Reduced glutathionate seems likely to supersede cysteinate inside most, if not all, cells. Although commonly overlooked in the literature, it is undoubtedly these complexes, rather than the uncoordinated (aquated) zinc ion, which usually participate in physiological processes. The free metal ion concentrations of trace elements in vivo are simply too low to be relevant in either transport or exchange mechanisms.

Nevertheless, free metal ion concentrations in vivo are of interest because they provide a good measure of general (thermodynamic) availability of the metal ion. In this regard, zinc (II) occupies a position intermediate between calcium (II) and magnesium (II) on the one hand and copper (II), iron (II) and iron (III) on the other. Estimates based on equilibration with albumin indicate that, in blood plasma, $[Zn^{2+}]$ lies between 10^{-9} M and 10^{-10} M (May et al. 1977). Magneson et al. (1987), using an assay that depends on activation of phosphoglucomutase, have estimated that concentrations of free Zn^{2+} in equine plasma are 2×10^{-10} M. These values compare with levels of calcium (II) and magnesium (II) at 10^{-3}M and 10^{-4}M respectively. Although the figures for other trace metal ions are much less certain, $[Fe^{2+}]$ seems likely to lie several orders of magnitude below $[Zn^{2+}]$ and one would expect $[Cu^{2+}]$ to be considerably lower still, say about 10^{-17}M.

In general, the concentrations of the low-molecular-weight complexes parallel the relative levels of the various free metal ions. In plasma, low-molecular-weight zinc (II) complexes occur at about 10^{-7}M, which is 1% or less of the

Table 6.1. Percentage distribution of zinc amongst low-molecular-weight, naturally occurring complexes in human blood plasma as found by computer simulation

	Percentage
$[Zn(II).Cys_2]^{2-}$	40
$[Zn(II).Cys.His]^-$	24
$[Zn(II).His]^+$	4
$[Zn(II).Cys_2.H]^-$	3
$[Zn(II).Cys.His.H]$	2
$[Zn(II).His_2]$	2
$[Zn(II).Cys]$	2

From Berthon et al. (1978).
Cys, cysteinate; His, histidinate.

concentration of the high-molecular-weight fraction. Accordingly, the small zinc (II) species are involved in processes which exploit their kinetic advantages over the complexes formed by proteins. For the most part, these involve transport to or through membranes and exchange between high-molecular-weight species.

Chelating Agents

The marked avidity of proteins for zinc (II) makes it difficult even for synthetic chelating agents to compete effectively for the metal ion in vivo. A notable exception is provided by diethylenetriaminepentaacetic acid (DTPA), however. At high enough concentrations this agent is capable of withholding sufficiently large amounts of zinc (II) from plasma proteins to significantly affect the relative distribution of the metal. The effect of DTPA on zinc (II) is similar to that of ethylenediaminetetraacetic acid (EDTA) on calcium (II), where it is known that levels in plasma can be seriously depleted if the drug is administered as a sodium salt. No corresponding effect occurs with any of the first row transition elements. This fact also accounts for the marked superiority of Ca–DTPA over Zn–DTPA in the decorporation of many toxic metals (May and Bulman 1983).

It is important to realize that, because of its small size, the low-molecular-weight fraction can be substantially enlarged without much affecting the concentration of the high-molecular-weight components. It follows from the rarity of agents capable of binding zinc (II) more strongly than the transport and storage proteins, that most compounds which interact directly with the metal ion exert their effects by such an enlargement of the low-molecular-weight fraction.

Interactions of most of the conventional chelating agents with zinc (II) are known. In addition to the polyaminopolycarboxylic acids, many others give rise to enhanced urinary excretion which ultimately can lead to deficiency symptoms. Due to its widespread use in the treatment of rheumatoid arthritis (and the fact that simultaneous administration with zinc (II) is contraindicated), penicillamine is worth mentioning specifically.

Other Therapeutic Agents

Since the early 1980s computer models have been used increasingly to investigate the interactions of agents which are administered for purposes other than binding metal ions but which are nevertheless capable of doing so. The side-effects of many drugs can be attributed to the undesirable chelation of trace metals, especially zinc.

Some of the most comprehensive studies have been performed by Berthon and his colleagues. They have developed two main lines of research concerned with zinc (II). These are (a) the interactions of histamine in blood plasma and (b) the interactions of amino acids and other components contained in the fluids infused during total parenteral nutrition.

It is well established that zinc has good antihistamine properties. Walker et al. (1975), for example, have shown that injections of zinc can counteract the effects of histamine in mice put into anaphylactic shock by otherwise lethal doses of histamine. Berthon's models show that the complexes formed with zinc (II) by histamine and various naturally occurring amino acids in plasma are neutral and therefore unlike histamine itself (which is inherently polar and, under the prevailing pH, mainly in a charged form due to protonation) (Berthon and Kayali 1982; Berthon and Germonneau 1982). These neutral, mixed-ligand complexes are thus thought to be capable of passively diffusing through lipid membranes into tissue where the histamine can be catabolized. They may therefore be instrumental in reducing the massive surges of histamine which can suddenly be released into plasma in response to a variety of local stimuli or general toxins.

Berthon et al. (1980, 1984) have also applied computer simulation techniques in a series of studies aimed at quantifying the supplements of zinc and copper which ought to be included in fluids for total parenteral nutrition. Although much more is now known about the daily requirements for such patients than when these investigations commenced (Berthon et al. 1980), the calculations remain relevant because optimum doses depend on the specific composition of each nutritive fluid. They cannot be quantified in an entirely general way because constituents of the fluid, particularly cysteine, chelate zinc (both exogenously and endogenously) and hence promote its urinary excretion to differing extents, depending on their relative concentrations and zinc (II)-binding abilities.

Other studies have revealed significant interactions between zinc and various pharmaceuticals. Side-effects of treatment with the anti-tuberculosis drug, ethambutol, have been attributed to the complexation of zinc (II) by a metabolic degradation product (Cole et al. 1981). Investigation into the binding of zinc by the hydrolysis products formed in vivo from the anti-tumour drug, razoxane, and its homologues has provided evidence of a curious selectivity towards zinc (II) by some of the compounds (but whether or not this is related to their mechanism of action remains unclear) (Huang et al. 1982).

Computer simulation studies have also concluded that zinc (II) interactions are not involved in the action of a variety of other agents. These include tetracycline (Brion et al. 1985), cimetidine (Akrivos et al. 1984), Adriamycin (May et al. 1980), hydralazine and prizidilol (Al-Falahi et al. 1984) and isoniazid (Cole et al. 1983). As is often the case with scientific models, negative results of this kind can be as valuable as positive findings and they are usually more reliable.

Another interesting issue concerns the predominant low-molecular-weight ligand of zinc in milk. This was the subject of a considerable controversy in the early years of this decade (Anonymous 1986) but now seems to have been much-to-do-about-nothing. Identification was thought to be necessary to understand the difference in bioavailability of zinc between human and cows' milk. However, it now seems that protein interactions, particularly inactivation by casein, are the critical factors responsible for this difference (Roth and Kirchgessner 1985).

Indirect Effects

Many agents having systemic interactions with zinc exert their effects through mechanisms which do not directly involve chelation. This means that intimate chemical interactions are also unlikely, although not impossible. Such cases are inherently more difficult to elucidate than those discussed above. Generally, they have remained obscure at the molecular level.

The Relationship between Binding in the High- and Low-Molecular-Weight Fractions

The role of the ligands in the low-molecular-weight fraction provides some interesting parallels between zinc and iron regarding the mechanisms for buffering intracellular concentrations. In both cases, synthesis of the major storage-regulating metalloprotein (metallothionein for zinc and ferritin for iron) is triggered by increasing levels of the free metal ion in the cell (Zähringer et al. 1976; Hamer 1986). Initially, the incoming metal ion may be complexed by various specific and non-specific sites or proteins in the cell cytoplasm or attached to surrounding membranes. This binding provides the most labile reservoir of the metal and the first pool to be depleted by any sudden demand. Metallothionein itself may participate in this buffering through limited exchange reactions (see below) but its main homeostatic role is accomplished by setting up a steady state in which the metalloprotein is continuously synthesized and degraded. Synthesis of the protein is probably triggered by the dramatic increase in free metal ion that would tend to occur once all the labile protein-binding sites have become saturated. Such a mechanism would permit the cell to regulate zinc levels in a way that is both highly efficient and also capable of very considerable capacity.

Another aspect of metal ion binding in vivo which now is largely resolved but which greatly worried early researchers concerns the way particular metal ions are incorporated into certain binding sites when other metals have higher inherent affinities for those sites. Although there are some notable exceptions, most organic donor groups follow the Irving–Williams series, having affinities for the first row, divalent metal ions in the order $Mn^{2+} < Fe^{2+} < Co^{2+} < Ni^{2+} < Cu^{2+} > Zn^{2+}$. Clearly zinc (II) can be pushed into metalloproteins by maintaining it at a higher concentration than, say, copper (II). However, this seems to imply a very delicate biological balance. Even more problematically, there would be difficulty in adjusting the concentrations of all metal ions so as to suit all the permutations required by the great spectrum of metalloenzymes. In fact, the problem is only one of perception: it is *because* of the Irving–Williams order of binding that the relative levels of the free metal ions in biological fluids turn out approximately in the way that they do. Indeed, this is a natural consequence of binding in this kind of multicomponent solution (i.e. one which contains a number of metal ions in the presence of a variety of ligands). At equilibrium, the lowest free concentrations are achieved by those metal ions which are generally complexed most strongly. Biological selectivity can thus be accomplished by relatively slight binding preferences compared with the norm.

Binding of Zinc to High-Molecular-Weight Ligands

Only a small proportion of the zinc present in plasma and tissue is present in the form of low-molecular-weight complexes. Most of the metal is invariably associated with protein and other components of high molecular weight. Zinc is an essential component of many enzymes and is involved either in the expression of enzyme activity or in the maintenance of protein conformation, as in carbonic anhydrase and superoxide dismutase respectively (Galdes 1982). No detailed quantitative information is available on the proportion of tissue zinc associated with enzymes, although it can be calculated that 10%–20% of zinc in rat liver could be associated with superoxide dismutase and carbonic anhydrase.

Isomorphous replacement of zinc in enzymes by other metals is readily accomplished in vitro. Such substitution has been invaluable in spectroscopic studies on the elucidation of enzyme structure and identification of metal-binding sites. It is often accompanied by loss of enzyme activity and it is suspected that similar substitutions may occur in vivo. For example, the reduced activity in lead-poisoned animals of erythrocyte $\triangle$-aminolevulinic acid dehydratase, which is a zinc-dependent enzyme, has been attributed to displacement of zinc from the active site in the enzyme by lead (Finelli et al. 1975). Increased exposure to cadmium results in reduced activity of renal leucine aminopeptidase, which is also a zinc-dependent enzyme (Washko and Cousins 1975). However, it can sometimes be difficult to establish whether such substitutions occur directly on the enzyme or whether the primary effect of the antagonist metal is to restrict the absorption of zinc and its subsequent incorporation into the enzyme.

Surprisingly little is known of the interaction of zinc with non-enzymatic tissue components of high molecular weight. The fact that clinical signs of zinc deficiency appear very quickly when animals are given low-zinc diets indicates that storage capacity for zinc is normally very low. In accord with this view, tissue zinc concentrations are not greatly decreased in zinc-deficient animals. However, zinc concentrations are still relatively high when compared with concentrations of copper and other metals, which suggests that the residual zinc is bound quite avidly to a variety of tissue proteins and is not readily utilized, except when tissue catabolism is induced by treatments such as restriction of food intake or feeding diets of low protein or calcium content (Hurley 1981; Masters et al. 1983).

Few of these zinc-binding macromolecules have been identified and there are indications that the metal is bound perhaps non-specifically to a range of proteins. Thus fractionation of tissue cytosol on Sephadex reveals a close similarity in the elution profiles for protein and zinc, which indicates that zinc is associated with most of the protein fractions separated by gel filtration procedures. This distribution is not fixed, however, as the pattern of zinc elution was found to vary markedly with age in foetal lambs (Bremner et al. 1977) and with changes in copper and sulphur status in adult lambs (Bremner 1982, 1987).

Metallothionein

One of the few non-enzymatic zinc-binding proteins which has been studied in detail is metallothionein (Hamer 1986; Bremner 1987). This is a widely distributed protein which is believed to play important roles in the control of the metabolism of zinc and also of copper and certain non-essential elements. It is characterized by its low molecular weight (about 6200), high cysteine content (about 30% of amino acid residues) and high metal-binding capacity (7 g atoms/mole). Metallothionein is found in variable amounts in most tissues and particularly in liver, pancreas, kidneys and intestinal mucosa. It occurs under normal physiological conditions as a zinc- and/or copper-binding protein.

Concentrations of metallothionein in tissues depend to a large extent on zinc status. They are often reduced to non-detectable levels in zinc-deficient animals and are increased after parenteral or dietary administration of zinc (Bremner and Davies 1975; Richards and Cousins 1976). Indeed there is usually a close relationship between the concentrations of zinc and of metallothionein in liver so that, as zinc concentrations increase, most of the additional metal is bound to metallothionein. This can be attributed to the efficient induction of metallothionein synthesis by zinc, by a process involving increased transcription of the metallothionein genes (Shapiro et al. 1978). This process is mediated by interactions between upstream regulatory DNA sequences and other cellular factors (see Palmiter 1987). Although specific zinc-responsive factors have yet to be identified, nuclear factors which respond to cadmium have been detected in recent work on induction of metallothionein synthesis in mouse L cells (Seguin and Hamer 1987). When metallothionein synthesis is induced by zinc in vivo, the time course of the changes in metallothionein mRNA concentrations is relatively short since these reach a maximum after 5 h in liver and decrease to control levels after about 24 h (Shapiro et al. 1978). Translation of the mRNA to produce metallothionein follows a similar but slightly delayed time course; maximum levels of hepatic metallothionein in zinc-injected rats occur after about 12 h (Bremner and Davies 1975).

These results are consistent with metallothionein performing an important regulatory role in the cell. There have been suggestions that it is involved in the homeostatic control of zinc absorption, in cellular detoxification or storage of zinc, in the control of differentiation and in direct activation of zinc-dependent enzymes (Bremner 1987). However, cell lines that do not synthesize metallothionein are fully viable and grow normally, indicating that the protein does not have an obligatory role in these processes (Compere and Palmiter 1981). The widespread occurrence of metallothionein in most tissues and the non-specific nature of its metal-binding properties imply a more general function and its most probable role in zinc metabolism is in the control of intracellular concentrations of ionic zinc. Zinc ions are potentially toxic and can interact, for example, with sulphydryl groups on membranes and enzymes with harmful consequences. Rapid induction of metallothionein synthesis in response to increased intracellular zinc concentrations is consistent with its action as a scavenger of metal ions. However, the binding constants for zinc–metallothionein are not so great that the metal cannot subsequently be released and utilized if necessary for normal metabolic purposes.

The transient nature of the association of zinc with metallothionein is illustrated by the half-life of only about 20 h for the protein in rat liver. In studies of the turnover of metallothionein in zinc-injected rats, the rates of disappearance of metal and of a ^{35}S label were almost identical, indicating that loss of metal and degradation are linked (Feldman and Cousins 1976). Indeed, comparison of the degradation rates of different iso- and metallo-forms of metallothionein indicates that removal of metal is an important rate-controlling step in the turnover of the protein (Bremner and Mehra 1983). Zinc appears to have a stabilizing effect on the protein, as zinc-induced metallothionein turns over more slowly than the dexamethasone-induced protein in HeLa cells (Karin et al. 1981), and degradation of copper- and cadmium-induced metallothionein is faster in zinc-deficient than in zinc-adequate rats (Bremner and Mehra 1983; Held and Hoekstra 1984).

Metallothionein in Zinc–Metal Interactions

On the basis of the hypothesis that metals with similar chemical and physical properties will interact biologically (Hill and Matrone 1970), interactions would be expected between zinc and copper and between cadmium and copper, as well as between many other transition metals. Examples of such mutual antagonisms are shown in Table 6.2, although in this chapter those that occur mainly at the absorption level, such as the iron–zinc interaction, are generally excluded.

It has been implied that this type of interaction involves isomorphous replacement of metals on functional proteins but the identity of these has rarely been established. However, as metallothionein can bind many metals in addition to zinc, it has been regarded as a probable site of metal–metal interactions. This has been confirmed in many experiments but effects of metals on the synthesis and turnover of the protein need to be considered in addition to simple competitive metal-binding reactions.

Zinc binds much less strongly than copper and cadmium to metallothionein, and it is therefore readily displaced by both these metals in vitro (Kagi and

Table 6.2. Examples of zinc–metal interactions

Susceptibility to cadmium toxicity increased in zinc deficiency
Tissue zinc concentrations changed after cadmium exposure
Placental transfer of zinc reduced after cadmium exposure
Cadmium inhibits zinc uptake by liver cells
Copper loading changes tissue zinc concentrations
Zinc supplements reduce copper absorption
Zinc protects against copper toxicity
High maternal copper intake exacerbates the teratogenic effects of zinc
  deficiency
Zinc supplementation of dam decreases foetal copper content

Kojima 1987). There is a great deal of circumstantial evidence that similar substitutions occur in vivo. For example the amount of zinc associated with hepatic metallothionein in copper-loaded animals is frequently less than expected from the liver zinc content, indicating that the zinc has been displaced by the excess copper (Bremner 1987). Similarly, when cadmium exposure occurs in situations where tissue zinc concentrations are already elevated, cadmium appears to displace zinc from liver and kidney metallothionein (Leber and Miya 1976). In such circumstances, the toxicity of cadmium is reduced, presumably because it is incorporated almost immediately into the non-toxic form of metallothionein, without the delay that occurs when cadmium has to induce transcription of the metallothionein genes. This important protective effect of endogenous zinc–metallothionein has been demonstrated in both cell culture and whole animal experiments (Leber and Miya 1976; Din and Frazier 1985).

In some circumstances, however, zinc is not displaced from metallothionein by the antagonist metals. Instead its binding to metallothionein is actually *increased*, which is contrary to the response predicted by the Hill and Matrone theory. This occurs if the exposure to the antagonist metal results in induction of metallothionein synthesis, as occurs after injection of cadmium or copper (Bremner et al. 1978; Winge et al. 1978). Zinc is then bound as a secondary metal and mixed cadmium, zinc- or copper, zinc-metallothioneins are produced. Indeed metallothionein which contains only cadmium, with no zinc, has never been isolated from a mammalian system, although it can be prepared in vitro. [In some extreme cases, such as after dosing with excessive amounts of lead, zinc may be the only metal associated with the induced protein (Ikebuchi et al. 1986).]

It was initially thought that the increased binding of zinc to metallothionein in copper- and cadmium-dosed animals resulted from its adventitious binding to vacant metal-binding sites on the protein. However it is now known that there is site-specific binding of metals in the two domains in metallothionein (Nielson and Winge 1985). Since admixture of Cd_7 and Zn_7 forms of metallothionein results in rapid and spontaneous rearrangement of the metals to yield the mixed (cadmium, zinc)–metallothionein, there must be some thermodynamic advantage in having zinc bound as a secondary metal in cadmium- or copper-induced metallothioneins (Nettesheim et al. 1985).

Many aspects of the copper–zinc interaction are known to involve metallothionein. This is particularly evident at the absorption level, where the decreased copper absorption in zinc-supplemented animals can be related to the induction of metallothionein synthesis in the intestinal mucosa by zinc (Hall et al. 1979). Since copper binds to the protein much more strongly than zinc, increased amounts of copper become associated with mucosal metallothionein, with a resultant decrease in the transfer of the metal across the serosal membrane into the blood. This interaction is of some clinical significance. Thus hypocupraemia has been reported in sickle-cell patients given zinc therapy (Prasad et al. 1978) and control of Wilson's disease is obtained by the administration of zinc supplements (Brewer et al. 1983).

Although the main effect of zinc in the treatment of Wilson's disease is probably in the reduction of copper absorption, mutual interactions between zinc and copper which involve metallothionein also occur in other tissues (Bremner 1987). For example, zinc supplementation can increase the proportion of hepatic copper which is bound to metallothionein and may therefore reduce its cytotoxic action. It is well established in animal studies that the binding of hepatic copper to metallothionein is very dependent on zinc status as no copper–metallothionein is present in the liver of zinc-deficient animals (Bremner et al. 1978). Indeed the binding of hepatic copper to metallothionein often seems to depend on prior induction of its synthesis by zinc or some other agents. Copper can also replace zinc on circulating metallothionein in plasma (Suzuki 1981).

Disturbances in zinc metabolism can occur in animals in which the supply of copper is varied, but this aspect of the copper–zinc interaction has received much less attention. As was described above, displacement of zinc from hepatic metallothionein can occur in copper-loaded liver, but this has not been associated with any major change in the excretion or cytotoxicity of the zinc. Even at the absorption level, only minor changes in uptake of zinc result from changes in the dietary supply of copper; these occur when copper supply is reduced from adequate to deficient levels (Hall et al. 1979). Similar increases in zinc uptake have been reported in foetal rats when maternal copper intake is reduced but whether the effect of copper is on maternal zinc status or specifically on placental transfer of zinc is unclear (Reinstein et al. 1984). Whether metallothionein is involved in these latter aspects of the copper–zinc interaction has yet to be established.

As is shown in Table 6.2, cadmium–zinc interactions also occur at the placental level. Administration of cadmium to pregnant animals results in decreased placental transport of zinc and causes teratogenic effects consistent in some cases with zinc deficiency (Webb and Samarawickrama 1981). The growth rate of the foetuses may be impaired and analogies have been drawn with the occurrence of intrauterine growth retardation in pregnant women who smoke (Van der Welde et al. 1983). It has been suggested that the increased inhalation of cadmium in these women may result in decreased transfer of zinc to the developing foetus and this is a topic which merits further investigation.

Metallothionein in Zinc–Hormone Interactions

Major disturbances in zinc metabolism occur in animals and humans subjected to infection and various types of physical or inflammatory stress. Plasma zinc concentrations are decreased as a consequence of increased hepatic uptake of the metal. Much of the additional zinc in the liver is bound to metallothionein, synthesis of which has been induced by the stress factors (Cousins 1985). It is assumed that this confers some advantage on the host and helps to combat the effects of the stress or infection, possibly by maintaining high cellular zinc reserves or low levels of circulating zinc.

Induction of metallothionein synthesis in stressed animals is mediated by a variety of hormones, including glucocorticoids (Failla and Cousins 1978), glucagon (DiSilvestro and Cousins 1984) and catecholamines (Brady and Helvig 1984) and by other factors such as interleukin-I, cyclic AMP and interferon (Palmiter 1987). Synthesis by these agents involves increased transcription of metallothionein genes. Promoter regions of the metallothionein gene which respond to glucocorticoids and do not correspond to the metal-regulatory regions have been identified (see Palmiter 1987). A third class of promoter region is involved in the induction of metallothionein synthesis which occurs after administration of bacterial endotoxin (Durnam et al. 1984).

Detailed consideration of hormonal induction of metallothionein synthesis is outwith the scope of this review but it does explain how the systemic metabolism of zinc is influenced by a wide variety of stress factors. Although it has been shown that glucocorticoids can stimulate zinc uptake by isolated cells, there is now little doubt that the primary effect of the mediating hormones is on the induction of metallothionein synthesis and that this is followed by hepatic uptake of zinc (Dunn et al. 1987). The fact that dexamethasone increases metallothionein mRNA levels in HeLa cells grown in zinc-deficient medium also shows that the synthesis of the protein is not preceded by the cellular uptake of zinc (Karin and Herschmann 1981).

Zinc–Vitamin Interactions

There have been many reports of interactions between zinc and a range of vitamins. However some of these, like the zinc–folate interaction, occur mainly at the absorption level and are outside the scope of this chapter. Other reports, such as that of reaction between zinc and pyridoxal phosphate (Ikeda et al. 1979), have not been substantiated. However the interaction between zinc and vitamin A has attracted a great deal of attention and may be of some clinical significance (Smith 1980).

Low serum vitamin A (retinol) levels occur in zinc-deficient animals (Smith et al. 1976) and humans with low plasma zinc levels (Halsted and Smith 1974). In severe zinc deficiency the dark adaptation threshold appears to be increased; this is possibly related to the decreased activity of the zinc-containing enzyme, retinal reductase (Sundaresan et al. 1977). Liver vitamin A levels are not affected, however, and supplementation of zinc-deficient rats with vitamin A does not restore plasma levels to normal. This may reflect reduced levels of retinol-binding protein, since this is required for mobilization of hepatic reserves of vitamin A (Smith et al. 1976). It is possible that some of the changes in plasma retinol levels in zinc-deficient animals are a secondary consequence of their reduced food intake (Smith 1980).

References

Akrivos F, Blais M-J, Hoffelt J, Berthon G (1984) An assessment of the physiological significance of cimetidine interactions with copper and zinc in biofluids as based on the computer-simulated distribution of the involved complexes at therapeutic levels of the drug. Agents Actions 15: 649–659

Al-Falahi H, May PM, Roe AM, Slater RA, Trott WJ, Williams DR (1984) Metal binding by pharmaceuticals. 4. A comparative investigation of the interaction of metal ions with hydralazine, prizidilol and related compounds. Agents Actions 14: 113–120

Anonymous (1986) Zinc bioavailability of human and cow's milk. Nutr Rev 44: 181–183

Berthon G, Germonneau P (1982) Histamine as a ligand in blood plasma. 6. Aspartate and glutamate as possible partner ligands for zinc and histamine to favour histamine catabolism. Agents Actions 12: 619–629

Berthon G, Kayali A (1982) Histamine as a ligand in blood plasma. 5. Computer simulated distribution of metal histamine complexes in normal blood plasma and discussion of the implications of a possible role of zinc and copper in histamine catabolism. Agents Actions 12: 398–407

Berthon G, May PM, Williams DR (1978) Computer simulation of metal–ion equilibria in biofluids. 2. Formation constants for Zn (II)–citrate-cysteinate binary and ternary complexes and improved models for low-molecular-weight zinc species in blood plasma. J Chem Soc Dalton Trans 1433–1438

Berthon G, Matuchansky C, May PM (1980) Computer simulation of metal ion equilibria in biofluids. 3. Trace metal supplementation in total parenteral nutrition. J Inorg Biochem 13: 63–73

Berthon G, Piktas M, Blais M-J (1984) Trace metal requirements in total parenteral nutrition. 6. A quantitative study of Cu(II)–histidine ternary complexes with leucine, glutamic acid, methionine, tryptophan and alanine, and final evaluation of the daily doses of copper and zinc specific to a nutritive mixture of a given composition. Inorg Chim Acta 93: 117–130

Berthon G, Hacht B, Blais M-J, May PM (1986) Copper–histidine ternary complex equilibria with glutamine, asparagine and serine. The implications for computer-simulated distributions of copper (II) in blood plasma. Inorg Chim Acta 125: 219–227

Brady FO, Helvig B (1984) Effect of epinephrine and norepinephrine on zinc thionein levels and induction in rat liver. Am J Physiol 247: E318–E322

Bremner I (1982) The effects of nutritional factors on the development of metal-induced renal damage. In: Bach PH, Bonner FW, Bridges JW, Lock EA (eds) Nephrotoxicity: assessment and pathogenesis. J. Wiley & Sons, Chichester, pp 280–295

Bremner I (1987) Interactions between metallothionein and trace elements. Prog Food Nutr Sci 11: 1–37

Bremner I, Davies NT (1975) The induction of metallothionein in rat liver by zinc injection and restriction of food intake. Biochem J 149: 733–738

Bremner I, Mehra RK (1983) Metallothionein: some aspects of its structure and function with special regard to its involvement in copper and zinc metabolism. Chem Scr 21: 117–121

Bremner I, Williams RB, Young BW (1977) Distribution of copper and zinc in the liver of the developing sheep foetus. Br J Nutr 28: 87–92

Bremner I, Hoekstra WG, Davies NT, Young BW (1978) Effect of zinc status on the synthesis and degradation of copper-induced thioneins. Biochem J 174: 883–892

Brewer GJ, Hill GM, Prasad AS, Cossack ZT, Rabbani P (1983) Oral zinc therapy for Wilson's disease. Ann Intern Med 99: 314–320

Brion M, Lambs L, Berthon G (1985) Metal ion tetracycline interactions in biological fluids. 5. Formation of zinc complexes with tetracycline and some of its derivatives and assessment of their biological significance. Agents Actions 17: 229–242

Cole A, May PM, Williams DR (1981) Metal binding by pharmaceuticals. 1. Cu(II) and Zn(II) interactions following ethambutol administration. Agents Actions 11: 296–305

Cole A, May PM, Williams DR (1983) Metal binding by pharmaceuticals. 3. Cu(II) and Zn(II) interactions with isoniazid. Agents Actions 13: 91–97

Compere SJ, Palmiter RD (1981) DNA methylation controls the inducibility of the mouse metallothionein gene in lymphoid cells. Cell 25: 233–240

Cousins RJ (1985) Absorption, transport and hepatic metabolism of copper and zinc: special reference to metallothionein and ceruloplasmin. Physiol Rev 65: 238–309

Din WS, Frazier JM (1985) Protective effect of metallothionein on cadmium toxicity in isolated rat hepatocytes. Biochem J 230: 295–402

DiSilvestro RA, Cousins RJ (1984) Mediation of endotoxin-induced changes in zinc metabolism. Am J Physiol 247: E436–E441

Dunn MA, Blalock TL, Cousins RJ (1987) Metallothionein. Proc Soc Exp Biol Med 185: 107–119

Durnam DM, Hoffman JS, Quaife CJ et al. (1984) Induction of mouse metallothionein-I mRNA by bacterial endotoxin is independent of metals and glucocorticoid hormones. Proc Natl Acad Sci USA 81: 1053–1056

Failla M, Cousins RJ (1978) Zinc accumulation and metabolism in primary cultures of rat liver cells: regulation by glucocorticoids. Biochim Biophys Acta 543: 293–304

Feldman SL, Cousins RJ (1976) Degradation of hepatic zinc-thionein following parenteral zinc administration. Biochem J 160: 583–588

Finelli VN, Klauder DS, Karaffa MA, Petering HG (1975) Interactions of zinc and lead on △-amino levulinic acid dehydratase. Biochem Biophys Res Commun 65: 303–311

Galdes A (1982) Zinc metalloenzymes. In: Inorganic Biochemistry, vol 3. Royal Society of Chemistry, London, pp 268–313

Hall AC, Young BW, Bremner I (1979) Intestinal metallothionein and the mutual antagonism between copper and zinc in the rat. J Inorg Biochem 11: 57–66

Halsted JA, Smith JC (1974) Night blindness and chronic liver disease. Gastroenterology 67: 193–194

Hamer D (1986) Metallothionein. Ann Rev Biochem 55: 913–951

Held DD, Hoekstra WG (1984) The effects of zinc deficiency on turnover of cadmium-metallothionein in rat liver. J Nutr 114: 2274–2282

Hill CH, Matrone G (1970) Chemical parameters in the study of in vivo and in vitro interactions of transition elements. Fed Proc 29: 1474–1481

Huang Z-X, May PM, Quinlan KM, Williams DR, Creighton AM (1982) Metal binding by pharmaceuticals. 2. Interactions of Ca(II), Cu(II), Fe(II), Mg(II), Mn(II) and Zn(II) with the intracellular products of the antitumour agent ICRF 159 and its inactive homologue ICRF 192. Agents Actions 12: 536–542

Hurley LS (1981) Teratogenic aspects of manganese, zinc and copper nutrition. Physiol Rev 61: 249–295

Ikebuchi H, Teshima R, Suzuki K, Terao T, Yamare Y (1986) Simultaneous induction of Pb-metallothionein-like protein and Zn-thionein in the liver of rats given lead acetate. Biochem J 233: 541–546

Ikeda M, Hosotani T, Ueda T, Kotake Y, Sakakibara B (1979) Observations of the concentrations of zinc and iron in tissues of vitamin B_6-deficient germ-free rats. J Nutr Sci Vitaminol (Tokyo) 25: 151–158

Jones DC, May PM, Williams DR, Reid MC, Sunderman FW Jr (1984) Antidotal efficacy of tetraethylenepentamine for acute nickel carbonyl poisoning in rats. Inorg Chim Acta 91: L51–L53

Jones MM, May PM (1987) The effect of kinetic factors on thermodynamic evaluations of therapeutic chelating agents. Inorg Chim Acta 138: 67–73

Kagi JHR, Kojima Y (1987) Chemistry and biochemistry of metallothionein. In: Kagi JHR, Kojima Y (eds) Metallothionein II. Birkhauser Verlag, Basel, pp 25–61

Karin M, Herschmann HR (1981) Induction of metallothionein in HeLa cells by dexamethasone and zinc. Europ J Biochem 113: 267–272

Karin M, Slater EP, Herschmann HR (1981) Regulation of metallothionein synthesis in HeLa cells by heavy metals and glucocorticoids. J Cell Physiol 106: 63–74

Leber AP, Miya TS (1976) A mechanism for cadmium- and zinc-induced tolerance to cadmium toxicity: involvement of metallothionein. Toxicol Appl Pharmacol 37: 403–414

McGrath SP, Sanders JR, Laurie SH, Tancock NP (1986) Experimental determinations and computer predictions of trace metal ion concentrations in dilute complex solutions. Analyst 111: 459–465

Magneson GR, Puvathingal JM, Ray WJ (1987) The concentrations of free Mg^{2+} and free Zn^{2+} in equine blood plasma. J Biol Chem 262: 11140–11148

Masters DG, Keen CL, Lönnerdal B, Hurley LS (1983) Zinc deficiency teratogenicity: the protective role of maternal tissue catabolism. J Nutr 113: 905–912

May PM, Bulman RA (1983) The present status of chelating agents in medicine. Prog Med Chem 20: 225–336

May PM, Williams DR (1977) Computer simulation of chelation therapy. FEBS Lett 78: 134–138

May PM, Linder PW, Williams DR (1977) Computer simulation of metal-ion equilibria in biofluids:

models for the low-molecular-weight complex distribution of Ca(II), Mg(II), Mn(II), Fe(III), Cu(II), Zn(II) and Pb(II) ions in human blood plasma. J Chem Soc Dalton Trans 588–595

May PM, Williams GK, Williams DR (1980) Speciation studies of Adriamycin, Quelamycin and their metal complexes. Inorg Chim Acta 46: 221–228

Nettesheim DG, Engeseth HR, Otvos JD (1985) Products of metal exchange reactions of metallothionein. Biochemistry 24: 6744–6751

Nielson KB, Winge DR (1985) Independence of the domains of metallothionein in metal-binding. J Biol Chem 260: 8698–8701

Palmiter RD (1987) Molecular biology of metallothionein gene expression. In: Kagi JHR, Kojima Y (eds) Metallothionein II. Birkhauser Verlag, Basel, pp 63–80

Prasad AS, Brewer GJ, Schoomaker EB, Rabbani P (1978) Hypocupremia induced by zinc therapy in adults. J Am Med Ass 240: 2166–2168

Reinstein NH, Lönnerdal B, Keen CL, Hurley LS (1984) Zinc–copper interactions in the pregnant rat: fetal outcome and maternal and fetal zinc, copper and iron. J Nutr 114: 1266–1279

Richards MP, Cousins RJ (1976) Metallothionein and its relationship to the metabolism of dietary zinc in rats. J Nutr 106: 1591–1599

Roth H, Kirchgessner M (1985) Utilization of zinc from picolinic or citric acid complexes in relation to dietary protein sources in rats. J Nutr 115: 1641–1649

Seguin C, Hamer DH (1987) Regulation in vitro of metallothionein gene binding factors. Science NY 235: 1383–1387

Shapiro SG, Squibb KS, Markowitz LA, Cousins RJ (1978) Cell-free synthesis of metallothionein directed by rat liver polyadenylated messenger ribonucleic acid. Biochem J 175: 833–841

Smith JC (1980) The vitamin A–zinc connection. A review. Ann NY Acad Sci 355: 62–75

Smith JC, Brown ED, McDaniel EG, Chen W (1976) Alterations in vitamin A metabolism during zinc deficiency and food and growth restriction. J Nutr 106: 569–574

Sundaresan PR, Cope FO, Smith JR (1977) Influence of zinc deficiency on retinal reductase and oxidase activities in rat liver and testes. J Nutr 107: 2189–2197

Suzuki Y (1981) Metallothionein and cadmium, zinc and copper distribution in blood of rats. Ind Health 18: 19–29

Van der Welde WJ, Copius Peereboom-Stegeman JHJ, Treffers PE, James J (1983) Structural changes in placentae of smoking mothers: a quantitative study. Placenta 4: 231–240

Walker WR, Reeves R, Kay DJ (1975) The role of Cu^{2+} and Zn^{2+} in the physiological activity of histamine in mice. Search 6: 134–135

Washko PW, Cousins RJ (1975) Effects of low dietary calcium on chronic cadmium toxicity in rats. Nutr Rep Int 11: 113–127

Webb M, Samarawickrama GP (1981) Placental transport and embryonic utilization of essential metabolites in the rat at the teratogenic dose of cadmium. J Appl Toxicol 1: 270–277

Winge DR, Premakumar R, Rajagopalan KV (1978) Studies on the zinc content of cadmium-induced thionein. Arch Biochem Biophys 188: 466–475

Zähringer J, Baliga BS, Munro HN (1976) Novel mechanism for translational control in regulation of ferritin synthesis by iron. Proc Natl Acad Sci USA 73: 857–861

Chapter 7

Biochemistry of Zinc in Cell Division and Tissue Growth

J.K. Chesters

Animal and Tissue Growth

Young rats offered a zinc-deficient diet show an abrupt reduction in growth after only a few days on the diet and before a major reduction in total body zinc content or concentration has occurred (Williams and Mills 1970). This coincides with an equally abrupt reduction in food intake to a level which, in pair-fed animals, severely restricts growth (Chesters and Quarterman 1970). However, when the intake of zinc-deficient rats is increased by gavage to that of controls offered the zinc-adequate diet ad lib, the animals fail to grow and become ill (Chesters and Quarterman 1970; Masters et al. 1983). The reduced tissue growth of a zinc-deficient rat results therefore from a zinc-responsive biochemical defect rather than from a physiological effect of loss of appetite.

This chapter will examine a hypothesis (Chesters 1978) that the crucial zinc required for growth is needed to facilitate alterations in the patterns of genetic expression of cells.

Protein Metabolism

Growth can be partitioned between increase in cell size and in cell number, only the latter being associated with net synthesis of DNA. Williams and Chesters (1970) showed that in liver, kidneys, spleen and testes, DNA synthesis was more severely affected by lack of zinc than was protein synthesis during the onset of zinc deficiency in rats. Similar results were obtained by Grey and Dreosti (1972) from a study of liver regeneration in zinc-deficient rats following partial hepatectomy. These observations suggest that DNA synthesis and cell division are more susceptible to lack of zinc than are increase in protein mass and cell size.

Table 7.1. Rates of protein synthesis (%/d) in muscles of zinc-deficient and control rats fed ad lib or restricted amounts of diet

Muscle	Dietary regimen				
	+Zn ad lib	−Zn ad lib	+Zn pair fed	−Zn restricted	+Zn restricted
Gastrocnemius	19.3	4.8	6.1	5.5	5.7
Soleus	27.7	8.6	11.4	10.7	10.3

After Giugliano and Millward (1987).

Recently, Giugliano and Millward (1987) found that the rate of protein synthesis in skeletal muscle was reduced in zinc-deficient rats compared with controls fed ad lib (Table 7.1). However, the effect was much less marked when pair-fed animals were used for the comparison. Furthermore, when the food intake of both zinc-deficient and control rats was restricted to a constant daily amount equivalent to the mean intake of zinc-deficient rats fed ad lib, the difference between control and deficient rats was abolished. Differences in food intake rather than in zinc intake appear to have been largely responsible for the different rates of protein synthesis in deficient and control rats.

Synthesis of Nucleic Acids

As well as restricting the growth of rats after birth, zinc deficiency in utero can cause severe teratogenic effects. These are associated with reduced rates of DNA synthesis in the foetus (Eckhert and Hurley 1977) and decreased DNA polymerase and thymidine kinase activities (Duncan and Hurley 1978). Loss of thymidine kinase activity has also been observed in the liver of zinc-deficient rats (Record and Dreosti 1979). However, in neither case was the loss of enzyme activity reversible by addition of zinc in vitro. The recognition that most nucleic acid polymerases contain zinc (Vallee and Galdes 1984) and the reduction in DNA polymerase activity in zinc-deficient rat foetuses and of RNA polymerase in zinc-deficient sucking rats (Terhune and Sandstead 1972) led to speculation that the growth retarding effects of zinc deficiency were due to decreased polymerase activities. On the other hand, thymidine kinase, the activity of which is also reduced by zinc deficiency, is not thought to contain zinc (Anderson 1973).

In hepatoma cells growing in zinc-deficient and control rats, lack of zinc inhibited both total DNA synthesis and thymidine incorporation into DNA (Baker and Duncan 1983). When the animals were treated with methotrexate, which is an inhibitor of the synthesis of thymidine de novo, there was a large rise in thymidine incorporation in the control rats due to induction of thymidine kinase. This did not occur in the zinc-deficient rats and a similar increase in DNA polymerase activity was also restricted to the control animals. This suggests a failure of enzyme induction in the zinc-deficient animals which may be a more general phenomenon of zinc deficiency since the activity of another inducible enzyme, hepatic ornithine decarboxylase, which normally increases following partial hepatectomy, was also significantly impaired when the rats were zinc deficient (Chesters 1971).

In addition to the above effects in hepatoma cells, lack of zinc resulted in closely correlated reductions of DNA polymerase and thymidine kinase in both cultured kidney cells and in foetal rats (Lieberman et al. 1963; Duncan and Hurley 1978), and in lymphocytes low zinc availability reduced both thymidine uptake and incorporation to similar extents (Chesters 1975). Although the latter relationship might possibly be expected, it is less obvious why DNA polymerase activity should be closely correlated with that of thymidine kinase. Furthermore, in regenerating rat liver following partial hepatectomy, in chick embryo cells in culture and in ascites cells in vivo, the inhibition of thymidine incorporation by zinc deficiency was proportional to the reduction in the number of cells which incorporated thymidine (Fujioka and Lieberman 1964; Rubin 1972; Sarayan et al. 1979). If reduced incorporation had resulted from low cellular activities of DNA polymerase or thymidine kinase because of a lack of zinc to synthesize functional enzyme proteins, then the fraction of the total cell population which incorporated thymidine would have been similar in the zinc-deficient and control groups but the rate of incorporation would have been lower in those lacking zinc. However, the experimental results suggest that individual cells either incorporate thymidine at a normal rate or not at all. This is more consistent with a requirement for zinc to facilitate induction of the group of enzymes which must be synthesized to allow the cells to progress from G_1 to the S phase of the cell cycle.

This hypothesis was strengthened by differences in the effects of zinc deficiency on DNA synthesis in normal cells and those from permanently culturable and virus-transformed cell lines. Chick embryo cells in culture required zinc for DNA synthesis but their dependence on zinc was greatly reduced after the cells were transformed with Rous sarcoma virus (Rubin 1972). Similarly, although DNA synthesis in primary cultures of kidney cells was highly sensitive to zinc deprivation, continuously culturable HeLa and L cells were relatively insensitive to lack of zinc (Lieberman and Ove 1962). Therefore, it is the change of genetic expression associated with a cell acquiring the ability to synthesize DNA that appears to be highly dependent on an adequate supply of zinc rather than the mechanics of DNA synthesis per se.

Cell Division and Differentiation

Although lack of zinc generally impairs DNA synthesis and hence cell division, there are situations in which it is associated with increased mitotic rates. In the zinc-deficient rat, Fell et al. (1973) found that the mitotic indices of the liver and epidermis were unchanged or decreased but those of the oesophagus and pancreas actually increased significantly (Table 7.2). In the buccal epithelium, lack of zinc results in a higher proportion of dividing cells and in an increase in those incorporating thymidine during a pulse label (Alvares and Meyer 1973; Chen 1986). These effects might have been caused by a slowing of the passage of the zinc-deficient cells through the S and M phases of the cell cycle. However, Chen (1986) showed that the duration of most phases was unaltered by zinc deficiency but that the G_1 phase (which included any G_0 component) was substantially shorter. Thus the cell cycle time was reduced in

Table 7.2. Mitotic indices of tissues of zinc-deficient and pair-fed control rats 6 h after colchicine injection.

	Mitoses/1000 cells			
	Hepatic parenchyma	Epidermis basal layer	Oesophagus germinal layer	Pancreas[a] zymogenic cells
−Zn ad lib	<0.2	1.8	163	2.7
+Zn pair fed	<0.2	5.0	12	<0.2

After Fell et al. (1973).
[a] Without colchicine.

the zinc-deficient epithelium and this was not caused by a freak accumulation of zinc in this tissue as analysis showed the zinc content to be very significantly lower than in the controls.

These effects are clearly not compatible with an impairment of DNA synthesis by zinc deficiency but could be explained by a failure of differentiation. The estimated duration of G_1 included any time spent by the cells in a non-proliferative or G_0 state and entry into a non-proliferative state can be considered to be the start of differentiation in cells of the basal layer which would otherwise continue to divide repetitively. Thus the differences observed might be attributable to a lower rate of differentiation in the zinc-deficient cells leaving a higher proportion of cells to continue dividing.

Zinc deficiency also impairs the differentiation of cells in the epiphyses of bone (Westmoreland 1971) and was associated with an immature appearance of fibroblasts in skin (Hsu et al. 1974). Furthermore in pregnant rats, the period of maximum risk of foetal abnormalities due to lack of zinc is during days 9 to 12 of pregnancy when tissue differentiation is maximal (Hurley et al. 1971). Zinc availabiilty, as indicated by plasma zinc concentration, is inversely related to a rat's food intake in the immediately preceding period (Chesters and Will 1973) and the requirement for readily available zinc during tissue differentiation is so critical that the extent of abnormalities can be predicted from the food intakes of the dams during days 9–12 of pregnancy (Record et al. 1985).

Malignancy

If zinc is required for cell division, one would expect zinc deficiency to restrict the growth of malignant tumours and numerous reports have confirmed that this occurs (De Wys and Pories 1972; McQuitty et al. 1970; Mills et al. 1981). In P388 leukaemia grown as an ascites tumour in mice, zinc deficiency reduced the fraction of the cells incorporating thymidine during a pulse label without apparently altering the rate of DNA synthesis in those cells active in incorporation (Barr and Harris 1973). In other studies on the growth of ascites tumorus in zinc-deficient mice, the zinc concentration in the ascites fluid and both thymidine uptake and incorporation were all reduced in proportion to the duration and severity of the deficiency (Minkel et al. 1979; Sarayan et al.

1979). However the patterns of thymidine uptake and incorporation with time were the same in zinc-deficient as in zinc-adequate cells. The authors suggest that this is inconsistent with a uniform reduction in activity in each of the zinc-deficient cells and more consistent with a higher proportion of the cells in the zinc-deficient animals being blocked in a $G_0–G_1$ phase.

In malignant as with normal oesophageal tissue, the usual reduction of cell division in zinc deficiency is not observed. Both here and in the "forestomach", tumours were induced more readily by methylbenzylnitrosamine in zinc-deficient rats. The proportion of malignant tumours after prolonged exposure to the carcinogen reached 100% while all the tumours in the controls remained benign (Fong et al. 1978, 1982). It seems likely that the elevated rate of cell division in these zinc-deficient tissues predisposes them to malignant conversion.

Mechanisms

Problems of Interpretation

Most of the experiments discussed so far relate to the effects of zinc deficiency in whole animals but the complexity of a living mammal is such that some investigators have turned to a simpler "animal", *Euglena*, for enlightenment (Vallee and Falchuk 1983). Others have worked with cultured mammalian cells (Lieberman and Ove 1962; Lieberman et al. 1963; Chesters 1972, 1975; Rubin 1972) in which zinc availabiilty has been restricted by addition of a chelator. Further information required to assemble the jigsaw of zinc deficiency in animals comes from the numerous reports of cell-free systems which require zinc. However valuable each of these approaches may be, it is often difficult to assess the significance of effects of zinc observed in cell cultures or in cell-free systems in relation to metabolic effects in whole animals. This problem arises from the different and generally unknown availabilities of Zn in vivo and in vitro.

As mentioned above, zinc deficiency rapidly inhibits the growth of animals before there is any major change in their total body zinc concentration. The zinc which is crucial for growth must therefore be a minor component which equilibrates readily with exchangeable zinc. Bremner and May (Chap. 6, this volume) suggest that low-molecular-weight complexes may be important and that the concentration of free zinc ions will give a good indication of zinc availability. They estimate that in plasma, low-molecular-weight complexes of zinc will account for 1% or less of the total zinc and the free ionic zinc will be between 10^{-9} and 10^{-10} M.

Cell Culture Studies

Much interesting information has come from the studies of zinc deficiency in *Euglena* (Vallee and Falchuk 1983) but certain of the major findings do not seem to be applicable to zinc deficiency in mammals. In *Euglena* which have ceased to grow because of lack of zinc, the DNA content of the cells doubles (Falchuk et al. 1975), the normal histones of the cell's chromatin are totally replaced by arginine-rich peptides (Vallee and Falchuk 1981) and the three normal RNA polymerases are replaced by a single enzyme which appears to be unique to zinc-deficient *Euglena* (Falchuk et al. 1985). In contrast, DNA concentrations in mammalian systems tend to be unchanged or reduced by zinc deficiency (Williams and Chesters 1970) and any changes in histones are of quantitatively minor nature (Castro et al. 1986). These major differences in response to zinc deficiency between *Euglena* and mammals seriously limit the extent to which conclusions drawn from the former can be applied to higher animals. However, the amounts and patterns of mRNA and protein synthesized by zinc-deficient *Euglena* differ from those in zinc-adequate cultures and Falchuk and Vallee (1985) conclude that in *Euglena*, zinc is primarily involved in the control of gene expression.

In a range of mammalian cultures low availability of zinc has been induced by addition of a chelator, normally EDTA, to the culture medium (Lieberman and Ove 1962; Lieberman et al. 1963; Chesters 1972, 1975; Rubin 1972; Rubin and Koide 1973). Computer simulation of zinc binding in lymphocyte cultures suggested that under optimal conditions for lymphocyte transformation, the free zinc concentration was about 2×10^{-10} M (Chesters 1972 and unpublished observations) and thus similar to that in plasma. Comparable estimates are not available for the other studies but the concentrations of EDTA and zinc used were similar. All the effects of EDTA reported here were reversible specifically by adding zinc.

Although stimulated uptake of nutrients is a common feature of the initial phase of growth in culture, neither the increase in K^+ accumulation by lymphocytes nor the uptake of 2-deoxyglucose and uridine in chick embryo cells was affected by addition of EDTA (Chesters 1978; Rubin and Koide 1973). Therefore the early alterations in membrane permeability do not seem to depend on readily available zinc. In each of the mammalian cultures, the reduction in DNA synthesis due to lack of zinc was much greater than that of RNA while protein synthesis was largely unaffected. However, the chelator had to have been present for some hours before DNA synthesis was measured for the effects of EDTA to be apparent. This again suggests that zinc is required for the metabolic steps which precede DNA synthesis rather than for DNA synthesis per se. The effects of zinc deprivation on the rate of DNA synthesis and the number of cells incorporating thymidine have been discussed above. Although ribosomal RNA synthesis in lymphocyte cultures was reduced by addition of EDTA, those 45S rRNA precursor molecules which were produced seemed to generate an initially normal ratio of 18S to (32S + 28S) rRNA (Chesters 1975). However, during maturation of these rRNA molecules there was a differential loss of 28S rRNA relative to 18S rRNA.

Transcription Factor IIIA

The failure of 28S rRNA maturation in zinc-deficient lymphocyte cultures did not appear to agree with previous conclusions that the primary function of zinc is to facilitate changes in gene expression. However, 5S rRNA, a constituent of the same ribosomal subunit as 28S rRNA, requires a protein transcription factor IIIA (TFIIIA) for its synthesis by RNA polymerase III (Lassar et al. 1983), and Hanas et al. (1983) have shown that zinc is required for binding of TFIIIA to the internal control region of the 5S rRNA gene. Furthermore, Wingender et al. (1984) found that concentrations of zinc which might occur in vivo were required for 5S rRNA synthesis by cell-free extracts of HeLa cells. Availability of zinc thus serves to regulate the transcription of the 5S rRNA gene in vitro, in line with the general hypothesis previously advanced. Subsequent studies of TFIIIA have suggested that it contains up to nine similar sites in which zinc binding stabilizes a finger-like loop of the protein and these "fingers" are thought to be responsible for the specificity of binding of TFIIIA to DNA (Fig. 7.1) (Miller et al. 1985). A number of other proteins have been shown to contain similar binding sites and without exception each is involved in the regulation of gene expression (Berg 1986).

Lack of zinc for TFIIIA binding to the 5S rRNA gene might therefore lead to inadequate 5S rRNA synthesis and reduced assembly of mature large ribosomal subparticles. Since these particles incorporate the 28S rRNA molecules, this in turn would leave the 28S rRNA relatively unprotected from ribonuclease attack and thus explain the differential loss of 28S rRNA from zinc-deficient lymphocytes. However, when the rates of synthesis of 5S rRNA were estimated in zinc-deficient and control rats there was no suggestion of a selective reduction in the rate of 5S rRNA synthesis in the zinc-deficient animals (Chesters, unpublished observations). In zinc-deficient lymphocytes, RNA synthesis required only a third of the concentration of zinc essential for optimal synthesis of DNA (Chesters 1978). It seems likely that while zinc is necessary for 5S rRNA synthesis in the rat, another essential function related to DNA synthesis has a higher requirement for zinc and that this limits the growth of the animals before free zinc levels fall sufficiently to affect 5S rRNA synthesis.

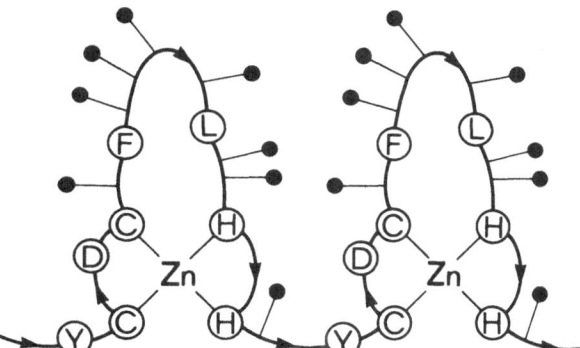

Fig. 7.1. Folding scheme for a linear arrangement of repeated domains, each centred on a tetrahedral arrangement of zinc ligands. Ringed residues are the conserved amino acids which include the Cys and His zinc ligands, the negatively charged Asp-11 and the three hydrophobic groups that may form a structural core. Black circles mark the most probable DNA-binding side chains. In the scheme drawn here the metal ion draws the ends of each unit together, leaving the central residues 14–25 to form a potential DNA-binding loop or "finger". An alternative but much less likely position for the zinc is between the His residues of one unit and the Cys residues of an adjacent one. Reproduced with permission from Miller et al. (1985).

Diadenosine Tetraphosphate

Another possible mechanism for the control of genetic expression by a zinc-mediated process involves the nucleotide diadenosine tetraphosphate (AP4A) whose synthesis by aminoacyl-tRNA synthetases is stimulated by zinc (Blanquet et al. 1983). Cellular levels of AP4A increase up to 1000-fold prior to and in proportion to DNA synthesis and are constitutively higher in cells which lack a defined G_1 phase or which have been virus transformed (Weinmann-Dorsch et al. 1984; Yamaguchi et al. 1985). Although some authors suggest that the concentrations of free zinc ions required to stimulate AP4A synthesis are probably above the physiological range (Goerlich and Holler 1984), others disagree and Grummt et al. (1986) have postulated that zinc acts as a "second messenger of mitotic induction". Such a role for zinc could explain many of the phenomena outlined above.

Tubulin

Zinc may also influence cell division through an interaction with tubulin, a cytoskeletal protein for which a variety of structural and functional roles have been suggested. In 1954, Fuji reported that zinc was associated with the mitotic spindle, and colchicine, which is known to bind to tubulin of the spindle, was more toxic to duodenal cells in zinc-deficient rats than in controls (Dinsdale and Williams 1977). In the testes of zinc-deficient rats, the few remaining spermatozoa were frequently seen to have defective microtubules within the sperm tails (Dinsdale and Williams 1980) but the atrophy of the testes was so severe that it is questionable whether the defects specifically originated from loss of microtubular organization. In brain extracts from pig and rat, zinc deficiency resulted in a reduced rate of tubulin aggregation in vitro and a lower level of free SH groups in crude microtubular preparations (Hesketh 1981). Zinc has been shown to bind to tubulin but only with a low affinity (Hesketh 1983) and zinc stimulates assembly of microtubules in vitro but only at concentrations of zinc which seem unlikely to be relevant in vivo (Haskins et al. 1980; Hesketh 1982).

Conclusion

Present evidence suggests that zinc is required in a highly available form for changes in the genetic expression of cells of which prime examples are the induction of the enzymes for DNA synthesis prior to entry into the S phase of the cell cycle and the induction of new proteins during tissue differentiation. However, the mechanism of its involvement in these processes is still unknown and remains a fascinating field for further research.

References

Alvares OF, Meyer J (1973) Thymidine uptake and cell migration in cheek epithelium of Zn-deficient rats. J Oral Pathol 2: 86–94

Anderson EP (1973) Nucleosides and nucleoside kinases. In: Boyer PD (ed) The enzymes, vol 9, 3rd edn. Academic Press, New York, London, pp 69–93

Baker GW, Duncan JR (1983) Possible site of Zn control of hepatoma cell division in Wistar rats. J Natl Cancer Inst 70: 333–336

Barr DH, Harris JW (1973) Growth of P388 leukaemia as an ascites tumor in Zn-deficient mice. Proc Soc Exp Biol Med 144: 284–287

Berg JM (1986) More metal binding fingers. Nature 319: 264–265

Blanquet S, Plateau P, Brevet A (1983) The role of Zn in 5′,5′ diadenosine tetraphosphate production by aminoacyl-tRNA synthetases. Mol Cell Biochem 52: 3–11

Castro CE, Alvares OF, Sevall JS (1986) Zinc deficiency decreases histone H1° in rat liver. Nutr Reports Int 34: 67–74

Chen S-Y (1986) Autoradiographic study of cell proliferation in acanthotic buccal epithelium of Zn-deficient rabbits. Arch Oral Biol 31: 535–539

Chesters JK (1971) Problems caused by variations in food intake in experiments on protein and nucleic acid metabolism. Proc Nutr Soc 30: 1–7

Chesters JK (1972) The role of zinc ions in the transformation of lymphocytes by phytohaemagglutinin. Biochem J 130: 133–139

Chesters JK (1975) Comparison of the effects of zinc deprivation and actinomycin D on ribonucleic acid synthesis by stimulated lymphocytes. Biochem J 150: 211–218

Chesters JK (1978) Biochemical functions of zinc in animals. World Rev Nutr Diet 32: 135–164

Chesters JK, Quarterman J (1970) Effects of zinc deficiency on food intake and feeding patterns of rats. Br J Nutr 24: 1061–1069

Chesters JK, Will M (1973) Some factors controlling food intake by zinc-deficient rats. Br J Nutr 30: 555–566

De Wys W, Pories W (1972) Inhibition of a spectrum of animal tumors by dietary Zn deficiency. J Natl Cancer Inst 48: 375–381

Dinsdale D, Williams RB (1977) The enhancement by dietary Zn deficiency of the susceptibility of the rat to colchicine. Br J Nutr 37: 135–142

Dinsdale D, Williams RB (1980) Ultrastructural changes in the sperm tail of Zn-deficient rats. J Comp Pathol 90: 559–566

Duncan JR, Hurley LS (1978) Thymidine kinase and DNA polymerase activity in normal and Zn-deficient developing rat embryos. Proc Soc Exp Biol Med 159: 39–43

Eckhert CD, Hurley LS (1977) Reduced DNA synthesis in zinc deficiency: regional differences in embryonic rats. J Nutr 107: 855–861

Falchuk KH, Vallee BL (1985) Zinc and chromatin structure, composition and function. In: Mills CF, Bremner I, Chesters JK (eds) Trace elements in man and animals – TEMA 5. Commonwealth Agricultural Bureaux, Slough, pp 48–55

Falchuk KH, Fawcett DW, Vallee BL (1975) Role of Zn in cell division of E. gracilis. J Cell Sci 17: 57–78

Falchuk KH, Mazus B, Ber E, Ulpino-Lobb L, Vallee BL (1985) Zinc deficiency and the E. gracilis chromatin: formation of an α-amanatin resistant RNA polymerase II. Biochemistry 24: 2576–2580

Fell BF, Leigh LC, Williams RB (1973) The cytology of various organs in Zn-deficient rats with particular reference to the frequency of cell division. Res Vet Sci 14: 317–325

Fong LYY, Swak A, Newberne PM (1978) Zinc deficiency and methylbenzyl nitrosamine-induced oesophageal cancer in rats. J Natl Cancer Inst 61: 145–150

Fong LYY, Lee JSK, Chan WC, Newberne PM (1982) Zn deficiency and the induction of oesophageal tumors in rats by benzylmethylamine and sodium nitrite. IARC Sci Publ 41: 679–683

Fuji T (1954) Presence of zinc in nucleoli and its possible role in mitosis. Nature 174: 1108–1109

Fujioka M, Lieberman I (1964) A Zn^{2+} requirement for synthesis of deoxyribonucleic acid by rat liver. J Biol Chem 239: 1164–1167

Giugliano R, Millward DJ (1987) The effects of severe Zn deficiency on protein turnover in muscle and thymus. Br J Nutr 57: 139–155

Goerlich O, Holler E (1984) Phenylalanyl-tRNA synthetase of E. coli K10. Effects of Zn(II) on partial reactions of diadenosine $5′,5′′′-P^1, P^4$ tetraphosphate synthesis, conformation and protein aggregation. Biochemistry 23: 182–190

Grey PC, Dreosti IE (1972) DNA and protein metabolism in Zn-deficient rats. J Comp Pathol 82: 223–228

Grummt F, Weinmann-Dorsch C, Schneider -Schaulies J, Lux A (1986) Zinc as a second messenger of mitogenic induction. Exp Cell Res 163: 191–200

Hanas JS, Hazuda DJ, Bogenhagen DF, Wu F Y-H, Wu C-W (1983) *Xenopus* transcription factor A requires Zn for binding to the 5S RNA gene. J Biol Chem 258: 14120–14125

Haskins KM, Zombola RR, Boling JM, Lee YC, Himes RH (1980) Tubulin assembly induced by cobalt and zinc. Biochem Biophys Res Commun 95: 1703–1709

Hesketh JE (1981) Impaired microtubule assembly in brain from Zn-deficient pigs and rats. Int J Biochem 13: 921–926

Hesketh JE (1982) Zinc stimulated microtubule assembly and evidence for Zn binding to tubulin. Int J Biochem 14: 983–990

Hesketh JE (1983) Zinc binding to tubulin. Int J Biochem 15: 743–746

Hsu JM, Kim KM, Anthony WL (1974) Biochemical and electron microscopic studies of rat skin during Zn deficiency. Adv Exp Med Biol 48: 347–388

Hurley LS, Gowan J, Swenerton H (1971) Teratogenic effects of short-term and transitory zinc deficiency in rats. Teratology 4: 199–204

Lassar AB, Martin PL, Roeder RG (1983) Transcription of class III genes: formation of pre-initiation complexes. Science 222: 740–748

Lieberman I, Ove P (1962) Deoxyribonucleic acid synthesis and its inhibition in mammalian cells cultured from the animal. J Biol Chem 237: 1634–1642

Lieberman I, Abrams R, Hunt N, Ove P (1963) Levels of enzyme activity and deoxyribonucleic acid synthesis in mammalian cells cultured from the animal. J Biol Chem 238: 3955–3962

McQuitty JT, De Wys WD, Monaco L et al. (1970) Inhibition of tumor growth by dietary zinc deficiency. Cancer Res 30: 1387–1390

Masters DG, Keen CL, Lönnerdal B, Hurley LS (1983) Zn deficiency teratogenicity: the protective effect of maternal tissue catabolism. J Nutr 113: 905–912

Miller J, McLachlan AD, Klug A (1985) Repetitive zinc-binding domains in the protein transcription factor IIIA from *Xenopus* oocytes. EMBO J 4: 1609–1614

Mills BJ, Broghamer WL, Higgins PJ, Lindeman RD (1981) A specific dietary Zn requirement for the growth of Walker 256/M1 tumor in the rat. Am J Clin Nutr 34: 1661–1669

Minkel DJ, Dolhoun PJ, Calhoun BL, Sarayan LA, Petering DH (1979) Zn deficiency and growth of Ehrlich ascites tumors. Cancer Res 39: 2951–2956

Record IR, Dreosti IE (1979) Effects of Zn deficiency on liver and brain thymidine kinase activities in the fetal rat. Nutr Rep Int 20: 749–755

Record IR, Dreosti IE, Manuel SJ, Buckley RA, Tulsi RS (1985) Teratological influence of the feeding cycle in Zn-deficient rats. In: Mills CF, Bremner I, Chesters JK (ed) Trace element metabolism in man and animals – TEMA 5. Commonwealth Bureaux of Nutrition, Slough, pp 210–213

Rubin H (1972) Inhibition of DNA synthesis in animal cells by EDTA and its reversal by Zn. Proc Natl Acad Sci 69: 712–716

Rubin H, Koide T (1973) Inhibition of DNA synthesis in chick embryo cultures by deprivation of either serum or zinc. J Cell Biol 56: 777–786

Sarayan LA, Minkel DT, Dolhoun PJ et al. (1979) Effects of Zn deficiency on cellular processes and morphology in Ehrlich ascites tumor cells. Cancer Res 39: 2457–2465

Terhune MW, Sandstead HH (1972) Decreased RNA polymerase activity in mammalian Zn deficiency. Science 177: 68–69

Vallee BL, Falchuk KH (1981) Zinc and gene expression. Philos Trans R Soc B 294: 185–197

Vallee BL, Falchuk KH (1983) Gene expression and zinc. In: Sarker B (ed) Biological aspects of metals and metal-related diseases. Raven Press, New York, pp 1–14

Vallee BL, Galdes A (1984) The metallobiochemistry of Zn enzymes. Adv Enzymol 56: 283–430

Weinmann-Dorsch C, Hedl A, Grummt I et al. (1984) Drastic rise in intracellular adenosine 5′ tetraphosphate 5′ adenosine correlates with onset of DNA synthesis in eukaryotic cells. Eur J Biochem 138: 179–185

Westmoreland N (1971) Connective tissue alterations in Zn deficiency. Fed Proc Fed Soc Exp Biol Med 30: 1001–1010

Williams RB, Chesters JK (1970) The effects of early zinc deficiency on DNA and protein synthesis in the rat. Br J Nutr 24: 1053–1059

Williams RB, Mills CF (1970) The experimental production of zinc deficiency in the rat. Br J Nutr 24: 989–1003

Wingender E, Dilloo D, Seifert KH (1984) Zinc ions are differentially required for the transcription of ribosomal 5S RNA and tRNA in a HeLa cell extract. Nucleic Acids Res 12: 8971–8985

Yamaguchi N, Kodama M, Ueda K (1985) Diadenosine tetraphosphate as a signal molecule linked with the functional state of rat liver. Gastroenterology 89: 723–731

Chapter 8

Zinc in Cell Division and Tissue Growth: Physiological Aspects

Barbara E. Golden

Introduction

Zinc's role in metabolism has been studied at many different levels, from its crucial presence at the active site of various enzymes to its more dubious efficacy in the treatment of various diseases. Somewhere between is its physiological role in normal growth. Cell biologists have established that zinc is essential for cell replication. In every animal species so far studied, growth failure is one of the earliest signs of experimental zinc deficiency. Paediatricians recognize failure to thrive as a major sign of zinc deficiency, inherited or dietary.

The aim of this chapter is to discuss zinc in relation to growth and vice versa. It will attempt to clarify the area of study between that of the biochemists and that of the clinicians, using the more precise, and, hopefully, accurate data from animal studies to add to the scantier data from human studies.

Figure 8.1 represents living tissue – a row of cells – in "zinc balance", that is, in the situation where zinc demand equals zinc supply. Zinc demand is increased in tissues with high rates of cell hyperplasia and hypertrophy or when zinc losses are high. Such tissues are therefore vulnerable to reduced zinc supply. They include (a) epidermis and intestinal mucosa, as is borne out clinically from the skin lesions and diarrhoea characteristic of acrodermatitis enteropathica (Hambidge et al. 1977); (b) tissues in the young which are undergoing normal growth; and (c) tissues undergoing very rapid or "catch-up" growth. In tissues with a high zinc concentration, zinc demand may also be increased for cell hypertrophy; losses from catabolism are also likely to be high.

Figure 8.2 shows zinc distribution in relation to gross body composition. The majority of body zinc is in lean tissue, twice as much in muscle as in bone,

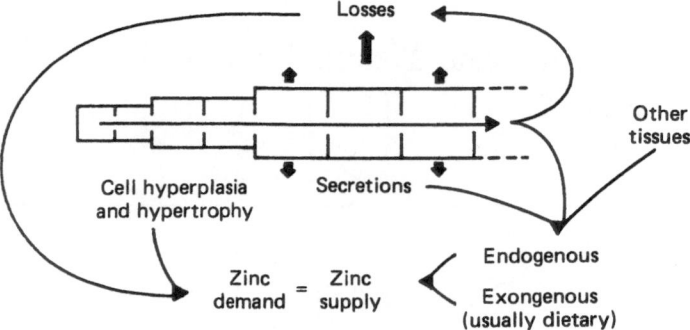

Fig. 8.1. Factors affecting zinc balance in living tissue (shown as a row of cells).

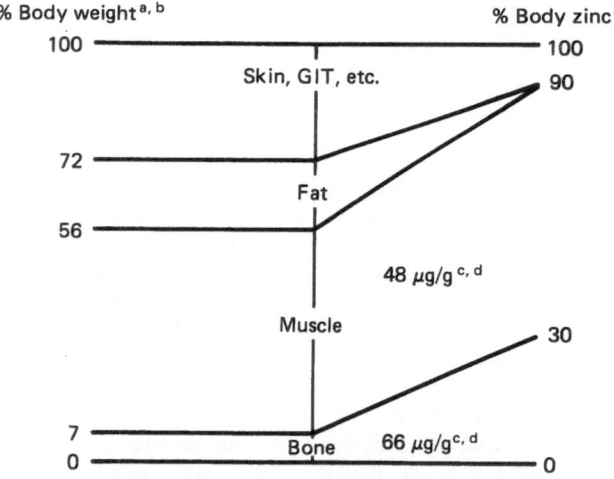

Fig. 8.2. Zinc in relation to gross body composition in adult man. *a*, Garn (1963); *b*, Diem and Lentner (1970); *c*, National Research Council (1979); *d*, μg zinc/g wet weight.

with relatively high zinc concentrations in each. Thus, of all tissues, muscle and bone together have by far the highest zinc demand during growth. Two hypotheses emerge:

1. Either muscle or bone may serve as a zinc store.
2. The rate or composition of growth in these tissues may be particularly dependent on zinc supply in the young or during "catch-up" growth.

Body Zinc Store

Unlike the energy store in adipose tissue, there is no specific zinc store. Thus, in all species studied, a marked reduction in dietary zinc intake is invariably followed quickly by harmful signs of zinc deficiency, e.g. skin lesions in calves, lambs (Mills et al. 1967) and baby pigs (Miller et al. 1968) and foetal malformations in pregnant rats (Swenerton and Hurley 1968). The other major clinical sign, growth failure, largely due to reduced food intake, should be viewed as a useful adaptation to an inadequate zinc supply. As there is no store from which to increase zinc supply, the body as a whole decreases its demand for zinc by reducing food intake and therefore growth rate. When Chesters and Quarterman (1970) and more recently, Flanagan (1984) and Park et al. (1986), force-fed weanling rats a zinc-deficient diet so that they were prevented from reducing their growth rates, the rats rapidly became ill and died. In contrast, rats fed similar diets ad libitum survived for several weeks longer, albeit with signs of zinc deficiency.

Several research groups have, however, argued cogently in favour of a functional, though limited, zinc store. Harland et al. (1975) demonstrated this phenomenon by pretreating young quail with excess dietary zinc. Their bone zinc concentrations increased. During an ensuing period of deficient dietary zinc, their bone zinc concentrations decreased again, and the expected reduction in growth rate due to zinc deficiency was significantly delayed. Calhoun et al. (1978) obtained similar results in weanling rats using intraperitoneal zinc as pretreatment. Only a gross excess of zinc prevented a reduction in growth rate during the following zinc-deficient period. Again, the excess zinc was apparently incorporated into, and then released from, bone. Brown et al. (1978), from the same laboratory, showed in weanling rats given a zinc-deficient diet without pretreatment, that femur zinc concentration still fell, and despite continued femur growth, total femur zinc content also fell significantly. Their data suggested that zinc was released from bone during periods of dietary zinc deficiency. Hurley and Tao (1972) found in pregnant rats that zinc released from bone reduced the effects of zinc deficiency on the developing foetuses, provided the dams were fed a calcium-deficient as well as a zinc-deficient diet. The calcium deficiency accelerated bone resorption and, by releasing zinc deposited in bone, ameliorated the pathological effects of a dietary deficiency of zinc.

More recently, Hurley's group (Masters et al. 1983) showed that dams fed a zinc-deficient diet ad libitum deposited more zinc in the products of conception than could be accounted for from maternal dietary zinc, though, of course, this was insufficient to prevent the profound effects of zinc deficiency. The dams reduced their food intake and lost (litter-free) weight. The extra zinc could be totally accounted for by tissue catabolism associated with the weight loss. This is another example of adaptation to a zinc-deficient diet. In this case, reduced food intake increased the endogenous zinc supply: in weanling rats, it had reduced the demand for exogenous zinc.

Zinc redistribution is another way of describing an increase in endogenous zinc supply in response to a decrease in exogenous zinc supply. Guigliano and Millward (1984) attempted to calculate the extent of zinc redistribution in zinc-deficient weanling rats. By measuring zinc concentrations and weight changes in different organs after 24 d of severe zinc deficiency, they estimated that the net gain of total muscle zinc, about 1600 μg, was less than 300 μg more than the net loss of total bone zinc. Dietary zinc intake during the same period was only 300 μg. Muscle had gained in weight but not altered its zinc concentration, whereas bone zinc concentration had fallen by two-thirds. Thus, zinc was apparently well conserved and redistributed.

Murray and Messer (1981) provided interesting data on bone zinc and calcium kinetics in weanling and older rats rendered zinc and/or calcium deficient. Bone resorption was estimated from the loss of previously incorporated ^{3}H-tetracycline. They showed that zinc release from bone was unaffected by zinc deficiency: thus bone zinc did not behave like a conventional store. However, zinc uptake by bone was markedly reduced during the same period of zinc deficiency. As a result, there was a slight net loss of zinc from the humerus, despite bone growth; in contrast there was a large net gain of bone zinc in zinc-supplemented controls. My interpretation of these data is that a relatively large net saving of zinc – from bone – became available for redistribution in the zinc-deficient rats. Thus the effects of zinc deficiency were buffered, indirectly, by bone turnover. The facility with which bone reduces its zinc uptake in response to dietary zinc deficiency is in stark contrast to the situation with muscle.

In conclusion, an effective zinc store does not exist. However, zinc pretreatment, tissue catabolism and even to a limited extent, normal bone, have all been shown to contribute endogenous zinc when the exogenous supply is deficient. More important perhaps, is the adaptive response of reduction in food intake to a zinc-deficient diet.

Tissue Growth and Composition

Zinc Deficiency

Numerous widely differing studies have been performed in order to delineate the effects of zinc deficiency in the young. These have ranged from zinc supplementation studies in children during rapid recovery from severe wasting, to force-feeding zinc-deficient diets to weanling rats. Probably the most physiological method is to simply reduce the zinc content of a diet fed ad libitum.

In weanling rats, among the earliest changes is a reduction in weight gained per gram of zinc-deficient diet ingested. This contributes to the reduction in growth rate which is largely due to reduced feed intake, as shown by comparison with pair-fed controls (Guigliano and Millward 1984). It also implies that either intestinal absorption is reduced, maintenance requirements are increased and/or different tissue is synthesized.

Several studies have demonstrated that different tissue is indeed synthesized. Thus, weanling rats fed either ad libitum (Guigliano and Millward 1984) or force-fed (Park et al. 1986) zinc-deficient diets gained relatively less muscle than overall weight. Weanling mice fed zinc-deficient diets following early undernutrition also had lower muscle gains than those fed zinc-sufficient diets (Morgan et al. 1987). Zinc deficiency limited muscle synthesis more than the other components of weight gain.

In growing humans, inducing experimental zinc deficiency is unethical. Acrodermatitis enteropathica is the best "natural" example of zinc deficiency: growth failure is a prominent feature (Hambidge et al. 1977).

In Colorado, Walravens and Hambidge sought and found infants on "low-zinc" infant formulas (Walravens and Hambidge 1976) and young children with low zinc intakes and low height for age (Walravens et al. 1983) who were mildly zinc deficient. In both groups, in double-blind studies, zinc supplementation resulted in an increased rate of height gain in the boys. These data alerted nutritionists and the food industry to the possibility of zinc deficiency affecting large sections of society in the developed world.

But what of the developing world, where malnutrition, with stunting, skin lesions and infections are still rife and often fatal?

Malnutrition

Low plasma zinc concentrations occur in many severely malnourished Jamaican children including all those with oedema (Golden and Golden 1979). Excluding the latter – oedema is a confounding factor – those with skin ulceration have particularly low values. The lesions respond specifically to local application of zinc (Golden et al. 1980). Also, their delayed hypersensitivity skin responses correlate with their plasma zinc concentrations and increase with local zinc application (Golden et al. 1978). These data indicate that zinc deficiency plays a part in their increased susceptibility and reduced resistance to infections. We also observed that percentage height for age correlates with plasma zinc (Golden and Golden 1979), and this is consistent with Hambidge's findings.

In other studies (Montgomery 1962; Hansen-Smith et al. 1979) markedly reduced muscle mass has been demonstrated in severely malnourished children, due to both atrophy and loss of fibres.

Recovery from Malnutrition

In experimental pigs (Widdowson 1974) and rats (Harris and Widdowson 1978) recovering rapidly from early severe undernutrition, excess fat was deposited.

The same phenomenon has been demonstrated in malnourished Jamaican children recovering on high-energy formulas. Thus, Ashworth (1969) measured percentage body fat and energy intake (above basal) per gram weight gained; Kerr et al. (1973) measured energy cost of growth (ECG); and Brook and Wheeler (1976) measured skin folds. Each set of data showed that malnourished children gained excess adipose tissue during their rapid recovery. MacLean and Graham (1980) in Peru also concluded from measurements of nitrogen content and energy cost of weight gained that "the impressive gains in weight made by recovering malnourished children are largely fat . . .".

As a corollary, lean tissue synthesis must be impaired. Cheek et al. (1970) and Reeds et al. (1978) showed that muscle cell mass only reached about 60% of "normal" when, respectively, Peruvian and Jamaican children reached their expected weight for height. Also, Hansen-Smith et al. (1979) demonstrated histologically that muscle fibre size of "recovered" Jamaican children nearly 14 months old was only 60% of that of "normal" 6-month-old infants.

Table 8.1 shows the approximate "normal" composition of the tissues principally synthesized during rapid recovery from severe wasting in infants. Summarizing, adipose tissue is largely energy in the form of fat, whereas muscle tissue is largely water and protein, with their "fellow travellers" such as zinc. Thus, the relative proportions of adipose and muscle tissue synthesized may be indirectly estimated from the amount of energy, water, nitrogen and zinc retained per gram of weight gained. Energy cost of growth [ECG = (Energy intake − Energy intake to maintain weight)/weight gain] approximates to energy retained. In recovering malnourished children, Jackson et al. (1977) showed that ECG correlated negatively ($r = -0.95$, $P < 0.01$) with increase in muscle mass and therefore was a useful measure of composition of tissue synthesized.

In Jamaica, the formulas used supply ample energy: adipose tissue synthesis appears to be limited only by volume of intake. The same formulas also apparently supply adequate protein: increasing the protein-to-energy ratio did not increase growth rate (Waterlow 1961) or protein synthesis rate (Golden et al. 1977). However, preliminary studies of increasing the zinc intake alone resulted in dramatic increases in growth rate in a few children (Golden and Golden 1981c).

In a retrospective study of malnourished children recovering on either cows' milk (CM) or soya-based (SB) high-energy formula, plasma zincs were low and inversely correlated with rates of weight gain (RWG). On the lower zinc SB formula, both plasma zincs and RWG were significantly less than on the CM formula. During early recovery, ECG was significantly higher on the SB formula; later, ECG was high on both formulas (Golden and Golden 1981a).

In a further study (Golden and Golden 1981b), the diets of similar children were supplemented with zinc during late recovery. Their RWG increased;

Table 8.1. Composition of adipose tissue and muscle in "normal" infants

	Adipose tissue	Muscle tissue
H_2O (%)	25[b]	78[a]
Protein (%)	7[b]	19[b]
Fat (%)	67[b]	<1[b]
Stored energy (kJ/g)	33[c]	5[c]
Zinc (μg/g)	Very low	48[d]

[a] Diem and Lentner (1970).
[b] Cheek et al. (1970).
[c] Jackson et al. (1977).
[d] National Research Council (1979).

change in RWG correlated with change in plasma zinc. Their ECG decreased; change in ECG correlated, negatively, with change in plasma zinc. These effects were greater in children on SB formula than on CM formula. These data implied that both rate and composition of weight gained on these formulas were limited by zinc deficiency.

However, if so, it is remarkable that such children had such good appetites. For as a result, they consumed vast quantities of energy and other nutrients so that tissue synthesis accelerated and therefore zinc demand increased. Thus, they aggravated their own zinc deficiency. Their lack of anorexia may be related to the zinc-to-protein ratios (Z/P) of the formulas. Experimental zinc-deficient diets have a low Z/P. Chesters and Will (1973) found that the appetites of zinc-deficient rats increased as their dietary Z/P increased, which was achieved by reducing protein content. The high-energy formulas we used had Z/P equal to that of "normal" infant formula, which is not high (Golden and Golden 1981a).

In order to define more accurately the effect of zinc on the composition of tissue synthesized, we gave three small groups of severely wasted children (11 children in all) high-energy SB formula supplemented with 0 (low), 5 (moderate) or 10 (high) mg zinc/kg feed during their recovery in weight for height. Their percentage reference weights for height on selection – 62%, 60% and 61% in low, moderate and high zinc groups respectively – were similar, as also were their ad libitum feed intakes. RWG was highest and ECG lowest in the high zinc group, but the differences were insignificant, probably due to the small number in each group (Table 8.2).

During mid and late recovery only, however, ECG was significantly higher in the low zinc group than in the moderate and high zinc groups ($t = 2.40$, $P = 0.03$).

Table 8.3 shows mean biceps and supra-iliac skinfolds as percentage Karlberg standards (Karlberg 1976) after 6 weeks' recovery.

The tendency to excess adipose tissue in these sites in both low and moderate zinc groups is apparent. When the skinfold data from four sites were used to estimate total body adipose tissue (Brooks 1971), it was again apparent that the low zinc group were synthesizing more than the high zinc group. Thus, in

Table 8.2. Rates of weight gain and energy costs of growth in previously malnourished children given differing intakes of zinc over the first 6 weeks of recovery

Zinc group	n	Weight gain (g/kg/d)		Energy cost of weight gain (kJ/g)	
Low[b]	4	10.1	(0.4)[a]	29.3	(5.4)
Moderate	4	11.6	(1.9)	24.8	(3.5)
High	3	11.7	(2.4)	25.0	(1.3)

[a] Mean (SD).
[b] Dietary zinc supplements: Low, nil; Moderate, 5 mg zinc/kg feed; High, 10 mg zinc/kg feed.

Table 8.3. Mean biceps (BSF) and supra-iliac (SIF) skinfolds in children previously malnourished, but given differing supplements of zinc during 6 weeks rehabilitation (as % Karlberg standards[a])

Zinc group	BSF	SIF
Low[b]	112	140
Moderate	118	154
High	99	99

[a] Karlberg (1976).
[b]|For key see Table 8.2.

the low, moderate and high zinc groups, respectively 48%, 42% and 40% of weight gained over 6 weeks comprised adipose tissue.

These data implied, very indirectly, that lean tissue synthesis was compromised in the low zinc group. More direct estimates of lean tissue synthesis were made, based on the information in Table 8.1. Total body water was measured during early, mid and late recovery in the 11 children (Golden and Golden 1985). Per kilogram of weight gain, it increased by 312 g in the low zinc group, 421 g in the moderate zinc group and 641 g in the high zinc group. The lower the dietary zinc intake, the less hydrated the new tissue, and thus the lower the lean tissue relative to adipose or total tissue synthesized.

Nitrogen and zinc retentions were also measured in balance experiments performed during early, mid and late recovery. As a proportion of nitrogen intake, which was similar in the three groups, nitrogen retention was significantly lower ($t = -2.60$, $P = 0.02$) in the low zinc group than in the high zinc group in mid and late recovery. Also, nitrogen retained per gram of weight gained was significantly lower in the low zinc group than in the high zinc group by late recovery ($t = -2.60$, $P < 0.05$).

Not surprisingly, the differences among the groups were greatest for zinc retention. Per gram of weight gained, 11 µg zinc was retained in the low zinc group, 34 µg zinc in the moderate zinc group, and 49 µg zinc in the high zinc group. Thus, the tissue synthesized by the low zinc group contained less water, nitrogen and zinc, but more energy than that synthesized by the high zinc group.

Protein turnover measurements were made using ^{15}N-glycine as a tracer and measuring ^{15}N enrichment of urinary ammonia. Rates of protein synthesis, and therefore lean tissue synthesis, were significantly less in the low zinc group than the high zinc group in both mid recovery ($t = -7.30$, $P < 0.002$) and late recovery ($t = -5.62$, $P = 0.002$).

Thus, many independent estimates of lean tissue synthesis in children recovering from severe wasting on similar intakes of high-energy formula but different intakes of zinc lead to similar conclusions, namely lean tissue synthesis is closely and positively related to dietary zinc intake. On a longer term basis, height increase is due to lean tissue synthesis: Hambidge's findings are entirely in keeping with ours.

Now that simpler methods of estimating body composition are becoming available, e.g. body impedance, mild degrees of zinc deficiency may become easier to diagnose and manage.

Conclusion

There is no readily available zinc store. However, bone appears to serve as a slowly mobilized endogenous zinc supply, a buffer, by reducing its zinc uptake during periods of zinc deficiency, when the body's demand for zinc exceeds its supply from other sources. In the same circumstances, muscle does not reduce its zinc uptake, nor its zinc concentration: it reduces its synthesis. This is clearly demonstrated in experimentally induced zinc deficiency in animals, and during recovery from severe malnutrition in childhood.

Acknowledgement. This work was supported by the UK Medical Research Council and by the Wellcome Trust, London.

References

Ashworth A (1969) Growth rates in children recovering from protein–calorie malnutrition. Br J Nutr 23: 835–845

Brook OG, Wheeler EF (1976) High energy feeding in protein–energy malnutrition. Arch Dis Child 51: 968–971

Brooks CGD (1971) Determination of body composition of children from skinfold measurements. Arch Dis Child 46: 182–184

Brown ED, Chan W, Smith JC (1978) Bone mineralization during a developing zinc deficiency. Proc Soc Exp Biol Med 157: 211–214

Calhoun NR, McDaniel EG, Howard MP, Smith JC (1978) Loss of zinc from bone during deficiency state. Nutr Rep Intern 17: 299–306

Cheek DB, Hill DE, Cordano A, Graham GG (1970) Malnutrition in infancy: changes in muscle and adipose tissue before and after rehabilitation. Pediatr Res 4: 135–144

Chesters JK, Quarterman J (1970) Effects of zinc deficiency on food intake and feeding patterns of rats. Br J Nutr 24: 1061–1069

Chesters JK, Will M (1973) Some factors controlling food intake by zinc-deficient rats. Br J Nutr 30: 555–566

Diem K, Lentner C (1970) Scientific tables, 7th edn. Geigy, Basle, pp 517, 518

Flanagan PR (1984) A model to produce pure zinc deficiency in rats and its use to demonstrate that dietary phytate increases the excretion of endogenous zinc. J Nutr 114: 493–502

Garn SM (1963) Human biology and research in body composition. In: Brozek J (ed) Body composition, part II. Ann NY Acad Sci 10: 429–446

Golden BE, Golden MHN (1979) Plasma zinc and the clinical features of malnutrition. Am J Clin Nutr 32: 2490–2494

Golden BE, Golden MHN (1981a) Plasma zinc, rate of weight gain, and the energy cost of tissue deposition in children recovering from severe malnutrition on a cow's milk or soya protein based diet. Am J Clin Nutr 34: 892–899

Golden BE, Golden MHN (1985) Effect of zinc supplementation on the composition of newly synthesized tissue in children recovering from malnutrition. Proc Nutr Soc 44: 110A

Golden MHN, Golden BE (1981b) Effect of zinc supplementation on the dietary intake, rate of weight gain, and energy cost of tissue deposition in children recovering from severe malnutrition. Am J Clin Nutr 34: 900–908

Golden MHN, Golden BE (1981c) Trace elements. Potential importance in human nutrition with particular reference to zinc and vanadium. Br Med Bull 37: 31–36

Golden M, Waterlow JC, Picou D (1977) The relationship between dietary intake, weight change, nitrogen balance and protein turnover in man. Am J Clin Nutr 30: 1345–1348

Golden MHN, Golden BE, Harland PSEG, Jackson AA (1978) Zinc and immunocompetence in protein-energy malnutrition. Lancet I: 1226–1228

Golden MHN, Golden BE, Jackson AA (1980) Skin breakdown in kwashiorkor responds to zinc. Lancet I: 1057 (letter)

Guigliano R, Millward DJ (1984) Growth and zinc homeostasis in the severely Zn-deficient rat. Br J Nutr 52: 545–560

Hambidge KM, Walravens PA, Neldner KH (1977) The role of zinc in the pathogenesis and treatment of acrodermatitis enteropathica. In: Brewer GJ, Prasad AS (eds) Zinc metabolism: current aspects in health and disease. Alan R. Liss Inc, New York, pp 329–340

Hansen-Smith FM, Picou D, Golden MH (1979) Growth of muscle fibres during recovery from severe malnutrition in Jamaican infants. Br J Nutr 41: 275–282

Harland BF, Spivey Fox MR, Fry BE Jr (1975) Protection against zinc deficiency by prior excess dietary zinc in young Japanese quail. J Nutr 105: 1509–1518

Harris PM, Widdowson EM (1978) Deposition of fat in the body of the rat during rehabilitation after early undernutrition. Br J Nutr 39: 201–211

Hurley LS, Tao S (1972) Alleviation of teratogenic effects of zinc deficiency by simultaneous lack of calcium. Am J Physiol 222: 322–325

Jackson AA, Picou D, Reeds PJ (1977) The energy cost of repleting tissue deficits during recovery from protein–energy malnutrition. Am J Clin Nutr 30: 1514–1517

Karlberg P (1976) The somatic development of children in a Swedish urban community. Acta Paediatr Scand Suppl 258: 48–55

Kerr D, Ashworth A, Picou D et al. (1973) Accelerated recovery from infant malnutrition with high calorie feeding. In: Gardner LI, Amacher P (eds) Endocrine aspects of malnutrition. The Kroc Foundation, California, pp 467–486

MacLean WC, Graham GG (1980) The effect of energy intake on nitrogen content of weight gained by recovering malnourished infants. Am J Clin Nutr 33: 903–909

Masters DG, Keen CL, Lönnerdal B, Hurley LS (1983) Zinc deficiency teratogenicity: the protective role of maternal tissue catabolism. J Nutr 113: 905–912

Miller ER, Luecke RW, Ullrey DE, Baltzer BV, Bradley BL, Hoefer JA (1968) Biochemical, skeletal and allometric changes due to zinc deficiency in the baby pig. J Nutr 95: 278–286

Mills CF, Dalgarno AC, Williams RB, Quarterman J (1967) Zinc deficiency and the zinc requirements of calves and lambs. Br J Nutr 21: 751–768

Montgomery RD (1962) Muscle morphology in infantile protein malnutrition. J Clin Pathol 15: 511–521

Morgan PN, Cardinet GC Jr, Keen CL, Lönnerdal B (1987) Effects of dietary zinc intake during recovery from undernutrition on mouse gastrocnemius muscle (GM). Fed Proc 46: 885 (abstract)

Murray EJ, Messer HH (1981) Turnover of bone zinc during normal and accelerated bone loss in rats. J Nutr 111: 1641–1647

National Research Council. Subcommittee on Zinc (1979) University Park Press, Baltimore, p 124

Park JHY, Grandjean CJ, Antonson DL, Vanderhoof JA (1986) Effects of isolated zinc deficiency on the composition of skeletal muscle, liver and bone during growth in rats. J. Nutr 116: 610–617

Reeds PJ, Jackson AA, Picou D, Poulter N (1978) Muscle mass and composition in malnourished infants and children and changes seen after recovery. Pediatr Res 12: 613–618

Swenerton H, Hurley LS (1968) Severe zinc deficiency in male and female rats. J Nutr 95: 8–18

Walravens PA, Hambidge KM (1976) Growth of infants fed a zinc supplemented formula. Am J Clin Nutr 29: 1114–1121

Walravens PA, Krebs NF, Hambidge KM (1983) Linear growth of low income preschool children receiving a zinc supplement. Am J Clin Nutr 38: 195–201

Waterlow JC (1961) The rate of recovery of malnourished infants in relation to the protein and calorie levels of the diet. J Trop Pediatr 7: 16–22

Widdowson EM (1974) Changes in pigs due to undernutrition before birth and for one, two and three years afterwards, and the effect of rehabilitation. Adv Exp Med Biol 49: 165–181

Chapter 9

Biochemical Pathologies of Zinc Deficiency

M.S. Clegg, C.L. Keen and Lucille S. Hurley†

Introduction

Todd et al. (1934) were the first to demonstrate that zinc was an essential component in the diet of mammals. Since then, it has been shown that animals which consume zinc-deficient diets eventually develop a variety of pathologies including anorexia, growth retardation, abnormal immune function, abnormal nitrogen metabolism, hypogeusia, impaired reproductive capacity, coarse and sparse hair growth, flaking seborrhoea of the skin, impaired connective tissue metabolism and behavioural defects.

If a deficiency of zinc occurs during early development, additional maladies are incurred. These include increased rates of pre- and postnatal death, abnormal development of the central nervous system and severe congenital malformations, as well as certain biochemical lesions (Hurley 1981). Thus, it is apparent that zinc deficiency affects virtually every organ system with the most negative effects occurring in those tissues which have a requirement for rapid cell division. It is therefore not surprising that young rapidly developing and growing animals show the most severe effects of a direct or induced zinc deficiency.

In addition to a role in cell division, zinc has been postulated to have a role in gene expression and protein synthesis, and it is an integral component of membrane proteins and enzymes where it is necessary for their proper structure and/or catalytic activity. In retrospect, this essential nutrient is required for virtually all aspects of normal cell metabolism, including such broad functions as the synthesis of DNA, the proper expression of DNA (transcription), the translation of mRNA into protein and finally, for the proper structure and functioning of the translated product. Of course, proper function at all these levels is required if the animal is to remain healthy.

In this review we will focus our comments on how a cellular zinc deficiency may lead to the various pathological effects associated with the consumption of a zinc-deficient diet. To accomplish this goal, we will review the role zinc plays in the following molecular events: (a) DNA synthesis/cell division; (b) chromatin structure and gene expression; (c) RNA and protein synthesis; and (d) protein structure and function. Prior to the discussion of these topics, a brief overview of the histological lesions associated with zinc deficiency is presented.

Histological Lesions of Zinc Deficiency

Despite the severe and widespread effects of zinc deficiency, there is a surprising paucity of information on the sequence of changes that occurs in the development of abnormal cell and tissue morphology. Diamond et al. (1971) observed that within 7 d of the introduction of a zinc-deficient diet there was evidence of testicular and oesophageal lesions in the weanling rat. The seminiferous tubules in zinc-deficient testes were slightly reduced in calibre, and there was a smaller number of maturing germ cells with a corresponding increase in lumen size; however, there was evidence of abundant mitotic activity in the deficient animals. The oesophageal lesions in deficient rats consisted of epithelial hyperplasia, hyperkeratosis and parakeratosis. These lesions were not noted in pair-fed animals, thus the described lesions were not the result of an energy deficit. Both the testicular and oesophageal lesions were reversible, completely disappearing after consumption of a zinc-adequate diet for 15 d.

As was noted by the authors, compared with the effects of most nutritional deficiencies, the relatively few days required for morphological evidence of the zinc deficiency state or recovery from it is surprising, and strongly suggests the absence of zinc stores in the animal which can be mobilized when zinc-deficient diets are fed. Oesophageal lesions similar to those observed in the zinc-deficient weanling rat have also been described in the zinc-deficient foetus at term (Diamond and Hurley 1970); the possible reversibility of these lesions with postnatal zinc repletion has not been investigated. This finding of hyperplasia in the oesophagus of the zinc-deficient foetus is particularly striking in that most other tissues of the zinc-deficient foetus are characterized by hypoplasia (Hurley and Shrader 1972). The mechanisms underlying the differential effects of zinc deficiency on foetal oesophageal tissue relative to other foetal tissues have not been identified.

For humans, the histology of the cutaneous lesions associated with zinc deficiency has been described in some detail by Gonzales et al. (1982) in their report on 12 cases of infants with acrodermatitis enteropathica (AE). These investigators were able to identify four different histological patterns which correlated with stages in the development of the cutaneous lesions normally associated with AE. Their findings are presented in detail below. According to the authors, the first stage of "deficiency" is characterized by a loss of the granular layer with a slight paleness of the upper one-third of the epidermis. Round to oval-shaped clear cells with pyknotic nuclei and "empty-looking cytoplasms" are found to replace the cells normally in this region. The spinous layer shows focal spongiosis and the cornified layer has a normal appearance in the outermost portion; the lower section is marked by confluent parakeratosis. In the second stage, paleness of the upper two-thirds of the epidermis is evident, and is accompanied by slight to fully developed psoriasiform hyperplasia. Slight spongiosis is evident and the cornified layer shows confluent parakeratosis. The upper part of the dermis is slightly oedematous and contains a sparse perivascular lymphohistiocytic infiltrate. The third stage of "deficiency" is characterized by psoriasiform hyperplasia and focal pallor in the upper part of the epidermis. The cornified layer is parakeratotic, and in the upper part of the dermis there is dilation and tortuosity of blood vessels, oedema and a perivascular infiltrate of mononuclear cells. In the fourth stage, there is confluent parakeratosis and slight epidermal hyperplasia. It is important to stress that while the above stages of "zinc deficiency" were given for putative AE patients, the lesions described are very similar to those described for cases of acquired zinc deficiency in human (O'kada et al. 1976; Ecker and Schroeter 1978; Van Vloten and Bos 1978).

From the above it is evident that zinc deficiency can have a rapid effect on cell division and morphology. Possible mechanisms by which zinc deficiency may lead to these changes are now discussed.

Cell Division and DNA Synthesis

A role for zinc in cell division was first hypothesized by Fujii (1954), who showed by histological methods that zinc was present in the nucleus of animal cells. His observations and the subsequent observations of others that zinc deficiency results in significant growth retardation and developmental abnormalities led a number of investigators to postulate that the primary locus of this nutrient deficiency was at the level of DNA synthesis. Following central dogma, this would result in less mRNA and protein and hence decreased anabolic processes by the animal. Accordingly, studies were designed that made use of whole animals or isolated cells to quantify DNA and/or to measure the incorporation of the radioactive DNA precursor, thymidine, into DNA under conditions of zinc deficiency. In these studies a consistent observation was that both total DNA and incorporation rates were reduced significantly as a consequence of the zinc deficiency (Fujioka and Lieberman 1964; Sandstead and Rinaldi 1969; Swenerton et al. 1969; Williams and Chesters 1970; Dreosti et al. 1972; Grey and Dreosti 1972; Sandstead et al. 1972; Fernandez-Madrid et al. 1973; Prasad and Oberleas 1973; Rubin 1973; Duncan and Dreosti 1974; Eckhert and Hurley 1977).

Subsequent studies were designed to determine the mechanism(s) underlying the observed reduction in DNA synthesis. Potential candidates for this role included reduced nucleotide precursor pools and/or a reduction in DNA synthesizing enzymes such as DNA polymerase and thymidine kinase. The DNA polymerases were considered because of (a) their essential and established role in DNA synthesis and (b) the prospect that they were zinc metalloenzymes (Slater et al. 1971; Springgate et al. 1973; Auld et al. 1975). However, doubts have recently arisen concerning whether DNA polymerases are true zinc metalloproteins or whether zinc is a co-purified contaminant (Wu and Wu 1987). Despite this ambiguity, Duncan and Hurley (1978) have shown in rats that zinc deficiency resulted in reduced activity of DNA polymerase.

Other investigators examined the potential relationship between zinc deficiency and thymidine kinase for reasons which are not entirely clear. Possibly this enzyme was initially studied because thymidine is the nucleotide of choice for DNA incorporation studies. Lieberman et al. (1963) and Lieberman and Ove (1962), in their pioneering studies on the events controlling the length of the G_1 period in the mammalian cell cycle, showed that decreased DNA synthesis is associated with decreased specific activity of both thymidine kinase and DNA polymerase, with the former being decreased more than the latter. Prasad and Oberleas (1973) reported that thymidine kinase activity was lower in rats fed a zinc-deficient diet for 6 d than in pair-fed controls. Additional papers have supported Prasad's initial investigations; hence, Swenerton et al. (1969), Record and Dreosti (1979) and Dreosti et al. (1980, 1985) all have shown reduced incorporation of thymidine into DNA as a consequence of zinc deficiency. However, while many groups have shown a persuasive decrease in liver thymidine kinase activity (Dreosti and Hurley 1975; Duncan and Hurley 1978) and in rates of thymidine incorporation into DNA (Swenerton et al. 1969), no evidence has supported the following possibilities: (a) that zinc is a necessary co-factor for thymidine kinase; (b) that zinc concentrations regulate the synthesis and/or degradation of thymidine kinase; or (c) that intracellular thymidine pool sizes and transport are similar between zinc-deficient and restricted-fed controls. These issues need to be clarified before the precise role of zinc in cell division can be delineated. The pathologies associated with impaired cell division are discussed below.

Associated with zinc-deficiency-induced reduction in cell division is impaired growth. This has been shown not only for various animal models (Todd et al. 1934; O'Dell et al. 1958; Miller 1970; Gupta et al. 1985), but also in humans (Prasad et al. 1961; Halsted et al. 1972; Hambidge et al. 1986). It has been suggested that a lower rate of growth makes more zinc available for zinc-requiring metabolic processes such as protein synthesis as well as for the structure and function of zinc metalloproteins. Another possible explanation for the slower growth rate is the suppression of appetite and food intake that occurs with zinc deficiency, hence the animal eats less, grows less and requires less zinc, thus reducing the metabolic costs associated with assimilating, processing and excreting unutilizable nutrients.

Anorexia is one of the more pronounced pathological effects associated with zinc deficiency. Zinc-deficiency-induced anorexia is seen in every animal species including humans (Williams and Chesters 1970; Bakan 1979; Essatara et al. 1984; Kawamoto et al. 1986; Kasarskis et al. 1986; Katz et al. 1987). In fact, it is often difficult for the investigator to separate those effects due to inanition from those unique to zinc deficiency.

Anorexia may represent an adaptation or, more accurately, an accommodation made by an animal exposed to environmental conditions which are not favourable to growth. Adaptation implies that an animal can function normally under these conditions for an extended period of time, while accommodation is the slow deterioration of a normal state with resultant adverse biochemical consequences including death. Anorexia by itself, and in association with zinc deficiency, can reduce the cell proliferation components required for growth; these include growth factors (i.e. nutrients and hormones), membrane receptors, second messengers, activated genes, messenger RNAs, proteins and enzymes. Even though several investigators have made an attempt to separate the phenomena of zinc deficiency and anorexia, it may be of only semantic importance because inanition itself is a consequence of zinc deficiency. Further comments concerning the potential biochemical lesions underlying zinc-induced anorexia are discussed below.

Abnormal bone development, growth and mineralization seen in zinc deficiency may also be a consequence of a zinc-mediated reduction in cell division. Follis (1941) reported that the number of osteoblasts is reduced in zinc-deficient rats. Others have shown reductions in chondrocyte number in epiphyseal cartilage as a consequence of zinc deficiency (Diamond and Hurley 1970; Bergman et al. 1970, 1972).

Beach et al. (1979, 1980a, b) have reported that zinc deficiency during early development can result in a marked alteration in immune function. These authors have shown that zinc deficiency, through its impairment of cellular division, can have a profound effect on the thymus and spleen. These organs, which are closely associated with immune function, can be reduced by as much as 90% relative to control animals. One functional outcome of this reduction in tissue mass is a smaller number of T and B lymphocytes. In addition, immunoglobulin profiles of zinc-deficient mice were abnormal and showed alterations in the concentrations of the various immunoglobulin subclasses. For example, IgA levels were undetectable while IgM and IgG levels were significantly lower or higher, respectively, than in controls. Repletion of the diet with zinc reversed many of these abnormalities; however, persistence of some of these defects could be detected well into adulthood (Fraker et al. 1984).

While the specific role of zinc in cell division is unknown, we will suggest one possible scenario. For this exercise, we will use a model whose framework has been discussed recently by Pardee et al. (1986). Pardee et al. suggest that prior to the incorporation of deoxynucleotides into DNA, a multiprotein "replicase" complex is formed consisting of such enzymes as DNA polymerase, thymidine kinase, thymidine synthetase, ribonucleotide reductase and others. Zinc may play a role in stabilizing this complex and without zinc, the spatial relationships among the various enzymes, co-factors, substrates and products may be altered to an extent prohibiting proper DNA synthesis. If this situation were true, it would take only a conformational alteration of one enzyme to decrease the activities of all other enzymes present in the complex. Pardee et al. have provided evidence that before the replicase can function, a final regulatory protein must be synthesized and integrated into the replicase complex. It is possible that a state of zinc deficiency may preclude or reduce the transcription of mRNA required for the eventual synthesis of adequate amounts of this protein. The potential role of zinc in gene expression is discussed later in this review.

There are several lines of evidence which suggest that zinc regulates both pre- and post-DNA synthesis events in the cell cycle. We have shown (Clegg et al. 1986) that a severe zinc deficiency results in a delayed rate of entry of cells into the S phase (Table 9.1). Falchuk et al. (1975) have also shown with zinc-deficient *Euglena gracilis* that entry of cells from the G_1 to S phase is

Table 9.1. Effects of maternal zinc deficiency on foetal brain-cell cycle. Analysis of gestation day 18

Diet	Stage of cell cycle (%)		
	G_0/G_1	S	G_2/M
100 μg zinc/g diet ad libitum	84.4±1.0	8.0±0.7	7.4±0.8
100 μg zinc/g diet restricted	85.7±1.2	7.4±0.9	6.8±1.0
0.7 μg zinc/g diet ad libitum	89.3±0.8[a]	3.4±0.6[a]	6.4±0.7

From Clegg et al. (1986).
[a] Significantly different from ad libitum and restricted controls, $P < 0.05$.

altered. It would be interesting to determine whether intracellular zinc concentrations rise, as is the case with magnesium ions (Walker 1986), prior to the onset of DNA synthesis. Fujioka and Lieberman (1964) have reported that only zinc but not other elements (including magnesium) can induce DNA synthesis in EDTA-arrested cell cultures.

Zinc regulation of the cell cycle may occur after DNA synthesis at the following levels: (a) chromatin decondensation preceding mitosis and (b) assembly of the mitotic spindle via its effects on microtubule assembly. With respect to the former, Sen and Crothers (1986) and Kvist and Björndahl (1985) have shown that zinc has the propensity to modulate the condensation/decondensation of chromatin. With respect to the latter, it has been known for some time that zinc can stimulate the polymerization of purified tubulin in vitro, and Morisawa and Mohri (1972) have shown in vivo that zinc is associated with sperm tail microtubules. Hesketh (1981) has reported that feeding a diet severely deficient in zinc (< 1 μg zinc/g diet) can reduce the in vitro rate of brain tubulin polymerization in adult rats and pigs. In our laboratory, Oteiza et al. (1988) have extended these observations by showing that feeding a marginally zinc-deficient diet (10 μg zinc/g diet) can also result in lower in vitro rates of brain tubulin polymerization in adult lactating rats. All of the above lines of evidence suggest that zinc plays a critical role in control of the cell cycle, and that a deficiency of this element can result in a disruption of normal cell synthetic processes. With regard to the pathology of zinc deficiency, this effect of zinc on cell replication may be particularly critical in the developing animal as it may lead to asynchrony in cell and tissue maturation.

Gene Expression

An exciting area of recent research has been studying the interaction of zinc with chromatin. Chromatin is the genetic material of eukaryotes and is composed of DNA, small amounts of mRNA and the histone and non-histone proteins. Early studies by Shin and Eichhorn (1968) were directed at studying the interaction of zinc with one component of chromatin, namely DNA. They showed that zinc can stabilize the DNA double helix with respect to thermal denaturation and promote re-annealing of single-stranded DNA into its native double-stranded form by its ability to bind to both the phosphate backbone and nucleoside bases of DNA. Zinc has been shown recently to promote the conformational transformation of DNA from the B to the Z form (Fazakerley 1984). That this transformation occurs in vivo has been demonstrated by Nordheim et al. (1981), who used antibodies specific to the left-handed Z-DNA in *Drosophila* polytene chromosomes. Castro et al. (1986a, b), Castro (1987) and Falchuk et al. (1986) have shown that zinc deficiency makes chromatin (DNA–RNA–protein complex) more resistant to micrococcal nuclease digestion. This implies either an increased ratio of nuclear protein to DNA and/or an alteration in the conformation of zinc-deficient chromatin. Either state would reduce the accessibility of the nuclease to its base recognition sites. Thus, access of RNA polymerase to various genes may be impeded by potential conformational changes occurring in zinc-deficient chromatin.

Earlier studies of Vallee and Falchuk (1981) and Mazus et al. (1984) demonstrated that zinc-deficient *Euglena gracilis* cells are almost totally lacking in histone proteins. Castro et al. (1986a, b) also have shown an alteration in the amounts of histone H1° present in zinc-deficient rat liver. Histones are quite stable by evolutionary standards; point mutations and/or other alterations in their structure and function would be expected to lead to dire consequences for the organism. Since histones are necessary for proper structure and functioning of chromatin (Igo-Kemenes et al. 1982), these alterations created by zinc deficiency may explain some of the aberrant metabolism associated with zinc-deficient organisms. However, at this time, the mechanism by which zinc impacts on histone metabolism is unknown.

Related to the interaction of zinc with DNA and chromatin is the interaction of zinc with various non-histone proteins responsible for gene expression. Falchuk et al. (1985) have shown that zinc-deficient *Euglena gracilis* have an altered expression of RNA polymerases. These authors showed that cells grown in zinc-sufficient media synthesized three distinct RNA polymerases, while cells grown in zinc-deficient media only synthesized one RNA polymerase, which is distinct from the zinc-sufficient polymerases. In this case, it is unknown whether zinc affected transcription of the RNA polymerase genes directly or indirectly.

Other researchers have been examining a direct role for zinc in gene transcription. Hanas et al. (1983, 1985), Wu and Wu (1987) and Wingender et al. (1984) have identified *Xenopus* transcription factor TFIIIA or Factor A as a zinc metalloenzyme. This protein of molecular weight 40 000 contains 2 moles zinc/mole protein and is involved in the regulation of 5S RNA synthesis in *Xenopus* oocytes. Factor A contains an ample amount of cysteine residues which, with histidine residues, are responsible for zinc binding (Miller et al. 1985; Klug and Rhodes 1987). Factor A must contain zinc atoms for proper functioning, as this protein incubated in the presence of the zinc chelator, o-phenanthroline, inhibits the in vitro transcription of the 5S gene. Berg (1986) has reported that, in addition to Factor A, several other nucleic-acid-binding proteins have zinc-binding domains (zinc-binding "fingers"). This suggests that there may be several genes which may be regulated in part or fully by zinc ions.

In summary, zinc has been shown both to be present and to interact with the various chromatin components, namely DNA, RNA, histone and non-histone proteins. Studies are needed to verify that nutritional modulation of in vivo zinc concentrations or specific zinc pools can directly influence the synthesis of nucleic-acid-synthesizing enzymes and/or can affect the regulation of other genes. However, given past and recent evidence, it seems reasonable to expect that at least part of the defects seen in zinc deficiency are a consequence of altered chromatin metabolism. This may be particularly important in the foetus where coordinate timing and expression among genes are essential for normal development.

RNA and Protein Synthesis

Scrutton et al. (1971) were the first to discuss the presence and possible role of zinc in RNA polymerases. Unlike DNA polymerases, RNA polymerases appear to contain more than 1 mole of zinc/mole protein. Hence, it is likely that zinc has both a structural and catalytic function in the RNA polymerases. It is therefore possible that a dietary zinc deficiency could result in reduced activity or function of RNA polymerases. In fact, Terhune and Sandstead (1972) have shown that RNA polymerase activity is significantly lower than normal in liver from the postnatally zinc-deficient rat. Vallee (1959) showed that zinc is a normal constituent of RNA, while Tal (1969) reported that metal ions, including zinc, play an important role in stabilizing ribosome conformation. Both of these macromolecules are present in mammalian pre-initiation and initiation complexes and therefore it is likely that zinc may play an essential role in stabilizing these structures.

These observations may explain why Fosmire et al. (1976) observed abnormal polysome profiles in zinc-deficient rat liver; alternatively, as suggested by the authors, abnormal free amino acid concentrations in plasma may account for the effect. Schneider and Price (1962) reported that *Euglena gracilis* grown in zinc-deficient media have decreased amounts of, and reduced synthetic rates for, RNA. Macapinlac et al. (1968) found reduced levels of RNA in zinc-deficient rat testes although they observed an unaltered ^{14}C adenine incorporation into RNA. Somers and Underwood (1969) also reported decreased RNA content of zinc-deficient testes. Authors of both reports suggested that the primary lesion responsible for lowered RNA content was an increased activity of ribonuclease and that this in turn was regulated by zinc or, in this case, the lack of it. On the other hand, Chesters and Will (1978) suggested that the alteration of tissue ribonuclease activities was part of the generalized response to alterations in growth seen in zinc-deficient animals.

Another possibility, discussed below in more detail, is that the increased ribonuclease activity is part of the alteration seen in nitrogen metabolism and is associated with the catabolic state of the zinc-deficient animal. In contrast, Fosmire et al. (1976) did not observe an increased activity of ribonuclease, yet they did report a lowered RNA content in zinc-deficient rat liver. In a zinc-deficient baby pig model, Prasad et al. (1971) also found lower than normal RNA levels in several tissues, and Falchuk et al. (1978) reported, in addition to abnormal synthesis of RNA polymerase, that the base composition of synthesized RNA was altered in a zinc-deficient *Euglena gracilis* model.

Taken together, this substantial body of information on abnormal gene expression and RNA metabolism in zinc deficiency supports a hypothesis that zinc deficiency does alter protein metabolism. Whereas total amount of protein in zinc-deficient animals is lower than normal because of a lack of growth (i.e. decreased number of cells), it has not been clearly demonstrated that zinc deficiency has a direct effect on protein synthesis at the level of translation. A number of investigators have reported that neither the rate of incorporation of various radioactive amino acids into protein nor the concentration of protein are decreased by a zinc-deficient state (O'Neal et al. 1970; Burke et al. 1981; Southon et al. 1985). However, Hicks and Wallwork (1987) have recently shown in a cell-free system that incorporation of ^{3}H-leucine into protein is reduced by zinc deficiency. They suggest that the defect occurs in the tRNA synthetase fraction and further suggest that one or more of the synthetases is zinc-dependent. It is also possible that one or more of the initiation or elongation factors may be zinc-dependent (i.e. via zinc-binding "fingers").

Hsu et al. (1969) found that the incorporation of ^{14}C-methionine into plasma, liver and kidney protein was reduced in zinc deficiency, although paradoxically, incorporation into pancreatic protein was increased. It is possible that the different hormonal environments in the various tissues result in the differing effects observed for protein synthetic rates. This concept is supported by the recent work done by Giugliano and Millward (1987), who observed different rates of protein synthesis in muscle and thymus tissue. These authors discussed the possible role of corticosterone in regulating protein synthetic and degradation rates in zinc-deficient tissues. Hsu and Woosley (1972) found decreased incorporation of methionine-^{35}S into skin proteins of zinc-deficient rats, while Wallwork and Duerre (1985; Duerre and Wallwork 1986) found higher catabolism and lower incorporation of [carboxyl-^{14}C]-methionine into histone and non-histone proteins of zinc-deficient rats than in pair-fed controls. However, these results must be interpreted with caution, as the authors found no difference in incorporation rates between ad libitum-fed controls and zinc-deficient animals. Duerre et al. (1977) found lower incorporation of ^{14}C-lysine into histones of zinc-deficient rat liver. Incorporation of L-[U-^{14}C]-proline into the epiphysis was significantly reduced in zinc deficiency (Suwarnasarn et al. 1982).

Morgan et al. (1988a, b) have shown that zinc intake plays an important role in determining the composition of gain in mice recovering from early malnutrition. These authors showed that accretion of lean body mass (hence muscle protein) is impaired when marginally zinc-deficient or zinc-deficient diets are fed to previously malnourished mice. A major requirement in all studies concerned with the incorporation and protein synthetic rates is that the specific activity of the precursor pool must remain constant during the period of the measurement. Few of the studies mentioned in this review actually measured the specific activity of the precursor during the period of incorporation. Given the general catabolic state that zinc deficiency induces in mammals, precursor pool sizes of both nucleotides and amino acids may not remain static over the duration of these various studies and therefore incorporation rates may not truly reflect synthesis rates.

Defects in protein synthesis would be detrimental to both the foetus and adult animal. Protein synthesis, in addition to its critical role in tissue maintenance and growth, is important in the regulation of amino acid pool size. Zinc deficiency results in a situation where not only is there a decreased utilization of amino acids for protein synthesis but there is also an increased flux of nitrogen coming from protein and nucleic acid catabolism (Giugliano and Millward 1987). This imposes a heavy nitrogen load on the animal, and this condition is likely to worsen if and when the animal consumes protein in its diet. Hsu and Anthony (1975) have shown that urea excretion is significantly elevated in zinc-deficient animals when compared with ad lib-fed and pair-fed control animals; this is consistent with the increased nitrogen catabolism postulated to occur in a zinc-deficient state. These authors also speculated that there may be a reduction in urea cycle function as a consequence of zinc deficiency. These combined defects may result in an intolerable production of circulating ammonia, which may itself be the signal that triggers the food intake reduction seen so consistently in zinc deficiency.

Elevated ammonia levels have been implicated in the control of food intake (Noda 1975), and Rabbani and Prasad (1977) have reported elevated blood ammonia concentrations in three zinc-deficient humans. Elevated levels and abnormal ratios of plasma amino acids have also been implicated in the regulation of food intake (Peng and Harper 1969; Leung et al. 1968, 1986). In fact, Reeves and O'Dell (1981) have shown abnormal ratios of tyrosine and tryptophan to total neutral amino acids in plasma of zinc-deficient animals, although others have not (Wallwork et al. 1979; Wallwork and Sandstead 1983). The putative elevated ratios may have a subsequent effect on the concentration of the brain monoamines including serotonin, epinephrine and norepinephrine, which are also potential candidates for the regulation of food intake (Halas et al. 1982; Wallwork et al. 1982). In contrast, Kasarskis et al. (1986) did not find any significant changes in tyrosine, dopamine, DOPA, norepinephrine, tryptophan, serotonin or 5-hydroxyindolacetic acid. Instead these authors hypothesize that the anorexic pathology seen in zinc deficiency is the result of a reduction in the activity of the receptor for norepinephrine.

Some interesting food choice experiments have been conducted with zinc-deficient rats to test their preference for protein and/or energy. Chesters and Will (1973) showed that zinc-deficient rats given a choice between a high- and low-protein diet will choose the low-protein diet. These data have been supported by the studies of Reeves and O'Dell (1981). Furthermore, Chesters and Will (1973) and Bettger et al. (1986) have shown that rats placed in a cold environment will eat significantly more of a zinc-deficient diet than if they are at ambient temperatures, presumably because of the increased energy requirements associated with decreased temperatures. However, zinc-deficient animals forced to increase their food intake as a result of cold temperatures show further reductions in plasma zinc when compared with zinc-deficient animals at ambient temperatures. Similarly, when zinc-deficient animals are force-fed zinc-deficient diets, they become sick and eventually die (Flanagan 1984). Chesters and Will (1973) conclude "that the cyclical patterns of food intake associated with zinc deficiency resulted from a slow but effective control of food intake by the energy balance of the animals".

Based on the above observations and arguments, we would like to propose our own hypothesis for the decrease in and cyclicity of food intake seen in zinc deficiency. Initially (approximately 2–3 d), when the animal is fed a zinc-deficient diet, it does not cycle or reduce its food intake. During this time, the animal's plasma zinc is dropping rapidly, but not enough to have a noticeable clinical effect (unless it is a foetus). The decrease in plasma zinc concentration is most likely a signal for the secretion of corticosterone, which is likely to slant the muscle protein synthesis/protein degradation relationship towards degradation, thereby putting the animal into a net catabolic state. This in turn results in an increased nitrogen load to the animal and, in association with hormones (i.e. corticosterone), induces the synthesis of urea cycle enzymes and ribonuclease. After the initial 2–3 d, the animal begins showing the decrease in and cycling of food intake. Protein synthesis is decreased (i.e. from defective development of polysomes) and protein and nucleic acid degradation are increased. The additional nitrogen coming from protein in the diet as the animal eats for energy results in an intolerable ammonia concentration and/or altered plasma amino acid profile. This situation is likely to result in a further drop in plasma zinc (and hence decreases zinc available for other essential functions), as the animal tries to defend protein synthesis which is being stimulated by other components in the diet (i.e. essential amino acids).

In this respect, zinc deficiency is similar to an essential amino acid deficiency in that when an animal is force-fed a diet lacking in one or more essential amino acids it invariably results in a further decrease in the plasma of the essential amino acid lacking in the diet. The decrease in the circulating essential amino acid in the case of the force-fed essential-amino-acid-deficient diet, and the decrease in circulating zinc seen in the case of a force-fed zinc-deficient animal increase the severity of deleterious effects of the diet to the animal and often result in rapid death.

The consequence of the negative events induced by a zinc deficiency is to reduce the animal's desire to eat while tissue catabolism continues to occur. The zinc released from the tissues (primarily muscle) is then used to support the essential protein synthesis required by the animal to survive. This situation may temporarily reprieve the animal with respect to ammonia levels and/or abnormal plasma amino acid ratios. However, the animal must again eat for energy and the cycle continues until essential activities can no longer be supported by the degradation of tissue. At this point the animal dies. Of course, the pair-fed animal is different. Despite its deficit of calories and essential nutrients, it is still receiving nutrients in the proper proportion and no essential nutrient is completely lacking. This allows this type of animal to have the higher feed efficiency ratio seen so often when compared with zinc-deficient animals.

The relationship between zinc deficiency and defective protein synthesis has not been well characterized in humans. However, Golden and Golden (1981a, b, 1985) have found that children recovering from malnutrition have an abnormal body composition. Namely, they show a greater gain in fat than in lean body mass. The Goldens suggest that the rapid accretion of body tissues following the introduction of "adequate diets" to malnourished children may place increased demands on available zinc. This potentially could lead to an induced zinc deficiency with subsequent decreased protein synthesis and lean body mass accrual. Zinc deficiency has been shown in animals to result in impaired wound healing. This may also be important in humans who have a marginal zinc status and/or are recovering from surgery where demands for zinc may be increased by the anabolic processes associated with replacing tissue. The lack of zinc at

the wound site may impair collagen synthesis (Prasad and Oberleas 1973) and/
or decrease cell proliferation required during the epithelialization process.

Other Zinc Metalloenzymes

While it is evident from the above that there are numerous mechanisms by
which zinc deficiency may affect nucleic acid metabolism, protein synthesis and
cell replication, it must also be appreciated that zinc is also known to be a
structural or catalytic component of over 200 other enzymes (Vallee and Galdes
1984). These enzymes cover a broad spectrum of metabolic functions and
include oxidoreductases, transferases, hydrolases, lyases, isomerases and ligases.
However, despite the large number of identified zinc metalloenzymes, to date
there is no clear-cut evidence that a reduction in the activity of any of these
enzymes is a factor underlying the problems associated with zinc deficiency.

Zinc and Membrane Stability

In addition to its catalytic and structural roles in enzymes and nucleic acids,
zinc is also thought to have a critical role in the stabilization of biomembranes.
Chvapil (1973) has suggested that zinc stabilizes membranes by binding to
sulphydryl groups and forming mercaptides. He found that treating a suspension
of erythrocytes in vitro with the zinc-chelating agent, 1,10-phenanthroline,
results in an increased susceptibility of the erythrocytes to haemolysis when
incubated with detergent. When he added zinc back to the medium, the fragility
of the erythrocyte membrane appeared normal. Based on these observations,
Chvapil has proposed that membrane-bound zinc alters the fluidity and
stabilization of membranes. Bettger et al. (1978) have suggested that zinc
deficiency may cause oxidative stress to membranes, structural strains, altered
activities of membrane-bound enzymes and changes in membrane receptors.
It has been suggested that some of the problems associated with zinc deficiency
are a consequence of a reduction in the zinc content in biomembranes, and
indeed, that a reduction in the membrane-bound zinc may represent one of
the earliest biochemical lesions associated with zinc deficiency. Similar to the
findings of Chvapil, O'Dell et al. (1987) have shown also that zinc deficiency
increases the sensitivity of erythrocytes to osmotic shock. Repleting the diet
with zinc quickly reverses this defect. However, in contrast to the findings of
Chvapil, in vitro addition of zinc to zinc-deficient red blood cells had no
protective effect. Recently, Dreosti (1987) has suggested that membrane lipid
damage may also be an important component in the teratogenic pathology of
zinc-deficient foetal brains. This hypothesis was based on the observations
using electron microscopy that in the neural tube of the day 11 zinc-deficient
rat embryo there is serious deterioration of cell membranes, especially those
of the mitochondrion, in the period just preceding cell death.

References

Auld DS, Kawaguchi H, Livingston DM, Vallee BL (1975) RNA-dependent DNA polymerase (reverse transcriptase) from avian myeloblastosis virus: a zinc metalloenzyme. Proc Natl Acad Sci USA 71: 2091–2094

Bakan R (1979) The role of zinc in anorexia nervosa: etiology and treatment. Med Hypotheses 5: 731–736

Beach RS, Gershwin ME, Hurley LS (1979) Altered thymic structure and mitogen responsiveness in postnatally zinc-deprived mice. Dev Comp Immunol 3: 725–738

Beach RS, Gershwin ME, Hurley LS (1980a) Growth and development in postnatally zinc deprived mice. J Nutr 110: 201–211

Beach RS, Gershwin ME, Makishima RK, Hurley LS (1980b) Impaired immunologic ontogeny in postnatal zinc deprivation. J Nutr 110: 805–815

Berg JM (1986) Potential metal-binding domains in nucleic acid binding proteins. Science 232: 485–486

Bergman B, Friberg U, Lohmander S, Oberg T (1970) Morphologic and autoradiographic observations on the effect of zinc deficiency on endochondral growth sites in the white rat. Odontol Rev 21: 379–389

Bergman B, Friberg U, Lohmander S, Oberg T (1972) The importance of zinc to cell proliferation in endochondral growth sites in the white rat. Scand J Dent Res 80: 486–492

Bettger WJ, Fish TJ, O'Dell BL (1978) Effects of copper and zinc status of rats on erythrocyte stability and superoxide dismutase activity. Proc Soc Exp Biol Med 158: 279–282

Bettger WJ, Wong LH, Paterson PG (1986) Effect of environmental temperature on food intake and deficiency signs in rats fed zinc-deficient diets. Nutr Behav 3: 241–249

Burke JP, Fenton MR, Miller ML, Tursi FD (1981) The effect of a zinc-deficient diet and the inflammatory response on rat liver mitochondrial protein synthesis. Biochem Med 25: 48–55

Castro CE (1987) Nutrient effects on DNA and chromatin structure. Ann Rev Nutr 7: 407–421

Castro CE, Alvares OF, Sevall J (1986a) Zinc deficiency decreases histone H1° in rat liver. Nutr Rep Intl 34: 67–74

Castro CE, Armstrong-Major J, Ramirez ME (1986b) Diet-mediated alteration of chromatin structure. Fed Proc 45: 2394–2398

Chesters JK, Quarterman J (1970) Effects of zinc deficiency on food intake and feeding patterns of rats. Br J Nutr 24: 1061–1069

Chesters JK, Will M (1973) Some factors controlling food intake by zinc-deficient rats. Br J Nutr 30: 555

Chesters JK, Will M (1978) Effect of age, weight and adequacy of zinc intake on the balance between alkaline ribonuclease and ribonuclease inhibitor in various tissues of the rat. Br J Nutr 39: 375

Chvapil M (1973) New aspects in the biological role of zinc: a stabilizer of macromolecules and biological membranes. Life Sci 13: 1041–1049

Clegg MS, Rogers JM, Zucker RM, Hurley LS, Keen CL (1986) Flow cytometry analysis of cell cycle stages in zinc deficient fetal rat brain. Fed Proc 45: 1086

Diamond I, Hurley LS (1970) Histopathology of zinc-deficient fetal rats. J Nutr 100: 325–329

Diamond I, Swenerton H, Hurley LS (1971) Testicular and esophageal lesions in zinc-deficient rats and their reversibility. J Nutr 101: 77–84

Dreosti IE (1987) Micronutrients, superoxide and the fetus. Neurotoxicology 8: 445–450

Dreosti IE, Hurley LS (1975) Depressed thymidine kinase activity in zinc-deficient rat embryos. Proc Soc Exp Biol Med 150: 161–165

Dreosti IE, Grey PC, Wilkins PJ (1972) Deoxyribonucleic acid synthesis, protein synthesis and teratogenesis in zinc-deficient rats. S Afr Med J 46: 1585–1588

Dreosti IE, Record IR, Manuel SJ (1980) Incorporation of ³H-thymidine into DNA and the activity of alkaline phosphatase in zinc-deficient fetal rat brains. Biol Trace Element Res 2: 21–29

Dreosti IE, Record IR, Manuel SJ (1985) Zinc deficiency and the developing embryo. Biol Trace Element Res 7: 103–122

Duerre JA, Wallwork JC (1986) Methionine metabolism in isolated perfused livers from rats fed zinc-deficient and restricted diets. Br J Nutr 56: 395–405

Duerre JA, Ford KM, Sandstead HH (1977) Effect of zinc deficiency on protein synthesis in brain and liver of suckling rats. J Nutr 107: 1082–1093

Duncan JR, Dreosti IE (1974) The effect of zinc deficiency on the timing of deoxyribonucleic acid synthesis in regenerating rat liver. S Afr Med J 48: 1697–1699

Duncan JR, Hurley LS (1978) Thymidine kinase and DNA polymerase activity in normal and zinc deficient developing rat embryos. Proc Soc Exp Biol Med 159: 39–43

Ecker RI, Schroeter AL (1978) Acrodermatitis and acquired zinc deficiency. Arch Dermatol 114: 937–939

Eckhert CD, Hurley LS (1977) Reduced DNA synthesis in zinc deficiency: regional differences in embryonic rats. J Nutr 107: 855–861

Essatara M'B, Levine AS, Morley JE, McClain CJ (1984) Zinc deficiency and anorexia in rats: normal feeding patterns and stress induced feeding. Physiol Behav 32: 469–474

Falchuk KH, Krishan A, Vallee BL (1975) DNA distribution in the cell cycle of *Euglena gracilis*. Cytofluorometry of zinc deficient cells. Biochemistry 14: 3439–3444

Falchuk KH, Hardy C, Ulpino L, Vallee BL (1978) RNA metabolism, manganese, and RNA polymerases of zinc-sufficient and zinc-deficient *Euglena gracilis*. Proc Natl Acad Sci USA 75: 4175–4179

Falchuk KH, Mazus B, Ber E, Ulpino-Lobb L, Vallee BL (1985) Zinc deficiency and the *Euglena gracilis* chromatin: formation of an α-amanitin-resistant RNA polymerase II. Biochemistry 24: 2576–2580

Falchuk KH, Gordon PR, Stankiewicz A, Hilt KL, Vallee BL (1986) *Euglena gracilis* chromatin: comparison of effects of zinc, iron, magnesium, or manganese deficiency and cold shock. Biochemistry 25: 5388–5391

Fazakerley GV (1984) Zinc Z-DNA. Nucleic Acids Res 12: 3643–3648

Fernandez-Madrid F, Prasad AS, Oberleas D (1973) Effect of zinc deficiency on nucleic acids, collagen, and noncollagenous protein of the connective tissue. J Lab Clin Med 82: 911–916

Flanagan PR (1984) A model to produce pure zinc deficiency in rats and its use to demonstrate that dietary phytate increases the excretion of endogenous zinc. J Nutr 114: 493–502

Follis RH Jr, Day HG, McCollum EV (1941) Histologic studies of the tissues of rats fed a diet extremely low in zinc. J Nutr 22: 223–233

Fosmire GJ, Fosmire MA, Sandstead HH (1976) Zinc deficiency in the weanling rat: effects on liver composition and polysomal profiles. J Nutr 106: 1152–1158

Fraker PJ, Hildebrandt K, Luecke KW (1984) Alteration of antibody-mediated responses of suckling mice to T-cell dependent and independent antigens by maternal zinc deficiency: restoration of responsivity by nutritional repletion. J Nutr 114: 170–177

Fujii T (1954) Presence of zinc in nucleoli and its possible role in mitosis. Nature 174: 1108–1109

Fujioka M, Lieberman I (1964) A Zn^{++} requirement for synthesis of deoxyribonucleic acid by rat liver. J Biol Chem 239: 1164–1167

Giugliano R, Millward DJ (1987) The effects of severe zinc deficiency on protein turnover in muscle and thymus. Br J Nutr 57: 139–155

Golden BE, Golden MHN (1981a) Plasma zinc, rate of weight gain, and energy cost of tissue deposition in children recovering from severe malnutrition on a cow's milk or soya protein based diet. Am J Clin Nutr 34: 892–899

Golden BE, Golden MHN (1985) Effect of zinc supplementation on the composition of newly synthesized tissue in children recovering from malnutrition. Proc Nutr Soc 44: 110A

Golden MHN, Golden BE (1981b) Effects of zinc supplementation on dietary intake, rate of weight gain, and energy cost of tissue deposition in children recovering from severe malnutrition. Am J Clin Nutr 34: 900–908

Gonzalez JR, Vazquez Botet M, Sanchez JL (1982) The histopathology of acrodermatitis enteropathica. Am J Dermatol 4: 303–311

Grey PC, Dreosti IE (1972) Deoxyribonucleic acid and protein metabolism in zinc-deficient rats. J Comp Pathol 82: 223–228

Gupta RP, Verma PC, Gupta PRK (1985) Experimental zinc deficiency in guinea pigs: clinical signs and some haematological studies. Br J Nutr 54: 421–428

Halas ES, Wallwork JC, Sandstead HH (1982) Mild zinc deficiency and undernutrition during prenatal and postnatal periods in rats: effects on weight, food consumption, and brain catecholamine levels. J Nutr 112: 542–551

Halsted JA, Ronaghy HA, Abadi P et al. (1972) Zinc deficiency in man. Am J Med 53: 277–284

Hambidge KM, Casey CE, Krebs NF (1986) Zinc. In: Mertz W (ed) Trace elements in human and animal nutrition, vol 2. Academic Press, Orlando San Diego New York Austin London Montreal Sydney Tokyo Toronto, pp 1–137

Hanas JS, Hazuda DJ, Bogenhagen DF, Wu FY-H, Wu C-W (1983) *Xenopus* transcription factor A requires zinc for binding to the 5 S RNA gene. J Biol Chem 258: 14120–14125

Hanas JS, Hazuda DJ, Wu C-W (1985) *Xenopus* transcription factor A promotes DNA reassociation. J Biol Chem 260: 13316–13320

Hesketh JE (1981) Impaired microtubule assembly in brain from zinc deficient pigs and rats. Int J Biochem 13: 921–926

Hicks SE, Wallwork JC (1987) Effect of dietary zinc deficiency on protein synthesis in cell-free systems isolated from rat liver. J Nutr 117: 1234–1240

Hsu JM, Anthony WL (1975) Effect of zinc deficiency on urinary excretion of nitrogenous compounds and liver amino acid-catabolizing enzymes in rats. J Nutr 105: 26–31

Hsu JM, Woosley RL (1972) Metabolism of L-methionine- ^{35}S in zinc-deficient rats. J Nutr 102: 1181–1186

Hsu JM, Anthony WL, Buchanan PJ (1969) Zinc deficiency and incorporation of ^{14}C-labeled methionine into tissue proteins in rats. J Nutr 99: 425–432

Hurley LS (1981) Teratogenic aspects of manganese, zinc, and copper nutrition. Physiol Rev 61: 249–295

Hurley LS, Shrader RE (1972) Congenital malformations of the nervous system in zinc-deficient rats. Int Rev Neurobiol [Suppl] 1: 7–51

Igo-Kemenes T, Hórz W, Zachau HG (1982) Chromatin. Annu Rev Biochem 51: 89–121

Kasarskis EJ, Sparks DL, Slevin JT (1986) Changes in hypothalamic noradrenergic systems during the anorexia of zinc deficiency. Biol Trace Element Res 9: 25–35

Katz RL, Keen CL, Litt IF, Hurley LS, Kellams-Harrison KM, Glader LJ (1987) Zinc deficiency in anorexia nervosa. J Adolesc Health Care 8: 400–406

Kawamoto JC, Castonguay TW, Keen CL, Stern JS, Hurley LS (1986) Age, sex and reproductive status alter the severity of anorexia in zinc deficient rats. Physiol Behav 38: 485–493

Klug A, Rhodes D (1987) "Zinc fingers": a novel protein motif for nucleic acid recognition. TIBS 12: 464–469

Kvist U, Björndahl L (1985) Zinc preserves an inherent capacity for human sperm chromatin decondensation. Acta Physiol Scand 124: 195–200

Leung PMB, Rogers QR, Harper AE (1968) Effect of amino acid imbalance on dietary choice in the rat. J Nutr 95: 482–492

Leung PMB, Larson DM, Rogers QR (1986) Influence of taste on dietary choice of rats fed amino acid imbalanced or deficient diets. Physiol Behav 38: 255–264

Lieberman I, Ove P (1962) Deoxyribonucleic acid synthesis and its inhibition in mammalian cells cultured from the animal. J Biol. Chem 237: 1634–1642

Lieberman I, Abrams R, Hunt N, Ove P (1963) Levels of enzyme activity and deoxyribonucleic acid synthesis in mammalian cells cultured from the animal. J Biol Chem 238: 3955–3962

Macapinlac MP, Pearson WN, Barney GH, Darby WJ (1968) Protein and nucleic acid metabolism in the testes of zinc-deficient rats. J Nutr 95: 569–577

Mazus B, Falchuk KH, Vallee BL (1984) Histone formation, gene expression, and zinc deficiency in *Euglena gracilis*. Biochemistry 23: 42–47

Miller JA, McLachlan D, Klug A (1985) Repetitive zinc-binding domains in the protein transcriptive factor IIIA from *Xenopus* oocytes. EMBO J 4: 1609–1614

Miller WJ (1970) Zinc nutrition in cattle: a review. J Dairy Sci 53: 1123–1135

Morgan PN, Keen CL, Lönnerdal B (1988a) The effect of varying dietary zinc intake of weanling mouse pups during recovery from early undernutrition on tissue mineral concentrations, relative organ weights, hematological variables, and muscle composition. J Nutr (in press)

Morgan PN, Keen CL, Calvert CC, Lönnerdal B (1988b) The effect of varying dietary zinc intake of weanling mouse pups during recovery from early undernutrition on growth, body composition and composition of gain. J Nutr (in press)

Morisawa M, Mohri H (1972) Heavy metals and spermatozoan motility. Exp Cell Res 70: 311–316

Noda K (1975) Possible effect of blood ammonia on food intake of rats fed amino acid imbalanced diets. J Nutr 105: 508–516

Nordheim A, Pardue ML, Lafer EM, Möller A, Stollar BD, Rich A (1981) Antibodies to left handed Z-DNA bind to interband regions of *Drosophila* polytene chromosomes. Nature 294: 417–422

O'Dell BL, Newberne PM, Savage JE (1958) Significance of dietary zinc for growing chickens. J Nutr 65: 503–524

O'Dell BL, Browning JD, Reeves PG (1987) Zinc deficiency increases the osmotic fragility of rat erythrocytes. J Nutr 117: 1883–1889

O'kada A, Takagi Y, Itakura T et al. (1976) Skin lesions during intravenous hyperalimentation: zinc deficiency. Surgery 80: 629–635

O'Neal RM, Pla GW, Spivey Fox MR, Gibson FS, Fry BE Jr (1970) Effect of zinc deficiency and restricted feeding on protein and ribonucleic acid metabolism of rat brain. J Nutr 100: 491–497

Oteiza PI, Hurley LS, Lönnerdal B, Keen CL (1988) Marginal zinc deficiency affects maternal brain microtubule assembly in rats. J Nutr (submitted)

Pardee AB, Coppock DL, Yang HC (1986) Regulation of cell proliferation at the onset of DNA synthesis. J Cell Sci [Suppl] 4: 171–180

Peng Y, Harper AE (1969) Amino acid balance and food intake. Effect of amino acid infusion on plasma amino acids. J Physiol 217: 1441–1445

Prasad AS, Oberleas D (1973) Thymidine kinase activity and incorporation of thymidine into DNA in zinc-deficient tissue. J Lab Clin Med 83: 634–639

Prasad AS, Halsted JA, Nadimi M (1961) Syndrome of iron deficiency anemia, hepatosplenomegaly, hypogonadism, dwarfism and geophagia. Am J Med 31: 532–546

Prasad AS, Oberleas D, Miller ER, Luecke RW (1971) Biochemical effects of zinc deficiency: changes in activities of zinc-dependent enzymes and ribonucleic acid and deoxyribonucleic acid content of tissues. J Lab Clin Med 77: 144–152

Rabbani P, Prasad AS (1977) Effect of zinc deficiency on blood urea, plasma ammonia, and liver ornithine carbamyl transferase activity in male rats. Fed Proc 36: 1139

Record IR, Dreosti IE (1979) Effects of zinc deficiency on liver and brain thymidine kinase activity in the fetal rat. Nutr Rep Int 20: 749–755

Reeves PG, O'Dell BL (1981) Short-term zinc deficiency in the rat and self-selection of dietary protein level. J Nutr 111: 375–383

Rubin H (1973) pH, serum and Zn^{++} in the regulation of DNA synthesis in cultures of chick embryo cells. J Cell Physiol 82: 231–238

Sandstead HH, Rinaldi RA (1969) Impairment of deoxyribonucleic acid synthesis by dietary zinc deficiency in the rat. J Cell Physiol 73: 81–84

Sandstead HH, Gillespie DD, Brady RN (1972) Zinc deficiency: effect on brain of the suckling rat. Pediatr Res 6: 119–125

Schneider E, Price CA (1962) Decreased ribonucleic acid levels: a possible cause of growth inhibition in zinc deficiency. Biochim Biophys Acta 55: 406–408

Scrutton ML, Wu CW, Goldwalt DA (1971) The presence and possible role of zinc in RNA polymerase obtained from E. coli. Proc Natl Acad Sci USA 68: 2497–2502

Sen D, Crothers DM (1986) Condensation of chromatin: role of multivalent cations. Biochemistry 25: 1495–1503

Shin YA, Eichhorn GL (1968) Interactions of metal ions with polynucleotides and related compounds. XI. The reversible unwinding and rewinding of deoxyribonucleic acid by zinc (II) ions through temperature manipulation. Biochemistry 7: 1026–1032

Slater JP, Milduan AS, Loeb LA (1971) Zinc in DNA polymerases. Biochem Biophys Res Commun 44: 27–32

Somers M, Underwood EJ (1969) Ribonuclease activity and nucleic acid and protein metabolism in the testes of zinc-deficient rats. Aust J Biol Sci 22: 1277–1282

Southon S, Livesey G, Gee JM, Johnson IT (1985) Intestinal cellular proliferation and protein synthesis in zinc-deficient rats. Br J Nutr 53: 595–603

Springgate CF, Milduan AS, Abramson R, Engle JL, Loeb LA (1973) Escherichia coli deoxyribonucleic acid polymerase I, a zinc metalloenzyme. Nuclear quadrupolar relaxation studies of the role of bound zinc. J Biol Chem 249: 5987–5991

Suwarnasarn A, Wallwork JC, Lykken GI, Low FN, Sandstead HH (1982) Epiphyseal plate development in the zinc-deficient rat. J Nutr 112: 1320–1328

Swenerton H, Shrader R, Hurley LS (1969) Zinc-deficient rat embryos: reduced thymidine incorporation. Science 166: 1014–1015

Tal M (1969) Metal ions and ribosomal conformation. Biochim Biophys Acta 195: 76–86

Terhune MW, Sandstead HH (1972) Decreased RNA polymerase activity in mammalian zinc deficiency. Science 177: 68–69

Todd WR, Elvehjem CA, Hart EB (1934) Zinc in the nutrition of the rat. Am J Physiol 107: 146–156

Vallee BL (1959) Biochemistry, physiology, and pathology of zinc. Physiol Rev 39: 443–490

Vallee BL, Falchuk KH (1981) Zinc and gene expression. Philos Trans R Soc Lond [Biol] 294: 185–197

Vallee BL, Galdes A (1984) The metallobiochemistry of zinc enzymes. In: Meister A (ed) Advances in enzymology, vol 56. John Wiley, New York Chichester Brisbane Toronto Singapore, pp 283–430

Van Vloten WA, Bos LP (1978) Skin lesions in acquired zinc deficiency due to parenteral nutrition. Dermatologica 156: 175–183

Walker GM (1986) Magnesium and cell cycle control: an update. Magnesium 5: 9–23

Wallwork JC, Duerre JA (1985) Effect of zinc deficiency on methionine metabolism, methylation reactions and protein synthesis in isolated perfused rat liver. J Nutr 115: 252–262

Wallwork JC, Sandstead HH (1983) Effect of zinc deficiency on appetite and free amino acid concentrations in rat brain. J Nutr 113: 47–54

Wallwork JC, Fosmire EJ, Sandstead HH (1979) Cyclic feeding patterns and plasma amino acid concentrations in zinc deficient rats. Fed Proc 38: 606

Wallwork JC, Boten JH, Sandstead HH (1982) Influence of dietary zinc on rat brain catecholamines. J Nutr 112: 514–519

Williams RB, Chesters JK (1970) The effects of early zinc deficiency on DNA and protein synthesis in the rat. Br J Nutr 24: 1153–1156

Wingender E, Dilloo D, Seifart KH (1984) Zinc ions are differentially required for the transcription of ribosomal 5S RNA and tRNA in a HeLa-cell extract. Nucleic Acids Res 12: 8971–8985

Wu FY-H, Wu C-W (1987) Zinc in DNA replication and transcription. Annu Rev Nutr 7: 251–272

Chapter 10

Zinc and Iron in Free Radical Pathology and Cellular Control

R.L. Willson

Die Gegenwart eines Zinksalzes verhindert die Farbenreaktion mit Eisen-chlorid, aber nicht mit Kupfer- oder Kobaltsalz

(Loven 1884)

In parts of New Zealand, the government is said to have saved millions of dollars by dosing sheep and cattle with zinc between January and April. In London, thanks to daily zinc supplements, an acrodermatitis enteropathica patient, who 30 years ago would not have been expected to live beyond infancy, recently gave birth to a healthy child (Brenton et al. 1981). In Switzerland, animals have survived normally lethal doses of radiation by previously being given zinc aspartate.

Zinc has been used as a medicine for centuries. Zinc oxide was mentioned as a constituent of a healing ointment by Pliny the Elder. Zinc carbonate (calamine) lotion is still popular for treating sunburn and zinc salts for eye inflammations. The epidermis, nails, hair and parts of the eye (tissues exposed to light) contain relatively high levels of zinc. High levels are also associated with spermatozoa.

In this chapter the possible free radical mechanisms underlying these observations will be discussed in the light of recent reports of the presence of carefully ordered zinc- and cysteine-containing domains in several regulatory proteins and the proposal presented at a Ciba Foundation Symposium in 1976, that zinc in conjunction with thiols has an important protective/stimulatory role in cellular control (Willson 1977a; Williams 1984) (Fig. 10.1).

Free Radicals and Disease

The possible importance of superoxide and other highly reactive free radicals in the development of disorders such as inflammation and cancer, and in tissue

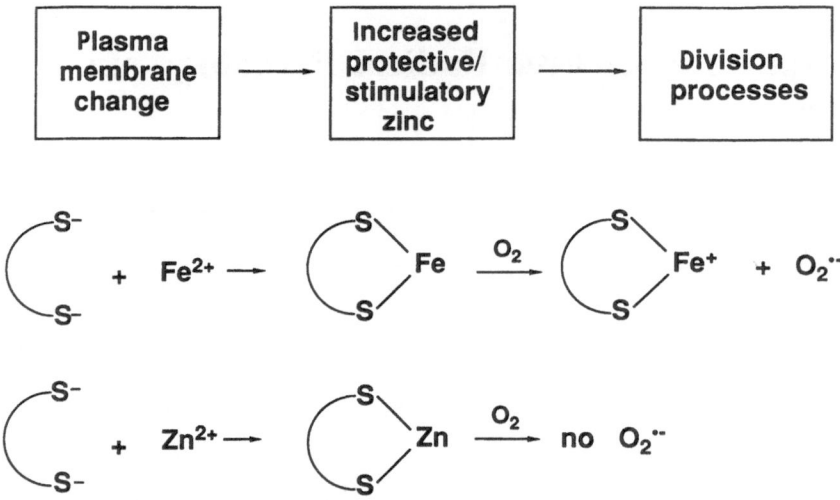

Fig. 10.1. Zinc as a protective/stimulatory factor in cell control (adapted from Willson 1977a).

injury following heart attack, transplant surgery or certain types of chemical poisoning, is presently attracting considerable interest. Although in the healthy body carefully controlled reactions catalysed by enzymes continue to be the rule, it does seem increasingly likely that in ill-health, non-enzymic iron-catalysed free radical oxidation processes can sometimes play a critical, if not a central, role.

Of course to nutritionists, such interest in non-enzymic oxidations is nothing new. The fact that unsaturated fats are particularly susceptible to oxidation by metal-catalysed free radical chain reactions and that foodstuffs often go rancid and deteriorate on storage has long been the bane of the food industry. Today, however, interest is focused more on the eater than on the eaten. We contain unsaturated fat, iron and oxygen so why don't we go rancid? Or put another way, why doesn't living matter go rancid until it in turn becomes food?

Such questions are providing much of the impetus for the current surge of interest in the role of free radical reactions, hydrogen peroxide and what has been variously described as "decompartmentalized", "ill-placed", "free" or "low-molecular-weight chelatable" iron. Fortunately, the healthy body does seem to have a fine armoury of defence mechanisms, continually alert to minimize the damaging effects of free radicals. Unfortunately, however, it is also apparent that such defences may sometimes be overwhelmed, perhaps through the action of light or radiation, a toxic chemical, infection or the breakdown of the cell's structural compartments. Serious pathological changes may then occur even in the previously healthiest of cells. Indeed, should the normal defence mechanism also be weakened by a nutritional or genetic deficiency then the onset of a damaging downward spiral of events may be greatly accelerated (Willson 1977b, 1987; Slater 1984).

How then might zinc be involved in free radical biochemistry and free radical pathology? What is the evidence for the metal having any significant protective and controlling role? To answer these questions let us first briefly remind ourselves of what free radicals are, how they are formed and what they can do.

What are Free Radicals?

In the present context a free radical is best considered simply as an atom or group of atoms containing an unpaired electron. In theory therefore, radicals might be formed readily by the homolytic cleavage of normal two-electron covalent bonds. For example, if as is the usual convention, the free radical is depicted with a bold superscript dot then:

$$H_2O = H^\cdot \text{ and } OH^\cdot$$
$$CH_4 = CH_3^\cdot \text{ and } H^\cdot$$
$$NADH = NAD^\cdot \text{ and } H^\cdot$$
$$H_2O_2 = HO_2^\cdot \text{ and } H^\cdot \text{ or } H_2O_2 = 2OH^\cdot$$

In living systems, however, conditions are such that simple homolytic bond cleavage does not take place unless systems are exposed to extremely high temperatures or to ultraviolet, ionizing or high-intensity microwave or ultrasonic radiation. Nevertheless, free radicals can still be formed by the addition or removal of a single electron, in other words by one-electron oxidation or one-electron reduction.

For example, with oxygen, with NAD^+ or with the herbicide and lung toxin paraquat (PQ^{2+}), one-electron reduction can readily occur in certain situations and it is by this route that many biological free radicals are formed:

$$O_2 + e \rightarrow O_2^{\cdot -}$$
$$NAD^+ + e \rightarrow NAD^\cdot$$
$$PQ^{2+} + e \rightarrow PQ^{\cdot +}$$

Free radicals, therefore, are not always neutral species. They may be positively or negatively charged; indeed like normal molecules they can take part in simple acid–base dissociation equilibria. For example, the superoxide radical ($O_2^{\cdot -}$) exists in equilibrium with its acid form, the hydroperoxyl free radical ($HO_2^\cdot$) (Bielski 1985):

$$O_2^{\cdot -} + H^+ \rightarrow HO_2^\cdot \quad pK = 4.8$$

How are Biological Free Radicals Formed?

The biological one-electron reductants and one-electron oxidants able to convert normal molecules into free radicals can be of two types: enzymic and non-enzymic. Although the net loss or gain of an even number of reducing equivalents, be they electrons or hydrogen atoms, is the norm in many enzyme-catalysed reactions, there is growing evidence that in some enzyme systems one-electron equivalent free radical intermediates are formed which under some circumstances can enter into diversionary non-enzymic pathways.

Enzymic Free Radical Formation by One-Electron Donation

$$\text{xanthine} + 2O_2 + H_2O \xrightarrow{\text{xanthine oxidase}} \text{urate} + 2O_2{}^{\cdot-} + 2H^+$$

$$\text{NADPH} + 2O_2 \xrightarrow{\text{neutrophil NADPH oxidase}} \text{NADP}^+ + 2O_2{}^{\cdot-} + H^+$$

$$\text{NADPH} + 2\text{ quinone} \xrightarrow{\text{cytochrome P450 reductase}} \text{NADP}^+ + 2\text{ semiquinone}^{\cdot-} + H^+$$

Chemists have long known that a large number of organic molecules can be readily oxidized or reduced by those transition metals such as iron, copper, cobalt or manganese whose ions can readily undergo one-electron oxidation or reduction between stable oxidation states. In the body the redox properties of such metals are usually carefully controlled through their complexation to specially designed proteins. However there are strong indications that in the case of iron, situations may occur where a metal complex ($Fe^{2+}X$ or $Fe^{3+}X$) reacts with oxygen, a peroxide (H_2O_2, RO_2H, $ROOR$) or some other electron acceptor and as a result free radicals are formed. Whether or not such reactions do occur at a significant rate depends on the nature of the acceptor and of the complexing ligand(s), X.

Non-enzymic Free Radical Formation by One-Electron Reduction

$$Fe^{2+}X + O_2 \quad \rightarrow Fe^{3+}X + O_2{}^{\cdot-}$$
$$Fe^{2+}X + H_2O_2 \rightarrow Fe^{3+}X + OH^{\cdot} + OH^-$$
$$Fe^{2+}X + ROOH \rightarrow Fe^{3+}X + RO^{\cdot} + OH^-$$
$$Fe^{2+}X + ROOR \rightarrow Fe^{3+}X + RO^{\cdot} + RO^-$$

What can Free Radicals Do?

Because of their unpaired electrons free radicals are, with few notable exceptions, highly reactive. As a result they may:

1. react with another free radical with whom they can share odd electrons (covalent binding or dimerization when both radicals are the same);
2. react with another free radical and donate or receive an odd electron (dismutation when both radicals are of the same type);
3. react with a normal molecule and either donate their odd electron or accept one, possibly in conjunction with the transfer of a proton;
4. add on to a normal molecule so forming a larger and chemically different species: the rapid addition of oxygen to many organic radicals and the addition of the hydroxyl free radical to unsaturated ethylenic and aromatic double bonds are perhaps the most important examples of this type of reaction;

5. break down immediately they are formed, as occurs with the electron adduct radicals of CCl_4 or H_2O_2 (dissociative electron capture).

Typical Free Radical Reactions

a: Radical–Radical Covalent Binding

$$CH_3{}^{\cdot} + CH_3{}^{\cdot} \rightarrow C_2H_6$$
$$\text{protein chain}{}^{\cdot} + \text{DNA}^{\cdot} \rightarrow \text{DNA} - \text{protein (cross-linking)}$$

b: Radical–Radical Electron or Hydrogen Transfer

$$RO_2{}^{\cdot} + O_2{}^{\cdot\ -} \rightarrow RO_2{}^- + O_2$$
$$O_2{}^{\cdot -} + O_2{}^{\cdot\ -} \rightarrow O_2 + O_2{}^{2-} \text{ (dismutation)}$$
$$CH_2OH^{\cdot} + CH_2OH^{\cdot} \rightarrow \quad CH_3OH + HCHO \text{ (dismutation)}$$

c: Radical–Molecule Electron Transfer (Oxidizing Radical)

$$OH^{\cdot} + RS^- \rightarrow OH^- + RS^{\cdot}$$
$$CCl_3O_2{}^{\cdot} + \text{aminopyrine} \rightarrow CCl_3O_2{}^- + \text{aminopyrine}^{\ +}$$

d: Radical–Molecule Electron Transfer (Reducing Radical)

$$NAD^{\cdot} + O_2 \rightarrow NAD^+ + O_2{}^{\cdot\ -}$$
$$O_2{}^{\cdot -} + \text{cyt III C} \rightarrow O_2 + \text{cyt II C}$$

e: Radical–Molecule Hydrogen Transfer (Abstraction)

$$OH^{\cdot} + C_2H_5OH \rightarrow H_2O + C_2H_4OH^{\cdot}$$
$$\text{glucose}(-H)^{\cdot} + GSH \rightarrow \text{glucose} + GS^{\cdot}$$

f: Radical–Molecule Addition

$$CH_3{}^{\cdot} + O_2 \rightarrow CH_3O_2{}^{\cdot}$$
$$OH^{\cdot} + \text{DNA base} \rightarrow \text{DNA base(OH)}^{\cdot}$$

g: Dissociative Electron Capture

$$H_2O_2 + e^- \rightarrow H_2O_2{}^{\cdot} - \rightarrow OH^{\cdot} + OH$$
$$CCl_4 + e^- \rightarrow CCl_4{}^{\cdot\ -} \rightarrow CCl_3{}^{\cdot} + Cl^-$$

Clearly, once free radicals are formed in biological systems, a host of different free radical reactions can take place. As to which reaction(s) predominate will depend on the concentration of the solutes present and the respective rate constants of reaction with the particular free radical in question.

Whether any of the above reactions results in biological damage will depend on how vital a particular damaged molecule is and the subsequent reactions the new radical derived from it can undergo. Damage to some molecules such as DNA or perhaps a key protein, be it an enzyme, an enzyme inhibitor or another cellular factor whose normal rate of synthesis is comparatively slow, may be very significant. On the other hand, the destruction of a few molecules of glucose is hardly likely to be noticed unless the products of such reactions are in themselves toxic.

However it must not be forgotten that in organic chemistry when one free radical reacts with another organic molecule, another free radical is generally formed and the reactions of this radical must be taken into consideration. Only when a free radical reacts with another free radical or with a transition metal able to donate or accept an odd electron is the free radical sequence terminated.

$$O_2{}^{\cdot-} + Fe^{3+}X \rightarrow O_2 + Fe^{2+}X$$
$$O_2{}^{\cdot-} + Cu^{2+}X \rightarrow O_2 + Cu^{+}X$$

Indeed, if a subsequent radical itself reacts with the original type of molecule then an extensive chain reaction may result with the initial damage being amplified considerably and far outstripping that anticipated from the initial radical concentration (Fig. 10.2).

If iron is present then oxidation and reduction reactions with the metal are also possible, reactions with peroxide products further amplifying the situation by causing more free radicals to be created.

Although relatively stable molecules will eventually be formed, on occasions these may in themselves be toxic. As they are likely to be able to diffuse greater distances than their free radical forebears, a widening of the overall damage may result. Some products of lipid peroxidation, for example hydroxynonenal, have been shown to be extremely cytotoxic (Slater 1984). Others may have clastogenic or chemotactic properties (Petrone et al. 1980; Emerit et al. 1982).

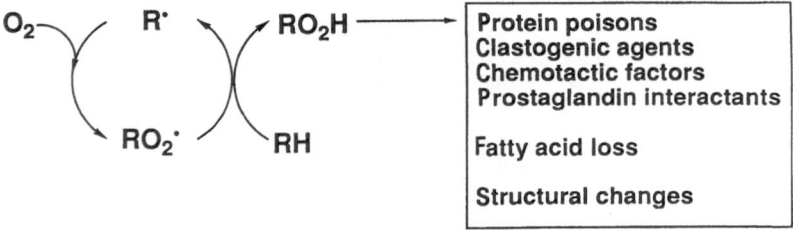

Fig. 10.2. Lipid peroxidation: formation of lipid hydroperoxides and subsequent products.

Protection against Free Radical Damage

In view of the above disturbances which free radicals can cause, it is perhaps not surprising that the cell seems to have developed a number of different protective measures to pre-empt or combat any free radical onslaught. These can be divided into four main types:

1. those aimed at *free radical prevention* including the maintenance of cellular structure, its compartmentalization and the prevention of geographical association between substances that might react to form free radicals; the effective control of iron distribution and the innocuous destruction of peroxides by catalase or by the selenium-containing glutathione peroxidase are included in this category;
2. those aimed at *free radical scavenging* including the maintenance of effective levels of anti-oxidants such as vitamin E, vitamin C, β-carotene and glutathione, as well as of the enzyme superoxide dismutase;
3. those aimed at *free radical repair* in particular the maintenance of effective levels of glutathione;
4. those involved in *nucleic acid repair* such as the polymerase enzymes.

Does zinc have a role in such defences? In biological systems, zinc itself cannot accept or donate reducing equivalents and therefore, unlike iron or copper, cannot interact directly in this manner with free radicals. However, its ubiquitous nature, its presence in over 100 enzymes and its widespread importance as a structural component does mean that it will undoubtedly have at least an indirect role in protecting against free radicals. But does it have a more specific role? There seem to be five possibilities. It may have a role:

1. as a factor in the maintenance of appropriate levels of metallothionein, an SH-rich protein thought to be a good free radical scavenger and repair agent;
2. as an essential component of superoxide dismutase, the free radical-destroying enzyme;
3. as a protective "mask" for thiol or other groups preventing their interaction with iron which could result in direct free radical formation or subsequent decompartmentalization of the metal;
4. as an essential component of nucleic acid repair enzymes;
5. as a stimulatory, triggering factor controlling cell processes until such a time that any possibility of damaging free radical reactions taking place is at a minimum.

Zinc and Metallothionein in Protection against Free Radical Damage

It is well-known that levels of the low-molecular-weight protein, metallothionein, are closely interwoven with zinc metabolism. Metallothionein is widely distributed within the body. High levels are present in the liver, kidney and intestine. Formation of the protein can be induced by increasing the zinc intake and is reduced in zinc deficiency. Since it contains an abundance of thiol groups the possibility of it scavenging oxygen free radicals or repairing organic free radicals in a similar manner to glutathione, must be considered (Thornalley and Vasak 1985; Greenstock 1987). Typical "sacrificial" reactions summarized below illustrate the opportunities for protection of [RH] by free radical attack upon metallothionein [metSH]:

$$O_2^{\cdot -} + H^+ \rightarrow HO_2^{\cdot}$$
$$HO_2^{\cdot} + RH \rightarrow H_2O_2 + R^{\cdot} \qquad \text{damage}$$
$$HO_2^{\cdot} + metSH \rightarrow H_2O_2 + metS^{\cdot} \qquad \text{protection}$$

$$OH^{\cdot} + RH \rightarrow H_2O + R^{\cdot} \qquad \text{damage}$$
$$OH^{\cdot} + metSH \rightarrow H_2O + metS^{\cdot} \qquad \text{protection}$$

$$R^{\cdot} + O_2 \rightarrow RO_2^{\cdot}$$
$$RO_2^{\cdot} + RH \rightarrow RO_2H + R^{\cdot} \qquad \text{damage}$$
$$RO_2^{\cdot} + metSH \rightarrow RO_2H + metS^{\cdot} \qquad \text{protection}$$
$$R^{\cdot} + metSH \rightarrow RH + metS^{\cdot} \qquad \text{repair}$$

Recently it has been suggested that zinc thiolate clusters in metallothionein are particularly efficient at scavenging hydroxyl free radicals (Thornalley and Vasak 1985). This is in agreement with other studies showing that thiol-containing compounds do react rapidly with OH$^{\cdot}$. However, the high general reactivity of OH$^{\cdot}$ with most biological molecules means that such scavenging may not be as significant as that by metallothionein of HO$_2^{\cdot}$ or RO$_2^{\cdot}$ since fewer biological modules react rapidly with these species (Willson 1985). Such scavenging of organic peroxy radicals by metallothionein and the induction of the protein by zinc are supported by recent studies with rat hepatocyte cultures (Coppen et al. 1985, 1986; Cousins 1986). The extent of lipid peroxidation and the extent of free radical formation occurring on incubation of the cells with t-butyl hydroperoxide or 3-methyl indole was found to be markedly lowered by the presence of increasing amounts of zinc in the culture medium and to vary inversely with the cellular metallothionein content. However, it was pointed out that the levels of cytochrome c reductase decreased with increasing zinc concentration thus raising the possibility that changes in this enzyme were somehow suppressing free radical activity.

Zinc and Superoxide Dismutase in Protection against Free Radical Damage

Since the first report (Mann and Keilin 1939) that the copper present in the blood was associated with a protein (haemocuprein), a large family of cupreins with closely related structures have been identified. Much of this research was stimulated by the surprise finding that erythrocuprein catalysed the dismutation of superoxide free radicals (McCord and Fridovich 1969; Bannister et al. 1987).

Chemical evidence that erythrocuprein also contained zinc was confirmed later by both neutron activation analysis and emission spectroscopy (Carrico and Deutsch 1970). Two atoms of copper and two atoms of zinc are present per molecule of protein and all four atoms are required for optimal superoxide dismutase activity. Both metals are relatively tightly bound. Dialysing against 1,10 phenanthroline at pH = 5.5 lowers the zinc content by only 13% and against EDTA at pH = 7.4 by only 10%. All the zinc and copper can be removed, however, by dialysis against KCN (0.05 M) and tris.HCl (0.05 M) or by dialysis through an EDTA-equilibrated Sephadex G-25 column at pH = 3.8 for 8–10 h (Weser et al. 1972).

Superoxide dismutase (SOD) is considered to be one of the most stable of globular proteins even when compared to the extreme thermophiles: the conformational melting temperature is 83 °C. The dimeric structure is not dissociated by 4% sodium dodecyl sulphate and the enzyme is still active in 8 M urea.

Detailed structures of the bovine erythrocyte enzyme from x-ray crystallographic data first at 5.5 Å, then at 3 Å and subsequently at 2 Å resolution, are now available along with the complete amino acid sequences of baker's yeast and of human copper/zinc superoxide dismutase (Thomas et al. 1974; Richardson et al. 1975; Barra et al. 1980; Steinman 1980; Tainer et al. 1982, 1983). The crystallographic characterization of a recombinant human copper/zinc superoxide dismutase expressed in yeast has also recently been published (Parge et al. 1986).

It is now known that the enzyme is a dimer with two identical subunits each containing some 153 amino acids, many of which are conserved in hologous positions in the proteins from a variety of species (Flohe et al. 1984; Steffens et al. 1986).

The actual catalytic mechanism of dismutation is thought to involve only copper, with the zinc atoms having a mainly structural role.

$$O_2^{\cdot -} + ECu(II) \xrightarrow{2H^+} O_2 + E(CuI)$$

$$O_2^{\cdot -} + E(CuI) \xrightarrow{2H^+} H_2O_2 + E(CuII)$$

$$\text{net } O_2^{\cdot -} + O_2^{\cdot -} \xrightarrow{2H^+} O_2 + H_2O_2$$

The fact that the enzyme affords protection to a variety of biochemical systems in vitro is undisputed. Its exact protective role, however, is still a matter for debate. When the superoxide dismutase activity of the protein was originally characterized it was thought that this reaction prevented otherwise damaging reactions of the superoxide free radical. The enzyme would therefore be of major importance in protecting all aerobic cells against the potentially damaging effects of oxygen.

$$O_2^{\cdot -} + \text{biomolecule} \rightarrow \text{bioradical} \quad \text{(damage)}$$

$$O_2^{\cdot -} + O_2^{\cdot -} \xrightarrow{2H^+} O_2 + H_2O_2 \quad \text{(protection)}$$

The major conundrum has been, however, that in many in vitro systems the superoxide free radical, $O_2^{\cdot -}$, is relatively unreactive towards biological molecules, particularly nucleic acid. One possibility therefore considered was that SOD prevents $O_2^{\cdot -}$ from helping to maintain decompartmentalized iron in the iron(II) state. In doing so it prevents the iron reacting with hydrogen peroxide, thereby generating hydroxyl free radicals, OH·, by the well-known Fenton reaction:

$$O_2^{\cdot -} + Fe^{3+}(X) \rightarrow Fe^{2+}(X) + O_2$$
$$Fe^{2+}(X) + H_2O_2 \quad \rightarrow Fe^{3+}(X) + OH^- + OH^{\cdot} \quad \text{(Fenton reaction)}$$
$$OH^{\cdot} + RH \quad \rightarrow H_2O \quad + R^{\cdot} \qquad \text{(damage)}$$

Another possibility, however, is that $O_2^{\cdot -}$, although itself relatively unreactive, can sometimes react by way of its protonated form, $HO_2^{\cdot}$. Although at pH = 7 only some 1% of the radical will be in its acid form, this species is a stronger oxidizing agent, reacting much more readily for example with sulphur groups and probably with polyunsaturated fatty acids.

In summary, there is good evidence that superoxide dismutase does have a role in protection against free radicals. The possibility that zinc deficiency could lead to a dangerously low SOD activity therefore cannot be ruled out. However, its relative importance clearly would depend on whether or not other vital zinc levels earlier had fallen dangerously. In a study with regenerating liver of zinc-deficient rats it was found that 8 h after the operation the lability of lysosomes was slightly decreased but increased slightly after 16 h although at this time the tissue SOD level was 30% less than normal (Dreosti and Record 1978). Finally, the fact that a manganese-containing protein with SOD activity but not containing zinc has also been identified in mammalian tissue may complicate such zinc deficiency studies. Although the subcellular distribution of this enzyme seems to differ from that of the copper–zinc enzyme, being principally located in the mitochondria rather than the cytosol, the possibility that levels of the manganese enzyme might increase during zinc deficiency and thereby compensate for any loss of the copper–zinc enzyme seems worthy of investigation.

"Normal" Zinc Levels: Natural Protection against Deleterious Iron–Thiol Interactions

The catalytic effect of iron salts on the oxidation of thiols is well-known with studies dating back to the last century (Baumann 1884; Loven 1884; Mathews and Walker 1909; Lamfrom and Nielsen 1957; Wills and Wilkinson 1967; Misra 1974). Even in the very early studies it had been noted that zinc inhibited the iron-catalysed oxidation of thioglycolate or cysteine and that a vivid purple-red colour seen in the presence of iron and oxygen was absent if zinc was present (Loven 1884; Mathews and Walker 1909). It is perhaps suprising, therefore, that it was not until the 1970s that the possibility that zinc may have a natural role in protecting cells against the deleterious effects of the iron-catalysed degradation of thiol groups, began to be considered in detail. Two findings were to prove particularly stimulating in this respect. First was the finding that zinc stabilized lysosomal membranes and protected microsomes from damage induced by carbon tetrachloride (Chvapil 1973; Chvapil et al. 1972 a,b,c, 1973; Chvapil and Zukoski 1974). The second was that iron could catalyse the reaction of the drug metronidazole (Flagyl) with cysteine and that zinc could inhibit the drug's biological reduction (Willson and Searle 1975; Willson 1976b, 1977a).

Zinc had been used by cytologists for some time in the isolation of intact cell plasma membrane from animal cells (Warren et al. 1966; Chvapil et al. 1972a). The addition of zinc chloride to saline-washed fibroblast cells appeared to strengthen the membrane and cause the cytoplasm to become firm, non-dispersible and tightly attached to the nucleus. When the cells were homogenized, the membrane came off and the nucleus and the cytoplasm remained intact. Thus there was relatively little debris in the homogenate. It was thought that a simple chemical interaction with sulphydryl groups was involved since the stabilization was independent of temperature and also occurred with sulphydryl reagents. Further support for the involvement of membrane thiol groups was provided by the finding that sulphydryl reagents decreased zinc uptake by HeLa cells (Cox and Ruckenstein 1970). The involvement of thiols in mercaptide formation had also been used to explain the stabilizing effect of zinc on neurotubules from the rat brain (Nickolson and Veldstra 1972).

Interest in the effects of zinc on metronidazole metabolism had stemmed from the finding that the drug's constituent nitro- group could be reduced exponentially by the anaerobic flora of the rat caecum provided oxygen was absent (Searle and Willson 1976). Reduction was inhibited by various thiol reagents, by the iron-complexing agent, 4,4,4-trifluoro-1,2-thienylbutane-1,3-dione, by nitrite and cyanide and also to a considerable extent, by zinc. It was concluded that non-haem iron and sulphur compounds were involved in the reduction and that zinc might be effectively competing with iron for cellular thiols.

Associated studies with simple solutions of cysteine showed that iron effectively catalysed the reduction of the drug with no reduction being detectable in its absence. As with the iron–cysteine and thioglycollated solutions studied nearly a century earlier, a purple colour was observed immediately on mixing the solutions; however, in this case it was also present in the absence of oxygen. Stopped-flow and related stationary-state kinetic studies indicated that the colour was due to an iron–cysteine–drug complex which subsequently decayed by reaction with cysteine to give the free radicals, $RNO_2^{\cdot-}$ and $RS^{\cdot}$ which also then subsequently decayed (Willson and Searle 1975). Zinc again inhibited the reduction. A subsequent reading of the literature, particularly the writings of Albert (1973), led to the inevitable conclusion that the safe compartmentalization of iron is critical to the life of the cell but that in ill health (possibly due to the action of a foreign agent, perhaps a chemical, a virus or changes induced by radiation) decompartmentalization of iron might occur and free radical reactions might be initiated. This would be particularly so if the iron or a reactive complex came into contact with hydrogen peroxide or sites containing labile sulphur such as those present in cysteine residues.

On this basis it was anticipated that relatively high concentrations of zinc might be present in those tissues vulnerable to oxidation, such as the hair, skin, eye and spermatozoa. When this was confirmed by further reading, the following corollaries were proposed (Willson 1976b, 1977a):

1. "in healthy cells, vital molecules are protected from the action of decompartmentalized iron by the presence of zinc";
2. "normal cells are designed in such a way that division is not initiated until the zinc concentration at critical sites within the cell is sufficient to protect them from decompartmentalized iron that might normally be present. Zinc thus plays a protective and stimulatory role."

The need for increased levels of zinc at the time of division is in agreement with the general concept that the total anti-oxidant levels might be expected to be high at this time. The fact that zinc is an essential component of the cell that cannot be synthesized or degraded may also be of advantage in its action as a controlling factor. If the progeny of a cell are to be the same as their parent, the zinc levels must be carefully monitored to ensure that they double within the cells' lifetime.

With respect to decompartmentalized iron, subsequent experiments have very much reinforced the arguments then presented, not only in relation to cancer and anti-cancer drugs such as metronidazole, misonidazole and bleomycin, but also in relation to inflammation and diseases such as malaria (Lown and Sim 1977; Bahnemann et al. 1978; Gutteridge 1979; Blake et al. 1981; Clark et al. 1984; Halliwell and Gutteridge 1984; Samuni et al. 1986). Indeed the observations that the levels of decompartmentalized (low-molecular-weight chelatable) iron are increased substantially in the brains of animals subject to cardiac arrest and subsequent resuscitation, is causing particular excitement amongst those interested in ischaemia and reperfusion injury (McCord 1985; Nayini et al. 1985; Bulkley 1987).

With respect to zinc and thiols, the fact that hydroxyl free radicals also have now been identified by the electron spin resonance spin trapping technique in cysteine solutions containing iron but not when excess zinc was present, has been particularly encouraging (Searle and Tomasi 1982). Even more encouraging has been the long-anticipated report that the fungal toxin sporidesmin, responsible for the debilitating facial excema and liver damage in New Zealand cattle and sheep, can be reduced to a dithiol which in the presence of copper or iron but not zinc, can be reoxidized by oxygen with the formation of superoxide free radicals (Mrs G. Reid, private communication 1978; Munday 1982, 1984a, b). However, it is perhaps in the molecular biology of cellular control that the most important application of zinc's proposed protective/stimulatory role may materialize.

Zinc, Thiols and Iron in Cell Control and Carcinogenesis

The evidence in 1976 that zinc in conjunction with thiols had an important protective/stimulatory role in cellular control was indeed circumstantial. Happily, however, a similar conclusion has since been reached along much the same lines. "It [zinc] is a protective/trigger unit in pumped storage systems in association with small chelates or proteins often containing thiols. It is a trigger in that it prevents reactions of DNA or RNA or proteins until they are required" (Williams 1984).

In 1976 it was argued that when normal cells cease proliferating on contact, changes in the plasma membrane inevitably occurred. It was considered not unreasonable to infer, therefore, that if zinc does have an important stimulatory/protective role, then the plasma membrane directly or indirectly controls the concentration of zinc in those sites which need both protection and stimulation for division to begin. When contact with a neighbouring cell is lost, for example after injury, it was suggested that the plasma membrane changes in such a way that the concentrations of protective and stimulatory zinc increase. Division processes then begin. If the concentration of decompartmentalized iron in critical sites is normal, the increased zinc concentration will be sufficient to protect the sites from iron-catalysed free-radical-induced damage. Should however the levels of decompartmentalized iron in critical sites be abnormally high when division is initiated, the zinc concentration may be insufficient to afford protection and free radical chain reactions may ensue. If these are sufficiently extensive the cell or its subsequent progeny will die. If on the other hand, oxidation is limited, due to a low oxygen tension or the presence of a nearby anti-oxidant in the form of a carcinogen, the cell may survive. The genetic apparatus of the cell may be modified, however, in such a way that the gene which controls the ability of the plasma membrane to regulate stimulatory zinc concentrations is damaged. As a result the subsequent progeny cannot prevent the increase of stimulatory zinc after cellular contact. Further division subsequently takes place and uncontrolled proliferation occurs (Fig. 10.3).

In 1976 the nature of the protective/stimulatory zinc factors of the controlling plasma membrane and of the cell division controlling system remained very much a matter for conjecture. All that could be said was that they were somehow related and that zinc–thiol complexes were likely to be involved. Techniques were not then available to detect zinc or iron at the atomic level within the cell and it was the zinc and iron levels in the microenvironment of critical thiol groups and not the tissue, serum or even the overall cellular levels that were considered important. Fortunately, 10 years later some of these ideas can now be re-examined in the light of the exciting developments in nucleic acid and protein sequencing techniques and the development of EXAFS (extended x-ray absorption fine structure) equipment which is providing valuable information concerning the distribution of metals at the atomic level. Two areas of study seem to hold out particular promise in helping to confirm or refute the proposed model.

First there has been the characterization of the plasma-membrane-associated protein kinase C which is now known to have a crucial role in signal transduction for a variety of biologically active substances which affect cellular functions and proliferation. It is also the major receptor for the tumour-promoting agent, phorbol ester (Takai et al. 1977; Nishizuka 1984, 1986). Second has been the identification of zinc–cysteine domains in a variety of regulatory proteins intimately associated with nucleic acid transcription, gene expression and hormone binding (Baudier and Gerard 1983; Miller et al. 1985; Klug and Rhodes 1987). Such domains, "zinc fingers", have been found to contain either two cysteine and two histidine residues, or alternatively four cysteine residues, bound to zinc, and to occur at defined intervals along the protein chains.

Similar cysteine-rich domains have also been shown to be present in protein kinase C (Coussens et al. 1986; Parker et al. 1986; Ono and Kikkawa 1987). Furthermore zinc, along with calcium, is now thought to be involved in the protein's regulation (Murakami et al. 1987). The possibility of assigning protein kinase C as the site for the plasma membrane changes responsible for increases in stimulatory/protective zinc in the above hypothesis and for assigning the production of "zinc fingers" as the control mechanism for cell division is therefore very tempting.

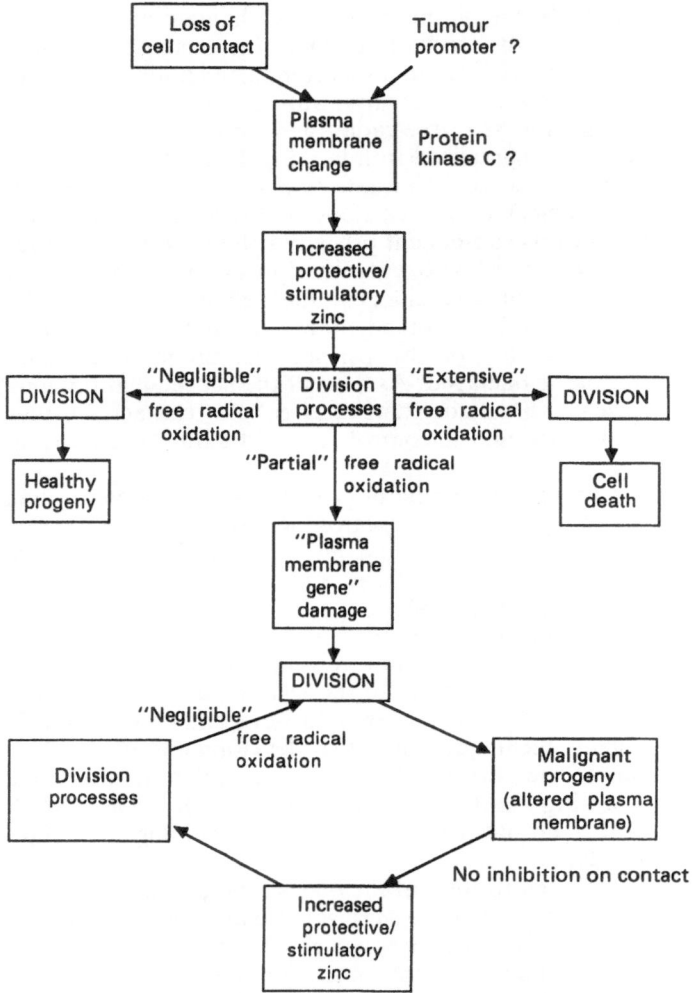

Fig. 10.3. A zinc/free radical model of cell division and carcinogenesis (adapted from Willson 1977b).

Encouragingly, in a recent paper discussing the possible regulatory role of "zinc fingers", Klug and Rhodes (1987) ask the question "Why zinc?". They go on to suggest that one reason might be that unlike iron or copper, zinc lacks redox chemistry and might be used in situations where the presence of redox reactions would lead to damaging radicals.

"Abnormal" Zinc Levels: Metabolic Disturbances and Cell Sensitivity

So far we have discussed how zinc could be involved in protecting the cell against free radical damage. When it comes to studying the relationship between zinc and free radical biochemistry, however, we must also consider the effect any changes in zinc metabolism may have on those cellular constituents most involved in free radical formation or in the cells' susceptibility to them. Of these the effects of zinc on free-radical-generating enzymes or on enzymes supplying reducing equivalents to them, on normal iron and copper metabolism and on the levels of unsaturated fatty acids and thiols, could be particularly important.

Abnormal Zinc Effects on Free-Radical-Generating Enzymes

The NADPH-dependent cytochrome P450 and cytochrome P450 reductase systems known to be involved in the generation of free radicals from a variety of substrates have both been shown to be inhibited by low concentrations of zinc (Chvapil et al. 1975; Hammermueller et al. 1986; Coppen et al. 1986).

Abnormal Zinc Effects on Iron and Copper Metabolism

As we have seen, both iron and copper can have a major effect on free radical reactions. The control of both metals at the correct levels is therefore vital. Zinc excess or deficiency has long been known to affect both copper and iron metabolism (Cox and Harris 1960; Seelig 1973; Solomons and Jacob 1981). Any disturbances leading to increased decompartmentalized iron or copper or decreased superoxide dismutase could clearly be significant. Zinc has been reported to interfere with ferritin iron and with iron turnover and can be transported by transferrin (Settlemire and Matrone 1967a,b; Evans and Winter 1975). When rats were chronically treated with calcium EDTA, the liver zinc significantly decreased with no change in iron and the capacity of the liver to peroxidize increased (Chvapil et al. 1974a). In another study, however, with rats fed only half the usual amount of dl α-tocopherol, changes in dietary zinc did not affect the content of copper or iron in any of the tissues studied (Chvapil et al. 1974b).

In a recent study in humans, a diet containing an iron/zinc sulphate ratio of 1 slightly inhibited zinc absorption and ratios of 2 and 3 did so substantially, as reflected by reduced plasma zinc levels. No effect on zinc absorption was observed if iron in the form of haem chloride was used. It was concluded that the possibility that intrinsic iron in formulas for feeding infants and in vitamin mineral supplements might inhibit the absorption of zinc justified concern about the iron/zinc ratio in the formulation of these products (Solomons and Jacob 1981).

Abnormal Zinc Effects on Unsaturated Fatty Acid Distribution and Prostaglandins

The ease of fatty acid oxidation increases with increased unsaturation (arachidonic > linolenic > linoleic). Thus changes in fatty acid content and distribution, induced by changes in zinc status, could have a marked effect on membrane susceptibility to free radical damage (Baird et al. 1980; Clejan et al. 1981; Cunnane and Wahle 1981; Hamilton et al. 1981; Sas and Pethes 1981; Ayala and Brenner 1983; Burke and Fenton 1985). Furthermore, arachidonic acid, free radicals, lipid peroxidation and prostaglandins are now known to be closely interrelated (Panganamala et al. 1974; Taylor et al. 1983). Zinc has been reported to inhibit prostaglandin synthesis and analogous effects of zinc deficiency and toxicity to aspirin, a proven prostaglandin inhibitor, has led to the suggestion that zinc does have a role in prostaglandin metabolism (Nugteren et al. 1966; O'Dell et al. 1977).

Zinc in Free-Radical-Related Disorders: Is there any Evidence?

Zinc is now thought to be involved with over 100 enzymes. Isn't this sufficient to explain the metal's biological importance? If zinc is so important in preventing free radical disorders wouldn't one expect to see an increase in such disorders in zinc deficiency? Conversely shouldn't an increased zinc intake protect or even ameliorate such disorders?

Regrettably, the only reply seems to be "not necessarily". Apart from the zinc required in the maintenance of general cell structure, the zinc in superoxide dismutase and the zinc associated with metalothionein and possibly with nucleic acid repair enzymes, it must be remembered that:

1. it is the levels of zinc in the microenvironment of sensitive functional groups that seem likely to be important in natural protection and not the tissue, serum or even overall cellular concentrations;
2. decreases in these normal natural protective layers in the critical microenvironments may be the last to be affected by zinc deficiency;
3. increases in serum or even dietary levels per se may not significantly affect these microenvironmental levels;
4. other deleterious effects of zinc deficiency may take place before any free radical damage is expressed pathologically;
5. in many disorders it is likely that free radical disturbances will be only one of many factors.

Bearing in mind these caveats, it seems useful to refer briefly, in the remaining space available, to some of the disorders in which there are at least suggestions that free radical biochemistry and zinc metabolism may be related.

Liver Injury

There is good evidence that free radicals are involved in many types of liver injury (Slater 1978, 1984, 1987) and zinc levels have been found to be low in many cirrhotic patients (Vallee et al. 1956; Prasad et al. 1965; Halsted et al. 1968; Sullivan and Heaney 1970). Serum lipoperoxide levels have been found to be significantly increased (Suematsu et al. 1977). In rats, the protective effect of zinc on carbon tetrachloride-induced liver damage has already been mentioned (Chvapil et al. 1972c, 1973). More recently, zinc deficiency has been reported to increase hepatic microsomal lipid peroxidation (Sullivan et al. 1980). Increased alcohol ingestion is also known to increase susceptibility to peroxidation. Recently it has been reported that acute ethanol toxicity in mice is diminished by administration of zinc aspartate (Floersheim 1985).

Inflammation

Ever since the characterization of superoxide dismutase and the suggestion that superoxide free radicals might be involved in inflammation, evidence has steadily grown indicating that oxygen-containing free radicals are widely involved (Flohe et al. 1985; Lunec et al. 1987). Stimulation of the protein kinase C of neutrophils causes the activation of NADPH oxidase and the subsequent generation of superoxide free radicals. Originally it was thought that these radicals themselves may cause damage to surrounding tissues when released as, for example, in the synovial cavity of the rheumatoid joint (McCord 1974). Alternatively however, through the action of decompartmentalized iron, the probably more damaging hydroxyl free radicals may be formed. Lipid peroxidation may also be initiated.

In one study the serum concentrations of zinc in patients suffering from rheumatoid arthritis had been found to be considerably below normal. The results of a preliminary trial indicated that oral administration of zinc salts was beneficial (Neidermeir and Griggs 1971; Simkin 1976).

Essential Fatty Acid Deficiency and Acrodermatitis Enteropathica

The dietary essential fatty acid in man is linoleic acid which is elongated and desaturated twice to form arachidonic acid (Fleming et al. 1976).

Because of similarities in the effects of zinc and essential fatty acid deficiency in both humans and animals (O'Dell et al. 1977; Bettger et al. 1979; Manku et al. 1979), the possibility of an interaction between both nutrients was suggested (Bettger et al. 1979; Cunnane et al. 1979). Patients with acrodermatitis enteropathica, the genetic disease observed in infants after weaning and typified by dermatitis, diarrhoea and growth failure, have been reported to have aberrant fatty acid metabolism (Cash and Berger 1969; White and Montalvo 1973; Neldner 1974). Some improvement on administration of cottonseed oil has been observed (Cash and Berger 1969). Fortunately the disease responds dramatically to oral zinc therapy (Moynahan and Barnes 1973). Whether free radicals are involved in the disorder remains a matter for conjecture. Since zinc is known to affect the unsaturated fatty acid content of lipid membranes it may be that zinc deficiency simply causes an increased incorporation of easily oxidized fatty acids into such membranes which subsequently peroxidize at the earliest opportunity. Thus not only are essential fatty acids lost but structural damage, interference with prostaglandin metabolism and the formation of toxic products may occur.

Loss of Spermatozoa Viability

The presence of zinc in relatively high concentrations in human semen is well-known (Mawson and Fischer 1953). Unsaturated fatty acids which readily peroxidize in the presence of iron and ascorbate have also been shown to be present in human spermatozoa (Jones and Mann 1976). High levels of superoxide dismutase have also been detected (Mennella and Jones 1980). Recently it has been found that spermatozoa from infertile men, which are unable to fuse with oocytes, are very readily oxidized with a production of reactive oxygen species some 40 times greater than normal (Aitken and Clarkson 1987). The extent of oxidation is further increased by the presence of the calcium ionophore A23187. It is suggested that a calcium-sensitive superoxide-generating system (very similar to the NADPH oxidase found in neutrophils) which is also linked to protein kinase C is present within the human spermatozoon. The system is thought to play an important role in the fusion process but in defective sperm the controlling system is disrupted, free radical processes are initiated and loss of function follows. On the basis of these findings clinical trials in which vitamin E is given to men with such defective spermatozoa are beginning. With respect to zinc, the fact that protein kinase C is again thought to be involved is interesting. Whether such spermatozoa also have an abnormal zinc distribution is not known.

Radiation Damage

A zinc asparate complex has been shown to protect mice against the lethal effects of radiation exposure (Floersheim and Floersheim 1986). Although free radicals have long been known to be involved in radiation injury, only a few compounds have been found to protect cells from radiation exposure. Of these, thiol compounds, particularly cysteine and cysteamine, are among the most efficient, but as with zinc aspartate, the protection is far from complete. It was in fact the possibility that metronidazole (Flagyl) sensitized cells to radiation by reacting with endogenous cellular thiols that led to the observation of its iron-catalysed reduction and subsequently to interest in the role of iron and zinc in radiation injury (Willson 1976a,b; Bahnemann et al. 1978).

The skin of patients addicted to alcohol had been reported to be relatively sensitive to radiation damage (Ellis 1950). Patients with postalcoholic cirrhosis excreted abnormally large amounts of zinc, and alcohol increased the rate of absorption of iron (II) chloride in normal subjects (Charlton et al. 1964; Sullivan and Lankford 1965). Administration of the chemotherapeutic drug bleomycin, now known to act by transporting iron into the vicinity of nucleic acid (Lown et al. 1977), increased the sensitivity of the skin to radiation damage: the onset of skin lesions correlated well with a decrease in serum zinc levels (Leith et al. 1975). Recent studies have shown that irradiation of alcohol dehydrogenase in vitro can result in the release of zinc from the protein, presumably due to direct or indirect damage to the metal-binding cysteine and/ or histidine residues (Abelidis et al. 1987). How supplementary zinc might protect against radiation injury in vivo, however, remains an open question.

Cancer

There is now considerable evidence that free radicals are involved in the development of several forms of cancer at both the initiation and promotional stages (Witz et al. 1980; McBrien and Slater 1982; Goldstein et al. 1983; Cerutti 1985; Kensler and Taffe 1986; Borek 1987).

Unfortunately the very manner in which cancers develop over a protracted time scale has meant that the identification of any role for free radicals in cancer development in vivo is extremely difficult. As in the case of viruses, the presence or absence of free radicals in tumour tissue does not imply that they are or are not involved in any stages of neoplastic growth. That said, the fact that some tumour tissues have been shown to peroxidize relatively slowly compared with related normal tissue is intriguing (Bartoli and Galeotti 1979; Cheeseman et al. 1986). This may be due to a variety of factors such as differences in levels of unsaturated fatty acids or anti-oxidants such as α-tocopherol or superoxide dismutase (Sykes et al. 1978; Oberley and Buettner 1979; Bize et al. 1980; Galeotti et al. 1980; Bartoli et al. 1983).

Measurements of zinc levels during the last half century have shown no consistent differences between levels in normal and malignant tissues nor between the levels in cancer and other patients. General nutritional factors associated with disease, the wide variety of malignant classifications and the fact that zinc is essential for any growing tissue makes literature information concerning zinc levels and cancer extremely difficult to interpret (Cristol 1922; Vikbladh 1951; Olson et al. 1958; Addink and Frank 1959; Davies et al. 1968; McQuitty et al. 1970; Wright and Dormandy 1972; Barr and Harris 1973; Dhar et al. 1973; Murthy et al. 1973; Mathur 1978; Andronikashvili et al. 1974; Kew and Mallet 1974; Dagher et al. 1977; Cavdar et al. 1980).

High dietary zinc has been associated with reduced incidences of gastric cancer in parts of Japan and Britain and of oesophageal cancer in parts of Africa (Hiryama 1962; Stocks and Davies 1964; McGlashan 1967). A high zinc intake in animals has been found to reduce the incidence of chemically induced tumours in the hamster cheek pouch (Poswillo and Cohen 1971) and the rat submandibular glands (Ciapparelli et al. 1972) although later experiments with the cheek pouch model failed to confirm the findings (Edwards 1976).

In another study both zinc-deficient and supplemented diets increased the incidence of chemically induced tumours in mice (Duncan and Dreosti 1975). Zinc-deficient diets have also been reported to cause a marked reduction in virally induced sarcomas in mice (Beach et al. 1981). Thus the role of zinc in actual carcinogenesis is far from clear. Matters are complicated further by reports that direct injection of simple zinc salts can be carcinogenic (Falin 1940; Guthrie and Guthrie 1974) and that high levels of zinc may be selectively cytotoxic to melanoma cells (Boravansky and Riley 1983; Borovansky et al. 1985).

Erythrocyte-Related Disorders

There is now considerable evidence for the occurrence of erythrocytes with abnormal membranes in patients with sickle cell anaemia, thalassaemia and Duchenne muscular dystrophy (see Rice-Evans and Dunn 1982 and refs cited). Since there is also now good evidence that zinc is able to stabilize erythrocyte membranes, the possibility arises that these disorders may be at least partly due to particular disturbances in zinc metabolism (Brewer and Kruckeberg 1979; Chvapil et al. 1979; Girotti et al. 1985, 1986). Indeed there are reports of improvements in adult patients with sickle cell anaemia following zinc treatment (see Brewer et al. 1976 and refs cited).

Concluding Remarks

In summary, there now seems to be a considerable amount of circumstantial evidence suggesting that zinc deficiency may lead to an increased susceptibility to free-radical-related disorders. Whether or not zinc supplementation at dietary levels over and above that presently recommended (15 mg per d) will reduce the incidence or affect the outcome of such disorders, however, remains very much an open question.

Acknowledgements. I am grateful to my zinc and iron collaborators, Dr. Andrew Searle, Ms Sonja Jonas and Professor Patrick Riley and to Professor Trevor Slater and the Cancer Research Campaign without whose encouragement and support for free radical biochemistry, my interest in zinc would probably never have materialized.

References

Abelidis, S, Moore JF, Chakravarty A (1987) Zinc release from irradiated yeast alcohol dehydrogenase. Int J Radiat Biol 52: 413–418

Addink NWH, Frank LJP (1959) Remarks apropos of analysis of trace elements in human tissues.Cancer NY 12: 544–551

Aitken RJ, Clarkson JS (1987) Cellular basis of defective sperm function and its association with the genesis of reactive oxygen species by human spermatozoa. J Reprod Fertil 81: 459–469

Albert A (1973) Selective toxicity, 5th edn. Chapman and Hall, London

Andronikashvili EL, Mosulishvili LM, Belokobilski AI, Kharabadze NE, Tevzieva TK, Efremova EY (1974) Content of some trace elements in sarcoma M-1 DNA in dynamics of malignant growth. Cancer Res 34: 271–274

Ayala S, Brenner RR (1983) Essential fatty acid status in zinc deficiency. Effect on lipid and fatty acid composition, desaturation activity and structure of microsomal membranes of rat liver and testes. Acta Physiol Lat Am 33: 193–204

Bahnemann D, Basaga H, Dunlop JR, Searle AJF, Willson RL (1978) Metronidazole (Flagyl), misonidazole (Ro 07–0582), iron, zinc and sulphur compounds in cancer therapy. Br J Cancer [Suppl III] 37: 16–19

Baird MB, Birnbaum LS, Sfeir GT (1980) NADPH-driven peroxidation in rat liver nuclei and nuclear membranes. Arch Biochem Biophys 200: 108–115

Bannister JV, Bannister WH, Rotillio G (1987) Aspects of the structure, function and applications of superoxide dismutase. (To be published)

Barr DH, Harris JW (1973) Growth of the P388 leukemia as an asciters tumor in zinc-deficient mice. Proc Soc Exp Biol Med 144: 284–287

Barra D, Martini F, Bannister JV et al. (1980) The complete amino acid sequence of human Cu/Zn superoxide dismutase. FEBS Lett 120: 53–56

Bartoli GM, Galeotti T (1979) Growth-related lipid peroxidation in tumour microsomal membranes and mitochondria. Biochim Biophys Acta 574: 537–541

Bartoli GM, Minotti G, Borrello S, Galeotti T (1983) A supposed role of superoxide dismutase in the control of tumor growth. In: Greenwald RA, Cohen G (eds) Oxy radicals and their scavenger systems, vol II. Elsevier Science Publications, New York, pp 179–184

Baudier J, Gerard P (1983) Ion binding to S100 proteins: structural changes induced by calcium and zinc on S100a and S100b proteins. Biochemistry 22: 3360–3369

Baumann E (1884) Zur Frage der Jodbestimmung im Harne. Z Physiol Chem 282–290

Beach RS, Gershwin ME, Hurley LS (1981) Dietary zinc modulation of moloney sarcoma virus oncogenesis. Cancer Res 41: 552–559

Bettger WJ, Reeves PG, Moscatelli EO, Reynolds G, O'Dell B (1979) Interaction of zinc and essential fatty acids in the rat. J Nutr 109: 480–488

Bielski BHJ (1985) Fast kinetic studies of dioxygen-derived species and their metal complexes. Philos Trans R Soc Lond [Biol] 311: 473–482

Bize IB, Oberley LW, Morris HP (1980) Superoxide dismutase and superoxide radical in Morris hepatomas. Cancer Res 40: 3686–3693

Blake DR, Hall ND, Bacon PA, Dieppe PA, Halliwell B, Gutteridge JMC (1981) The importance of iron in rheumatoid disease. Lancet II: 1142–1144

Borek C (1987) Radiation and chemically induced transformation: free radicals, antioxidants and cancer. Br J Cancer [Suppl VIII] 55: 74–86

Borovansky J, Riley PA (1983) The effect of divalent cations on Cloudman melanoma cells. Eur J Cancer Clin Oncol 19: 91–99

Borovansky J, Riley PA, Vrankova E, Necas E (1985) The effect of zinc on mouse melanoma growth in vitro and in vivo. Neoplasma 32: 401–406

Brenton DP, Jackson MJ, Young A (1981) Two pregnancies in a patient with acrodermatitis enteropathica treated with zinc sulphate. Lancet II: 500–502

Brewer GJ, Kruckeberg WC (1979) The anticalcium and erythrocyte membrane effects of zinc and their potential value in the treatment of sickle cell anaemia. In: Rosa J, Benzard Y, Hercules J (eds) INSERM Symposium No 9. Elsevier Biomedical Press, Holland, pp 195–204

Brewer GJ, Prasad AS, Oelshlegel FJ Jr, Schoomaker EB, Ortega J, Oberleas D (1976) Zinc and sickle-cell anaemia. In: Trace elements in human health and disease, vol 1. Academic Press, New York, pp 283–294

Bulkley GB (1987) Free radical-mediated reperfusion injury: a selective review. Br J Cancer [Suppl VIII] 55: 66–73

Burke JP, Fenton MR (1985) Effect of a zinc-deficient diet on lipid peroxidation in liver and tumor subcellular membranes (42083). Proc Soc Exp Biol Med 179: 187–191

Carrico RJ, Deutsch HF (1970) The presence of zinc in human cytocuprein and some properties of the apoprotein. J Biol Chem 245: 723–727

Cash R, Berger CK (1969) Acrodermatitis enteropathica: defective metabolism of unsaturated fatty acids. J Pediatr 74: 717–729

Cavdar AO, Babacan E, Arcasoy A, Erten J, Ertem U (1980) Zinc deficiency in Hodgkin's disease. Eur J Cancer 16: 311–321

Cerutti PA (1985) Prooxidant states and tumor promotion. Science 227: 375–381

Charlton RW, Jacobs B, Seftel H, Bothwell TH (1964) Effect of alcohol on iron absorption. Br Med J 1427–1429

Cheeseman KH, Collins M, Proudfoot K et al. (1986) Studies on lipid peroxidation in normal and tumour tissue: the Novikoff rat liver tumour. Biochem J 235: 507–514

Chvapil M (1973) New aspects on the biological role of zinc: a stabilizer of macromolecules and biological membranes. Life Sci 13: 1041–1049

Chvapil M, Zukoski CF (1974) New concept on the mechanism(s) of the biological effect of zinc. In: Pories WJ, Strain WH (eds) Clinical approaches of zinc metabolism. Thomas, Springfield, Illinois, pp 75–86

Chvapil M, Ryan JN, Brada Z (1972a) Effects of selected chelating agents and metals on the stability of liver lysosomes. Biochem Pharmacol 21: 1097–1105

Chvapil M, Ryan JN, Zukoski CF (1972b) The effect of zinc and other metals on the stability of lysosomes. Proc Soc Exp Biol Med 140: 642–646

Chvapil M, Ryan JN, Zukoski CF (1972c) Effect of zinc on lipid peroxidation in liver microsomes and mitochrondria. Proc Soc Exp Biol Med 141: 150–153

Chvapil M, Ryan JN, Elias SL, Peng YM (1973) Protective effect of zinc on carbon-tetrachloride-induced liver injury in rats. Exp Mol Pathol 19: 186–196

Chvapil M, Aronson AL, Peng YM (1974a) Relation between zinc and iron and peroxidation of lipids in liver homogenate in CaEDTA-treated rats. Exp Mol Pathol 20: 216–227

Chvapil M, Peng YM, Aronson AL, Zukoski C (1974b) Effect of zinc on lipid peroxidation and metal content in some tissues of rats. J Nutr 104: 434–443

Chvapil M, Sipes IG, Ludwig JC, Halladay SC (1975) Inhibition of NADPH oxidation and oxidative metabolism of drugs in liver microsomes by zinc. Biochem Pharmacol 24: 917–919

Chvapil M, Montgomery D, Ludwig JC, Zukoski CF (1979) Zinc in erythrocyte ghosts. Proc Soc Exp Biol Med 162: 480–484

Ciapparelli L, Retief DH, Fatti LP (1972) The effect of zinc on 9,10-dimethyl-1,2-benzanthracene (DBMA) induced salivary gland tumours in the albino rat – a preliminary study. S Afr J Med Sci 37: 85–90

Clark IA, Cowden WB, Hunt NH, Maxwell LE, Mackie EJ (1984) Activity of divicine in Plasmodium vinckei-infected mice has implications for favism and epidemiology of G-6-P deficiency. Br J Haematol 57: 479–487

Clejan S, Maddaiah VT, Castro-Magana M, Collipp PJ (1981) Zinc deficiency-induced changes in the composition of microsomal membranes and in the enzymatic regulation of glycerolipid synthesis. Lipids 16: 454–460

Coppen DE, Cousins RJ, Richardson DE (1985) Effect of zinc on chemically-induced peroxidation in rat liver parenchymal cells in primary culture. Fed Proc (Abs) 44: 6404

Coppen DE, Richardson DE, Cousins RJ (1986) Suppression of lipid peroxidation in rat hepatocytes in primary culture by supplemental zinc. Fed Proc 45: 1083

Cousins RJ (1986) Towards a molecular understanding of zinc metabolism. Clin Physiol Biochem 4: 20–30

Coussens L, Parker PJ, Rhee L et al. (1986) Multiple distinct forms of bovine and human protein kinase C suggest diversity in cellular signalling pathways. Science 233: 859–866

Cox DH, Harris DL (1960) Effect of excess dietary zinc on iron and copper in the rat. J Nutr 70: 514–520

Cox RP, Ruckenstein A (1971) Studies on the mechanism of hormonal stimulation of zinc uptake in human cell cultures: hormone–cell interactions and characteristics of zinc accumulation. J Cell Physiol 77: 71–82

Cristol P (1922) Zinc et cancer. Comp Rend Acad Sci 174: 887–889

Cunnane SC, Wahle KWJ (1981) Zinc deficiency increases the rate of Δ^6 desaturation of linoleic acid in rat mammary tissue. Lipids 16: 771–774

Cunnane SC, Horrobin DF, Ruf KB, Sella G (1979) Prevention of dietary effects of zinc deficiency by administration of essential fatty acids. Proc Physiol Soc 296: 83P–84P

Dagher RK, Ellis GD, Tanski DT et al. (1977) Prostate cancer: zinc levels in normal and metastasized cancerous lung tissue. IRCS Med Sci 5: 509–510

Davies IJT, Musa M, Dormandy TL (1968) Measurements of plasma zinc. J Clin Pathol 21: 359–365

Dhar NK, Goel TC, Dube PC, Chowdhury AR, Kar AB (1973) Distribution and concentration of zinc in the subcellular fractions of benign hyperplastic and malignant neoplastic human prostate. J Exp Mol Pathol 19: 139–142

Dreosti IE, Record IR (1978) Lysosomal stability, superoxide dismutase and zinc deficiency in regenerating rat liver. Br J Nutr 40: 133–137

Duncan JR, Dreosti IE (1975) Zinc intake, neoplastic DNA synthesis and chemical carcinogenesis in rats and mice. J Natl Cancer Inst 55: 195–196

Edwards MB (1976) Chemical carcinogenesis in cheek pouch of Syrian hamsters receiving supplementary zinc. Arch Oral Biol 21: 133–135

Ellis F (1950) Discussion on the chemical factors modifying radiotherapeutic response. Proc R Soc Med 43: 399–405

Emerit I, Keck M, Levy A, Feingold J, Michelson AM (1982) Activated oxygen species at the origin of chromosome breakage and sister–chromatid exchanges. Mutat Res 103: 165–172

Evans GW, Winter TW (1975) Zinc transport by transferrin in rat portal blood plasma. Biochem Biophys Res Commun 66: 1218–1224

Falin LI (1940) Experimental teratoma in fowl. Am J Cancer 38: 199–211

Fleming CR, Smith LM, Hodges RE (1976) Essential fatty acid deficiency in adults receiving total parenteral nutrition. Am J Clin Nutr 29: 976–983

Floersheim GL (1985) Protection against acute ethanol toxicity in mice by zinc aspartate, glycols, levulos and pyritinol. Agents Actions 16: 580–584

Floersheim GL, Floersheim P (1986) Protection against ionising radiation and synergism with thiols by zinc aspartate. Br J Radiol 59: 597–602

Flohe L, Gunzler WA, Kim S-MA et al. (1984) The phylogenetic position of the Cu-Zn-SOD of *P. leiognathi*. In: Bors W, Saron M, Tait D (eds) Oxygen radicals in chemistry and biology. Walter de Gruyter, Berlin, pp 793–801

Flohe L, Beckmann R, Giertz H, Loschen (1985) Oxygen-centered free radicals as mediators of inflammation. In: Sies H (ed) Oxidative stress. Academic Press, London New York

Galeotti T, Borello S, Seccia A, Farallo E, Bartoli GM, Serri F (1980) Superoxide dismutase content in human epidermis and squamous cell epithelioma. Arch Dermatol Res 267: 83–86

Girotti AW, Thomas JP, Jordan JE (1985) Inhibitory effect of zinc(II) on free radical lipid peroxidation in erythrocyte membranes. J Free Rad Biol Med 1: 395–401

Girotti AW, Thomas JP, Jordan JE (1986) Xanthine oxidase-catalysed crosslinking of cell membrane proteins. Arch Biochem Biophys 251: 639–653

Goldstein B, Witz G, Zimmerman J, Gee C (1983) Free radicals and reactive oxygen species in tumor promotion. In: Greenwald RA, Cohen G (eds) Oxy radicals and their scavenger systems. Elsevier Science 2: 321–325r

Greenstock CL, Jinot CP, Whitehouse RP, Sargent MD (1987) DNA radiation damage and its modification by metallothionein. Free Rad Res Comms 2: 233–239

Guthrie J, Guthrie O (1974) Embryonal carcinomas in Syrian hamsters after intratesticular inoculation of zinc chloride during reasonal testicular growth. Cancer Res 34: 2612–2614

Gutteridge JMC (1979) Identification of malondialdehyde as the TBA-reactant formed by bleomycin-iron damage to DNA. FEBS Lett 105: 278–282

Halliwell B, Gutteridge JMC (1984) Oxygen toxicity, oxygen radicals, transition metals and disease. Biochem J 219: 1–14

Halsted JA, Hackley B, Rudzki Ç, Smith JC Jr (1968) Plasma zinc concentrations in liver diseases. Gastroenterology 54: 1098–1105

Hamilton RM, Gillespie CT, Cook HW (1981) Relationships between levels of essential fatty acids and zinc in plasma of cystic fibrosis patients. Lipids 16: 374–376

Hammermueller JD, Bray TM, Bettger WJ (1986) Effect of zinc deficiency on NADPH and cytochrome P-450 dependent active oxygen generation in rat lung and liver. Fed Proc 45: 1083

Hiryama T (1962) Quoted in S Afr Cancer Bull 6: 114

Ho S-Y, Catallanotto FA, Lisak RP, Dore-Duffy P (1986) Zinc in multiple sclerosis. II. Correlation with disease activity and elevated plasma membrane-bound zinc in erythrocytes from patients with multiple sclerosis. Ann Neurol 20: 712–715

Jones R, Mann T (1976) Lipid peroxides in spermatozoa: formation, role of plasmologen and physiological significance. Proc R Soc Lond [Biol] 193: 317–333

Kensler TW, Taffe BG (1986) Free radicals in tumor promotion. Adv Free Rad Biol Med 2: 347–387

Kew MC, Mallet RC (1974) Hepatic zinc concentrations in primary cancer of the liver. Br J Cancer 29: 80–83

Klug A, Rhodes D (1987) "Zinc fingers": a novel protein motif for nucleic acid recognition. Trends Biochem Sci 464–469

Lamfrom H, Nielsen SO (1957) The iron catalysis of thioglycolate oxidation by oxygen. J Am Chem Soc 79: 1966–1970

Leith JT, Lewinsky BB, Schilling WA (1975) Modification of the response of mouse skin to X-irradiation by bleomycin treatment. Radiat Res 61: 100–109

Loven JM (1884) Ueber die Thiomilchsauren und die Thiodilactylsauren. J f prakt Chem 29: 366–378

Lown JN, Sim S (1977) The mechanism of the bleomycin-induced cleavage of DNA. Biochem Biophys Res Commun 77: 1150–1157

Lunec J, Griffiths HR, Blake DR (1987) Oxygen radicals in inflammation. ISI Atlas of Science: Pharmacology 1: 45–48

McBrien DCH, Slater TF (eds) (1982) Free radicals, lipid peroxidation and cancer. Academic Press, London New York

McCord JM (1974) Free radicals and inflammation. Protection of synovial fluid by superoxide dismutase. Science 185: 529–531

McCord JM (1985) Oxygen derived free radicals in post ischemic injury. N Engl J Med 312: 159–163

McCord JM, Fridovich I (1969) Superoxide dismutase. An enzymic function for erythrocuprein (hemocuprein). J Biol Chem 244: 6049–6055

McGlashan ND (1967) Zinc and oesophageal cancer. Lancet I: 578

McQuitty JT, DeWys WD, Monaco L et al. (1970) Inhibition of tumor growth by dietary zinc deficiency. Cancer Res 30: 1387–1390

Manku MS, Horrobin DF, Karmazyn M, Cunnane SC (1979) Prolactin and zinc effects on rat vascular reactivity: possible relationship to dihomo-γ-linoleic acid and to prostaglandin synthesis. Endocrinology 104: 774–779

Mann T, Keilin D (1939) Haemocuprein and hepatocuprein, copper–protein compounds of blood and liver in mammals. Proc R Soc Biol 128: 303

Mathews AP, Walker S (1909) The action of metals and strong salt solutions on the spontaneous oxidation of cystein. J Biol Chem 6: 299–312

Mathur A (1978) The role of zinc in experimental and human oral cancer. Dissertation, University of Lund, Sweden

Mawson CA, Fischer MI (1953) Zinc and carbonic anhydrase in human semen. Biochem J 55: 696–700

Mennella MRF, Jones R (1980) Properties of spermatozoal superoxide dismutase and lack of involvement of superoxides in metal-ion-catalysed lipid-peroxidation reactions in semen. Biochem J 191: 289–297

Miller J, McLachlan AD, Klug A (1985) Repetitive zinc-binding domains in the protein transcription factor IIIA from Xenopus oocytes. EMBO J 4: 1609–1614

Misra HP (1974) Generation of superoxide free radical during the autoxidation of thiols. J Biol Chem 249: 2151–2155

Moynahan EJ, Barnes PM (1973) Zinc deficiency and a synthetic diet for lactose intolerance. Lancet I: 676–677

Munday R (1982) Studies on the mechanism of toxicity of the mycotoxin sporidesmin: 1. Chem Biol Interact 41: 361–374

Munday R (1984a) Studies on the mechanism of toxicity of the mycotoxin sporidesmin: 2. J Appl Toxicol 4: 176–181

Munday R (1984b) Studies on the mechanism of toxicity of the mycotoxin sporidesmin: 3. J Appl Toxicol 4: 182–186

Murakami K, Whitely MK, Routtenberg A (1987) Regulation of protein kinase C activity by cooperative interaction of Zn^{2+} and Ca^{2+}. J Biol Chem 262: 13902–13906

Murthy ASK, Vawte GF, Kopito L, Rossen E (1973) Biochemical studies on liver tumors of children. Arch Pathol 96: 48–52

Nayini NR, White BC, Aust SD et al. (1985) Post resuscitation iron delocalization and malondialdehyde production in the brain following prolonged cardiac arrest. J Free Rad Biol Med 1: 111–116

Neidermeir W, Griggs JH (1971) Trace metal composition of synovial fluid and blood serum of patients with rheumatoid arthritis. J Chronic Dis 23: 527–536

Neldner KH, Hagler L, Wise WR, Stifel FB, Lufkin EG, Herman RH (1974) Acrodermatitis enteropathica: comparison of an infant with her family. J Pediatr 83: 999–1006

Nickolson VJ, Veldstra H (1972) The influence of various cations on the binding of colchicine by rat brain homogenates. Stabilization of intact neurotubules by zinc and cadmium ions. FEBS Lett 23: 309–313

Nishizuka Y (1984) Studies and perspective of protein kinase C. Science 233: 305–312

Nishizuka Y (1986) The role of protein kinase C in cell surface signal transduction and tumour promotion. Nature 308: 693–698

Nugteren DH, Beerthuis RK, Van Dorp DA (1966) The enzymic conversion of all-cis 8,11,14-eicosatrienoic acid into prostaglandin E_1. Recl Trav Chim Pays-Bas Belg 85: 405–419

Oberley LW, Buettner GR (1979) Role of superoxide dismutase in cancer: a review. Cancer Res 39: 1141–1149

O'Dell BL, Reynolds G, Reeves PG (1977) Analogous effects of zinc deficiency and aspirin toxicity in the pregnant rat. J. Nutr 107: 1222–1228

Olson KB, Heggen GE, Edwards CF (1958) Analysis of 5 trace elements in the liver of patients dying of cancer and non-cancerous disease. Cancer 11: 554–561

Ono Y, Kikkawa U (1987) Do multiple species of protein kinase C transduce different signals? Trends Biochem Sci 421–423

Panganamala RV, Sharma HM, Sprecher JC, Cornwell DG (1974) Evaluation of superoxide anion and singlet oxygen in the biosynthesis of prostaglandins from eicosa-8,11,14-trenoic acid. Prostaglandins 7: 21–28

Parge HE, Getzoff ED, Scandella CS, Hallewell RA, Tainer JA (1986) Crystallographic characterization of recombinant human CuZn superoxide dismutase. J Biol Chem 261: 16215–16218

Parker PJ, Coussens L, Totty N et al. (1986) The complete primary structure of protein kinase C – the major phorbol ester receptor. Science 233: 853–859

Petrone WF, English DK, Wong K, McCord JM (1980) Free radicals and inflammation: superoxide-dependent activation of a neutrophil chemotactic factor in plasma. Proc Natl Acad Soc 77: 1159–1163

Poswillo DE, Cohen B (1971) Inhibition of carcinogenesis by dietary zinc. Nature 231: 447–448

Prasad AS, Oberleas D, Halsted JA (1965) Determination of zinc in biological fluids by atomic absorption spectrophotometry in normal and cirrhotic subjects. J. Lab Clin Med 66: 508–516

Rice-Evans CA, Dunn MJ (1982) Erythrocyte deformability and disease. Trends Biochem Sci 7: 282–286

Richardson JS, Thomas KA, Rubin BH, Richardson DC (1975) Crystal structure of bovine Cu, Zn superoxide dismutase at 3 A resolution: chain tracing and metal ligands. Proc Natl Acad Sci 72: 1349–1353

Samuni A, Bump EA, Mitchell JB, Brown JM (1986) Enhancement of misonidazole cytotoxicity by iron. Int J Radiat Biol 49: 77–83

Sas B, Pethes G (1981) Influence of zinc deficiency on the stability of subcellular membranes and on the Zn incorporation into metallothionein. Acta Vet Acad Sci Hung 29: 441–450

Searle AJF, Tomasi A (1982) Hydroxyl free radical production in iron-cysteine solutions and protection by zinc. J Inorg Biochem 17: 161–166

Searle AJF, Willson RL (1976) Metronidazole (Flagyl); degradation by the intestinal flora. Xenobiotica 6: 457–464

Seelig MS (1973) Proposed role of copper–molybdenum interaction in iron-deficiency and iron-storage diseases. Am J Clin Nutr 26: 657–672

Settlemire CT, Matrone G (1967a) In vivo interference of zinc with ferritin iron in the rat. J Nutr 92: 153–158

Settlemire CT, Matrone G (1967b) In vivo effect of zinc on iron turnover in rats and lifespan of the erythrocyte. J Nutr 92: 159–164

Simkin PA (1976) Oral zinc sulphate in rheumatoid arthritis. Lancet I: 539–542

Slater TF (ed) (1978) Biochemical mechanisms of liver injury. Academic Press London New York

Slater TF (1984) Free radical mechanisms in tissue injury. Biochem J 222: 1–15

Slater TF (1987) Free radicals and tissue injury: fact and fiction. Br J Cancer [Suppl VIII] 55: 5–10

Solomons NW, Jacob RA (1981) Studies on the availability of zinc in humans: effects of heme and nonheme iron on the absorption of zinc. Am J Clin Nutr 34: 475–482

Steffens GJ, Michelson AM, Otting F, Puget K, Strassburger W, Flohe L (1986) Primary structure of Cu–Zn superoxide dismutase of brassica oleracea proves homology with corresponding enzymes of animals, fungi and prokaryotes. J Biol Chem 367: 1007–1016

Steinman HM (1980) The amino acid sequence of copper–zinc superoxide dismutase from bakers' yeast. J Biol Chem 255: 6758–6765

Stocks P, Davies RI (1964) Zinc and copper content of soils associated with incidence of cancer of the stomach and other organs. Br J Cancer 18: 14–24

Suematsu T, Kamada T, Abe H, Kinkuchi S, Yogi K (1977) Serum lipoperoxide level in patients suffering from liver diseases. Clin Chem Acta 79: 267

Sullivan JF, Heaney RP (1970) Zinc metabolism in alcoholic liver disease. Am J Clin Nutr 23: 170–177

Sullivan JF, Lankford HG (1965) Zinc metabolism and chronic alcoholism. Am J Clin Nutr 17: 57–63

Sullivan JF, Jetton MM, Hahn HKJ, Burch RE (1980) Enhanced lipid peroxidation in liver microsomes of zinc-deficient rats. Am J Clin Nutr 33 51–56

Sykes JA, McCormak FX, O'Brien TJ (1978) A preliminary study of the superoxide dismutase content of some human tumors. Cancer Res 38: 2759–2762

Tainer JA, Getzoff ED, Beem KM, Richardson JS, Richardson DC (1982) Determination and analysis of the 2 A structure of copper, zinc superoxide dismutase. J Mol Biol 160: 181–217

Tainer JA, Getzoff ED, Richardson JS, Richardson DC (1983) Structure and mechanism of copper, zinc superoxide dismutase. Nature 306: 284–290

Takai Y, Kishimot A, Inoue M, Nishizuka Y (1977) Studies on a cyclic nucleotide-independent protein kinase and its proenzyme in mammalian tissues. 1. Purification and characterization of an active enzyme from bovine cerebellum. J Biol Chem 252: 7603–7609

Taylor L, Menconi MJ, Polgar P (1983) The participation of hydroperoxides and oxygen radicals in the control of prostaglandin synthesis. J Biol Chem 258: 6855–6857

Thomas KA, Rubin BH, Bier CJ, Richardson JS, Richardson DC (1974) The crystal structure of bovine Cu^{2+}, Zn^{2+} superoxide dismutase at 5.5-A resolution. J Biol Chem 249: 5677–5683

Thornalley PJ, Vasak M (1985) Possible role for metallothionein in protection against radiation-induced oxidative stress. Kinetics and mechanism of its reaction with superoxide and hydroxyl radicals. Biochim Biophys Acta 827: 36–44

Vallee BL, Wacker WEC, Bartholomay AF, Robin ED (1956) Zinc metabolism in hepatic dysfunction 1. Serum zinc concentration in Laennecs cirrhosis and their validation by sequential analysis. N Engl J Med 255: 403–408

Vikbladh I (1951) Studies on zinc in blood: II. Scand J Clin Lab Invest [Suppl 2] 3: 5–73

Warren L, Glick MC, Nass MK (1966) Membranes of animal cells. I. Methods of isolation of the surface membrane. J Cell Physiol 68: 269–287

Weser U, Barth G, Djerassi C et al. (1972) A study on purified apo-erythrocuprein. Biochim Biophys Acta 278: 28–44

White HB Jr, Montalvo JM (1973) Serum fatty acids before and after recovery from acrodermatitis enteropathica: comparison of an infant with her family. J Pediatr 83: 999–1006

Williams RJP (1984) Zinc: what is its role in biology? Endeavour New Series 8: 65–70

Wills ED, Wilkinson AE (1967) The effect of irradiation on subcellular particles. Destruction of sulphydryl groups. Int J Radiat Biol 13: 45–55

Willson RL (1976a) Metronidazole and iron in cancer therapy. Lancet I: 304–305

Willson RL (1976b) Metronidazole and tissue zinc/iron ratio in cancer therapy. Lancet I: 1407

Willson RL (1977a) Iron, zinc, free radicals and oxygen in tissue disorders and cancer control in iron metabolism. In: Porter R (ed) Ciba Foundation Symposium, 51, Elsevier Excerpta Medica, pp 333–354

Willson RL (1977b) Zinc: a radical approach to disease. New Scientist 558–560

Willson RL (1985) Organic peroxy free radicals as ultimate agents in oxygen toxicity. In: Sies H (ed) Oxidative stress. Academic Press, London, pp 41–72

Willson RL (1987) Vitamin, selenium, zinc and copper interactions in free radical protection against ill-placed iron. Proc Nutr Soc 46: 27–34

Willson RL, Searle AFJ (1975) Metronidazole (Flagyl): iron catalysed reaction with sulphydryl groups and tumour radiosensitisation. Nature (Lond) 255: 498–500

Witz G, Goldstein BD, Amoruso M, Stone DS, Troll W (1980) Retinoid inhibition of superoxide anion radical production by human polymorphonuclear leukocytes stimulated with tumor promoters. Biochem Biophys Res Commun 97: 883–888

Wright EB, Dormandy TL (1972) Liver zinc in carcinoma. Nature 237: 166

Chapter 11

Zinc Status and Food Intake

B.L. O'Dell and P.G. Reeves

Introduction

Zinc deficiency induces a complex of pathological signs in all animal species studied. Among them are reduced growth rate and food intake, alopecia, skin lesions including parakeratosis and hyperkeratinization, impaired skeletal development and abnormal gait and stance. Of these visual signs, decreased food consumption is one of the first, if not *the* first, to be exhibited as deficiency develops. Reduced food intake in turn has dramatic effects on many other physiological and biochemical parameters so that it is difficult to distinguish the effects of reduced food intake from those specific to zinc deficiency.

Zinc Status, Growth Rate and Food Intake

From the earliest research on the nutritional requirement of young animals for zinc, depressed growth rate has been delineated commonly as the cardinal sign of deficiency. Growth retardation was first described in rats (Stirn et al. 1935) and later in many other species, including pigs (Tucker and Salmon 1955), chicks (O'Dell et al. 1958), calves (Miller and Miller 1962), lambs (Ott et al. 1964), monkeys (Macapinlac et al. 1967), guinea pigs (Alberts et al. 1977), mice (Fraker et al. 1977), immature men (Prasad et al. 1963) and children (Walravens et al. 1976; Golden and Golden 1981). In many of these and innumerable subsequent studies, particularly in the rat, it has been shown that growth depression is largely the result of reduced food intake. This is evident in as much as limiting the intake of zinc-adequate controls (pair feeding) to the ad libitum intake of rats consuming a low-zinc diet results in comparable weight gain over 2- to 5-week periods (Chesters and Quarterman 1970; Faraji and Swenseid 1983; Flanagan 1984; Kasarkis et al. 1986; Wallwork et al. 1981, 1982; Wallwork and Sandstead 1983). In these studies the rate of weight gain of rats fed low-zinc diets (< 1 ppm) ranges from 0% to 40% of ad libitum-fed controls. In some studies the pair-fed controls gained weight significantly faster than those fed the deficient diet, but in others there was no difference. It may be concluded that zinc status has little effect on the efficiency of food utilization. Almost all of the energy consumed by severely deficient young animals, and their pair-fed controls, is used for maintenance.

The growth rate of rats consuming a low-zinc diet can be maintained near normal for a short period by forced feeding of the diet (Faraji and Swenseid 1983; Flanagan 1984) although force-feeding can lead quickly to severe illness and death (Chesters and Quarterman 1970). It is clear that severe zinc deprivation has a rapid and dramatic effect on appetite and consequently on the growth rate of most, if not all, species. As deficiency progresses, food intake is barely sufficient to maintain body mass and the ratio of weight gain to food consumption becomes extremely low, as has been pointed out in many species in which the period, but not daily, food intake was measured.

Daily or more frequent measurement of food intake has been limited largely to growing rats. In this species cyclical daily food intake begins within 4 to 5 d after a low-zinc (< 1 ppm) diet is fed (Williams and Mills 1970; Chesters and Quarterman 1970). Less dramatic cyclical feeding has been observed in guinea pigs (Quarterman and Humphries 1983; Gordon and O'Dell 1983). Recently cyclical feeding has been observed in growing chicks fed a low-zinc diet (J.D. Browning, J.M. Hempe, J.E. Savage, B.L. O'Dell, unpublished observations; see Fig. 11.1). Variable food intake in zinc-deprived human patients has not been described, but zinc supplementation (4 mg/d) of children, 2–6 years of age, who exhibited signs of mild zinc deficiency, significantly increased energy consumption of the males but not of the females (Krebs et al. 1984). However, infants approximately 1 year of age, who were recovering from severe malnutrition and had low plasma zinc levels, responded to zinc therapy with increased weight gain without substantial increase in appetite (Golden and Golden 1981). The failure to observe an appetite effect may be related to the low protein content of the diet consumed, in as much as appetite is depressed in animals fed low-zinc diets only when the protein is at or above the level required for maximal growth rate.

Appetite depression by zinc deprivation is not restricted to young growing animals. It occurs also when nearly adult rats are fed severely zinc-deficient diets. Kasarkis et al. (1986) observed reduced and cyclical feeding in adult male rats, and Record et al. (1986) reported similar patterns of food intake in pregnant rats, which began within 2–3 d after they started consuming a low-zinc diet; the intake of the females ranged from 1 to 21 g/d with a 4-d cycle. Food intake was inversely related to serum zinc and the incidence of congenital malformations correlated with food intake on days 8 and 9 of gestation, a critical period for organogenesis.

Although there is no report of appetite depression in adult men deprived of zinc, there is suggestive evidence of either decreased intake or energy utilization in subjects who consumed experimental diets. Prasad et al. (1978) fed four adult men low-zinc diets (about 3 mg/d) with a constant caloric intake for 6

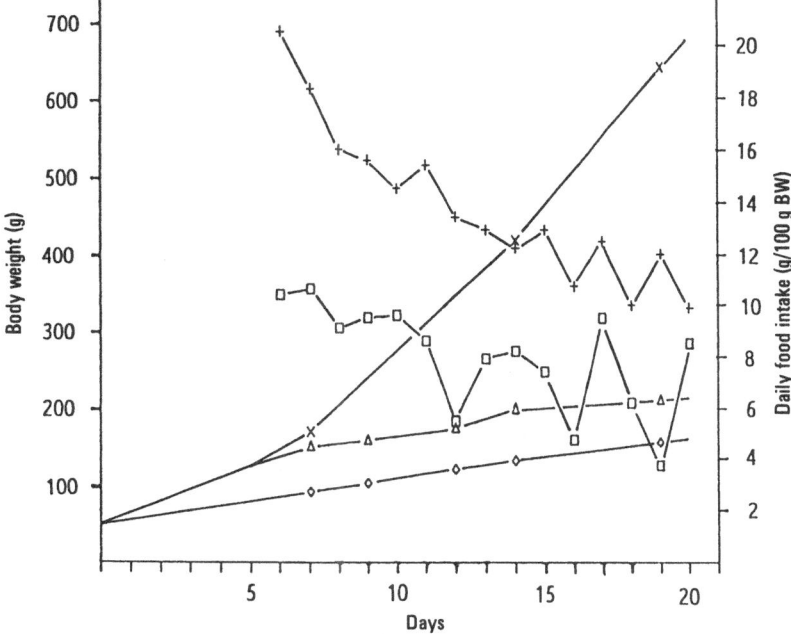

Fig. 11.1. Growth rate and food intake of chicks fed egg-white-based diets containing low or adequate levels of zinc. Groups of broiler strain chicks were fed the respective diets from hatching and food consumption measurements (+, 50 ppm zinc; □, 5 ppm zinc) and restricted feeding were started on day 5. The growth rate of the 5 ppm group (◇) was severely depressed compared with the group fed ad libitum (×) but similar to that of the pair-fed (△) controls. Group synchronization of food intake, depicting the cyclical daily pattern, occurred by chance in this trial.

months and observed continuous weight loss. Baer et al. (1985) fed six men a low-zinc diet (0.28 mg/d) for 5 weeks and found it necessary to increase caloric intake during the experiment to maintain body weight. In man the strong psychological aspects of food intake may override the mechanisms that control feeding in zinc-deficient animals. The apparent species differences may also relate to the fact that the human diets consumed were not as deficient in zinc as those commonly used in animal studies.

Behavioural Patterns of Food Intake in Zinc Deprivation

The typical cyclical feeding pattern of individual animals may be obscured by the group mean intake unless by chance, or otherwise, all in the group are synchronized in their behaviour. Nevertheless, the variation around the mean intake of the zinc-deprived rats is much larger than that of controls (Chesters and Will 1973). The typical cycle in the growing rat is 3–4 d when the diet contains a 20% or higher level of protein; reduction of the protein to 5% increases total intake and eliminates the cyclical pattern (Chesters and Quarterman 1970; Griffith and Alexander 1972). Zinc-deficient rats eat on fewer occasions during the day than controls but consume approximately the same amount at a given meal.

Obviously an animal cannot gain weight unless it consumes more than the maintenance requirement of food, but it is not clear whether or not failure of zinc-deprived rats to eat, and thus to grow, is a direct effect of zinc on appetite or the indirect result of growth failure due to impaired cell division.

If cells cannot divide and grow, the cellular building blocks and their metabolites may feedback to inhibit appetite. The higher food consumption of zinc-deprived rats fed low protein, compared with normal protein, diets led Chesters and Will (1973) to conclude that appetite is not impaired by zinc deprivation per se but results from adaptation to a change in energy balance. Of course, low-protein diets would not support growth, but neither would protein metabolites accumulate. In any case, the food intake of zinc-deficient rats responds within 1–2 h after zinc supplementation of a protein-adequate diet. This rapid response is related to the plasma zinc concentration. There is an inverse relationship between the plasma zinc level of rats in the morning and their food intake during the previous 24-h period (Chesters and Will 1973).

The protein content of low-zinc diets fed to rats affects both their food intake and plasma zinc concentration. Furthermore, the zinc status of rats affects their selection of energy sources. In a 9-d experiment zinc-deficient rats given a choice of two diets, one containing 10% protein and the other 50% protein, consumed 8% less total food than zinc-adequate controls but selected only 15% of their calories as protein while controls selected 30% protein calories (Reeves and O'Dell 1981a, b). The choice of a low-protein diet occurs within 1 d after zinc is removed from the diet and as little as 2 ppm zinc stimulates protein intake (Chesters and Quarterman 1970; Reeves and O'Dell 1981a, b). It has been suggested (Ashley and Anderson 1975; Gibson and Wurtman 1977) that plasma and brain amino acid concentrations determine protein and carbohydrate selection by normal rats. A high ratio of plasma tryptophan to total neutral amino acids correlates with low protein intake. Also in zinc-deficient rats, selection for protein is inversely correlated with high ratios of both tryptophan and tyrosine to total neutral amino acids in plasma (Reeves and O'Dell 1981a, b).

There are conflicting reports that taste perception is impaired in zinc deficiency and conceivably this could have a bearing on food intake. A complete review of this literature is outside the scope of the present chapter, but the reader may consult more extensive reviews of the subject (Henkin 1984; Anonymous 1981). While it is not possible to measure accurately taste acuity in experimental animals, zinc-deficient rats consume excessive quantities of dilute salt solutions and even solutions which, normally, they avoid, such as quinine and hydrochloric acid (Catalanotto 1979). Regardless of the effect of zinc deficiency on taste acuity and perception, there is no experimental evidence that, in the long term, hypogeusia impairs food intake in man or animals.

Possible Mechanisms by which Zinc Status Affects Food Intake

The actual biochemical basis of the decreased food intake observed in zinc deficiency is unknown, but several promising mechanisms have been explored. The hypothalamus has long been recognized as playing a key role in appetite regulation, but it is probably only part of a complex regulatory system. It is now believed that specific areas of the hypothalamus coordinate chemical mediators that stimulate or inhibit efferent pathways at neuronal synapses (Meyers 1975). Amino acids and their metabolites appear to play a key role in the regulatory process.

The fact that zinc deficiency depresses food intake only when dietary protein is adequate to support growth suggests that amino acid metabolism is involved in the regulatory mechanism. The dietary protein and amino acid effect is associated largely with the essential amino acids, particularly methionine, phenylalanine, threonine and tryptophan (Chesters and Will 1973). Griffith and Alexander (1972) found that the plasma amino acid pattern is changed in zinc-deficient rats regardless of the dietary protein level. The levels of non-essential amino acids, including glycine, serine, cystine and tyrosine, are elevated. This relationship in zinc deficiency is reminiscent of the depressed food intake associated with amino acid imbalance. Out of a large body of work on amino acid nutriture came the suggestion that non-utilizable amino acids serve as a satiety signal (Peng et al. 1969). In zinc-deficient rats, only plasma tyrosine levels correlate with the daily fluctuation in food intake; however, there is no change in plasma tyrosine levels when rats consume a low-protein, low-zinc diet (Griffith and Alexander 1972).

Although plasma amino acid concentrations may serve as a food regulation signal, the plasma zinc concentration is also a viable candidate even though cyclical feeding does not occur when the diet is low in protein. Bettger (1985) has shown that the level of protein or of an amino acid mixture in a low zinc diet has a significant effect on the plasma zinc level within 16 h after the diet is fed to fasted rats. When zinc is limiting, high amino acid or protein intake lowers plasma zinc whereas diets containing low or no protein have no effect. Fasting also maintains a normal plasma zinc concentration.

The ratios of tryptophan and tyrosine to the total neutral amino acids in plasma correlate inversely with the intake of a normal protein diet (Reeves and O'Dell 1981a, b; Wallwork et al. 1981). Reeves and O'Dell (1981a, b) found no differences in brain levels of tyrosine, tryptophan, the catecholamines or serotonin. However, Wallwork and Sandstead (1983) observed an inverse correlation of the brain tyrosine to neutral amino acid ratio and food intake in zinc-deficient rats; there was not a correlation with the ratio of tryptophan to neutral amino acids. If the ratio of tyrosine to neutral amino acids is elevated in the plasma of zinc-deficient rats, it is reasonable to assume that brain tyrosine is increased at some stage of the cyclical feeding. This could have a direct effect on catecholamine levels and function, as will be discussed below.

Recent work implicates the cholinergic neurons of the hypothalamus as a major component in the regulation of food ingestion (Leibowitz 1980).

Norepinephrine can inhibit or stimulate the feeding response depending on the area of the hypothalamus in which it is injected (Grossman 1962). The paraventricular nucleus in the anterior hypothalamus is particularly sensitive to exogenous norepinephrine (Leibowitz and Brown 1980). Since norepinephrine is predominantly an inhibitory transmitter in the brain, present evidence suggests that release of norepinephrine inhibits neurons which in turn activate the feeding response (Baker et al. 1971). There is also evidence that dopamine is involved in the control of food intake (Heffner et al. 1980).

Wallwork et al. (1982) found the concentration of norepinephrine to be elevated in the whole brains of zinc-deficient compared with control rats, while Reeves and O'Dell (1981a, b) found no difference in the catecholamine levels in the whole brain but significant differences in the hypothalamus. Rats fed a zinc-deficient diet for only 48 h had elevated norepinephrine levels in the anterior hypothalamus compared with pair-fed controls. Kasarkis et al. (1986) also found elevated levels of norepinephrine in the hypothalamuses of rats fed a low-zinc diet for 10 d; there was no difference after 4 d. These workers found elevated levels of tyrosine and dihydroxyphenylalanine in this tissue and proposed that impaired norepinephrine recognition sites result from zinc deficiency. Reeves and O'Dell (1984) postulated that the increased level of norepinephrine in the anterior hypothalamus is part of the trigger mechanism that initiates food intake depression in zinc-deprived rats. They found that rats fed for 19 d on low-zinc diets containing no tyrosine and a low level of phenylalanine had higher food intake than that of comparable animals fed higher levels of these amino acids. This was coupled with lower levels of both norepinephrine and dopamine, as well as of tyrosine, in the anterior hypothalamus.

Numerous neuroactive compounds besides the catecholamines and serotonin, including the opiate and intestinally active peptides, exert a function in appetite control. Essatara et al. (1984a, b) have studied the effects, in zinc-deficient rats, of various intracerebroventricularly injected agents known to induce feeding. Included were norepinephrine, muscimol, a γ-aminobutyrate agonist, bromergocryptine, a dopamine agonist and dynorphin, an endogenous opiate. In general the zinc-deficient rats exhibited resistance to feeding induced by all of these agents. Twice as much norepinephrine and muscimol was required to increase intake above baseline and the response was one-half or less that of controls. Zinc-deficient rats required ten times more dynorphin to stimulate a feeding response than did controls. The authors made the reasonable suggestion that the anorexia of zinc deficiency results from defective opiate or other receptors.

Future Research Needs

A dietary deficiency of many nutrients depresses appetite, but zinc deprivation causes an unusually rapid onset that is readily reversible. The zinc-deficient rat in particular provides an excellent model for exploration of food intake regulatory mechanisms and research in this area deserves support. The brain and intestinal peptides should be investigated in zinc-deprived animals in as much as both opiates and cholecystokinin exert a marked effect on food intake (Baile et al. 1986). Equally, the food intake depression associated with zinc deficiency deserves the attention of researchers in order to provide insight into the basic physiological and biochemical function of this important nutrient. An understanding of the mechanisms involved would also be useful in the design of experiments to define the other biological roles of zinc.

With the commonly used techniques of pair feeding or pair weighing, it is not possible to separate adequately the effects of zinc itself from those of low food intake. Other approaches to this problem have been made, and they, as well as others, should be explored further. The usual pair-fed animal rapidly becomes a meal-feeder and fasts most of the day. Thus, pair feeding is an inadequate control for many metabolic studies involving zinc status. Quarterman et al. (1970) attempted to circumvent this problem by using a piece of apparatus that automatically supplies small amounts of food uniformly and continuously. This technique has not been widely adopted, and would require computer programming to simulate the normal feeding pattern, which in the case of rats occurs mostly at night and as discrete meals. Force-feeding, as used by Flanagan (1984) and Park et al. (1986), offers promise for short-term food intake control, but fails after about 8 d in rats fed extremely low-zinc diets. Another approach is to make all experimental animals (zinc deficient and controls) meal feeders (Reeves and O'Dell 1983). All of the techniques presently available suffer from shortcomings to the extent that new ideas and procedures are much needed.

Conclusions

Reduced food intake is one of the first signs of zinc deprivation in most species studied. Not only is food consumption in rats dramatically and quickly decreased, but on a daily basis it is highly cyclical. There have been few food intake studies involving children and adult human subjects who have consumed low-zinc diets, but the growth rate of low zinc status infants and children is markedly improved by zinc supplementation. The cyclical daily food intake observed in zinc deficiency correlates with both zinc and tyrosine concentrations in plasma. Whether plasma zinc plays a direct regulatory role or acts indirectly through other regulatory agents, such as brain catecholamines and peptides, and plasma amino acids, is unclear.

References

Alberts JC, Lang JA, Reyes PS, Briggs GM (1977) Zinc requirement of the young guinea pig. J Nutr 107: 1517–1527
Anonymous (1981) Decreased taste acuity in chronic renal patients. Nutr Rev 39: 207–210
Ashley DVM, Anderson GH (1975) Correlation between the plasma tryptophan to neutral amino acid ratio and protein intake in the self-selecting weanling rat. J Nutr 105: 1412–1421

Baer MT, King JC, Tamura T et al. (1985) Nitrogen utilization, enzyme activity, glucose intolerance and leukocyte chemotaxis in human experimental zinc depletion. Am J Clin Nutr 41: 1220–1235

Baile CA, McLaughlin CL, Della-Fera MA (1986) Role of cholecystokinin and opioid peptides in control of food intake. Physiol Rev 66: 172–234

Baker JL, Crayton JW, Nicoll RA (1971) Noradrenaline and acetylcholine responses of supraoptic neurosecretory cells. J Physiol 218: 19–32

Bettger WJ (1985) Effect of dietary protein or amino acids on the rapid change in plasma zinc concentration in rats fed zinc deficient diets. Nutr Res 5: 1153–1159

Catalanotto FA (1979) Alterations of short-term tastant-containing fluid intake in zinc deficient adult rats. J Nutr 109: 1079–1085

Chesters JK, Quarterman J (1970) Effects of zinc deficiency on food intake and feeding patterns of rats. Br J Nutr 24: 1061–1069

Chesters JK, Will M (1973) Some factors controlling food intake by zinc-deficient rats. Br J Nutr 30: 555–566

Essatara M'B, McClain CJ, Levine AS, Morley JE (1984a) Zinc deficiency and anorexia in rats: the effect of central administration of norepinephrine, muscimol and bromergocriptine. Physiol Behav 32: 479–482

Essatara M'B, Morley JE, Levine AS, Elson MK, Shafer RB, McClain CJ (1984b) The role of endogenous opiates in zinc deficiency anorexia. Physiol Behav 32: 475–478

Faraji B, Swenseid ME (1983) Growth rate, tissue zinc levels and activities of selected enzymes in rats fed a zinc-deficient diet by gastric tube. J Nutr 113: 447–455

Flanagan PR (1984) A model to produce pure zinc deficiency in rats and its use to demonstrate that dietary phytate increases the excretion of endogenous zinc. J Nutr 114: 493–502

Fraker PJ, Haas SM, Luecke RW (1977) Effect of zinc deficiency on the immune response of the young adult A/J mouse. J Nutr 107: 1889–1895

Gibson CJ, Wurtman RJ (1977) Physiological control of brain catechol synthesis by brain tyrosine concentration. Biochem Pharmacol 26: 1137–1142

Golden MHN, Golden BE (1981) Effect of zinc supplementation on the dietary intake, rate of weight gain and energy cost of tissue deposition in children recovering from severe malnutrition. Am J Clin Nutr 34: 900–908

Gordon PR, O'Dell BL (1983) Zinc deficiency and impaired platelet aggregation in guinea pigs. J Nutr 113: 239–245

Griffith PR, Alexander JC (1972) Effect of zinc deficiency on amino acid metabolism of the rat. Nutr Repts Intl 6: 9–20

Grossman SP (1962) Direct adrenergic and cholinergic stimulation of hypothalamus mechanisms. Am J Physiol 202: 872–882

Heffner TG, Hartman JA, Seiden LS (1980) Feeding increased dopamine metabolism in the rat. Science 208: 1168–1170

Henkin RI (1984) Zinc in taste function. Biol Trace Element Res 6: 263–280

Kasarkis EJ, Sparks DL, Slevin JT (1986) Changes in hypothalamic noradrenergic systems during the anorexia of zinc deficiency. Biol Trace Element Res 9: 25–35

Krebs NF, Hambidge KM, Walravens PA (1984) Increased food intake of young children receiving a zinc supplement. Am J Dis Child 138: 270–273

Leibowitz SF (1980) Control of feeding and drinking behavior and water electrolyte excretion. In: Morgane PJ, Panksepp J (eds) Behavioral studies on the hypothalamus, vol 3, part A. Marcel Dekker, New York, pp 229–437

Leibowitz SF, Brown LL (1980) Histochemical and pharmacological analysis of noradrenergic projections to the paraventricular hypothalamus in relation to feeding stimulation. Brain Res 201: 289–314

Macapinlac MP, Barney GH, Pearson WN, Darby WJ (1967) Production of zinc deficiency in the squirrel monkey. J Nutr 93: 499–510

Miller JK, Miller WJ (1962) Experimental zinc deficiency and recovery of calves. J Nutr 76: 467–474

Meyers RD (1975) Handbook of drug and chemical stimulation of the brain. Van Nostrand Reinhold, New York

O'Dell BL, Newberne PM, Savage JE (1958) Significance of dietary zinc for the growing chicken. J Nutr 65: 503–523

Ott EA, Smith WH, Stob M, Beeson WM (1964) Zinc deficiency syndrome in the young lamb. J Nutr 82: 41–50

Park JHY, Grandjean CJ, Hart MH, Erdman SH, Pour P, Vanderhoof JA (1986) Effect of pure zinc deficiency on glucose tolerance and insulin and glucagon levels. Am J Physiol 251: E273–E278

Peng Y, Benevenga NJ, Harper AE (1969) Amino acid imbalance and food intake: effect of previous diet on plasma amino acids. Am J Physiol 216: 1020–1025

Prasad AS, Miale A, Farid Z, Sandstead HH, Schulert AR, Darby WJ (1963) Biochemical studies on dwarfism, hypogonadism, and anemia. Arch Int Med 111: 407–428

Prasad AS, Rabbani P, Abbassi A, Bowersox E, Spivey Fox MR (1978) Experimental zinc deficiency in human. Ann Intern Med 89: 483–490

Quarterman J, Humphries WR (1983) The production of zinc deficiency in the guinea pig. J Comp Pathol 93: 261–270

Quarterman J, Williams RB, Humphries WR (1970) An apparatus for the regulation of the food supply to rats. Br J Nutr 24: 1049–1051

Record IR, Dreosti IE, Tulsi RS, Manuel SJ (1986) Maternal metabolism and teratogenesis in zinc-deficient rats. Teratology 33: 311–317

Reeves PG, O'Dell BL (1981a) Short-term zinc deficiency in the rat and self-selection of dietary protein level. J Nutr 111: 375–383

Reeves PG, O'Dell BL (1981b) Regulation of protein intake in zinc deficient rats. In: Howell JMcC, Gawthorne JM, White CL (eds) Trace element metabolism in animals and man. Australian Academy of Science, Canberra, pp 338–341

Reeves PG, O'Dell BL (1983) The effect of zinc deficiency on glucose metabolism in meal-fed rats. Br J Nutr 49: 441–452

Reeves PG, O'Dell BL (1984) The effect of dietary tyrosine levels on food intake in zinc deficient rats. J Nutr 114: 761–767

Stirn FE, Elvehjem CA, Hart EB (1935) The indispensability of zinc in the nutrition of the rat. J Biol Chem 109: 347–359

Tucker HF, Salmon WD (1955) Parakeratosis or zinc deficiency disease in the pig. Proc Soc Exp Biol Med 88: 613–616

Wallwork JC, Sandstead HH (1983) Effect of zinc deficiency on appetite and free amino acid concentrations in rat brain. J Nutr 113: 47–54

Wallwork JC, Fosmire GJ, Sandstead HH (1981) Effect of zinc deficiency on appetite and plasma amino acid concentrations in the rat. Br J Nutr 45: 127–136

Wallwork JC, Botnen JH, Sandstead HH (1982) Influence of dietary zinc on rat brain catecholamines. J Nutr 112: 514–519

Walravens PA, Hambidge KM (1976) Growth of infants fed a zinc supplemented formula. Am J Clin Nutr 29: 1114–1121

Williams RB, Mills CF (1970) The experimental production of zinc deficiency in the rat. Br J Nutr 24: 989–1003

Chapter 12

Zinc and Reproduction: Effects of Deficiency on Foetal and Postnatal Development

C.L. Keen and Lucille S. Hurley†

Introduction

The first evidence that zinc was an essential nutrient for animals was provided by Todd et al. in 1934 using the rat as an experimental model. Twenty-five years later, Turk et al. (1959) showed that chicks hatched from dams fed a zinc-deficient diet were weak and died within 4 d. Subsequent studies in the chick demonstrated that severe maternal zinc deficiency resulted in skeletal defects in the offspring, including agenesis of the limbs, skull and beak deformities, dorsal curvature of the spine and shortened and fused vertebrae. Other anomalies included brain abnormalities, microphthalmia, micromelia and herniation of viscera (Blamberg et al. 1960; Keinholz et al. 1961).

Congenital Malformations in Zinc-Deficient Mammals

Severe Zinc Deficiency: Gross Malformations

In 1966, Hurley and Swenerton, using the Sprague-Dawley rat as a model, demonstrated that dietary zinc deficiency is teratogenic in mammals. Dams fed diets severely deficient in zinc (< 1 μg zinc/g diet) throughout pregnancy gained less weight than controls and had significantly fewer live foetuses at term than did controls. Surviving foetuses from the deficient dams were significantly lighter than controls, despite the smaller number of live foetuses per litter. In addition these foetuses were characterized by a variety of congenital defects including skeletal abnormalities, delayed ossification of the bones and skull, misshapen heads, fused or missing digits and micrognathia. In addition to the external malformations, foetuses from zinc-deficient dams showed internal soft tissue malformations affecting the brain, heart, lungs and urogenital system. That the observed defects were a direct result of the maternal zinc deficiency was suggested by the observation that the foetuses from the zinc-deficient dams contained significantly less zinc (content and concentration) than control foetuses (Hurley and Swenerton 1966; Hurley 1967).

Since the mid-1960s it has been shown that for the rat, virtually every developing organ system can be affected by prenatal zinc deficiency (Hurley 1969, 1981; Mills et al. 1969; Apgar 1972; Dreosti et al. 1972; Warkany and Petering 1972; Rogers et al. 1985a, b, 1987). Typical foetal defects associated with severe maternal zinc deficiency are shown in Fig. 12.1. In addition to the wide range of defects observed, it is important to note that the frequency of these defects within an entire litter, and within individual foetuses, is very high (Table 12.1). Overall, it has been conservatively estimated that up to 90% of all implantation sites are affected adversely by severe maternal zinc deficiency in the rat (Hurley 1981; Rogers et al. 1985a).

As in the rat, gestational zinc deficiency in sows, ewes and mice also results in a marked increase in the frequency of foetal defects (Palludan and Wegger 1976; Hackman and Hurley 1983; Apgar and Fitzgerald 1985; Sato et al. 1985; Dreosti et al. 1986).

In addition to gross structural defects, severe gestational zinc deficiency has also been shown to affect adversely the biochemical development of the lung and pancreas. Maturation of the lung in preparation for birth requires the synthesis of pulmonary surfactant. In a study by Vojnik and Hurley (1977), offspring of zinc-deficient dams had low lung concentrations of phosphatidylcholine and phosphatidylethanolamine compared with controls, indicating a defect in the production of pulmonary surfactant. Daston (1982) observed similar low levels of pulmonary surfactants in foetuses of cadmium-exposed rats, and ascribed the defect to a cadmium-induced foetal zinc deficiency.

In the developing foetal pancreas, maternal zinc deficiency resulted in low zymogen granulation of acinar cells near the end of gestation. Procarboxypeptidase A and chymotrypsinogen were 30% lower per cell than in controls (Robinson and Hurley 1981a). An additional observation of this work was that zinc-deficient foetuses taken from large litters of seven or more had lower levels of procarboxylase A and chymotrypsinogen in the placenta than did zinc-deficient foetuses taken from smaller litters (Robinson and Hurley 1981a). Pancreatic insulin and glucagon levels in zinc-deficient foetuses were also lower than those found in control foetuses. Based on results from pair-fed controls, it was determined that the low insulin levels were in part due to maternal food restriction, while the low glucagon levels were independent of maternal food intake (Robinson and Hurley 1981b).

Timing and Rapid Effects of Severe Maternal Zinc Deficiency

Using the rat as a model, the timing of maternal zinc deficiency has been shown to be critical in its impact on development. Hurley and Swenerton

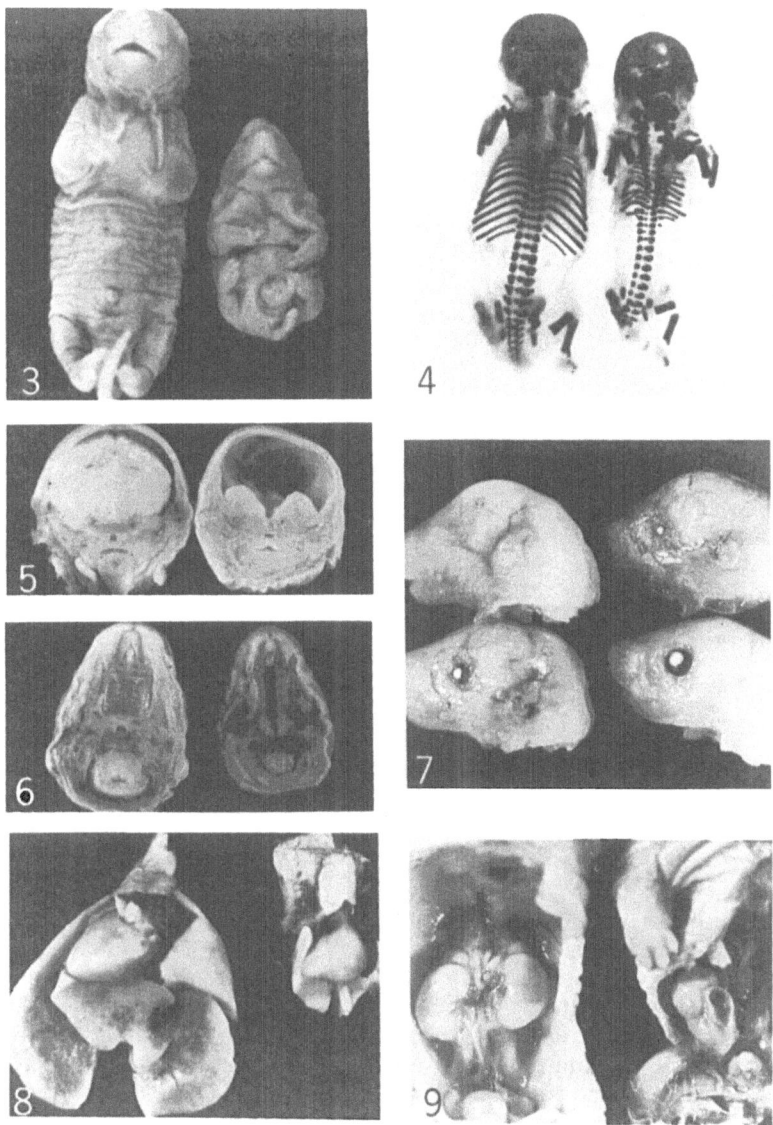

Fig. 12.1. Typical abnormalities observed in zinc-deficient foetuses from rats fed 0.5 µg zinc/g diet. (3) *Left*: control foetus, day 21. *Right*: day 21 zinc-deficient foetus. Note severe growth retardation, syndactyly, omphalocoele and short tail. (4) Alizarin-red skeletal preparations. *Left*: control foetus, day 21. *Right*: day 21 zinc deficient. Note short wavy ribs and poorly ossified occiputs and vertebral centra. (5) *Left*: razor section through the cerebral region of a control foetus, day 21. *Right*: cerebral region of a day 21 zinc-deficient foetus, showing severe hydrocephalus. (6) *Left*: palatal region of a control foetus, day 21. *Right*: cleft palate in a day 21 zinc-deficient foetus. (7) Day 19 zinc-deficient foetuses from a single litter, showing varying severity of microphthalmia. (8) *Left*: lungs and heart of a control foetus, day 21. *Right*: lungs and heart of a day 21 zinc-deficient foetus. Lung size is severely affected, but all lobes are present. (9) *Left*: dorsal abdominal region of a control foetus, day 21, with normal kidneys. *Right*: day 21 zinc-deficient foetus. Note horseshoe kidney and hydroureter. Reprinted with permission, Rogers et al. 1985b.

Table 12.1. Malformations in rat foetuses from dams fed 0.5 μg zinc/g diet during gestation

Malformation	Litters examined		Foetuses examined	
	No.	(% affected)	No.	(% affected)
External malformations				
Cranioschisis or exencephaly	16	(81)	102	(47)
Microphthalmia	25	(68)	184	(38)
Cleft palate	17	(94)	127	(87)
Micrognathia	16	(12)	102	(6)
Omphalocoele	16	(50)	102	(23)
Arthrogryposis	16	(88)	102	(67)
Syndactyly	16	(94)	102	(58)
Short or kinky tail	16	(94)	102	(85)
Internal malformations[a]				
Hydrocephalus	16	(75)	61	(36)
Small lungs	16	(94)	61	(66)
Diaphragmatic hernia	16	(6)	61	(2)
Hydronephrosis and/or hydroureter	16	(63)	61	(26)
Spina bifida	16	(31)	61	(15)
Skeletal defects[b]				
Cranial defects	–		21	(57)
Vertebral defects	–		21	(71)
Missing or wavy rib	–		21	(90)
Sternal defects	–		21	(90)
Missing tibia and fibula	–		21	(9)
Missing digits	–		21	(48)

Table adapted with permission from data presented in Rogers et al. (1985b).
[a] Revealed by razor sectioning.
[b] Twenty-one foetuses from six litters examined using Alizarin-red staining.

(1966) reported that when a zinc-deficient diet was fed prior to mating, there was a severe disruption of the oestrous cycle and mating did not take place. Similar observations have been made for the bonnet monkey (Swenerton and Hurley 1980), mice and hamsters (Watanabe et al. 1983). For this reason, a diet severely deficient in zinc is usually not introduced experimentally before day 0 of pregnancy, the day that mating is confirmed by the presence of sperm plugs or sperm in vaginal smears. While zinc deficiency throughout pregnancy in the rat (21 d) results in a high frequency of foetal malformations, it is important to recognize that even transitory periods of zinc deficiency during pregnancy can result in foetal malformations. When rats were fed a zinc-deficient diet from day 0 to day 10 of pregnancy, 22% of the foetuses were malformed; for rats fed zinc-deficient diets from day 0 to day 12, 56% of the foetuses were malformed. When the deficient diet was fed from day 6 to day 14 of pregnancy, 46% of the foetuses were malformed (Hurley et al. 1971).

Warkany and Petering (1972) reported that even 3 d of zinc deficiency, from day 10 to day 12 of pregnancy, were sufficient to result in brain abnormalities and that even brains which appeared grossly normal were characterized by microscopic lesions. When zinc deficiency was imposed only during the last third of pregnancy, foetal brains were smaller in absolute terms than were those of controls, but were normal relative to body weight, although livers were smaller in both bases (McKenzie et al. 1975). It should be noted that the type of malformations induced with different periods of zinc deficiency is dependent on the developmental events occurring when the deficiency is imposed. For example, eye and brain defects are associated with early zinc deficiency, while cleft palate is normally observed only when the deficiency is imposed after day 12 of pregnancy (Rogers et al. 1985b).

The rapid and severe effects of transitory periods of zinc deficiency on foetal development strongly suggest the absence of mobilizable zinc pools in the dam which can be utilized during brief periods of zinc deficiency. Consistent with this idea are reports that plasma zinc concentrations in the rat can decrease by over 50% in the first 24 h following the introduction of a zinc-deficient diet (Dreosti et al. 1968; Hurley et al. 1982). It is reasonable to suggest that the rapid effect of zinc deficiency on the embryo/foetus is due to the rapid reduction in maternal plasma zinc as this is the primary source of zinc for developing mammalian embryos. Additional evidence that it is the reduction in plasma zinc that is critical for the embryo is the observation that if tissue catabolism is increased in the zinc-deficient mother by either simultaneously withholding dietary calcium (Hurley and Tao 1972; Tao and Hurley 1975; Masters et al. 1986) or by reducing the intake of the zinc-deficient diet (Masters et al. 1986), the teratogenicity of the zinc deficiency is partially alleviated. In both cases it has been shown that the increased maternal tissue catabolism (bone and muscle) results in a release of zinc into the plasma, and foetal zinc concentrations are increased relative to those in foetuses from zinc-deficient dams with normal rates of tissue catabolism.

A rapid effect of zinc deficiency on preimplantation embryos has also been shown (Hurley and Shrader 1975). In eggs flushed from the oviducts of control rats on day 3 of pregnancy, all of the eggs appeared normal and were at the eight or 16 cell stages. in contrast only 70% of the eggs collected from dams fed a zinc-deficient diet during the first 3 d of pregnancy were found to be normal and all the eggs were at the eight cell stage. The remaining eggs had either failed to cleave and were undergoing necrosis with disruption and fragmentation of cellular contents or consisted of three to seven blastomeres of variable size and orientation. In eggs collected from dams on day 4 of pregnancy, less than 25% of the eggs from zinc-deficient dams were at the normal morula or blastocyst stage, while 98% of the eggs collected from control dams were at a normal blastocyst stage. Typical abnormalities observed in eggs collected from zinc-deficient dams are shown in Fig. 12.2.

The only sources of zinc for the preimplantation mammalian embryo are maternal oviductal and uterine fluids. Similar to the findings for maternal plasma, uterine fluid zinc concentrations were 50% lower in dams fed zinc-deficient diets for 4 d than in controls (Gallaher and Hurley 1980). (The zinc concentration of uterine fluid was about five times higher than that of plasma in both groups.) Thus the reduction in maternal plasma zinc concentration following the introduction of a zinc-deficient diet is reflected in other body fluids, indicating that the extent of the decrease in plasma zinc is functionally significant.

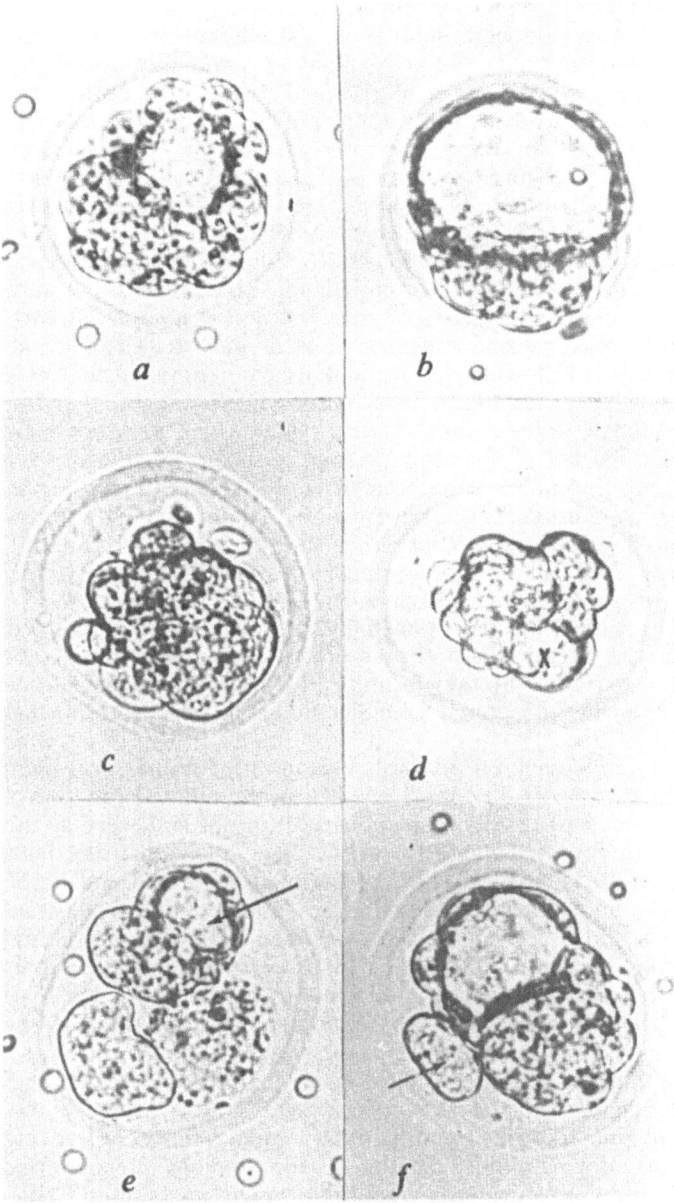

Fig. 12.2. Examples of embryos obtained from the uterine horns of control (**a** and **b**) and zinc-deprived (**c–f**) female rats on day 4 of gestation. Note the presence of an intact zona pellucida in all examples. All embryos of both control and deficient dams were stained successfully with neutral red following photography (demonstrating viability). **a** Control, early blastula stage. **b** Control, normal blastula. **c** Abnormal morula showing variation in blastomere size. **d** Early blastocyst from zinc-deficient female showing derangement of blastomeres and beginning of blastocoele (*X*). **e** Abnormal blastulation with dissociated blastomeres and a degenerating mass of cytoplasm. Blastocoele indicated by *arrow*. **f** Abnormal, lateral continuation of blastocoele cavity (*arrow*). Reprinted with permission, Hurley and Shrader 1975.

Mechanisms of Teratogenicity of Zinc Deficiency in Experimental Animals

It is evident from the foregoing studies that severe maternal zinc deficiency can result in embryonic and foetal abnormalities; however, the mechanisms underlying these defects are not well understood. A fundamental question is whether the effects of zinc deficiency on the embryo/foetus are due directly to a deficiency of zinc in embryonic cells, or do they occur in part through an indirect effect of zinc deficiency on the metabolism of the mother. The observation that feeding a zinc-deficient diet results in lower than normal maternal plasma and uterine fluid zinc concentrations, as well as lower embryonic and foetal zinc concentrations, is suggestive of a direct effect of zinc deficiency on the embryo/foetus, but does not prove a causative relationship between low foetal zinc and malformations.

For example, since one effect of zinc deficiency is a marked reduction in food intake (Chesters and Quarterman 1970; Kawamoto et al. 1986), it could be argued that some of the teratogenic effects of zinc deficiency are secondary to maternal food restriction. This possibility has been tested using pair-fed controls. Although pair-fed females gained almost no weight during pregnancy (in contrast to ad libitum-fed controls), and their foetuses were smaller than those of ad libitum-fed controls, normal formations were observed. Thus, the gross teratogenic effects of zinc deficiency are not related to the depressed food intake produced by this condition (Hurley et al. 1971). However, it is important to point out that while it is evident that gross structural defects associated with prenatal zinc deficiency are not due to maternal food restriction, more subtle biochemical lesions may be, in part, due to this effect of zinc deficiency. Thus, it is important that pair-fed controls be used in investigations of the biochemical lesions underlying the teratogenicity of zinc deficiency.

One recent approach to the investigation of the direct effects of zinc deficiency on the embryo has been the use of embryo culture systems. In these systems the potential teratogenic effects of zinc-induced changes in maternal metabolism can be eliminated. Using the embryo culture system, embryos are removed from control dams during early development (typically from day 8 to day 9.5 in the rat) (New 1966; New et al. 1973). The embryos are then placed in culture tubes containing rat serum and distilled water, cultured up to 48 h and examined. The strength of this system is that an investigator can use serum collected from either control or zinc-deficient rats. Mieden et al. (1986) using this system have reported that rat embryos grown on serum collected from zinc-deficient dams developed abnormally. Abnormal embryos exhibited a single malformation complex; the heads were small, the optic sulci narrow and the optic placode ill-defined (Fig. 12.3). In addition, the embryos were characterized by a shorter crown–rump length, a small head/body ratio and a lower protein content than shown by embryos grown on control serum. While these data strongly suggest that some of the effects of zinc deficiency on the embryo are direct, they do not rule out the possibility that other factors in maternal serum were altered by the feeding of the zinc-deficient diet and that it is these alterations that are producing the teratogenic effects.

To clarify this issue, Mieden et al. (1986) added zinc to the zinc-deficient serum so that its total concentration of zinc was equal to that found in control rat serum. Embryos grown on the zinc-deficient sera which were supplemented with zinc developed normally, supporting the idea that the teratogenicity of the initial zinc-deficient sera was the direct result of the low zinc concentrations (Fig. 12.3). Similarly, defects in neural tube closure have also been observed in early chick embryo explants cultivated in a zinc-deficient medium (Iniguez et al. 1978).

In contrast to the findings of Mieden et al. (1986), Record et al. (1985a) were unable to produce the teratogenic effects of zinc deficiency in the rat embryo culture system using methodologies very similar to those employed by Mieden et al. (1986). These investigators reported that in their experiments, embryos collected from control dams developed normally for 48 h when grown on sera collected from deficient dams, although the yolk-sac protein content was lower in the embryos grown on the deficient sera than in controls. The

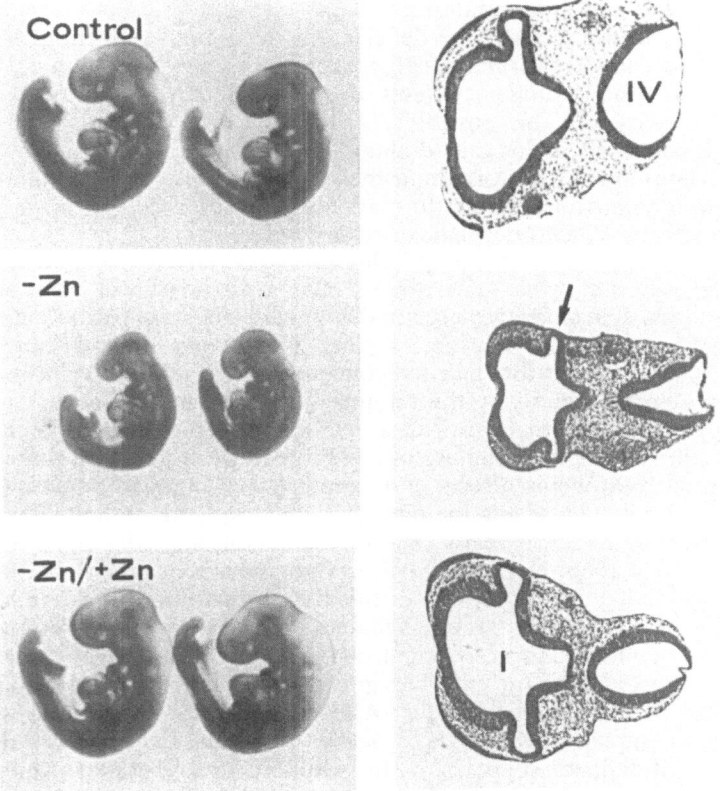

Fig. 12.3. Cultured 48-h whole rat embryos (×12) and horizontal sections (×50) through the optic vesicles of respective embryos. Control represents embryos cultured on control serum. −Zn, embryos cultured on zinc-deficient serum. These embryos are abnormal, having small heads, narrow optic sulci and ill-defined otic placodes. The same serum samples used to culture the above abnormal −Zn embryos were then supplemented with zinc, and the resulting cultured embryos are shown in −Zn/+Zn. The serum teratogenicity was overcome; the embryos were similar to controls in all respects. *Arrow* indicates optic sulci; *I*, lateral ventricle; *IV*, 4th ventricle. Reprinted with permission, Mieden et al. 1986.

explanation for the difference in results between the two rat embryo culture studies has yet to be determined.

While Record et al. (1985a) did not document abnormal development in control embryos grown on zinc-deficient serum, they did observe abnormal development in 9.5-day-old embryos collected from zinc-deficient dams which were then cultured on zinc-deficient sera for 48 h. A particularly interesting observation was that the embryos from the zinc-deficient dams fell into two broad morphological classes. The first group appeared normal at day 9.5 when viewed under a dissecting microscope, while the second group was generally smaller and the embryonic pole of the egg cylinder appeared to be developmentally abnormal and/or retarded. Culture of the first group of embryos in either zinc-deficient or zinc-supplemented serum produced morphologically normal embryos; in contrast, those that appeared abnormal at day 9.5 were grossly malformed after 48 h incubation in either serum (Fig. 12.4). Future studies directed at investigating the biochemical differences between these two subgroups of embryos collected from zinc-deficient dams should provide valuable new information on the biochemical lesions underlying zinc deficiency teratogenicity.

With regard to potential biochemical lesions which may underlie in part the negative effects of zinc deficiency on embryonic and foetal development, several ideas have received attention. Some of the postulated biochemical

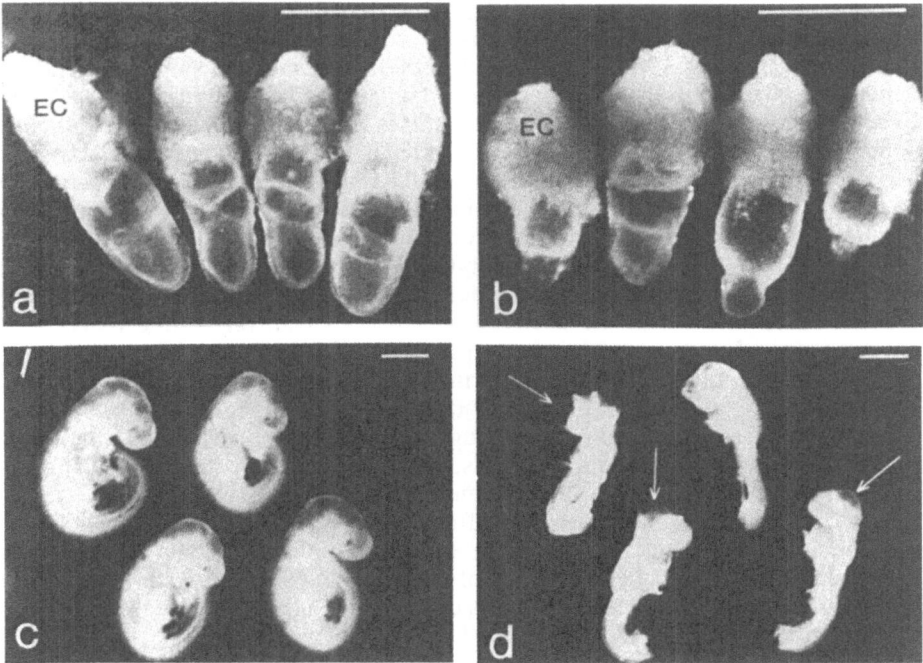

Fig. 12.4. Zinc-deficient morphologically normal (**a**) and abnormal (**b**) egg cylinders removed on day 9.5 of gestation; 11.5-day-old embryos grown from zinc-deficient egg cylinders that were apparently normal (**c**) or morphologically abnormal (**d**). *EC,* ectoplacental cone. Bar = 1 mm. Reprinted with permission, Record et al. 1985.

lesions are briefly discussed below. For additional treatment of this subject the reader is directed to other recent reviews (Hurley 1981; Dreosti et al. 1985; Hambidge et al. 1986).

Considerable attention has been given to the idea that one of the basic defects underlying the abnormalities observed in zinc deficiency is abnormal nucleic acid metabolism. DNA synthesis is significantly depressed in zinc-deficient embryos and foetuses compared with controls, and this lower rate of synthesis has been linked to depressed activities of DNA polymerase and thymidine kinase (Swenerton et al. 1969; Dreosti and Hurley 1975; Eckhert and Hurley 1977; Dreosti et al. 1980). Biochemical abnormalities of chromatin have also been reported in zinc-deficient *Euglena gracilis* (Falchuk et al. 1986) and in zinc-deficient adult and weanling rats (Castro et al. 1986; Clegg et al. 1987). If the impairment in nucleic acid synthesis or expression is sufficient, then this could result in alterations in the differential rates of cellular growth necessary for normal morphogenesis. Evidence that the rate of cellular growth is altered in zinc-deficient embryos is provided by the finding that the mitotic index in the brains of zinc-deficient embryos is higher than in control brain (Hurley and Shrader 1972; Eckhert and Hurley 1977). The resulting asynchrony in cellular growth could then result in structural defects.

In addition to abnormalities in nucleic acid metabolism, several other metabolic defects may also be occurring as a result of cellular zinc deficiency. For example, there is considerable evidence that cellular membranes and cytoskeletal integrity can be affected by zinc deficiency. Tubulin assembly has been shown to be impaired in brains of zinc-deficient adult, weanling and foetal rats (Hesketh 1981; Oteiza et al. 1987). The mechanism by which zinc deficiency affects tubulin polymerization has yet to be established; however, since the concentration of tubulin is similar in brain supernatants of zinc-deficient and control rats (Oteiza et al. 1987), it can be postulated that zinc plays a direct role in the regulation of tubulin polymerization in vivo. Zinc has been shown to stabilize neurotubules in vitro, possibly through the formation of zinc mercaptide bridges between the tubulin dimer subunits (Nicholson and Veldstra 1972). Abnormal tubulin polymerization resulting from cellular zinc deficiency could result in significant defects in cellular endo- and exocytosis. For example, a defect in the normal exocytosis of transferrin due to reduced microtubule polymerization has been proposed as one mechanism underlying the hepatic accumulation of iron associated with zinc deficiency (Rogers et al. 1987). Defects in tubulin polymerization have also been postulated as a potential biochemical lesion underlying chromosomal aberrations, including gaps, fragments and terminal deletions, that are found in the tissues of zinc-deficient foetuses (Bell et al. 1975; Keen and Hurley 1987). Similarly, maternal zinc deficiency has been reported to affect foetal brain cell cycle kinetics with the deficiency resulting in a block in the G_0/G_1 phase (Clegg et al. 1986). While this block could be the result of an impairment in nucleic acid and/or protein synthesis, it could also arise as a result of decreased tubulin polymerization (Keen and Hurley 1987).

Finally, it has been reported that in both maternal and foetal liver of zinc-deficient rats, there are significantly elevated levels of malondialdehyde, an indicator of lipid peroxidation (Dreosti 1987). The increase in levels of lipid perodixation products in the zinc-deficient foetus may reflect a role of zinc in membrane stabilization (Chvapil 1976), or an elevation in the pool of unsaturated fatty acids available for peroxidation (Odutuga 1982). Evidence that the increased malondialdehyde products observed in zinc-deficient foetuses reflect cellular membrane lipid perodixation has been provided by the studies of Harding et al. (1987). These investigators, using electron microscopy, observed that in the 11 d zinc-deficient foetus there is severe deterioration of cell membranes, especially those of the mitochondrion, in the period just preceding the appearance of cell death in the neural tube. Previous work by this group has shown that in the zinc-deficient foetus there is extensive cell necrosis in those areas of the embryos undergoing rapid cell division, prior to the development of identifiable malformations (Record et al. 1985b). Similar observations of large numbers of pyknotic cells and cellular debris in the wall and lumen of cells in all regions of the developing brain in zinc-deficient embryos examined on days 12–14 of gestation have been reported by Rogers et al. (1985b). Thus it is reasonable to suggest that an increased rate of cellular lipid peroxidation in the zinc-deficient foetus could result in membrane damage with resultant cellular necrosis and cell death which leads to asynchrony in the development of the affected tissue primordia.

Effect of Zinc Deficiency on Parturition

Zinc deficiency during late gestation in the rat results in severe problems in parturition. Delayed and prolonged labour, excessive bleeding, failure to eat the afterbirth or care for pups and increased dam and pup postnatal mortality are common abnormalities observed (Apgar 1972, 1977; O'Dell et al. 1977; Mutch and Hurley 1980). Zinc deficiency is relatively specific in its effects on parturition, as similar phenomena are not observed in pregnant rats fed calorie-restricted diets, or diets deficient in protein, thiamin, copper or manganese (Apgar 1975). An effect of zinc deficiency on parturition in rats is evident even when the deficient diet is introduced as late as day 18 of gestation; alternatively, zinc repletion on day 19 of pregnancy results in normal delivery (Apgar 1973). Marginal zinc deficiency also has been reported to have an adverse effect on parturition in ewes and rhesus monkeys (Apgar and Travis 1979; Golub et al. 1984b). Sharp decreases in body temperature, blood pressure and serum corticosterone have been noted in zinc-deficient rats at, or immediately following, parturition (O'Dell et al. 1977; Reeves et al. 1977). The biochemical lesions underlying the stress associated with delivery in zinc-deficient dams have not been identified firmly; however, it has been suggested that the biochemical lesion may involve a reduction in oestrogen release and/or binding which subsequently results in a delay in the normal development of myometrial gap junctions essential for normal delivery (Bunce et al. 1983; Dylewski et al. 1986; Lytton and Bunce 1986).

Marginal Deficiencies of Zinc During Development in Experimental Animals

Non-human Primates

As described above, the effects of a severe deficiency of zinc on development are dramatic, but its occurrence in human populations under normal conditions is relatively rare. This is not to imply that studies of severe deficiency are inappropriate, as they serve as valuable tools in investigating the role of zinc in development. In a different category, however, is marginal zinc deficiency, which can occur in large groups of people. Thus, investigations on the effects of marginal zinc deficiency on reproductive outcome are essential for better understanding the potential effects of inadequate dietary zinc intake on human pregnancy outcome.

One animal model used to study the effects of marginal zinc deficiency is the rhesus monkey. This model is particularly useful with regard to the extrapolation of pregnancy outcome results to the human population as the metabolic demands of pregnancy are considerably less in large mammals, and in primates in particular (Payne and Wheeler 1968), than in rodents. Thus the level of dietary zinc which constitutes a "marginal" or "severe" deficiency of the element may be considerably lower in humans than in rodents. While a link between zinc deficiency and poor pregnancy outcome has been strongly suggested by some human studies, the relative importance of zinc deficiency as a risk factor in human pregnancy has not been defined. Below we discuss in some detail a long-term study designed to define the consequences of zinc deprivation in pregnant rhesus monkeys and their offspring which has been conducted at the University of California, Davis.

In these studies, the control monkeys were fed a diet containing 100 μg zinc/g while the deficient monkeys were fed a diet containing 4 μg zinc/g. The 4 μg zinc/g diet does not produce overt clinical signs of zinc deficiency in non-pregnant monkeys; however, it does result in a 20%–30% decrease in plasma zinc, a level which has been defined as "marginal deficiency" (Golub et al. 1982). All of the monkeys received the control diet for a few weeks prior to mating, and were then assigned to their respective diets at the time of conception. The diets were fed during gestation and lactation (to 5.5 months postnatal). The studies also included a group of monkeys which were fed the control diet in amounts equal to those consumed by the marginal zinc group to correct for the possible adverse effects of the anorexia induced by zinc deficiency on the outcome of pregnancy.

Maternal Effects

Overt signs of zinc deficiency were not observed in the zinc-deprived monkeys during the first half of pregnancy (Golub et al. 1984a). However, by mid-pregnancy, deficiency signs, including reduced plasma zinc levels, anorexia and dermatitis were evident in many of the deprived monkeys. The appearance of the zinc deficiency syndrome in late pregnancy was accompanied by a drop in plasma zinc below 0.65 μg/ml, a level which can be considered to represent clinically relevant deficiency in humans (Swanson and King 1982). It is interesting to note that during late pregnancy the plasma zinc levels in the control monkeys, as in humans (Swanson and King 1982), declined despite an adequate zinc intake (Fig. 12.5).

During the third trimester, the anorexia was such that maternal weight gain was significantly lower in the marginal zinc group than in controls. However, it is important to note that this lower rate of maternal weight gain in the marginal zinc group was due mainly to a subgroup characterized by severe

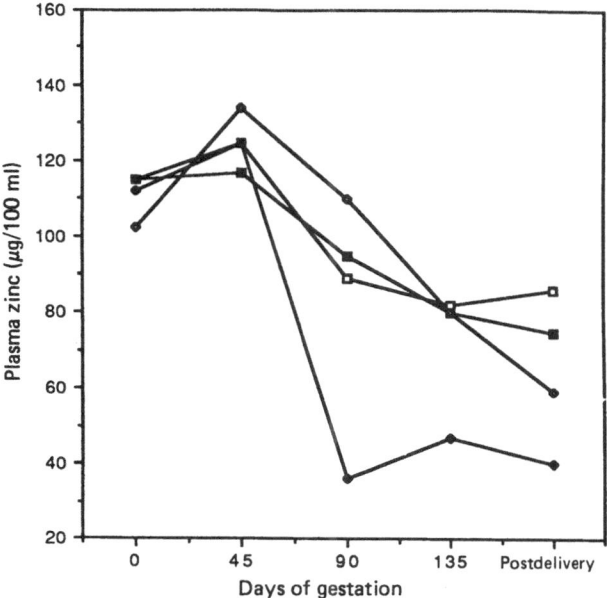

Fig. 12.5. Plasma zinc concentrations in pregnant rhesus monkeys. (□) = animals fed ad libitum control diet (100 ppm zinc); (■) = animals fed restricted amounts of control diet; (○) animal fed zinc-deficient diet (4 ppm zinc) that developed anorexia; (●) animal fed zinc-deficient diet that did not develop anorexia. All diets were fed throughout pregnancy. Figure adapted with permission from Golub et al. 1984a.

anorexia. Consistent with this observation, those monkeys in the pair-fed group which were paired to severely anorexic mothers also showed lower weight gains during the third trimester. Thus the lower maternal weight gain in the marginal zinc group was mainly due to the anorexia associated with the zinc deficiency (Golub et al. 1984a).

An interesting observation of this study was that within the marginal zinc group, plasma zinc was inversely related to the degree of anorexia; during late pregnancy, monkeys that were not anorexic had significantly lower plasma zinc levels than did controls, while anorexic monkeys had plasma zinc levels similar to those of control monkeys (Fig. 12.5). This observation is undoubtedly related to the effects of food intake on maternal tissue catabolism.

Maintenance of "normal" plasma zinc levels for an extended period of time during pregnancy in anorexic, zinc-deprived monkeys suggests that inadequate zinc nutriture during pregnancy can be masked by concurrent anorexia when zinc status is assessed by plasma zinc measurement alone. It should be recognized that while one interpretation of the variable expression of anorexia in the marginal zinc group is that the more affected monkeys (i.e. those with the greater degree of zinc deficiency) are the ones most likely to develop severe anorexia, the release of zinc from maternal tissue pools during periods of catabolism, and the resultant rise in maternal plasma zinc concentrations, should lessen the severity of the zinc deficiency for the embryo. Consistent with this idea, Masters et al. (1983) have reported that when the food intake of pregnant rats fed a zinc-deficient diet was restricted, foetal outcome was better than in ad libitum-fed zinc-deficient dams, whereas increasing the food intake of zinc-deficient dams by gavaging resulted in a marked increase in the number and severity of foetal malformations normally associated with zinc deficiency in the rat. Also consistent with these findings, Record et al. (1985b) have reported that abnormal cellular necrosis was more evident in day 9.5–11.5 foetuses from zinc-deficient dams that consumed, ad libitum, large quantities of diet compared with embryos from zinc-deficient dams that consumed smaller quantities of the diet. Masters and Fels (1985) have reported that maternal tissue catabolism in ewes during periods of zinc deficiency can similarly protect the developing lamb.

It can therefore be speculated that the development of an anorexic response to zinc deficiency during pregnancy is a positive response of the mother with regard to pregnancy outcome. Consistent with this idea is the recent report by Dreosti et al. (1986) that C57BL/6J mice fed low zinc diets (5 ppm) during pregnancy did not exhibit anorexia, and for these animals, the 5 ppm zinc diet was highly teratogenic. This is in marked contrast to other strains studied in which the feeding of a 5 ppm zinc diet would not normally result in the production of gross congenital defects. The authors suggested that the C57BL/6J mice may be particularly sensitive to zinc deficiency teratogenicity due to their failure to develop an anorexic response to the low-zinc diet which would result in tissue catabolism and zinc mobilization. In terms of human pregnancy, this idea raises the question of whether periods of anorexia may in some instances be linked to conditions of marginal zinc deficiency, and if this is the case whether they serve a protective function for the foetus.

During the third trimester, monkeys in the zinc-deprived group showed a variety of metabolic defects. Haematological changes similar to those of iron-deficiency anaemia [reduced packed cell volume (PCV), haemoglobin (Hb) and mean corpuscular haemoglobin concentration (MCHC)] were observed in the deficient group, and these animals had a high incidence of abnormal red blood cell morphology (poikilocytosis, anisocytosis, leptocytosis) (Golub et al. 1984a). The observation of refractory anaemia in the zinc-deficient monkeys is consistent with the suggestion of Jameson (1976) that anaemia may be a complication of marginal zinc deficiency in pregnant women. It has been suggested that anaemia may be a consequence of zinc deficiency due to either an increased rate of erythrocyte destruction (Dash et al. 1974) or a reduction in the rate of haemoglobin synthesis (Garnica 1981).

Peripheral lymphocyte response to mitogens [Con A and phytohaemagglutinin (PHA)] was found to be lower in the deprived mothers than in controls (Golub et al. 1984a). This finding of reduced immune responsiveness is consistent with a variety of reports showing immunodeficiency in zinc-deficient states in all species studied to date (Gershwin et al. 1985). A third metabolic defect observed in the zinc-deficient mothers was an alteration in vitamin A metabolism (Baly et al. 1984; Golub et al. 1984a). As plasma zinc concentrations declined during pregnancy, there was a positive correlation to plasma vitamin A and retinol-binding protein concentrations. This effect of zinc deficiency on vitamin A metabolism in pregnant monkeys is consistent with similar observations in pregnant rodent models (Duncan and Hurley 1978; Peters et al. 1986), and in non-pregnant humans (Morrison et al. 1978; Palin et al. 1979).

Pregnancy Outcome

Pregnancy outcome (stillbirths, abortions, delivery complications) was significantly worse in the zinc-deficient and pair-fed control monkeys than in the ad libitum-fed controls (Golub et al. 1984b; Table 12.2). Like the mothers, some of the zinc-deficient infants showed microcytic anaemia. Gross malformations were not noted in any of the offspring, an observation consistent with the finding that plasma zinc concentrations in the deficient monkeys were "normal" during the period of organogenesis. However, male zinc-deficient infants showed lower birth weight, shorter crown–rump and femur lengths, lower plasma zinc and iron concentrations and poorer muscle tonus than did control male infants. In contrast to the male infants, female zinc-deficient infants were not growth retarded. Similar sex-specific effects of zinc deficiency on growth have been reported for rats (Mutch and Hurley 1974) and suggested for humans (Butrimovitz and Purdy 1978; Nishi et al. 1980).

That males may be more susceptible to zinc deficiency during early development is also supported by observations that an influence of zinc intake on postnatal growth in humans is more apparent in males than in females (Ronaghy and Halsted 1975; Walravens and Hambidge 1976), and that the majority of cases of dwarfism attributed to zinc deficiency in Iran and Egypt have been males (Ronaghy and Halsted 1975). In the study by Golub et al. (1984b) the greater impact of zinc deficiency on male monkeys could be attributed to the normally greater growth rate of male foetuses as evidenced by a higher birth weight, or to the differential body composition of males and females as evidenced by a higher muscle : fat ratio in males than in females (Cheek 1975). Regardless of the explanation for the greater sensitivity of the males to zinc deficiency, it is evident that the monkey model is consistent with

Table 12.2. Pregnancy outcome in rhesus monkeys fed control (100 mg zinc/kg) or zinc-deficient (4 mg zinc/kg) diets throughout pregnancy

	Controls		ZD
	AL	PF	
Viability			
Conceptions	9	13	18
Abortion	1	1	1
Stillborn	0	2	3
Neonatal death	0	1	2
Loss ratio	11%	31%	33%
Delivery complications			
Premature	0	2	2
Postmature	0	2	2
Prolapse (uterine–vaginal)	0	1	2
Retained placenta	0	1	2
Infection (breast, endometrial)	0	0	2

Reprinted, with permission, from Golub et al. (1984a).
AL, ad libitum control; PF, pair-fed control; ZD, zinc deficient.

reports on zinc status and growth in other species, including man, and demonstrates the necessity of evaluating male and female data separately in studies on growth and zinc status.

A strong negative correlation ($r = -0.72$) was found in the monkey study between maternal plasma zinc (day 90 gestation) and infant birth weight in the zinc-deprived group, while a positive correlation ($r = +0.42$) was seen for the ad libitum-fed and restricted intake groups (Golub et al. 1984b). Negative correlations between maternal plasma zinc and infant birth weight have been reported for human populations (Metcoff et al. 1981; Prema 1980; McMichael et al. 1982), strongly supporting the idea that maternal zinc status may be a significant predictor of infant birth weight in humans. It is important to note that in the zinc-deficient group, third trimester food intake was a negative determinant of plasma zinc. Thus in the pregnant zinc-deficient monkey, a decline in food intake will result in an increase in maternal tissue catabolism and an increase in the amount of zinc available to the foetus and thus it could be argued that foetal growth should be increased. However, overall nutrient supply to the foetus may be compromised due to the general reduction in maternal food intake in this group, thus reducing the rate of foetal growth. To what extent maternal tissue catabolism, and subsequent zinc release, offset the overall reduction in maternal nutrient intake, is currently unclear. However, taken together, the above data support the idea that information on maternal calorie and zinc intake during pregnancy, in addition to data on maternal plasma zinc levels during pregnancy, are essential if we are to clarify the influence of maternal zinc status on infant outcome in human populations.

Postnatal Outcome

Investigations of the postnatal development of rhesus monkey infants from the study discussed above have resulted in several findings. It is important to note that in these investigations the mothers continued to receive their respective diets after pregnancy. Infants in the zinc-deprived group were characterized by a pattern of rapid postnatal catch-up in body weight, a small but persistent lag in linear growth throughout the 1st year of life, low plasma zinc levels from birth to 9 months of age, delayed skeletal maturation and mineralization, a marked increase in subcutaneous fat after weaning, diminished weight gain associated with reduced food intake, a reduced food : efficiency ratio, reduced taste sensitivity at 1 year of age, reduced immunocompetence and behavioural defects (Golub et al. 1984c, 1985; Leek et al. 1984, 1988; Haynes et al. 1985).

An intriguing observation in this study was that the intrauterine growth-retarded male infants in the zinc-deficient group gained weight at a rate five times that of control male infants during the 1st month of postnatal life (Golub et al. 1984c). The finding of rapid postnatal growth in the growth-retarded zinc-deficient male monkeys is consistent with previous reports of rapid catch-up growth in low birth weight monkeys (Sackett 1981) and humans (Tanner 1981) where there was no evidence that the cause of the low birth weight was zinc deficiency. As the diet of the zinc-deficient mothers was unchanged during the postnatal period, it is evident that the impact of marginal maternal zinc deficiency was less during neonatal than during prenatal development. Possible explanations for this difference in sensitivity of the infant to maternal zinc intake include: (a) a lower requirement of the mother for zinc during lactation than during pregnancy; (b) a lower "dietary" requirement of the infant for zinc during the 1st month postnatal than during in utero development; and (c) lack of an effect of maternal marginal zinc deficiency on milk zinc concentration. It is important that future studies address these issues as information on these points is essential to clarify the relative influence of maternal zinc intake on prenatal versus postnatal development in human infants.

Plasma zinc concentrations in the infant monkeys were found, in general, to be inversely correlated with weight gain during the 1st year of life (Golub et al. 1984c). For example, plasma zinc levels in the zinc-deficient monkeys were lowest during month 1, the period of rapid catch-up growth (zinc-deprived infants 0.63 µg/ml; control infants 1.0 µg/ml), and rose to control values (0.98–1.25 µg/ml) when growth rates slowed towards the end of year 1. Since the plasma zinc drop was marked in the zinc-deficient infant monkeys during the period of rapid growth, it suggests that the zinc status of these infants may be suboptimal with regard to other developmental processes during this time. The observation of an inverse relation between plasma zinc and growth rate in the infant monkey is consistent with similar data on nutritionally deprived children (Butrimovitz and Purdy 1978). While the observation that plasma zinc levels in the zinc-deficient monkey infants were similar to control infants at 1 year of age is suggestive of an adequate zinc status in these infants, during this time period the zinc-deficient monkey infants showed reduced taste sensitivity to quinine (Golub et al. 1984c). Since hypoguesia is thought to be a common expression of zinc deficiency in humans (Henkin 1984), this observation suggests a functional zinc deficiency in the zinc-deprived infants despite apparently normal plasma levels of the element. These data underscore the difficulty of using plasma zinc measurements as indicators of infant zinc status during periods of rapid growth.

From birth to 1 year of age, skeletal maturation (defined on the basis of epiphyseal ossification centres) in the zinc-deficient infant monkeys was significantly retarded compared with controls, and the deficient infants were characterized by a picture of defective endochondral bone mineralization that was very similar to human rachitic syndromes (Leek et al. 1984). The defects in skeletal maturation observed in the zinc-deficient infant monkeys persisted up to 3 years of age (all monkeys were weaned to the same diet as that consumed by their mothers), although bone mineralization in the deficient monkeys increased during years 2 and 3 to values which were only slightly below those of controls (Leek et al. 1988).

The finding of abnormalities of skeletal growth in the zinc-deficient monkeys is consistent with reports of skeletal defects in zinc-deficient humans, rats, pigs and cows (Ronaghy et al. 1974; Hickory et al. 1979; Hurley 1981; Apgar 1985). The mechanism of abnormal bone mineralization in the presence of zinc deficiency is unknown. The concentration of zinc in bone is very high, leading some to suggest a structural function, i.e. that the element may be an essential component of the calcified matrix (Calhoun et al., 1974). Others have suggested that osteoblast maturation may be limited by zinc deficiency (Haumont and McLean 1966; Bergman et al. 1972; Suwarnasarn et al. 1982; Yamaguchi et al. 1982; Yamaguchi and Yamaguchi 1986), and that calcium metabolism is abnormal in zinc deficiency (Hurley et al. 1969). Recently it has also been suggested that zinc and vitamin D may act synergistically with regard to bone formation. Yamaguchi and Sakashita (1986) have reported that while both vitamin D and zinc supplementation resulted in increased femoral diaphyseal DNA content and alkaline phosphatase activity in rats, the greatest increase occurred when the supplements were given together. It has also been reported that zinc is required for the effect of vitamin D on intestinal calcium uptake (Sivakumar and Belavady 1975).

In addition to its putative interaction with vitamin D, zinc is a component of alkaline phosphatase which is essential for normal bone formation. Defective activity of bone alkaline phosphatase such as that seen in humans with hypophosphatasia (Rasmussen 1983) produces radiographic findings similar to those observed in zinc-deficient monkeys during the 1st year of life (Leek et al. 1984). Finally, it has been suggested that the reduction in bone growth observed with postnatal zinc deficiency may be secondary to a generalized reduction in growth, possibly due to a lower output of a growth-promoting factor such as growth hormone or somatomedin. It is obvious that future studies need to be directed towards the investigation of the biochemical defects underlying bone abnormalities in zinc deficiency.

Measurements of skinfold thickness during the 1st year of life showed that the zinc-deficient monkey infants had a higher rate of fat/muscle deposition than did the control infants (Golub et al. 1984c). A similar relative hypertrophy of fat tissue has been reported in children recovering from protein–calorie malnutrition (Graham et al. 1969; Ravelli et al. 1976) and this has been linked to a condition of marginal zinc deficiency in these children (Golden and Golden 1981a, b). Based on information from a mouse model for protein–calorie malnutrition, it has been suggested that the increased fat/protein deposition observed in marginally zinc-deficient infants is the result of a zinc-deficiency-induced impairment in protein synthesis (Morgan et al. 1988a, b). Studies directly evaluating the rates of protein and fat synthesis in marginally zinc-deficient infants have not been reported.

Consistent with observations made on the zinc-deficient mothers (Golub et al. 1984a), infant zinc-deficient monkeys showed defective immunocompetence compared with controls (Haynes et al. 1985). Between birth and 1 year of age major immunological findings in the zinc-deficient infants included depressed mitogenesis to Con A, PHA-P (phytohaemagglutinin, purified) and pokeweed mitogen (PWM), alterations in serum IgM levels and impaired polymorpho-nuclear leucocyte (PMN) function as shown by reduced chemotaxis and phagocytic activity.

The findings of reduced immunocompetence in the zinc-deficient monkey infants is consistent with reports that zinc deficiency in children can lead to alterations in immunocompetence as manifested by thymic hypoplasia, reduced cellular immunity, defective monocyte and polymorphonuclear leucocyte function and increased susceptibility to infectious agents (Golden et al. 1977, 1978; Weston et al. 1977; Castillo-Duran et al. 1987). That zinc is required for the normal function and ontogeny of the immune system in mammals is now well established (Gershwin et al. 1985), although the biochemical lesions underlying the effects of zinc deficiency on the immune system have yet to be identified. Whether immune defects persist in the zinc-deficient monkey after zinc supplementation has not been established.

Finally, in addition to the structural and biochemical defects described above, the zinc-deficient infant monkeys demonstrated behavioural defects (Golub et al. 1985). During the 1st year, postnatal zinc-deficient infants showed normal development of reflexes, motor patterns and social behaviours. A transient hypotonia was observed in the zinc-deficient infants, but this cleared up rapidly. It is likely that the hypotonia was the result of the metabolic condition or dystocia of the mother and is not clinically relevant (Dubowitz 1969). The zinc-deficient infants were able to learn and perform simple delayed response and visual discrimination tasks. However, these infants showed a marked deficit in the amount and variety of spontaneous behaviour relative to controls. Significantly, the decrease in spontaneous behaviour was positively correlated with rapid growth and negatively correlated with plasma zinc levels. Thus, the observation of reduced spontaneous activity in the zinc-deficient infants may be interpreted as a concurrent effect of zinc deficiency and should therefore be responsive to zinc therapy.

The observation of behavioural deficits in the zinc-deficient infant monkeys is consistent with a number of reports documenting abnormal behaviour as a result of either marginal or severe dietary zinc deficiency in rats and mice (Bradford et al. 1981; Hughes and Horsburgh 1982; Golub et al. 1984d; Gordon 1984; Halas et al. 1986; Tininus et al. 1986; Prohaska 1987). Lethargy and apathy have also been suggested to be associated with zinc deficiency in young humans (Walravens et al. 1978) and adults (Halsted et al. 1972; Henkin et al. 1975; Kay et al. 1976). In most cases, the behavioural deficits have been reported to improve upon zinc supplementation; however, Sandstead et al. (1978) have described residual behavioural effects in infant monkeys following a brief, severe dietary zinc deficiency (the deficient diet contained 1 μg zinc/g) during the third trimester of pregnancy (day 110 to day 150). Infants from the deprived mothers showed reduced play times and a reduction in exploratory behaviour at the time of weaning. Residual behavioural defects in rats due to prenatal zinc deficiency have also been reported (Halas and Sandstead 1975; Halas et al. 1976; Hughes and Horsburgh 1982; Kawamoto and Halas 1984); however, there is considerable variation in the type of defects reported. Based on the reports to date, it is evident that the extent and severity of the zinc deficiency is an important predictor with regard to the persistence of the associated behavioural abnormalities. Biochemical lesions underlying the behavioural defects associated with zinc deficiency have not been firmly identified. (Studies of the influence of zinc deficiency on behaviour are considered in more detail in Chap. 14).

At least five major conclusions can be drawn from the monkey studies described above. First, dietary levels of zinc that do not produce overt signs of deficiency in non-pregnant monkeys can lead to a severe deficiency syndrome by the end of pregnancy. Second, the metabolic impact of zinc deficiency depends in part on whether or not anorexia is induced, with severe anorexia resulting in a masking of maternal zinc deficiency and partial protection of the conceptus. Third, a syndrome of anaemia and reduced immune responsiveness can develop in zinc-deficient mothers that have low plasma zinc concentrations. Fourth, inadequate dietary zinc during gestation can result in a reduction in foetal growth and an increase in complications of delivery. Fifth, postnatal development of marginally zinc-deficient infants is characterized by periods of low plasma zinc concentrations, abnormalities in growth rate and tissue composition, skeletal defects, reduced immunocompetence and behavioural abnormalities.

In the previous section we have described some of the effects of feeding a marginally zinc-deficient diet during pregnancy and lactation on foetal and infant development in the rhesus monkey model. While this model is useful for the investigation of the effects of zinc deficiency on development, it is compromised for the extrapolation of the results to the human condition by the introduction of the marginal zinc diet at the onset of pregnancy. In human populations it is more likely that the consumption of a marginal zinc diet would be of longstanding duration, occurring prior to, as well as during, pregnancy. This is an important point as it could be argued that long-term exposure to marginal zinc diets could result in a significant adaptation such that the impact of these diets during pregnancy would be limited. For example, it has been reported that rats fed marginally zinc-deficient diets (10 μg zinc/g) can adapt in part to the diets by reducing zinc excretion (Fairweather-Tait et al. 1985).

To address this issue Haynes et al. (1987a, b) have initiated studies on the effects of chronic (including prepregnancy) marginal zinc deficiency on reproduction in the rhesus monkey. In these studies, the marginal zinc diet (4 μg zinc/g) was introduced 1 year prior to breeding. After receiving marginal zinc diets for 1 year, the non-pregnant female monkeys showed low plasma zinc levels. In addition, several immune variables, including serum IgM and IgG levels and polymorphonuclear function, were depressed in the deficient animals compared with controls (Haynes et al. 1987a). To date, information has been published only on the effects of the zinc deficiency up to mid-gestation (Haynes et al. 1987b). It was reported that foetal growth, as evaluated by ultrasound, was significantly retarded in the marginal zinc group compared with controls. These results suggest that even if the non-pregnant rhesus monkey can adapt in part to the consumption of a marginal-zinc diet, the adaptation is insufficient to allow for normal pregnancy.

Rodent Models

Pregnancy outcome in rodent models, like that in the rhesus monkey, is markedly affected by marginal zinc deficiency. One of the more dramatic effects of transitory, marginal zinc deficiency on development is its effect on the immune system. Beach et al. (1982, 1983) have developed an experimental model in which pregnant mice receive a control diet (100 μg zinc/g) for the 1st week of pregnancy, followed by a diet marginally deficient in zinc (5 μg/g) for the last 2 weeks of pregnancy. After parturition the dams are again fed the control diet. When the offspring of these mice were examined they showed a profound suppression of serum IgM levels with impaired direct plaque-forming cell responsiveness to heterologous erythrocytes early in life. This immunodeficiency persisted even at 6 months of age in the F_1 progeny. Furthermore, the immunodeficiency was passed on to the F_2 and F_3 generations, although it was of a lesser magnitude than that seen in the first generation offspring. It is important to point out that this impaired immunocompetence persisted even after there was apparent rehabilitation of the spleen and thymus of the offspring deprived of zinc in utero (Beach et al. 1982). Keller and Fraker (1986) have similarly reported that in mice, gestational zinc deficiency results in a retardation of B-cell development in utero which persists into the young adult period despite nutritional rehabilitation. The observation that immune defects of the same type occur in marginally zinc-deficient infant monkeys (Haynes et al. 1985) underscores the potential significance of these results for human populations. The mechanisms whereby defective immunocompetence is passed on to subsequent generations after in utero deprivation of zinc are unknown, but it is evident that such multigenerational effects on immune function may be a serious potential public health problem for populations marginally deficient in zinc.

It is important to note that the persistence of immune defects due to an in utero nutritional insult are not specific for zinc. Chandra (1975) reported similar immunosuppression defects in F_1 and F_2 mice as a result of caloric restriction in F_0 dams. Thus an understanding of the mechanisms underlying prenatal zinc-deficiency-associated immune defects may also lead to a better understanding of the effects of other nutrient deficiencies on the immune system during development.

Marginal prenatal zinc deficiency may also have a direct persistent effect on zinc metabolism itself. Vruwink et al. (1987) have reported that in mice, transitory prenatal zinc deficiency resulted in a marked change in the response to a zinc challenge at 10 weeks of postnatal age. While both control mice and those prenatally deprived of zinc accrued similar concentrations of liver zinc following a zinc challenge, the prenatally deprived mice showed markedly elevated levels of liver metallothionein (MT) compared with the controls. This result suggests that prenatal zinc deficiency resulted in a molecular lesion which postnatally was expressed in part by an amplification of MT synthesis (or a reduction in MT turnover). An increase in the rate of MT synthesis could occur if prenatal zinc deficiency resulted in amplification of the MT gene. Metallothionein gene amplification has been demonstrated in Chinese hamster ovary cells exposed to excess cadmium (Crawford et al. 1985). It has been suggested that MT gene amplification can occur as a result of chromosomal aneuploidy and rearrangements (Hamer 1986). Since it is known that severe prenatal zinc deficiency can result in chromosomal lesions (Bell et al. 1975), it may be speculated that transitory marginal zinc deficiency may also be affecting chromosomal integrity.

An alternative possibility is that prenatal zinc deficiency resulted in a persistent hypomethylation of the MT gene, as hypomethylation of this gene has been shown to cause amplification of its synthesis. While the observation that prenatal zinc deficiency can affect postnatal MT metabolism is intriguing, the functional significance of this observation with regard to postnatal zinc homeostasis has not been determined. It is interesting to speculate that the alteration in MT metabolism may be one of the factors underlying the intergenerational effects of prenatal zinc deficiency.

Disease-Induced Deficiencies of Zinc During Pregnancy

In addition to simple dietary deficiencies of zinc it is important to recognize that zinc deficiency may also occur as a result of some disease states. For example, it has been suggested that zinc deficiency may be a complication of diabetes in some patients (Pidduck et al. 1970; Kumar and Rao 1974) and a diabetes-induced alteration in maternal zinc metabolism has been suggested as a potential biochemical lesion underlying the high incidence of teratogenicity associated with this disease (Simpson et al. 1983; Uriu-Hare et al. 1985; Wibell et al. 1985; Eriksson 1984). The hypothesis that the teratogenicity of diabetes is in part due to a diabetes-induced foetal zinc deficiency has recently been examined by Uriu-Hare et al. (1985, 1987). These investigators found that, like the non-pregnant diabetic rat (Failla et al. 1985), pregnant diabetic rats are characterized by markedly elevated liver zinc and MT concentrations. However, in contrast to their mothers, foetuses from the diabetic dams showed lower levels of liver zinc and metallothionein than those found in controls.

Similarly, Eriksson (1984) has reported that whole-body zinc levels were lower in foetuses from diabetic dams than in control foetuses. One interpretation of these results is that the diabetic state in the mother results in a stimulation of the synthesis of MT in maternal liver. Such an increase could cause a redistribution of maternal zinc pools, resulting in an embryonic/foetal zinc deficiency which is expressed in part by low concentrations of liver zinc. Evidence that the diabetes-induced alteration in foetal zinc metabolism is functionally significant is suggested by the observation that zinc supplementation of diabetic dams resulted in a significant improvement in foetal weight, length and skeletal ossification (Uriu-Hare et al. 1987).

A second example of a "disease" state which has been linked to potential disturbances in zinc metabolism is alcoholism. It is now well accepted that excess alcohol consumption during pregnancy is highly teratogenic in humans resulting in a syndrome referred to as foetal alcohol syndrome (FAS) (Jones et al. 1973; Rosett et al. 1983; Iosub et al. 1985). While the mechanisms underlying the teratogenicity of alcohol are undoubtedly multifactorial, one hypothesis is that ethanol functions as a teratogen partly by inducing maternal nutritional deficiencies, thereby affecting the nutrition of the embryo/foetus (Hurley 1979; Kumar 1982). Zinc in particular has received attention as a nutrient which may be deficient in some alcoholics (Henderson et al. 1980; Flynn et al. 1981; Gordon et al. 1981), and it has been suggested that one biochemical lesion underlying FAS is embryonic zinc deficiency (Hurley 1979; Flynn et al. 1981). Flynn et al. (1981) reported that plasma zinc levels were lower in pregnant alcoholic women than in control pregnant women, and that there was an inverse relationship between plasma zinc levels in these women and the expression of FAS.

In contrast with the report by Flynn et al. (1981), Ehalmesmaki and Ylikorkala (1985) did not find maternal plasma zinc levels to be lower in alcoholic pregnant women than in controls. However, these investigators did report that plasma zinc levels were lower in the infants of alcoholic women than in those of controls. Assadi and Ziai (1986) have also found abnormal zinc metabolism in FAS infants. Foetal alcohol syndrome infants ($n = 6$) were characterized by low plasma zinc concentrations and increased urinary zinc excretion compared with control infants ($n = 6$). The functional significance of the observed abnormal zinc metabolism in the FAS child is not known.

While the evidence that maternal zinc deficiency may be a significant factor in FAS in humans is strong, investigations of this hypothesis using experimental animals have yielded conflicting results. A reduction in rat foetal zinc levels and/or uptake due to maternal ethanol consumption was reported by Suh and

Firek (1982). and Ghisham et al. (1982). However, supplemental zinc feeding was not found to result in a reduction in the expression of FAS in rats (Ghisham and Greene 1983; Zidenberg-Cherr et al. 1988). While the lack of an effect of zinc supplementation on FAS expression in rats is discouraging, it seems imperative that additional studies on the role of abnormal zinc metabolism in the expression of FAS in humans be conducted, as FAS represents the major identifiable cause of mental retardation in the human population.

In addition to the possible direct effect of some diseases on zinc metabolism, zinc deficiency may also be induced during pregnancy as a result of certain drug therapies. Broadly speaking, drug–mineral interactions can be separated into two categories; in the first there is direct interaction between the drug and the element; in the second, the drug indirectly affects the metabolism of the element.

With regard to zinc, the first category of interactions is exemplified by drugs that have the potential to chelate zinc and reduce its bioavailability and/or increase its excretion. For example, ethylenediaminetetraacetic acid (EDTA) is a divalent chelating agent used for various purposes including the treatment of lead toxicity. Injecting EDTA into pregnant rats or feeding it in the diet at levels of 2% and higher results in foetuses with malformations similar to those found with zinc deficiency. That the teratogenicity of EDTA is due in part to a drug-induced embryonic/foetal zinc deficiency is suggested by the finding that supplemental dietary zinc prevents the teratogenicity of this drug (Swenerton and Hurley 1971; Kimmel and Sloan 1975). Furthermore, the chelating agents triethylenetetramine and D-penicillamine used in the treatment of Wilson's disease, rheumatoid arthritis and other connective tissue disorders are teratogenic in rats and evidence has also been presented that their teratogenicity is in part due to drug-induced changes in maternal and foetal zinc metabolism (Keen et al. 1983a, b, 1984). That the use of divalent chelating agents during pregnancy may be teratogenic should not be surprising and it seems reasonable to suggest that the zinc status of pregnant women receiving these drugs should be closely monitored throughout pregnancy.

The second category of zinc–drug interactions, however, is more difficult to predict, as simple examination of the drug's chemical structure may not allow an accurate prediction of its effect on zinc metabolism. A representative of this category of interaction is 6-mercaptopurine (6-MP) used in the treatment of leukaemia. In the rat model, it has been shown that the teratogenicity of 6-MP is associated with a reduction in foetal zinc concentration, and that the teratogenicity of the drug can be increased with marginal dietary zinc intake, and reduced with supplemental dietary zinc (Hirsch and Hurley 1978; Amemiya et al. 1986a). However, in contrast to the foetuses, 6-MP-treated dams are characterized by transitory increases in hepatic zinc concentration, although maternal plasma zinc concentrations are reduced (Amemiya et al. 1988).

These data suggest that 6-MP is in part teratogenic through an effect on maternal zinc metabolism resulting in an increase in maternal hepatic zinc stores with a consequent reduction in plasma zinc concentrations. One mechanism by which 6-MP could have this effect would be if the drug, directly or indirectly, stimulated the synthesis of maternal liver metallothionein. Increases in hepatic metallothionein concentrations following 6-MP adminis- tration have been reported in non-pregnant animals (Amemiya et al. 1985,

1986b). The mechanism by which 6-MP may induce metallothionein synthesis is unclear. However, it is known that 6-MP is hepatotoxic (Minow et al. 1976), and hepatic metallothionein synthesis is known to be stimulated under conditions which result in hepatic damage as part of the acute phase reaction (Cousins 1985). If indeed the effect of 6-MP on maternal zinc metabolism is secondary to the cytotoxicity of the drug, this suggests that any drug with pronounced cytotoxicity may indirectly affect maternal zinc metabolism. If the reduction in plasma zinc is of sufficient magnitude, then the foetus will be exposed to a zinc-deficient environment.

Other drugs which have been suggested to be teratogenic in part due to the induction of foetal zinc deficiency include thalidomide (Jackson and Schumacher 1979), salicylate (Kimmel et al. 1972; Koshakji and Schulert 1972; Hackman and Hurley 1984a, b) and acetazolamide (Hackman and Hurley 1983). Currently it is not known if the effect of these drugs on zinc metabolism is direct or indirect.

Zinc Deficiency in Human Pregnancy

The 1980 United States recommended dietary allowance (RDA) for zinc during pregnancy is 20 mg/d, an increase of 5 mg/d over the recommended allowance set for non-pregnant women (National Research Council 1980). However, dietary studies of pregnant women show that they rarely consume the RDA for zinc, and indeed intakes are often less than 10 mg/d (Hunt et al. 1979, 1983, 1987; Abraham 1982; Hambidge et al. 1983; Moser and Reynolds 1983; Turnlund et al. 1983; Abu-Assal and Craig 1984). Given, on one hand, the observations of dietary zinc intake during pregnancy often markedly lower than that suggested by health organizations and evidence on the other hand of teratogenic effects of severe or marginal zinc deficiency in experimental animals, it becomes important that we determine to what extent, if any, marginal zinc deficiency is a risk factor with regard to human pregnancy outcome.

It has been well documented that a significant decline in plasma zinc concentrations can occur during the course of normal pregnancy. Based on both longitudinal and cross-sectional studies it has been estimated that a 10%–20% reduction in plasma zinc concentration occurs in most women between early and either mid- or late pregnancy (Vir et al. 1981; Dreosti et al. 1982; Hambidge et al. 1983; Breskin et al. 1983; Hunt et al. 1984, 1987; Campbell-Brown et al. 1985; Qvist et al. 1986). The reduction in plasma zinc concentration during pregnancy is thought to be due in part to plasma volume expansion, the effect of rising plasma oestrogen and progesterone levels on liver zinc accumulation and the uptake of zinc by the foetal unit. Supplementation of pregnant women with zinc (10–20 mg/d) does not eliminate the pregnancy-associated decline in plasma zinc concentration (Breskin et al. 1983; Hunt et al. 1983, 1984); however, it has been reported to reduce the extent of the decline in some population groups (Hunt et al. 1983).

The observation that the extent of the decline in plasma zinc during pregnancy in some women can be influenced by dietary zinc supplementation suggests that while part of the decline may be a "normal" consequence of pregnancy (and thus presumably is not a risk factor for the foetus), the extent of the decline in some women may be of sufficient magnitude to represent a risk to the foetus. Evidence that zinc deficiency could be a teratogenic agent in humans was first provided by Sever and Emanuel (1973) who noted that there was a high frequency of neural tube defects in regions of the world where zinc deficiency was prevalent. Further support for this idea was provided by Hambidge and co-workers (1975) who summarized the literature of pregnancy outcome in women with the autosomal genetic recessive disorder acrodermatitis enteropathica (AE), a disease which mimics zinc deficiency. The seven pregnancies (three women) reviewed by Hambidge et al. (1975) occurred prior to the recognition of the therapeutic value of zinc. Of these pregnancies, one resulted in a spontaneous abortion and two resulted in the birth of babies with malformations; one was an anencephalic foetus and the other an achondroplastic dwarf. Two of the other pregnancies resulted in low birth weight infants. The incidence of abnormal births in this study (43%) represents a tenfold increase over the frequency of birth defects normally found in healthy populations. After the introduction of zinc therapy for AE patients, Brenton et al. (1981) reported that pregnant women with AE who are able to maintain normal plasma zinc levels via supplementation are characterized by normal deliveries and births.

The relationship between maternal zinc status and pregnancy outcome in humans has also been investigated in prospective studies. Jameson (1976) in Sweden reported that in a study of 316 pregnancies, a high proportion (60%) of the women who gave birth to infants with congenital defects showed low serum zinc concentrations in the first trimester. In addition, women who delivered before or after normal term had low serum zinc levels in the third trimester. Prema (1980) reproted that in a study of 291 Indian women, two mothers of three infants with major malformations showed very low serum zinc levels at delivery. Similarly, Cherry et al. (1981) in a study of 272 teenage pregnant women observed that six out of the ten women who delivered infants with malformations had low term plasma zinc levels.

Similar observations that low maternal plasma zinc may be associated with increased incidence of pregnancy complications and malformations at birth have been made by Damyanov and Dutz (1971) in Iran, Cavdar et al. (1977, 1980, 1985) in Turkey and Soltan and Jenkins (1982) in Ireland. More recently, in a study of 450 women followed during and after pregnancy, plasma zinc was reported to be an indicator of foeto-maternal complications in women in the lowest quartile for plasma zinc (Mukherjee et al. 1984). Buamah et al. (1984) reported that in a study of 259 pregnancies, mean serum zinc concentrations of women with anencephalic foetuses ($n = 9$) were significantly lower during the second trimester than those of women who delivered normal infants ($n = 253$). In contrast to the finding of low plasma zinc concentrations in the mothers of anencephalic infants, Bergmann et al. (1980) reported increased zinc concentrations in the hair of mothers of children with spina bifida. The authors interpreted this finding as an indication of abnormal zinc metabolism in the maternal/foetal unit.

Additional evidence suggesting a relationship between zinc deficiency in humans and the outcome of pregnancy has been provided by reports of significant correlations between birth weight and maternal plasma zinc levels (Jameson 1976; Crosby et al. 1977; Meadows et al. 1981; Ghosh et al. 1985; Simmer and Thompson 1985a, b). Similarly, it has been reported that maternal and infant leucocyte zinc levels at term are correlated with low birth weight (Meadows et al. 1981, 1983a, b; Patrick and Dervish 1984; Simmer and Thompson 1985a). It is important to note that the association between low maternal leucocyte zinc levels and low birth weights was only observed in mothers of intrauterine growth-retarded term infants; mothers of premature infants were found to have leucocyte zinc levels similar to those of women with normal term infants (Meadows et al. 1981, 1983a, b). Simmer and Thompson (1985b) observed that in mothers of low birth weight infants, there was a high incidence of smoking in the women characterized by low leucocyte zinc levels. These authors suggested that cadmium inhaled as part of the cigarette smoke may have been absorbed and subsequently inhibited cellular zinc uptake. Simmer et al. (1985) have reported that prostaglandin production is altered in the "zinc-deficient" leucocytes, suggesting that the observed low leucocyte zinc levels represent a functional maternal zinc deficiency. Recently, Wells et al. (1987) in a study of 70 women have reported that maternal leucocyte zinc levels during the third trimester were significantly associated with birth weight and suggested their measurement as a predictor of intrauterine growth retardation at this stage of pregnancy.

In contrast, McMichael et al. (1982), Metcoff et al. (1981) and Fehily et al. (1986) have reported an inverse relationship between maternal plasma zinc at mid-pregnancy and birth weight. One explanation for the finding of an inverse relationship between mid-pregnancy maternal zinc levels and infant birth weight is that the "high" maternal plasma zinc levels reflect defective foetal uptake of the element (McMichael et al. 1982). A second explanation for the difference in findings with regard to maternal plasma zinc concentrations and birth weight is that the catabolic/anabolic state of the mothers was different in the two sets of studies. Finally, some investigators have reported no association between indicators of maternal zinc status and infant birth weight (Campbell-Brown et al. 1985; Tuttle et al. 1985); however, given the foregoing reports of both positive and negative correlations between maternal plasma zinc and birth weight, it is difficult to interpret these results.

In addition to reports suggesting an association between low maternal plasma zinc levels and poor foetal outcome, maternal morbidity has been reported to be more frequent in women with low plasma zinc compared with women with normal to high plasma zinc. Jameson (1976) observed in a study of 84 well-nourished pregnant adult women that low plasma zinc concentrations during mid-pregnancy were associated with prolonged labour, atonic bleeding and postdate delivery. Similar observations of an association between low maternal serum zinc and both prolonged pregnancy and premature delivery have been made by others (Antoniou et al. 1982; Kiilholma et al. 1984). McMichael et al. (1982) observed an association of low mid-pregnancy plasma zinc levels with an increased risk of intrapartum haemorrhaging. Low plasma zinc during pregnancy has also been associated with pregnancy-induced hypertension. Cherry et al. (1981) and Hunt et al. (1984) observed that hypertensive pregnant teenagers are characterized by low plasma zinc concentrations compared with normotensive pregnant teenagers. Similarly, Zimmerman et al. (1984) followed 46 well-nourished adult women throughout pregnancy and the five women who developed pre-eclampsia had lower serum zinc concentrations during early pregnancy than women who did not develop pre-eclampsia.

Bassiouni et al. (1979), in a study of 52 women with pre-eclampsia and 20 women whose pregnancy progressed normally, have also observed a correlation between low maternal zinc levels and pre-eclampsia. Brophy et al. (1985) reported low placental zinc concentrations in women who suffered pre-eclampsia compared with women with normal pregnancies. Hunt et al. (1984) did not observe an association between low plasma zinc concentrations and pregnancy-induced hypertension in adult women; however, the incidence of hypertension was lower in pregnant women given zinc supplements (20 mg/d) than in controls. In contrast, Mukherjee et al. (1984) in a study of 450 adult women observed no correlation between plasma zinc concentrations and pregnancy-induced hypertension, premature delivery or abnormal bleeding, although women with the lowest plasma zinc concentrations had a high incidence of infections.

Taken together, the clinical reports summarized above may be viewed as offering strong support for the idea that maternal zinc status may be a significant determinant of pregnancy outcome in many population groups. However, it is important to recognize that in most cases it is not clear if low maternal plasma (or leucocyte) zinc levels reflect a cause or effect of the factors underlying poor pregnancy outcome. For example, it has been well documented that a large variety of stresses, ranging from tissue injury to hyperthermal and hypothermal stress, will result in lower plasma zinc levels as a result of liver sequestration of the element (Cousins 1985; Hambidge et al. 1986). However, even if low maternal plasma zinc levels are secondary to a primary insult (such as liver disease), the low plasma zinc levels may still represent a primary teratogenic insult to the embryo/foetus. This is an important point as it suggests that maternal zinc supplementation may be of value in a large number of cases where the pregnant mother is suffering from disease or chronic stress.

While it is evident from experimental animal studies that the consumption of a zinc-deficient diet during pregnancy is teratogenic, documentation of a strong correlation between inadequate maternal dietary zinc intake and poor pregnancy outcome in human population groups has been surprisingly limited. As mentioned above, the current RDA for zinc for pregnant women is 20 mg/d, an amount which few reach without supplementation. Indeed, it has been estimated that in many population groups dietary zinc intake during pregnancy is more of the order of 10–12 mg/d. While superficially this large difference between actual dietary zinc intake and the RDA suggests that maternal primary zinc deficiencies should be common, epidemiological studies have not shown strong associations between maternal dietary zinc intake and pregnancy outcome, even when some diets contain very low (~8 mg/d) amounts of zinc (Hunt et al. 1979). As discussed in Chaps. 18 and 21 one explanation for the lack of documentation of an effect of maternal dietary zinc intake on pregnancy outcome is that the current RDA for zinc for pregnant women may be set at a level in excess of the true requirement.

Balance studies indicate that non-pregnant adults must absorb about 2 mg of zinc/d to replace endogenous losses and maintain zinc equilibrium (Hess et al. 1977; Baer and King 1984). Swanson and King (1987) have calculated that at its peak, the additional maternal demand for zinc during pregnancy would be of the order of 0.6 mg/d to accommodate the synthesis of additional maternal and foetal tissues. It is currently thought that this additional zinc must be provided by the diet since women (Swanson et al. 1983), in contrast to rodents (Davies and Williams 1977), apparently do not increase their fractional gastrointestinal uptake of zinc during pregnancy. Swanson et al. (1983) reported that pregnant women fed 16 mg of zinc absorbed 25%, or 4 mg, of the dose. Thus, as pointed out by these authors, given a 25% bioavailability, an intake of 10.5 mg/d should provide the metabolic requirement for zinc of the maternal–foetal unit.

Based on the calculations of Swanson and King (1987), it is tempting to suggest that dietary zinc intakes in excess of 10 mg/d should be adequate for most pregnant women. However, as pointed out by these authors and others (Chaps. 4, 22), it is important also to consider other factors in the diet which may negatively affect the bioavailability of the element. For example, Cavdar et al. (1980) have suggested that the high incidence of neural tube defects in parts of Turkey may be linked to the consumption by pregnant women of a high cereal diet with low zinc bioavailability. Because of the potential competitive interaction between iron and zinc (Hill and Matrone 1970), it has been suggested that excessive iron supplementation during pregnancy may be a risk factor for the precipitation of maternal zinc deficiency (Solomons 1986). Breskin et al. (1983) reported that iron supplements in excess of 30 mg/d were correlated with low plasma zinc levels during early pregnancy. Hambidge et al. (1983) observed in a study of pregnant women that the degree of prenatal iron supplementation was negatively correlated with plasma zinc levels despite similar intakes of zinc during pregnancy. Similarly, Campbell-Brown et al. (1985) reported an association of low maternal plasma zinc concentrations with high iron supplementation (> 100 mg/d) during pregnancy. In contrast Sheldon et al. (1985) observed no effect of iron supplementation (160 mg/d) on maternal plasma zinc levels from pregnancy to 12 weeks postpartum.

Recently, there has been considerable interest in the idea that folic acid supplements may also reduce the bioavailability of zinc. Mukherjee et al. (1984) found a weak negative association between maternal zinc and folic plasma levels, and increased maternal–foetal complications were correlated with low plasma zinc levels. Consistent with these findings, Milne et al. (1984) observed decreased urinary zinc excretions, and increased faecal zinc excretions in adult males given 400 μg of folate every other day. Combined iron–folate supplements also decreased intestinal zinc absorption in healthy male volunteers (Meadows et al. 1983a, b). Simmer et al. (1987) reported that both combined iron–folate supplements (100 mg and 350 μg, respectively), and folate supplements alone (350 μg), reduced zinc absorption in healthy pregnant women during the second trimester. Furthermore, these investigators observed an inhibitory effect of both iron and folate when the supplements were given 24 h prior to, as well as during, the zinc absorption test, suggesting a mucosal rather than a luminal effect of the supplements. In contrast, Keating et al. (1987) observed no effect of folate supplements (10 mg) on zinc absorption in adult men. It is obvious that additional studies investigating the effects of folic acid supplements on maternal zinc status are urgently needed, especially in view of the recent interest in the idea that folic acid supplements during pregnancy may reduce the incidence of neural tube defects (Smithells et al. 1983; Wild et al. 1986).

While, as discussed above, there is a lack of papers showing a correlation between low dietary zinc intake and poor pregnancy outcome, dietary zinc supplementation trials have provided additional support for the concept that maternal zinc status may be a significant predictor of human pregnancy outcome. Jameson (1982) reported that zinc supplementation (45 mg/d) of 64 women characterized by low plasma zinc in the first trimester of pregnancy resulted in a lower frequency of prolonged labour, extended gestation and postmature babies than in a similar group of non-supplemented women. In a study of 46 pregnant women, ten of whom were supplemented with 15 mg zinc/d, zinc supplementation was associated with higher maternal plasma alkaline phosphatase activities and a lower incidence of dysfunctional labour patterns than in the non-supplemented controls (Hambidge et al. 1983).

Hunt et al. (1984) reported that adult low-income Mexican–American women supplemented with 20 mg of zinc daily had a lower frequency of pregnancy-induced hypertension than non-supplemented women; however no other differences in pregnancy complications were noted. Zinc supplementation had no marked effects on pregnancy outcome in low-income teenage Mexican–Americans (Hunt et al. 1984). Kynast and Saling (1986) reported that in a prospective study in which 179 apparently healthy pregnant women were supplemented with zinc (20 mg zinc aspartate) between the 12th and 34th week of pregnancy, there was a lower frequency of pregnancy complications and low birth weight in the supplemented group than in 345 unsupplemented women selected from a birth register.

Conclusions

Pregnancy is associated with extraordinary metabolic demands on both the mother and developing foetus. In this review evidence has been presented that adequate maternal zinc nutriture is essential for normal embryogenesis. Studies in laboratory animals and non-human primates have demonstrated that maternal zinc deficiency produces effects ranging from infertility, prolonged labour, intrauterine growth retardation and teratogenesis, to embryonic or foetal death. Postnatal complications of prenatal zinc deficiency can also occur, as is evidenced by impaired immunocompetence and abnormal behaviour patterns in adult animals that were prenatally deprived of zinc.

As with the findings from experimental animals, numerous studies have suggested a strong effect of maternal zinc status on pregnancy outcome in humans. Significantly, reports are now being published that zinc supplementation of apparently healthy pregnant women can result in significant improvements in pregnancy outcome. Thus, there is considerable evidence from human epidemiological and intervention studies that marginal to severe maternal zinc deficiency may be a complication of human pregnancies. In some situations, the maternal zinc deficiency may arise as a result of a primary dietary deficiency of the element; however, it is possible that in other cases maternal zinc deficiency (as defined by low plasma zinc levels), may be, in part, secondary to tissue injury and/or stress.

Future studies need to be directed at (a) a better understanding of the mechanisms underlying the teratogenicity of zinc deficiency; (b) improvement of techniques for the determination of maternal and infant zinc status; (c) determination of the effects of modest zinc supplementation during pregnancy on the risk of pregnancy complications and foetal outcome in large population groups; and (d) development of optimal therapies for the treatment of infants who were prenatally zinc deprived.

References

Abraham R (1982) Trace element intake by Asians during pregnancy. Proc Nutr Soc 41: 261–265

Abu-Assal MJ, Craig WJ (1984) The zinc status of pregnant vegetarian women. Nutr Rep Int 29: 485–494

Amemiya K, Hurley LS, Keen CL (1985) Effects of the anticarcinogenic drug 6-mercaptopurine on mineral metabolism in the mouse. Toxicol Lett 25: 55–62

Amemiya K, Keen CL, Hurley LS (1986a) 6-Mercaptopurine induced alterations in mineral metabolism and teratogenesis in the rat. Teratology 34: 321–334

Amemiya K, Hurley LS, Keen CL (1986b) The effect of 6-mercaptopurine on mineral and metallothionein metabolism in the mouse. Biol Trace Element Res 11: 161–175

Amemiya K, Hurley LS, Keen CL (1988) The effect of 6-mercaptopurine on ^{65}Zn distribution in the pregnant rat. Teratology (in press)

Antoniou K, Vassilaki-Grimmani M, Lolis D, Grimanis AP (1982) Concentrations of cobalt, rubidium, selenium and zinc in maternal and cord blood serum and amniotic fluid of women with normal and prolonged pregnancies. J Radioanal Chem 70: 77–84

Apgar J (1972) Effect of zinc deprivation from day 12, 15, or 18 of gestation on parturition in the rat. J Nutr 102: 343–348

Apgar J (1973) Effect of zinc repletion late in gestation on parturition in the zinc-deficient rat. J Nutr 103: 973–981

Apgar J (1975) Effects of some nutritional deficiencies on parturition in rats. J Nutr 105: 1553–1561

Apgar J (1977) Effects of zinc deficiency and zinc repletion during pregnancy on parturition in two strains of rats. J Nutr 107: 1399–1403

Apgar J (1985) Zinc and reproduction. Annu Rev Nutr 5: 43–68

Apgar J, Fitzgerald JA (1985) Effect on the ewe and lamb of low zinc intake throughout pregnancy. J Anim Sci 60: 1530–1538

Apgar J, Travis HF (1979) Effect of a low zinc diet on the ewe during pregnancy and lactation. J Anim Sci 48: 1234–1238

Assadi FK, Ziai M (1986) Zinc status of infants with fetal alcohol syndrome. Pediatr Res 20: 551–554

Baer MT, King JC (1984) Tissue zinc levels and zinc excretion during experimental zinc depletion in young men. Am J Clin Nutr 39: 556–570

Baly DL, Golub MS, Gershwin ME, Hurley LS (1984) Studies of marginal zinc deprivation in rhesus monkeys. III. Effects on vitamin A metabolism. Am J Clin Nutr 40: 199–207

Bassiouni BA, Foda AI, Rafei AA (1979) Maternal and fetal plasma zinc in pre-eclampsia. Eur J Obstet Gynecol Reprod Biol 9: 75–80

Beach RS, Gershwin ME, Hurley LS (1982) Gestational zinc deprivation in mice: persistence of immunodeficiency for three generations. Science 218: 469–471

Beach RS, Gershwin ME, Hurley LS (1983) Persistent immunological consequences of gestational zinc deprivation. Am J Clin Nutr 38: 579–590

Bell LT, Branstrator M, Roux C, Hurley LS (1975) Chromosomal abnormalities in maternal and fetal tissues of magnesium or zinc deficient rats. Teratology 12: 221–226

Bergman B, Friberg U, Lohmander S, Oberg T (1972) The importance of zinc to cell proliferation in endochondral growth sites in the white rat. Scand J Dent Res 80: 486–492

Bergmann KE, Makosch G, Tews KH (1980) Abnormalities of hair zinc concentration in mothers of newborn infants with spina bifida. Am J Clin Nutr 33: 2145–2150

Blamberg DL, Blackwood WB, Supplee WC, Combs CF (1960) Effect of zinc deficiency in hens on hatchability and embryonic development. Proc Soc Exp Biol Med 104: 217–220

Bradford LD, Oner G, Lederis K (1981) Effects of zinc deficiency on locomotor and stereotypic behavior in rats. Pharmacologist 23: 142

Brenton DP, Jackson MJ, Young A (1981) Two pregnancies in a patient with acrodermatitis enteropathica treated with zinc sulphate. Lancet II: 500–502

Breskin MW, Worthington-Roberts BS, Knopp RH et al. (1983) First trimester serum zinc concentration in human pregnancy. Am J Clin Nutr 38: 943–953

Brophy MH, Harris NF, Crawford IL (1985) Elevated copper and lowered zinc in the placenta of pre-eclamptics. Clin Chem Acta 15: 107–111

Buamah PK, Russell M, Bakes M, Milford Ward A, Skillen AW (1984) Maternal zinc status: a determination of central nervous system malformations. Br J Obstet Gynaecol 91: 788–790

Bunce GE, Wilson GE, Mills CF, Klopper A (1983) Studies on the role of zinc in parturition in the rat. Biochem J 210: 761–767

Butrimovitz GP, Purdy WC (1978) Zinc nutrition and growth in a childhood population. Am J Clin Nutr 31: 1409–1412

Calhoun NR, Smith JC, Becker KL (1974) The role of zinc in bone metabolism. Clin Orthop 103: 212–234

Campbell-Brown M, Ward RJ, Haines AP, North WRS, Abraham R, McFayden IR (1985) Zinc and copper in Asian pregnancies – is there evidence for a nutritional deficiency? Br J Obstet Gynaecol 92: 875–885

Castillo-Duran C, Heresi G, Fisberg M, Uauy R (1987) Controlled trial of zinc supplementation during recovery from malnutrition: effects on growth and immune function. Am J Clin Nutr 45: 602–608

Castro CE, Alvares OF, Sevall JS (1986) Zinc deficiency decreases histone H1⁰ in rat liver. Nutr Rep Int 34: 67–74

Cavdar AO, Arcasoy A, Baycu T, Himmetoglu O (1977) Zinc deficiency and anencephaly in Turkey. Teratology 22: 141

Cavdar AO, Babacan E, Arcasoy A, Ertem U (1980) Effect of nutrition on serum zinc concentration during pregnancy in Turkish women. Am J Clin Nutr 33: 542–544

Cavdar AO, Babacan E, Asik S et al. (1985) Neural tube defects and zinc. Nutr Res [Suppl] I: 331–334

Chandra RK (1975) Antibody formation in first and second generation offspring of nutritionally deprived rats. Science 190: 289–290

Cheek DB (1975) Fetal and postnatal cellular growth. John Wiley, New York

Cherry FF, Bennett EA, Bazzano GS, Johnson LK, Fosmire GJ, Barson HK (1981) Plasma zinc in hypertension/toxemia and other reproductive variables in adolescent pregnancy. Am J Clin Nutr 34: 2367–2375

Chesters JK, Quarterman J (1970) Effects of zinc deficiency on food intake and feeding patterns of rats. Br J Nutr 24: 1061–1069

Chvapil M (1976) Effects of zinc on cells and biomembranes. Med Clin North Am 60: 799–812

Clegg MS, Rogers JM, Zucker RM, Hurley LS, Keen CL (1986) Flow cytometry analysis of cell cycle stages in zinc deficient fetal rat brain. Fed Proc 45: 1086

Clegg MS, Hurley LS, Keen CL (1987) Effects of Zn deficiency on liver nonhistone profiles in weanling male rats. Fed Proc 46: 597

Cousins RJ (1985) Absorption, transport, and hepatic metabolism of copper and zinc: special reference to metallothionein and ceruloplasmin. Physiol Rev 65: 238–309

Crawford BD, Enger MD, Griffith BB et al. (1985) Coordinate amplification of metallothionein I and II genes in cadmium-resistant Chinese hamster cells: implications for mechanisms regulating metallothionein gene expression. Mol Cell Biol 5: 320–329

Crosby WM, Metcoff J, Costiloe JP et al. (1977) Fetal malnutrition: an appraisal of correlated factors. Am J Obstet Gynecol 128: 22–31

Damyanov I, Dutz W (1971) Anencephaly in Shiraz, Iran. Lancet I: 82

Dash S, Brewer GJ, Oelshlegel FJ (1974) Effect of zinc on hemoglobin binding by red blood cell membranes. Nature 250: 251–252

Daston GP (1982) Fetal zinc deficiency as a mechanism for cadmium induced toxicity to the developing rat lung and pulmonary surfactant. Toxicology 24: 55–63

Davies NT, Williams RB (1977) The effect of pregnancy and lactation on the absorption of zinc and lysine by the rat duodenum in situ. Br J Nutr 38: 417–423

Dreosti IE (1987) Zinc deficiency and the developing embryo. Neurotoxicology 8: 369–378

Dreosti IE, Hurley LS (1975) Depressed thymidine kinase activity in zinc deficient rat embryos. Proc Soc Exp Biol Med 150: 161–165

Dreosti IE, Tao S, Hurley LS (1968) Plasma zinc and leukocyte changes in weanling and pregnant rats during zinc deficiency. Proc Soc Exp Biol Med 127: 169–174

Dreosti IE, Grey PC, Wilkins PJ (1972) Deoxyribonucleic acid synthesis, protein synthesis and teratogenesis in zinc-deficient rats. S Afr Med J 46: 1585–1588

Dreosti IE, Record IR, Manuel SJ (1980) Incorporation of ^{3}H-thymidine into DNA and the activity of alkaline phosphatase in zinc-deficient fetal rat brains. Biol Trace Element Res 2: 21–29

Dreosti IE, McMichael AJ, Gibson GT, Buckley RA, Hartsthorne JH, Colley DP (1982) Fetal and maternal copper and zinc levels in human pregnancy. Nutr Res 2: 591–602

Dreosti IE, Record IR, Manuel SJ (1985) Zinc deficiency and the developing embryo. Biol Trace Element Res 7: 103–122

Dreosti IE, Buckley RA, Record IR (1986) The teratogenic effect of zinc deficiency and accompanying feeding patterns in mice. Nutr Res 6: 159–166

Dubowitz V (1969) The floppy infant. Spastics International Medical Publications, Lavenham, England

Duncan JR, Hurley LS (1978) An interaction between zinc and vitamin A in pregnant and fetal rats. J Nutr 108: 1432–1438

Dylewski DP, Lytton FDC, Bunce GE (1986) Dietary zinc and parturition in the rat. II. Myometrial gap junctions. Biol Trace Element Res 9: 165–175

Eckhert CD, Hurley LS (1977) Reduced DNA synthesis in zinc deficiency: regional differences in embryonic rats. J Nutr 107: 855–861

Ehalmesmaki E, Ylikorkala O (1985) Concentrations of zinc and copper in pregnant problem drinkers and their newborn infants. Br Med J 291: 1470–1471

Eriksson UJ (1984) Diabetes in pregnancy: retarded fetal growth, congenital malformations and feto-maternal concentrations of zinc, copper and manganese in the rat. J Nutr 114: 477–484

Failla ML, Caperna TJ, Dougherty JM (1985) Influence of pancreatic and adrenal hormones on altered trace metal metabolism in the STZ-diabetic rat. In: Mills CF, Bremner I, Chesters JK (eds) Trace elements in man and animals – TEMA 5. Commonwealth Agricultural Bureaux, Farnham Royal, England, pp 330–333

Fairweather-Tait SJ, Wright AJA, Cooke J, Franklin J (1985) Studies of zinc metabolism in pregnant and lactating rats. Br J Nutr 54: 401–413

Falchuk KH, Gordon RR, Stankiewicz A, Hilt KL, Vallee BL (1986) *Euglena gracilis* chromatin: comparison of effects of zinc, iron, magnesium, or manganese deficiency and cold shock. Biochemistry 25: 5388–5391

Fehily D, Fitzsimmons B, Jenkins D, Cremin FM, Flynn A, Soltan MH (1986) Association of fetal growth with elevated maternal plasma zinc concentration in human pregnancy. Hum Nutr Clin Nutr 40C: 221–227

Flynn A, Martier SS, Sokol RJ, Miller SI, Golden NL, Villano BC (1981) Zinc status of pregnant alcoholic women: a determinant of fetal outcome. Lancet I: 572–574

Gallaher D, Hurley LS (1980) Low zinc concentration in rat uterine fluid after four days of dietary deficiency. J Nutr 110: 591–593

Garnica AD (1981) Trace metals and hemoglobin metabolism. Ann Clin Lab Med 11: 220–228

Gershwin ME, Beach RS, Hurley LS (1985) Nutrition and immunity. Academic Press, New York

Ghisham F, Greene HL (1983) Fetal alcohol syndrome: failure of zinc supplementation to reverse the effect of ethanol on placental transport of zinc. Pediatr Res 17: 529–531

Ghisham FK, Patwardhan R, Greene HL (1982) Fetal alcohol syndrome: inhibition of placental zinc transport as a potential mechanism for fetal growth retardation in the rat. J Lab Clin Med 100: 45–52

Ghosh A, Fong LYY, Wan CW, Laing ST, Woo JSK, Wong V (1985) Zinc deficiency is not a cause for abortion, congenital abnormality and small-for-gestational age infant in Chinese women. Br J Obstet Gynaecol 92: 886–891

Golden BE, Golden MHN (1981a) Plasma zinc, rate of weight gain, and the energy cost of tissue deposition in children recovering from severe malnutrition on a cow's milk or soya protein based diet. Am J Clin Nutr 34: 892–899

Golden MHN, Golden BE (1981b) Effect of zinc supplementation on the dietary intake, rate of weight gain, and energy cost of tissue deposition in children recovering from severe malnutrition. Am J Clin Nutr 34: 900–908

Golden MHN, Jackson AA, Golden BE (1977) Effect of zinc on thymus of recently malnourished children. Lancet II: 1057–1059

Golden MHN, Golden BE, Harland PSEG, Jackson AA (1978) Zinc and immunocompetence in protein-energy malnutrition. Lancet I: 1226–1230

Golub MS, Gershwin ME, Hurley LS, Hendrickx AG, Baly DL (1982) Induction of marginal zinc deficiency in female rhesus monkeys. Am J Primatol 3: 299–305

Golub MS, Gershwin ME, Hurley LS, Baly DL, Hendrickx AG (1984a) Studies of marginal zinc deprivation in rhesus monkeys. I. Influence on pregnant dams. Am J Clin Nutr 39: 265–280

Golub MS, Gershwin ME, Hurley LS, Baly DL, Hendrickx AG (1984b) Studies of marginal zinc deprivation in rhesus monkeys. II. Pregnancy outcome. Am J Clin Nutr 39:879–887

Golub MS, Gershwin ME, Hurley LS, Saito WY, Hendrickx AG (1984c) Studies of marginal zinc deprivation in rhesus monkeys. IV. Growth of infants in the first year. Am J Clin Nutr 40: 1192–1202

Golub MS, Keen CL, Vijayan VK, Gershwin ME, Hurley LS (1984d) Early development of brain and behavior in mice fed a marginally zinc deficient diet. In: Frederickson CJ, Howell BA, Kasarskis EJ (eds) The neurobiology of zinc. Part B: Deficiency, toxicity and pathology. Alan R. Liss, New York, pp 65–76

Golub MS, Gershwin ME, Hurley LS, Saito WY (1985) Studies of marginal zinc deprivation in rhesus monkeys. VII. Infant behavior. Am J Clin Nutr 42: 1229–1239

Gordon EF (1984) Behavioral correlates of experimental zinc deficiency. In: Frederickson CJ, Howell BA, Kasarskis EJ (eds) The neurobiology of zinc. Part B: Deficiency, toxicity and pathology. Alan R. Liss, New York, pp 77–90

Gordon EF, Gordon RC, Passal DB (1981) Zinc metabolism: basic, clinical and behavioral aspects. J Pediatr 99: 341–349

Graham GG, Cordano A, Blizzard RM, Cheek DB (1969) Infantile malnutrition: changes in body composition during rehabilitation. Pediatr Res 3: 579–589

Hackman RM, Hurley LS (1983) Interaction of dietary zinc, genetic strain and acetazolamide in teratogenesis in mice. Teratology 28: 355–368

Hackman RM, Hurley LS (1984a) Drug–nutrient interactions in teratogenesis. In: Roe DA, Campbell TC (eds) Drugs and nutrients. The interactive effects. Marcel Dekker, New York, pp 299–329

Hackman RM, Hurley LS (1984b) Interactions of salicylate, dietary zinc, and genetic strain in teratogenesis in rats. Teratology 30: 225–236

Halas ES, Sandstead HH (1975) Some effects of prenatal zinc deficiency on behavior of the adult rat. Pediatr Res 9: 94–97

Halas ES, Rowe MC, Johnson OR, McKenzie JM, Sandstead HH (1976) Effects of intrauterine zinc deficiency on subsequent behavior. In: Prasad AS (ed) Proceedings of the international symposium on trace elements in human health and disease. Nutrition Foundation Monograph. Academic Press, New York, pp 327–343

Halas ES, Hunt CD, Eberhardt MJ (1986) Learning and memory disabilities in young adult rats from mildly zinc deficient dams. Physiol Behav 37: 451–458

Halsted JA, Ronaghy HA, Abadi P et al. (1972) Zinc deficiency in man: the Shiraz experiment. Am J Clin Med 53: 277–284

Hambidge KM, Nelder KH, Walravens PA (1975) Zinc, acrodermatitis enteropathica, and congenital malformations. Lancet I: 577–578

Hambidge KM, Krebs NF, Jacobs MA, Favier A, Guyette L, Ikle DN (1983) Zinc nutritional status during pregnancy: a longitudinal study. Am J Clin Nutr 37: 429–442

Hambidge KM, Casey CE, Krebs NF (1986) Zinc. In: Mertz W (ed) Trace elements in human and animal nutrition, 5th edn. Academic Press, New York, pp 1–137

Hamer DH (1986) Metallothionein. Annu Rev Biochem 55: 913–951

Harding AJ, Dreosti IE, Tulsi RS (1987) Zinc deficiency in the 11 day rat embryo: a scanning and transmission electron microscope study. Life Sci 42: 889–896

Haumont S, McLean FC (1966) Zinc and the physiology of bone. In: Prasad AS (ed) Zinc metabolism. Charles C. Thomas, Springfield, pp 169–186

Haynes DC, Gershwin ME, Golub MS, Cheung ATW, Hurley LS, Hendrickx AG (1985) Studies of marginal zinc deprivation in rhesus monkeys. VI. Influence on the immunohematology of infants in the first year. Am J Clin Nutr 42: 252–262

Haynes DC, Golub MS, Gershwin ME, Cheung ATW, Hurley LS, Hendrickx AG (1987a) Long-term marginal zinc deprivation in rhesus monkeys. I. Effects on adult female breeders before conception. Am J Clin Nutr 45: 1492–1502

Haynes DC, Golub MS, Gershwin ME, Hurley LS, Hendrickx AG (1987b) Long-term marginal zinc deprivation in rhesus monkeys. II. Effects on maternal health and fetal growth at midgestation. Am J Clin Nutr 45: 1503–1513

Henderson GI, Hoyumpa AM, Rothschild MA (1980) Effect of ethanol and ethanol-induced hypothermia on protein synthesis in pregnant and fetal rats. Alcohol Clin Exp Res 4: 165–177

Henkin RI (1984) Zinc in taste function. A critical review. Biol Trace Element Res 6: 263–280
Henkin RI, Patten BM, Re PK, Bronzert DA (1975) A syndrome of acute zinc loss. Arch Neurol 32: 745–751
Hesketh JE (1981) Impaired microtubule assembly in brain from zinc-deficient pigs and rats. Int J Biochem 13: 921–926
Hess FM, King JC, Margen S (1977) Zinc excretion in young women on low zinc intake and oral contraceptive agents. J Nutr 107: 1610–1620
Hickory W, Nanda R, Catalanotto FA (1979) Fetal skeletal malformations associated with moderate zinc deficiency during pregnancy. J Nutr 109: 883–891
Hill CH, Matrone G (1970) Chemical parameters in the study of in vivo and in vitro interactions of transition elements. Fed Proc 29: 1474–1481
Hirsch KS, Hurley LS (1978) Relationship of dietary zinc to 6-mercaptopurine teratogenesis and DNA metabolism in the rat. Teratology 17: 303–314
Hughes RN, Horsburgh RJ (1982) Some behavioral effects of prenatal marginal zinc deficiency in young rats. Nutr Res 2: 513–520
Hunt IF, Murphy NJ, Shroads J, Smith JC Jr (1979) Dietary zinc intake of low-income pregnant women of Mexican descent. Am J Clin Nutr 32: 1511–1518
Hunt IF, Murphy NJ, Cleaver AE et al. (1983) Zinc supplementation during pregnancy: zinc concentration of serum and hair from low-income women of Mexican descent. Am J Clin Nutr 37: 572–582
Hunt IF, Murphy MJ, Cleaver AE et al. (1984) Zinc supplementation during pregnancy: effect on selected blood constituents and on progress and outcome of pregnancy in low-income women of Mexican descent. Am J Clin Nutr 40: 508–521
Hunt IF, Murphy NJ, Martner-Hewes PM et al. (1987) Zinc, vitamin B-6, and other nutrients in pregnant women attending prenatal clinics in Mexico. Am J Clin Nutr 46: 563–569
Hurley LS (1967) Studies on nutritional factors in mammalian development. J Nutr 91: 27–39
Hurley LS (1969) Nutrients and genes: interactions in development. Nutr Rev 27: 3–10
Hurley LS (1979) The fetal alcohol syndrome: possible implications of nutrient deficiencies. In: Li TK, Schenker J, Lumeng L (eds) Alcohol and nutrition. National Institute of Alcohol Abuse and Alcoholism, Rockville, Maryland, pp 367–379
Hurley LS (1981) Teratogenic aspects of manganese, zinc, and copper nutrition. Physiol Rev 61: 249–295
Hurley LS, Shrader RE (1972) Congenital malformations of the nervous system in zinc-deficient rats. In: Pfeiffer CC (ed) Neurobiology of the trace metals zinc and copper. Academic Press, New York, pp 7–51
Hurley LS, Shrader RE (1975) Abnormal development of preimplantation rat eggs after three days of maternal zinc deficiency. Nature (Lond) 254: 427–429
Hurley LS, Swenerton H (1966) Congenital malformations resulting from zinc deficiency in rats. Proc Soc Exp Biol Med 123: 692–697
Hurley LS, Tao S (1972) Alleviation of teratogenic effects of zinc deficiency by simultaneous lack of calcium. Am J Physiol 222: 322–325
Hurley LS, Gowan J, Milhaud G (1969) Calcium metabolism in manganese-deficient and zinc-deficient rats. Proc Soc Exp Biol Med 130: 856–860
Hurley LS, Gowan J, Swenerton H (1971) Teratogenic effects of short-term and transitory zinc deficiency in rats. Teratology 4: 199–204
Hurley LS, Gordon P, Keen CL, Merkhofer L (1982) Circadian variation in rat plasma zinc and rapid effect of dietary zinc deficiency. Proc Soc Exp Biol Med 170: 48–52
Iniguez C, Casas J, Carreres J (1978) Effects of zinc deficiency on the chick embryo blastoderm. Acta Anat 101: 120–129
Iosub S, Fuchs M, Bingol N, Stone RF, Gromishch DS, Wasserman E (1985) Incidence of major congenital malformations in offspring of alcoholics and polydrug abusers. Alcohol 2: 521–523
Jackson AJ, Schumacher HJ (1979) The teratogenic activity of a thalidomide analogue EM_{12} in rats on a low-zinc diet. Teratology 19: 341–344
Jameson S (1976) Effects of zinc deficiency in human reproduction. Acta Med Scand 593: 5–89
Jameson S (1982) Zinc status and pregnancy outcome in humans. In: Prasad AS, Dreosti IE, Hetzel BS (eds) Clinical applications of recent advances in zinc metabolism. Alan R. Liss, New York, pp 39–52
Jones KL, Smith DW, Ulleland CN (1973) Pattern of malformation in the offspring of chronic alcoholic mothers. Lancet I: 1267–1271
Kawamoto JC, Halas ES (1984) Lasting morphological effects of mild perinatal zinc deficiency in the adult hippocampal formation. In: Frederickson CJ, Howell GA, Kasarskis EF (eds) The

neurobiology of zinc. Part B: Deficiency, toxicity, and pathology. Alan R. Liss, New York, pp 33–48

Kawamoto JC, Castonguay TW, Keen CL, Stern JS, Hurley LS (1986) Age, sex and reproductive status alter the severity of anorexia in zinc deficient rats. Physiol Behav 38: 485–493

Kay RG, Tasman-James C, Pybus J, Whiting R, Black H (1976) A syndrome of acute zinc deficiency during total parenteral alimentation in man. Ann Surg 4: 331

Keating JN, Wada L, Stokstad ELR, King JC (1987) Folic acid: effect on zinc absorption in humans and in the rat. Am J Clin Nutr 46: 835–839

Keen CL, Hurley LS (1987) Effects of zinc deficiency on prenatal and postnatal development. Neurotoxicology 8: 379–388

Keen CL, Cohen NL, Lönnerdal B, Hurley LS (1983a) Teratogenesis and low copper status resulting from triethylenetetramine in rats. Proc Soc Exp Biol Med 173: 598–605

Keen CL, Mark-Savage P, Lönnerdal B, Hurley LS (1983b) Teratogenic effects of D-penicillamine in rats: relation to copper deficiency. Drug–Nutrient Interactions 2: 17–34

Keen CL, Cohen NL, Hurley LS, Lönnerdal B (1984) Molecular localization of copper and zinc in rat fetal liver in dietary and drug-induced copper deficiency. Biochem Biophys Res Commun 118: 697–703

Keinholz EW, Turk DE, Sunde ML, Hoekstra WG (1961) Effects of zinc deficiency in the diets of hens. J Nutr 75: 211–221

Keller PR, Fraker PJ (1986) Gestational zinc requirement of the A/J mouse: effects of a marginal zinc deficiency on in utero B-cell development. Nutr Res 6: 41–50

Kiilholma P, Gronroos M, Erkkola R, Pakarinen P, Nanto V (1984) Role of calcium, copper, iron, and zinc in preterm delivery and premature rupture of fetal membranes. Gynecol Obstet Invest 17: 194–201

Kimmel CA, Sloan CS (1975) Studies on the mechanism of EDTA teratogenesis. Teratology 12: 330–331

Kimmel CA, Butcher RE, Schumacher HJ (1972) Salicylates and nutrition: pre- and postnatal effects. Anat Rec 172: 345

Koshakji RP, Schulert AR (1972) Biochemical mechanisms of salicylate teratology in the rat. Biochem Pharmacol 22: 407–416

Kumar SP (1982) Fetal alcohol syndrome. Mechanisms of teratogenesis. Ann Clin Lab Sci 12: 254–257

Kumar S, Rao KSJ (1974) Blood and urinary zinc levels in diabetes mellitus. Nutr Metab 17: 231–235

Kynast G, Saling E (1986) Effect of oral zinc application during pregnancy. Gynecol Obstet Invest 21: 117–123

Leek JC, Vogler JB, Gershwin ME, Golub MS, Hurley LS, Hendrickx AG (1984) Studies of marginal zinc deprivation in rhesus monkeys. V. Fetal and infant skeletal defects. Am J Clin Nutr 40: 1203–1212

Leek JC, Keen CL, Vogler JB et al. (1988) Long-term marginal zinc deprivation in rhesus monkeys. IV. Effects on skeletal growth and mineralization. Am J Clin Nutr (in press)

Lytton FDC, Bunce GE (1986) Dietary zinc and parturition in the rat. I. Uterine pressure cycles. Biol Trace Element Res 9: 151–164

McKenzie JM, Foxmire GJ, Sandstead HH (1975) Zinc deficiency during the latter third of pregnancy: effects on fetal rat brain, liver, and placenta. J Nutr 105: 1466–1475

McMichael AJ, Dreosti IE, Gibson GT, Hartshorne JM, Buckley PA, Colley DP (1982) A prospective study of serial maternal zinc levels and pregnancy outcome. Early Hum Dev 7: 59–69

Masters DG, Fels HE (1985) Zinc supplements and reproduction in grazing ewes. Biol Trace Element Res 7: 89–93

Masters DG, Keen CL, Lönnerdal B, Hurley LS (1983) Zinc deficiency teratogenicity: the protective role of maternal tissue catabolism. J Nutr 113: 905–912

Masters DG, Keen CL, Lönnerdal B, Hurley LS (1986) Release of zinc from maternal tissues during zinc deficiency or simultaneous zinc and calcium deficiency in the pregnant rat. J Nutr 116: 2148–2154

Meadows NJ, Ruse W, Smith MF et al. (1981) Zinc and small babies. Lancet II: 1135–1137

Meadows NJ, Grainger SL, Ruse W, Keeling PWN, Thompson RPH (1983a) Oral iron and the bio-availability of zinc. Br Med J 287: 1013–1014

Meadows N, Ruse W, Keeling PWN, Scopes JW, Thompson RPH (1983b) Peripheral blood leucocyte zinc depletion in babies with intrauterine growth retardation. Arch Dis Child 58: 807–809

Metcoff J, Cottilo JP, Crosby W et al. (1981) Maternal nutrition and fetal outcome. Am J Clin Nutr 34: 708–721

Mieden GD, Keen CL, Hurley LS, Klein NW (1986) The effects on whole rat embryos cultured on serum from zinc and copper deficient rats. J Nutr 116: 2424–2431

Mills CF, Quarterman J, Chester JK, Williams RB, Dalgarno AC (1969) Metabolic role of zinc. Am J Clin Nutr 22: 1240–1249 (Symp Zinc Metab)

Milne DB, Canfield WK, Mahalko JR, Sandstead HH (1984) Effect of oral folic acid supplements on zinc, copper, and iron absorption and excretion. Am J Clin Nutr 40: 535–540

Minow RA, Stern MM, Casey JM, Rodriguez V, Luna MA (1976) Clinico-pathologic correlation of liver damage in patients with 6-mercaptopurine and adriamycin. Cancer 38: 1524–1528

Morgan PN, Keen CL, Calvert CC, Lönnerdal B (1988a) The effects of varying dietary zinc intake on weanling mouse pups during recovery from early undernutrition on growth, body composition and composition of gain. J Nutr (in press)

Morgan PN, Keen CL, Lönnerdal B (1988b) The effect of varying dietary zinc intake on weanling mouse pups during recovery from early undernutrition on tissue mineral concentrations, relative organ weights, and hematological indices. J Nutr (in press)

Morrison SA, Russell RM, Carney EA, Oaks EV (1978) Zinc deficiency: a cause of abnormal dark adaptation in cirrhotics. Am J Clin Nutr 31: 275–281

Moser PB, Reynolds RD (1983) Dietary zinc intake and zinc concentrations of plasma, erythrocytes, and breast milk in antepartum and postpartum lactating and nonlactating women; a longitudinal study. Am J Clin Nutr 38: 101–108

Mukherjee MD, Sandstead HH, Ratnaparki MV, Johnson LK, Milne DB, Stelling HP (1984) Maternal zinc, iron, folic acid and protein nutriture and outcome of human pregnancy. Am J Clin Nutr 40: 496–507

Mutch PB, Hurley LS (1974) Effect of zinc deficiency during lactation on postnatal growth and development of rats. J Nutr 104: 828–842

Mutch PB, Hurley LS (1980) Mammary gland function and development: effect of zinc deficiency in rat. Am J Physiol 238: 26–31

National Research Council (1980) Recommended dietary allowances, 9th edn. National Research Council, Washington DC

New DAT (1966) The culture of vertebrate embryos. Logos Press, London, pp 18–46

New DAT, Coppola PT, Terry S (1973) Culture of explanted rat embryos in rotating tubes. J Reprod Fertil 35: 135–138

Nicholson VJ, Veldstra H (1972) The influence of various cations on the binding of colchicene by rat brain homogenates. FEBS Lett 23: 309–313

Nishi Y, Lifshitz F, Bayne MA, Daum F, Silverberg M, Aiges H (1980) Zinc status and its relation to growth retardation in children with chronic inflammatory bowel disease. Am J Clin Nutr 33: 2613–2621

O'Dell BL, Reynolds G, Reeves PG (1977) Analogous effects of zinc deficiency and aspirin toxicity in the pregnant rat. J Nutr 107: 1222–1228

Odutuga AA (1982) Effects of low-zinc status and essential fatty acid deficiency on growth and lipid composition of rat brain. Clin Exp Pharmacol Physiol 9: 213–221

Oteiza PI, Keen CL, Lönnerdal B, Hurley LS (1987) Marginal Zn deficiency affects maternal brain microtubule assembly in rats. Fed Proc 46: 596

Palin HD, Underwood BA, Denning CR (1979) The effect of oral zinc supplementation on plasma levels of vitamin A and retinol-binding protein in cystic fibrosis. Am J Clin Nutr 32: 1253–1259

Palludan B, Wegger I (1976) Importance of zinc for foetal and post-natal development in swine. In: Nuclear techniques in animal production and health. International Atomic Energy Agency, Vienna, pp 191–205

Patrick J, Dervish C (1984) Leucocyte zinc in the assessment of zinc status. CRC Crit Rev Clin Lab Sci 20: 95–114

Payne PR, Wheeler EF (1968) Comparative nutrition in pregnancy and lactation. Proc Nutr Sci 27: 129–138

Peters AJ, Keen CL, Lónnerdal B, Hurley LS (1986) Zinc–vitamin A interaction in pregnant and fetal rats: high vitamin A does not compensate. J Nutr 116: 1765–1771

Pidduck HG, Wren PJJ, Evans DAP (1970) Hyperzincuria of diabetes mellitus and possible genetical implications of this observation. Diabetes 19: 240–247

Prema K (1980) Predictive value of serum copper and zinc in normal and abnormal pregnancy. Indian J Med Res 71: 554–560

Prohaska JR (1987) Functions of trace elements in brain metabolism. Physiol Rev 67: 858–901

Qvist I, Abdulla M, Jagerstad M, Svensson S (1986) Iron, zinc, and folate status during pregnancy and two months after delivery. Acta Obstet Gynecol Scan 65: 15–22

Rasmussen H (1983) Hypophosphatasia. In: Stanbury JB, Wyngaarden JB, Frederickson DS, Goldstein JL, Brown MS (eds) The metabolic basis of inherited disease, 5th edn. McGraw-Hill, New York, pp 1493–1507

Ravelli GP, Stein ZA, Susser MW (1976) Obesity in young men after famine exposure in utero and early infancy. N Engl J Med 295: 349–353

Record IR, Dreosti IE, Tulsi RS (1985a) In vitro development of zinc-deficient and replete rat embryos. Aust J Exp Biol Med Sci 63: 65–71

Record IR, Tulsi RS, Dreosti IE, Fraser FJ (1985b) Cellular necrosis in zinc-deficient rat embryos. Teratology 32: 397–405

Reeves PG, Frissell SG, O'Dell BL (1977) Response of serum corticosterone to ACTH and stress in the zinc-deficient rat. Proc Soc Exp Biol Med 156: 500–504

Robinson LK, Hurley LS (1981a) Effect of maternal zinc deficiency or food restriction on rat fetal pancreas. 1. Procarboxypeptidase A and chymotrypsinogen. J Nutr 111: 858–868

Robinson LK, Hurley LS (1981b) Effect of maternal zinc deficiency or food restriction on rat fetal pancreas. 2. Insulin and glucagon. J Nutr 111: 869–877

Rogers JM, Keen CL, Hurley LS (1985a) Zinc deficiency in pregnant Long-Evans hooded rats: teratogenicity and tissue trace elements. Teratology 31: 89–100

Rogers JM, Keen CL, Hurley LS (1985b) Zinc, copper, and manganese deficiencies in prenatal and neonatal development, with special reference to the central nervous system. In: Gabay S, Harris J, Ho BT (eds) Metal ions in neurology and psychiatry. Alan R. Liss, New York, pp 3–34

Rogers JM, Lönnerdal B, Hurley LS, Keen CL (1987) Increased iron concentration in zinc-deficient rat fetuses: timing, localization, and maternal ^{59}Fe absorption. J Nutr (in press)

Ronaghy HA, Halsted JA (1975) Zinc deficiency occurring in females. Report of 2 cases. Am J Clin Nutr 28: 831–836

Ronaghy HA, Reinhold JG, Mahloudji M, Ghavami P, Spivey Fox MR, Halsted JA (1974) Zinc supplementation of malnourished school boys in Iran: increased growth and other effects. Am J Clin Nutr 27: 112–121

Rosett HL, Weiner L, Lee A, Zuckerman B, Dooling E, Oppenheimer E (1983) Patterns of ethanol consumption and fetal development. Obstet Gynecol 61: 539–546

Sackett GP (1981) A nonhuman primate model for studying causes and effects of poor pregnancy outcomes. In: Friedman S, Sigman M (eds) Preterm birth and psychological development. Academic Press, New York, pp 41–63

Sandstead HH, Strobel DA, Logan GM Jr, Marks EO, Jacob RA (1978) Zinc deficiency in pregnant rhesus monkeys: effects on behavior of infants. Am J Clin Nutr 31: 844–849

Sato F, Watanabe T, Hoshi E, Endo A (1985) Teratogenic effect of maternal zinc deficiency and its co-teratogenic effect with cadmium. Teratology 31: 13–18

Sever LE, Emanuel I (1973) Is there a connection between maternal zinc deficiency and congenital malformations of the central nervous system in man? Teratology 7: 117

Sheldon WL, Aspillaga MO, Smith PA, Lind T (1985) The effect of oral iron supplementation on zinc and magnesium levels during pregnancy. Br J Obstet Gynaecol 92: 892–898

Simmer K, Thompson RPH (1985a) Zinc in the fetus and newborn. Acta Paediatr Scand [Suppl] 319: 158–163

Simmer K, Thompson RPH (1985b) Maternal zinc and intrauterine growth retardation. Clin Sci 68: 395–399

Simmer K, Punchard NA, Murphy G, Thompson RPH (1985) Prostaglandin production and zinc deficiency in human pregnancy. Pediatr Res 19: 697–700

Simmer K, Iles C, James C, Thompson RPH (1987) Are iron-folate supplements harmful? Am J Clin Nutr 45: 122–125

Simpson JL, Elias S, Martin AO, Palmer MS, Ogata ES, Radvany RA (1983) Diabetes in pregnancy, Northwestern University Series (1977–1981). Am J Obstet Gynecol 146: 263–270

Sivakumar B, Belavady B (1975) Effect of zinc on vitamin D-dependent calcium uptake in rat intestine. Indian J Biochem Biophys 12: 386–388

Smithells RW, Seller MJ, Harris R et al. (1983) Further experience of vitamin supplementation for prevention of neural tube defect recurrences. Lancet I: 1027–1030

Solomons NW (1986) Competitive interaction of iron and zinc in the diet: consequences for human nutrition. J Nutr 116: 927–935

Soltan MH, Jenkins DM (1982) Maternal and fetal plasma zinc concentration and fetal abnormality. Br J Obstet Gynaecol 89: 56–58

Suh SM, Firek AF (1982) Magnesium and Zn deficiency and growth retardation in offspring of alcoholic rats. J Am Coll Nutr 1: 193–201

Suwarnasarn A, Wallwork JC, Lykken GI, Low FN, Sandstead HH (1982) Epiphyseal plate development in the zinc-deficient rat. J Nutr 112: 1320–1328

Swanson CA, King JC (1982) Zinc utilization in pregnant and nonpregnant women fed controlled diets providing the zinc RDA. J Nutr 112: 697–707

Swanson CA, King JC (1987) Zinc and pregnancy outcome. Am J Clin Nutr 46: 763–771

Swanson CA, Turnlund JR, King JC (1983) Effect of dietary zinc sources and pregnancy on zinc utilization in adult women fed controlled diets. J Nutr 113: 2557–2567

Swenerton H, Hurley LS (1971) Teratogenic effects of a chelating agent and their prevention by zinc. Science 173: 62–64

Swenerton H, Hurley LS (1980) Zinc deficiency in rhesus and bonnet monkeys, including effects on reproduction. J Nutr 110: 575–583

Swenerton H, Shrader R, Hurley LS (1969) Zinc-deficient embryos: reduced thymidine incorporation. Science 166: 1014–1015

Tanner JM (1981) Catch-up growth in man. Br Med Bull 3: 233–238

Tao SH, Hurley LS (1975) Effect of dietary calcium deficiency during pregnancy on zinc mobilization in intact and parathyroidectomized rats. J Nutr 105: 220–225

Tininus TP, Beckwith BE, Halas ES (1986) Effects of mild perinatal zinc deficiency on passive avoidance. Nutr Behav 3: 163–168

Todd WR, Elvehjem CA, Hart EB (1934) Zinc in the nutrition of the rat. Am J Physiol 107: 146–156

Turk DE, Sunde ML, Hoekstra WG (1959) Zinc deficiency experiments with poultry. Poultry Sci 38: 1256

Turnlund JR, King JC, Wahbeh CJ, Ishkanian I, Tannous RI (1983) Zinc status and pregnancy outcome of pregnant Lebanese women. Nutr Res 3: 309–315

Tuttle S, Aggett PJ, Campbell D, MacGillivray I (1985) Zinc and copper nutrition in human pregnancy: a longitudinal study in normal primigravidae and in primigravidae at risk of delivering a growth retarded baby. Am J Clin Nutr 41: 1032–1041

Uriu-Hare JY, Stern JS, Reaven GM, Keen CL (1985) The effect of maternal diabetes on trace element status and fetal development in the rat. Diabetes 34: 1031–1040

Uriu-Hare JY, Stern JS, Keen CL (1987) Dietary zinc reduces the expression of diabetes-induced teratogenicity in the rat. Fed Proc 46: 595

Vir SC, Lover AHG, Thompson W (1981) Zinc concentration in hair and serum of pregnant women in Belfast. Am J Clin Nutr 34: 2800–2807

Vojnik C, Hurley LS (1977) Abnormal prenatal lung development resulting from maternal zinc deficiency in rats. J Nutr 107: 862–872

Vruwink K, Gershwin ME, Hurley LS, Keen CL (1987) Persistent effects of gestational zinc (Zn) deficiency on metallothionein (MT) induction in mice. Fed Proc 46: 885

Walravens PA, Hambidge KM (1976) Growth of infants fed a zinc supplemented formula. Am J Clin Nutr 29: 1114–1121

Walravens PA, Van Doorminck WJ, Hambidge KM (1978) Metals and mental function. J Pediatr 93: 535–541

Warkany J, Petering HG (1972) Congenital malformations of the central nervous system in rats produced by maternal zinc deficiency. Teratology 5: 319–334

Watanabe T, Sato F, Endo A (1983) Cytogenetic effects of zinc deficiency on oogenesis and spermatogenesis in mice. Yamagata Med J 1: 13–20

Wells JL, James DK, Luxton R, Pennock CA (1987) Maternal leucocyte zinc deficiency at start of third trimester as a predictor of fetal growth retardation. Br Med J 294: 1054–1056

Weston WL, Huff JC, Hambert JR, Hambidge KM, Neldner KH, Walravens PA (1977) Zinc correction of defective chemotaxis in acrodermatitis enteropathica. Arch Dermatol 113: 422–425

Wibell L, Gebre-Medhin M, Lindmark G (1985) Magnesium and zinc in diabetic pregnancy. Acta Paediatr Scand 320: 100–106

Wild J, Read AP, Sheppard S et al. (1986) Recurrent neural tube defects, risk factors and vitamins. Arch Dis Child 61: 440–444

Yamaguchi M, Sakashita T (1986) Enhancement of vitamin D_3 effect on bone metabolism in weanling rats orally administered zinc sulphate. Acta Endocrinol 111: 285–288

Yamaguchi M, Yamaguchi R (1986) Action of zinc on bone metabolism in rats. Increases in alkaline phosphatase activity and DNA content. Biochem Pharmacol 35: 773–777

Yamaguchi M, Mochizuki A, Okada S (1982) Stimulatory effect of zinc on bone growth in weanling rats. J Pharmacobiodyn 5: 619–626

Zidenberg-Cherr S, Benak PA, Hurley LS, Keen CL (1988) Altered mineral metabolism: a mechanism underlying the fetal alcohol syndrome in rats. Drug–Nutrient Interactions (in press)

Zimmerman AW, Dunham BS, Nochimson DJ, Kaplan BM, Clive JM, Kunkel SL (1984) Zinc transport in pregnancy. Am J Obstet Gynecol 149: 523–528

Chapter 13

A Note on Zinc and Immunocompetence

R.A. Good

In Prasad's initial description of the extraordinarily complex syndrome produced by profound zinc deficiency in humans, evidence of increased susceptibility to infection was one manifestation of this nutritional deficiency (Prasad et al. 1961). Holstein–Friesian cattle bearing the highly lethal A46 mutation also have severe zinc deficiency that is attributable to failure to absorb zinc normally from the gastrointestinal tract. These cattle have a profound immune deficiency disease and frequently die from infections. Their immunodeficiency disease is characterized by failure of development of normal T cell-mediated immunities, defective delayed hypersensitivity reactions, feeble allograft immunity, failure of normal development of the thymus, deficiencies of T lymphocytes in blood and thymus-dependent regions of the lymphoid tissues and defective defences against viruses, fungi and bacteria. In addition, these genetically defective cattle exhibit striking malfunctions of the gastrointestinal tract, and a pleurioroficial dermatitis characterized by parakeratoses. All of these manifestations, including all of the immunodeficiencies which occur in these zinc-deficient cattle, are corrected by administration of zinc orally or parenterally (Brummerstedt et al. 1971; Andresen et al. 1974).

Zinc is similarly prescribed as the perfect cure for the hereditary disease acrodermatitis enteropathica of human infants and children. In this syndrome, T cell numbers, thymic hormone levels and T cell-mediated cellular and humoral immunities are all deficient, zinc levels in the blood are low and zinc excretion in urine is deficient (Moynahan and Barnes 1973). Susceptibility to many infections is greatly increased in these children. In both the cattle with the autosomal recessive A46 mutation and the children with recessively transmitted acrodermatitis enteropathica, zinc absorption is grossly deficient, and both these diseases are completely cured by administration of adequate amounts of zinc.

Once the major manifestations of zinc deficiency had thus been outlined, first by Prasad's findings and also in these two "experiments of nature", laboratory experiments on the immunological influences of nutritional zinc restriction could be carried out to clarify the role of zinc in immunological functions (Fraker et al. 1977, 1978; Fernandes et al. 1979; Schloen et al. 1979; Fraker et al. 1986). In experiments employing rats, mice and human subjects, nutritional zinc restriction as a single nutritional variable has been shown regularly to produce deficiency of T cell numbers; reduced proliferative responses of T lymphocytes to the phytomitogens (phytohaemagglutinin, concanavalin A and pokeweed mitogen), allogeneic cells or common antigens; decreased T cell-dependent antibody production, e.g. in primary antibody responses; decreased tempo of skin allograft rejection; decreased natural killer cell activities; and decreased functional activity levels (Dardenne et al. 1982) of thymic hormones. For example, thymulin, one of the thymic hormones, is a peptide produced by the thymus that has been shown to require zinc for its functional activity, that is, promoting development and differentiation of T lymphocytes (Dardenne et al. 1982). Zinc restriction in animals results in rapid decline of thymulin functional levels in the circulating blood and restoration of zinc nutriture promptly brings the thymulin levels back to normal (Iwata et al. 1979).

Thus, zinc is a crucial nutritional component required for normal development and maintenance of the immune functions and critical to normal ability to resist infections caused by viruses, bacteria, fungi and protozoa.

Nutritional zinc deficiency, conditioned zinc deficiency and combinations of the two produce states of profound acquired immunodeficiency (Schloen et al. 1979; Hansen et al. 1982; Sandstead et al. 1982; Fraker et al. 1986); these deficiencies are seen frequently in hospitals and general medical practices. Such zinc-deficiency-dependent immunodeficiencies have been encountered, for example, in patients with a wide variety of advanced cancers (Garofalo et al. 1980), severe burns or traumas, chronic alcoholism, liver disease, intestinal disorders (e.g. intestinal fistulas), protein–calorie malnutrition and in aged persons who are undernourished or malnourished (Good and Gajjar 1986) and patients with sickle cell anaemia (Prasad et al. 1975; Prasad 1982). Indeed, among the most frequent and most correctable of the acquired immunodeficiency diseases are those that are based on nutritional or conditioned deficiencies of the element zinc (Sandstead et al. 1982; Good and Gajjar 1986).

Aspects of the relationships between zinc status and immune function are considered in further detail in the review by Fraker et al. (1986) and in Chaps. 5, 12 and 18 of this volume.

References

Andresen E, Basse A, Brummerstedt E, Flagstad T (1974) Lethal trait A46 in cattle. Additional genetic investigations. Nord Vet Med 26: 275–278

Brummerstedt E, Flagstad T, Basse A, Andresen E (1971) The effect of zinc on calves with hereditary thymus hypoplasia (lethal trait A-46). Acta Pathol Microbiol Scan (A) 79: 686–687

Dardenne M, Pleau JM, Nabarra B et al. (1982) Contribution of zinc and other metals to the biological activity of the serum thymic factor. Proc Natl Acad Sci USA 79: 5370–5373

Fernandes G, Nair M, Onoe K, Tanaka T, Floyd R, Good RA (1979) Impairment of cell-mediated immunity functions by dietary zinc deficiency in mice. Proc Natl Acad Sci USA 76: 457–461

Fraker PJ, Haas S, Luecke RW (1977) Effect of zinc deficiency on the immune response of the young adult A/J mouse. J Nutr 107: 1889–1895

Fraker PJ, DePasquale-Jardieu P, Zwickl CM, Luecke RW (1978) Regeneration of T-cell helper function in zinc-deficient adult mice. Proc Natl Acad Sci USA 75: 5660–5664

Fraker PJ, Gershwin ME, Good RA, Prasad A (1986) Interrelationships between zinc and immune function. Fed Proc 45: 1474–1479

Garofalo JA, Erlandson E, Strong EW et al. (1980) Serum zinc, serum copper, and the Cu/Zn ratio in patients with epidermoid cancers of the head and neck. J Surg Oncol 15: 381–386

Good RA, Gajjar AJ (1986) Diet, immunity and longevity. In: Hutchinson ML, Munro HN (eds) Nutrition and aging. Bristol-Myers nutrition symposia, vol 5. Academic Press, Orlando, pp 235–249

Hansen MA, Fernandes G, Good RA (1982) Nutrition and immunity: the influence of diet on auto-immunity and the role of zinc in the immune response. Ann Rev Nutr 2: 151–177

Iwata T, Incefy GS, Tanaka T et al. (1979) Circulating thymic hormone levels in zinc deficiency. Cell Immunol 47: 100–105

Moynahan EJ, Barnes PM (1973) Zinc deficiency and a synthetic diet for lactose intolerance. Lancet I: 676

Prasad AS (1982) Clinical disorders of zinc deficiency. In: Prasad AS, Dreosti IE, Hetzel BS (eds) Current topics in nutrition and disease: clinical applications of recent advances in zinc metabolism, vol 7. Alan R. Liss, New York, pp 89–120

Prasad AS, Halsted JA, Nadimi M (1961) Syndrome of iron deficiency anemia, hepatosplenomegaly, hypogonadism, dwarfism and geophagia. Am J Med 31: 532–546

Prasad AS, Schoomaker EB, Ortega J, Brewer GJ, Oberleas D, Oelschlegel FJ (1975) Zinc deficiency in sickle cell disease. Clin Chem 21: 582–587

Sandstead HH, Henriksen LK, Greger JL, Prasad AS, Good RA (1982) Zinc nutriture in the elderly in relation to taste acuity, immune response and wound healing. Am J Clin Nutr 36: 1046–1059

Schloen LH, Fernandes G, Garofalo JA, Good RA (1979) Nutrition, immunity and cancer – a review. II. Zinc, immune function and cancer. Clinical Bulletin, Memorial Sloan-Kettering Cancer Center 9: 63–75

Chapter 14

Zinc and Behaviour

Sandra E. File

Introduction

When reviewing the consequences of zinc deficiency in humans, Hambidge (1981) commented

> behavioural abnormalities are prominent in severe zinc deficiency (Walravens et al. 1978) and may also occur in less severe cases of zinc depletion. Irritability, lethargy and depression are evident even at an early stage in the clinical course. Improvement in hedonic tone and motivation to engage in the environment follow rapidly after the institution of zinc therapy.

He suggested that such observations indicated an important, although ill-defined, role for zinc in human brain function, but also emphasized the lack of any evidence of impaired intellectual capacity. Such comments are typical of many that have stimulated interest in experimental investigation of the behavioural consequences of zinc deficiency.

The purpose of this chapter is to review studies which have investigated the behavioural consequences of a low-zinc diet, and to examine critically the claims that such a diet leads to defects that can be specifically attributed to the zinc deficiency, rather than to general debilitation. Since the anorexia frequently developing during zinc deficiency may have its own non-specific behavioural consequences it has been recognized widely that experimental studies should include a group of animals with normal zinc status, but fed a diet restricted in calories to the intake of the zinc-deficient animals. Most investigators have therefore included a group of pair-fed animals (PF) as well as a group fed a zinc-adequate diet ad libitum (AL). In some studies low-zinc diets were administered prenatally or early postnatally and the lasting consequences of this were examined by testing after a period of nutritional rehabilitation. Such treatment protocols will be referred to as PZD (previous zinc deficiency). This is in order to differentiate these studies from those in which the behavioural testing took place at the time of zinc deficiency. These latter animals will be referred to as ZD (zinc deficient). There has been reasonable consistency amongst the experimental studies in the zinc concentrations in diets given to deficient and control groups, but both the developmental stage of the animals and the duration of the treatment have varied.

In order to attribute any behavioural changes to the specific effects of a low-zinc diet it is essential that as many other sources of difference as possible be excluded. Unfortunately, most of the studies have used inadequate experimental design and sometimes there is a lack of crucial methodological detail. Thus while the results from some studies suggest that behavioural differences result from a period of zinc deficiency, it is not possible to attribute these specifically to zinc. The final section of this chapter will summarize the main confounding factors and suggest the crucial controls that should be incorporated in future studies.

Before reviewing studies on learning, experiments that have measured other behaviours will be discussed. Thus the first section will cover studies on aggressive behaviour, stress, open-field behaviour and conditioned suppression. The changes found in these behaviours will be borne in mind when interpreting the experiments on learning.

Effects of Zinc Deficiency on Aggressive Behaviour, Stress, Open-Field Behaviour and Conditioned Emotional Response

Two studies (Halas et al. 1977; Peters 1978) have investigated the effects of a low-zinc diet late in pregnancy on the subsequent aggressive behaviour of Long-Evans hooded rats. Pregnant rats received a zinc-deficient diet (ZD) and were pair-fed (PF) or fed ad libitum (AL). After birth, the pups remained in litter groups with their own mothers and a zinc-supplemented diet was available ad libitum. Despite this, the offspring of the PZD females consumed less food than the other groups and grew more slowly.

The effects of this depletion–repletion regime on defensive aggression (measured by shock-elicited fighting) were investigated at 82 or 112 d of age (Halas et al. 1977). Female, but not male, rats from PZD dams showed a significantly higher percentage of aggressive responses than the PF or AL groups. Unfortunately, it is not specified what the aggressive score was a percentage of and therefore we do not know whether the increase was due to a decrease in other behaviours or to an increase in overall activity. Weight differences are known to change general activity level.

Peters (1978), using a similar experimental protocol, tested PZD, PF and AL male rats at day 112 in a T-maze in which each rat was given a choice of target rat. At the end of each arm was a rat from a different experimental group. The order of preference was AL > PF > PZD, and it was claimed therefore that, since rats prefer the less aggressive target, the PZD rats were more aggressive. The experimental and target rats were then trapped in the goal arms for 1 min and the aggressive behaviour of the experimental rat was scored. The PZD group had more aggressive encounters than the other groups, probably reflecting an increase in offensive attacks. As the behaviour of the target rats was not scored it is not possible to say whether the PZD rats also received more aggression. The results of these two studies indicate that defensive behaviour might be increased in PZD female rats and offensive behaviour in PZD males. However, as will be discussed later, the groups may well have differed in the extent of their early social interaction and this, rather than a direct action of zinc deficiency, may have produced changes in aggression.

Strobel and Sandstead (1984) examined the effects of a period of zinc deficiency during the last trimester of pregnancy in monkeys. The ZD mothers developed alopecia and dermatitis, and showed impaired weight gain. The PF animals adjusted poorly to the pattern of restricted feeding and three of the four mothers lost their infants. After birth, a zinc-supplemented diet was freely available. From 2 months of age the PZD monkeys showed more physical contact, spent longer nursing, played less, explored less and had lower activity levels than the AL infants. At 6 months the infants were separated from their mothers and social interaction was observed in peer groups of the same prenatal dietary group. The only significant difference was that AL monkeys spent a greater proportion of time in ventral–ventral clutching. At 1 year the monkeys were observed interacting with a monkey from a different prenatal dietary group and the only significant difference was that the PZD monkeys showed *less* withdrawal. The results of this study suggest that the prenatal restriction of zinc had little enduring effect on social interactions. It is possible that the decreased activity observed in infancy was counteracted by the increase in time spent with their mother.

In a second study, a zinc-deficient diet was imposed during lactation. The ZD mothers spent more time huddling in the first 3 months, but less in the 4th month than did the AL mothers. At 4 months the infants were separated from their mothers and given a normal diet. The PZD monkeys showed greater signs of depression (decreased locomotor activity, increased cooing vocalizations and an increase in passive behaviours) than did AL animals. However, there were no differences when the infants were introduced to a stranger monkey. When separated from their peers at 6 months of age PZD monkeys showed less environmental exploration. There were no effects on social behaviour of a low-zinc diet imposed on juvenile monkeys, but ZD monkeys maintained greater visual contact with objects introduced into their environment.

Essatara et al. (1984) assessed lasting effects of a low-zinc diet on stress-induced eating. Weanling rats were exposed to a zinc-deficient diet for 35 d and then rehabilitated for 40 d. Neither during nor after the period of zinc restriction were there any differences between the groups in the amount of food eaten in response to the stress of a tail-pinch. However, since the ZD rats were normally eating less and were lighter than the AL rats, in relation to their body weight the ZD rats actually consumed *more* in response to a tail-pinch. The authors concluded that stress could overcome the normal anorexia of zinc deficiency, and therefore that the ZD rats had a normal stress response. Results from a study with mice also suggest that zinc deficiency during the perinatal period does not influence the stress response. There were no differences between the basal corticosterone concentrations of PZD and AL mice, or in the increases shown in response to novelty stress (Golub et al. 1983).

In the open-field test, rats are stressed by exposure to an unfamiliar large circular arena and their activity walking around this arena is scored. Gordon et al. (1982) investigated the effects on open-field activity of low-zinc diets starting in adolescence (day 35) and lasting for 49 d, or starting in old-age (day 300) and lasting for 105 d. Both the young adult ZD rats, suffering from inflamed paws, skin lesions and urethritis, and the less debilitated, but still unwell, old rats waited significantly longer before leaving the centre of the arena and they crossed fewer sectors of the open field. It is likely that non-specific effects of debility were the cause of low activity.

Caldwell et al. (1976) chronically exposed female rats to zinc deficiency from day 25, repeatedly mated them at 14, 21 and 28 weeks and gave their offspring a supplemented diet from weaning until testing in the open field at day 45. Zinc depletion was associated with increased pre- and post-partum mortality. The mean litter sizes at weaning for the ZD mothers ranged from 0.6 to 1.9 pups, compared with 8.5–9.2 for the AL groups, and 90% of the weaned ZD pups died before day 45. The PZD pups from the 2nd and 3rd pregnancies were less active in the open field, but it must be questioned whether the scores of so few survivors were typical of the group as a whole. Additionally, differences in weight and the social group size also may have changed behaviour.

The results in the open field have been interpreted as showing that a period of zinc restriction makes rats more emotional. If this interpretation is correct one might expect to see an enhanced effect in the conditioned emotional response. In this situation animals are exposed to a series of tone–shock pairings in a classical conditioning paradigm. On a later occasion, usually while they are emitting some steady response such as lever pressing, they are presented with the tone. The extent to which the animals' lever pressing is disrupted is taken as a measure of the effects of the prior conditioning. Halas et al. (1979) examined the effects of a low-zinc diet from birth to weaning; the conditioning trials took place during the period of zinc depletion. Pups were given tone–shock pairings for 3 d, starting at postnatal days 11, 14, 17 and 20. The effects of this early conditioning were assessed 32 d later.

None of the pups that started conditioning on days 11 and 14 showed any evidence of conditioning, which is not surprising because an auditory conditioned stimulus was used and such young pups have limited auditory sensitivity (Crowley and Hepp-Raymond 1966). All the pups conditioned from day 20 showed significantly suppressed responses to the tone, but there were no significant differences between the groups. Differences did emerge, however, in the groups conditioned from day 17 and, on the first two-tone presentations, the PZD rats showed significantly less suppressed responses than the PF or AL animals.

The authors concluded that the zinc deficiency had retarded long-term *memory*. However, we do not know that the original *acquisition* was equal in all the groups. The pups differed significantly in weight at the time of conditioning and therefore the different groups received training at different developmental ages. The ZD pups would have been trained at a developmentally earlier age than the other groups. Since no conditioning was apparent prior to day 17 the effect of a low-zinc diet could have been simply due to a non-specific effect of slowing growth. In order to test for specific effects of zinc in this task all the groups should have been conditioned at equal weights. In a second experiment, Halas et al. (1979) restricted the diets of rats prenatally, and this period of zinc restriction did not significantly change conditioned suppression.

Avoidance Learning

Halas and Sandstead (1975) restricted the zinc intake of rats from days 15 to 20 of gestation. At day 60, male rats were trained in a two-way active avoidance task. Rats could avoid, or escape from, a shock by jumping to the other side of the apparatus. The conditioning days were then followed by two extinction days, during which the conditioned stimulus (CS), but not the shock, was presented. During avoidance conditioning the PZD group made fewer jumps and took longer to jump than the AL and PF groups, which did not differ from each other. The PF group showed faster extinction than the AL group, but the extinction of the ZD animals cannot be assessed because they were at zero response level before the extinction phase was entered. In addition to being less active in response to the CS, the PZD animals were also less active during the inter-trial interval, thus raising the possibility that changes in overall activity could account for performance differences in this task. There is evidence that the avoidance performance of female rats is not disrupted by prenatal zinc deficiency (Rowe 1984; Halas et al. 1979). Interestingly the lack of significant difference in avoidance response was accompanied by no differences in activity during the inter-trial interval, again suggesting the two effects might be linked.

Caldwell and Oberleas restricted the zinc intake of rats from day 30 to day 78, continuing throughout the period of behavioural testing. In a one-way active avoidance task the ZD rats were not significantly different overall, but on the first ten trials they were slower to cross to the unshocked side of the apparatus. However, they were lethargic in the home cage, less active in the open field and had clinical signs of zinc deficiency. The slight impairment was therefore likely to be secondary to a non-specific malaise. In a second study female rats were exposed to a less severely zinc-deficient diet from day 25 and were mated at 14, 21 and 28 weeks, as previously described. The few pups surviving in the second and third pregnancies of ZD mothers had longer avoidance latencies in a one-way active avoidance task.

Whilst a lower level of general activity would impair performance in an active avoidance task, the opposite applies to passive avoidance. Golub et al. (1983, 1984) restricted the zinc intake of mice from day 16 of gestation until postnatal day 15. At day 70, mice were placed in a two-compartment chamber and when they crossed to the other side they were shocked until they escaped; they were tested 30 min later. Acquisition and retention are indicated by the mice not crossing to the other compartment, or showing a long latency period before they do so. The PZD males showed shorter latencies before crossing to the other compartment and the females fed the mildly restricted diet, but not those fed the severely restricted diet, also showed significantly shorter latencies than the AL controls.

Maze Learning

Halas et al. (1976) exposed dams during gestation to a low-zinc diet under PF or AL conditions. On day 95 male offspring were tested in a multiple choice Tolman–Honzik maze which involved running for food reward. Previous dietary history did not influence errors or running time. When the period of zinc restriction was imposed during lactation the PZD group did not differ in running times, but did make more cul-de-sac entry errors (Lokken et al. 1973).

A more prolonged period of zinc restriction was imposed in a third study (Caldwell et al. 1976). Female rats were given a zinc-deficient diet from 25 d of age and first mated at 14 weeks. The zinc deficiency was maintained throughout lactation and for two further pregnancies. The pups were given a supplemented diet from weaning until day 55 when they were trained in a Lashley III water maze. The performance of the 10% of the PZD group that survived to the time of testing was not significantly different from the AL group.

The effects of a ZD diet imposed from day 30 and throughout testing were also studied using a Lashley III water maze (Caldwell and Oberleas), which has the advantage of not requiring food deprivation and therefore being less affected by changes in body weight or food motivation. The ZD rats took longer to swim the maze than the AL group and did not decrease their swimming time over trials. The AL group had significantly decreased its errors by trial 3, but the ZD rats had not. The ZD rats were debilitated at the time of testing, were lethargic in the home cage and had lower activity in the open field. This may well have contributed to the slower swimming speeds, but less readily accounts for a failure to reduce errors over trials.

Three studies using the radial arm maze examined the effects of various periods of zinc restriction on spatial learning and memory. Hungry rats are required to visit each of the baited arms of the maze. Whilst there is no prescribed order in which they have to enter the arms, an error is scored if the rat re-enters an arm or enters an arm that has not been baited. In order to identify the arms the rat is required to use extra-maze spatial cues.

Halas and Kawamoto (1984) exposed dams to a zinc-deficient diet throughout gestation and lactation. After dietary repletion, the PZD adult females were exposed for 15 d to a 17-arm maze with all arms baited. They made significantly more re-entry errors than the AL or PF groups, but only on the first 5 d and only on choices 10–17. In a repeat experiment, the PZD rats made more errors on all 15 d.

Halas et al. (1983) tested the lasting effects of a low-zinc diet during lactation on the radial maze performance of zinc-repleted adult male and female offspring. The rats were required to learn that eight out of the 17 arms were baited. In the first 20 acquisition trials there were no group differences in entries into unbaited arms (indicating unimpaired reference memory). Over trials 20–30 the PZD males and females made more errors, but by trials 30–40 there were again no group differences. Female, but not male, PZD rats made significantly more errors of working memory (indicated by re-entries into a baited arm) than PF or AL rats on trials 30–40. The animals were then retested after an interval of 16–42 d with the opposite arms baited, i.e. they were exposed to a reversal learning task. There were no group differences in the first ten trials and no differences throughout in the number of working memory errors. However, on trials 10–25 the PZD male rats made significantly more reference memory errors than did the AL or PF rats, but these differences disappeared in subsequent trials. In a final experiment on the female rats all 17 arms were baited and the percentage of errors on choices 10–17 showed the PZD rats made more re-entry errors than the AL rats for trials 5–20. This is similar to the result obtained by Halas and Kawamoto (1984) after a period of dietary restriction extending through gestation and lactation. It seems thus that restriction solely at the time of lactation is sufficient to change performance in this radial arm maze task.

Chafetz et al. (1984) trained adult male rats on an eight-arm maze with all the arms baited. They then imposed a zinc-deficient diet for 3 d and compared performance with a group of rats whose diet was restricted so that they were 70%–80% of their free-feeding weight. The arms were baited with either a zinc-deficient or a normal food pellet. Only when all arms were baited with the diet appropriate for each group did the ZD group make more re-entry errors. It is hard to interpret these results, but there was clearly no general learning impairment.

Cognitive Performance (Miscellaneous Tasks)

Massaro et al. (1982) fed rats a zinc-deficient diet from day 45 for 17 d; behavioural testing started on day 13 of the diet. Rats were trained to lever-press for food reward in a chamber that contained two levers, each with a light and a speaker above. In phase 1, one of the two lights was randomly illuminated and the rat was rewarded if it pressed the lever that corresponded to the light. Some animals (2/22 of rats on normal diet and 8/26 ZD rats) failed to reach the criterion of 80% correct for 50 consecutive trials. Of those that did achieve this level of correct responses there was no difference in the rate of learning.

In phase 2, each dietary group was divided into two training groups. One group received an auditory stimulus paired with the light for 100 trials. The right light was always associated with a particular tone that was emitted from the right speaker; the left light was associated with a different tone emitted from the left speaker. The second group simply received a further 100 trials with the visual stimuli alone. This procedure of pairing a redundant stimulus with an informative CS leads to the phenomenon of "blocking", whereby it is subsequently harder to condition to the redundant stimulus, compared with a group with no prior experience of the stimulus. The extent of blocking was then assessed in phase 3 by randomly presenting auditory stimuli in the absence of visual stimuli. In the groups simply transferring from the visual to the auditory stimuli there was no difference between the dietary groups, showing that cross-modal transfer was unaffected by a low-zinc diet. However, the ZD group showed significantly *enhanced* blocking. In the final phase of learning, instead of emitting a particular tone from a particular speaker, on any trial the rats received the tone stimulus from both speakers. Thus the rats were now required to discriminate on the basis of tone *frequency* alone; the cue of spatial localization was no longer available. Both the ZD groups learned this task faster than the normal diet groups, indicating that more of the ZD animals had attended to the frequency cue in phase 3. For both dietary groups, previous experience of the tone presented at the same time as the light had the effect of enhancing attention to the spatial location of the cue, at the expense of coding the frequency. Thus these groups showed slower learning in the last phase compared with the groups that had never received tone–light pairings (even though these rats had actually received fewer tone presentations).

This is a complex experiment and is made harder to interpret by the use of relearning measures in phases 3 and 4, which confound how much the animals had already learned with how rapidly they could learn afresh. However, the data indicate the ZD rats showed enhanced blocking and a greater attention to the tone frequency. There is absolutely no justification for the authors' conclusion that the ZD rats had a deficit in associative learning.

At 500 d of age the offspring from four ZD and four AL rhesus monkey mothers were tested in a colour-discrimination task (Strobel and Sandstead 1984). No differences were found either on the acquisition of the initial task or on subsequent reversals. When the monkeys were tested in an object-quality learning set task the ZD offspring were impaired at days 554 and 730, but not at 1000 d of age. Insufficient experimental details were provided for these results to be assessed. The monkeys were also tested on oddity problems at an unspecified age; no group differences were found. Although the details of these cognitive tasks are sparse it does seem that the monkeys were not suffering any general impairment.

Conclusions

Sadly, despite several animal studies on the behavioural effects of a low-zinc diet, little can be proffered by way of firm conclusions. When the animals were tested during the period of zinc restriction they were too unwell for specific changes in learning or emotionality to be inferred. The experiments that manipulated diet during early development and then tested after a period of rehabilitation avoided this problem. However, these studies face other interpretational difficulties. Because the pups were not fostered at birth to mothers that had received AL diets, genetic effects and any effects of zinc restriction on maternal behaviour were confounded with prenatal diet effects. The fostering should also have been done so that each mother received pups from every dietary condition. This would help to minimize any effects from changes in sibling behaviour. For example, if ZD pups are less active they will interact less in the litter and this may have lasting effects on their social behaviour or learning ability. This would be only an indirect effect of the low-zinc diet. Such indirect effects of a low-zinc diet could also occur when the restriction occurs postnatally, and would be difficult to detect unless the animals were observed at the time of restriction. From the studies in which zinc was restricted in the perinatal period, there is some evidence of changes in aggressive behaviours, in active avoidance responding, in passive avoidance and in performance in a radial arm maze. These changes have not been found consistently and none of the changes can be unequivocally interpreted as indicating impaired learning or memory. More importantly, in no case can it be definitely concluded that the behavioural change was a direct consequence of the reduced dietary zinc. If the changes are secondary to genetic, maternal or social factors then the results may not be applicable to the human situation.

Of course, just as the studies do not permit positive conclusions about any detrimental effects from zinc deficiency, nor do they permit a firm conclusion that no changes result.

Acknowledgements. I am grateful to Professor N.K. Mackintosh for detailed discussions of the experiments by Massaro et al. 1982. I am a Wellcome Trust Senior Lecturer.

References

Caldwell DF, Oberleas D, Prasad A (1976) Psychobiological changes in zinc deficiency. In: Prasad A, Oberleas D (eds) Trace elements in human health and disease. Academic Press, New York, pp 311–325

Chafetz MD, Abshire FM, Bernard DL (1984) Zinc deficiency in adult rats alters foraging in a radial arm maze. In: Frederickson CJ, Howell GA, Kasarskis EJ (eds) The neurobiology of zinc, part B: Deficiency, toxicity and pathology. Alan R. Liss, New York, pp 109–119

Crowley DE, Hepp-Raymond HC (1966) Development of cochlear function in the ear of the infant rat. J Comp Physiol Psychol 62: 427–432

Essatara MB, Levine AS, Morley JE, McClain CJ (1984) Zinc deficiency and anorexia in rats: normal feeding patterns and stress induced feeding. Physiol Behav 32: 469–474

Golub MS, Gershwin ME, Vijayan VK (1983) Passive avoidance performance of mice fed marginally or severely zinc deficient diets during post-embryonic brain development. Physiol Behav 30: 409–413

Golub MS, Keen CL, Vijayan VK, Gershwin ME, Hurley LS (1984) Early development of brain and behaviour in mice fed a marginally zinc deficient diet. In: Frederickson CJ, Howell GA, Kasarskis EJ (eds) The neurobiology of zinc, part B: Deficiency, toxicity and pathology. Alan R. Liss, New York, pp 65–75

Gordon EF, Bond JT, Gordon RC, Denny MR (1982) Zinc deficiency and behavior: a developmental perspective. Physiol Behav 28: 893–897

Halas ES, Sandstead HH (1975) Some effects of prenatal zinc deficiency on behavior of the adult rat. Pediatr Res 9: 94–97

Halas ES, Kawamoto JC (1984) Correlated behavioral and hippocampal effects due to perinatal zinc deprivation. In: Frederickson CJ, Howell GA, Kasarskis EJ (eds) The neurobiology of zinc, part B: Deficiency, toxicity and pathology. Alan R. Liss, New York, pp 91–107

Halas ES, Reynolds GM, Sandstead HH (1977) Intra-uterine nutrition and its effects on aggression. Physiol Behav 19: 653–661

Halas ES, Heinrich MD, Sandstead HH (1979) Long term memory deficits in adult rats due to postnatal malnutrition. Physiol Behav 22: 991–997

Halas ES, Eberhard MJ, Diers MA, Sandstead HH (1983) Learning and memory impairment in adult rats due to severe zinc deficiency during lactation. Physiol Behav 30: 371–381

Halas ES, Rowe MC, Johnson OR, McKenzie JM, Sandstead HH (1976) Effects of intrauterine zinc deficiency on subsequent behavior. In: Prasad AS (ed) Proceedings of the international symposium on trace elements in human health and disease. Nutrition Foundation Monograph. Academic Press, New York, pp 327–343

Hambidge KM (1981) Zinc deficiency in man: its origins and effects. Philos Trans R Soc Lond [Biol] 294: 129–144

Lokken PM, Halas ES, Sandstead HH (1973) Influences of zinc deficiency on behavior. PSEBM 144: 680–682

Massaro TF, Mohs M, Fosmire G (1982) Effects of moderate zinc deficiency on cognitive performance in young adult rats. Physiol Behav 25: 117–121

Peters DP (1978) Effects of prenatal nutritional deficiency on affiliation and aggression in rats. Physiol Behav 20: 359–362

Rowe MC (1984) Physiological and behavioral effects of zinc deficiency in rats. MA thesis, University of North Dakota, Grand Forks

Strobel DA, Sandstead HH (1984) Social and learning changes following prenatal or postnatal zinc deprivation in rhesus monkeys. In: Frederickson CJ, Howell GA, Kasarskis EJ (eds) The neurobiology of zinc, part B: Deficiency, toxicity and pathology. Alan R.Liss, New York, pp 121–138

Walravens PA, van Doorninck WJ, Hambidge KM (1978) Metals and mental function. J Pediatr 93: 535–541

Chapter 15

Neurobiology of Zinc

I.E. Dreosti

Introduction

Nowadays, the importance of zinc in neurobiology is widely acknowledged and has been reviewed recently by several authors (Dreosti 1983; Frederickson et al. 1984b; Sandstead 1985). However, little more than a decade ago such recognition was very limited, and interest in zinc and the brain centred mainly on the pioneering studies by Hurley and Swenerton (1966) concerning abnormal development of the central nervous system in zinc-deficient rat embryos, and on the early behavioural investigations by Apgar (1968) and Caldwell et al. (1970) with zinc-deprived experimental animals.

In the ensuing years, interest in zinc burgeoned as neuroscientists identified distinctive patterns in the distribution of the metal in various brain regions and nutritionists encountered a growing number of neurological disorders, both teratological and functional, associated with zinc impoverishment in man and animals. The extent and complexity of these emerging interactions generated the current interest in the neurobiology of zinc.

In temporal terms, the involvement of zinc in the central nervous system occurs throughout ontogenesis. In rodents, early prenatal zinc impoverishment seriously impairs embryonic neurulation and development of the primitive neural tube (Hurley and Shrader 1972; Warkany and Petering 1972), and results in a high incidence of exencephalus (Fig. 15.1a) and spina bifida, which together represent the most severe of the zinc-related neural defects.

Late prenatal and early postnatal zinc impoverishment mainly affect brain function (Sandstead et al. 1975), largely because of the involvement of zinc in several neurochemical processes, but possibly also due to a degree of dysmorphogenesis accompanying later brain development. Functional, postnatal neuropathy develops relatively slowly in zinc-depleted animals and the attendant behavioural disorders are very much less striking than the teratogenic outcome of an embryonic zinc deficit.

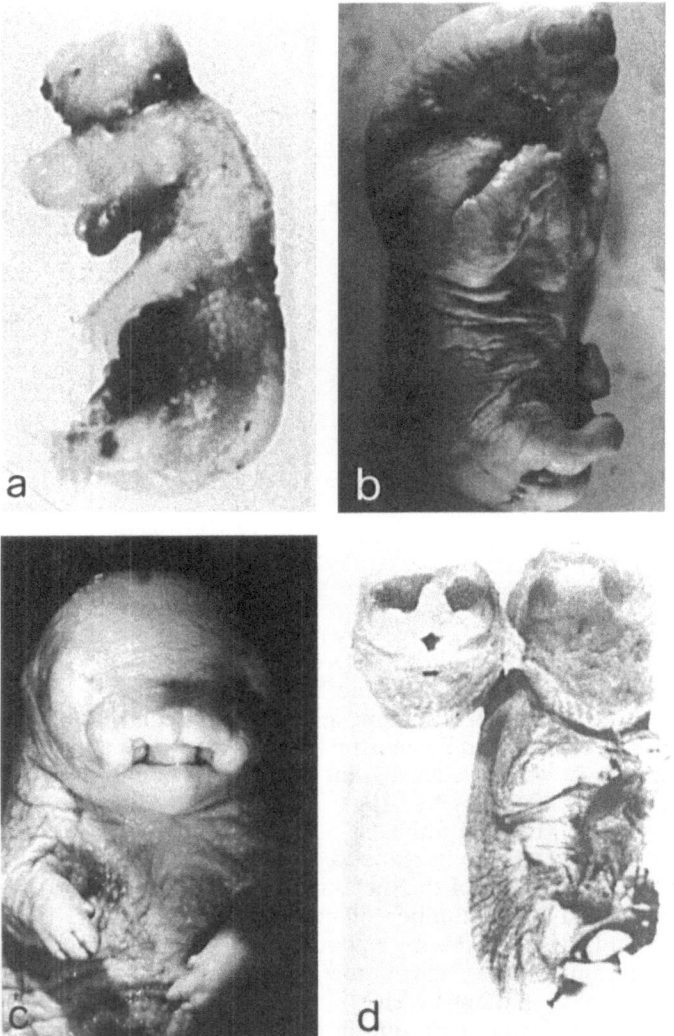

Fig. 15.1. Exencephalus (**a**), anophthalmia, ectrodactyly, talipes and missing tail (**b**), bilateral cheiloschisis and syndactyly (**c**) and hydrocephalus (**d**) in zinc-deficient rat foetuses.

Much of the recent interest in zinc and the brain has centred on the hippocampus, where zinc is found to accumulate in the mossy fibre pathway. While undoubtedly a useful research model, it should be stressed that zinc is present in all brain cells and the neurobiological importance of the metal almost certainly extends beyond its function in the hippocampus. The present review has therefore avoided excessive focus on any one brain region, and has sought instead to present the role of zinc in the brain in more general terms and, where possible, in relation to humans.

Zinc and Brain Development

The teratogenic outcome of maternal zinc impoverishment in animals is well documented, and the rapidity with which the condition affects the embryo has received considerable attention (Hurley et al. 1971; Record et al. 1986). The pattern of terata produced is greatly influenced by the time during pregnancy when maternal dietary zinc restriction is imposed (Dreosti et al. 1985). All organ systems may be affected (Fig. 15.1b and c), but development of the central nervous system appears to be especially vulnerable to inadequate zinc (Hurley 1981). Neural tube defects develop during the embryonic organogenic period, while stenosis of the aqueduct of Sylvius and hydrocephalus are evident in older embryos and foetuses (Fig. 15.1d).

In humans, unlike other animals, a direct causal relationship has not yet been established between zinc deficiency and congenital brain defects, although the possibility was alluded to in the late 1960s (Hurley 1968). Nevertheless, all current evidence suggests that the human foetus is no less vulnerable to zinc depletion than are the offspring of other species (Dreosti 1982). A speculative association was first proposed in the 1970s (Sever and Emanuel 1973; Prasad 1979), when it was suggested that the high incidence of anencephalus seen in the Middle East may be related to the low zinc status of the population in that region. Since then, further tentative associations have been drawn between various brain terata and zinc deficiency, by several workers in a number of countries (Table 15.1). Particularly revealing in evaluating the teratogenic potential of zinc deficiency in humans is the high incidence of birth defects, including anencephalus, observed in the offspring of women suffering from the genetic zinc deficiency disease, acrodermatitis enteropathica (Hambidge et al. 1975).

Biochemically, the teratogenicity of zinc deficiency has been widely attributed to impaired cell division during embryonic development (Swenerton et al. 1969; Dreosti et al. 1972), and to an accompanying asynchrony in histogenesis and organogenesis, which leads to a distortion of the differential rates of growth responsible for normal morphogenesis (Hurley and Shrader 1972). The metal is intricately involved in the processes of replication, transcription and translation and is therefore critically required during cell division. The precise locus of action has not yet been identified in metabolic terms, but substantial experimental evidence implicates diminished activity of the zinc-dependent

Table 15.1. Brain malformations and suspected zinc deficiency in humans

Malformation	Country	Reference
Anencephalus	Middle East	Damyanov and Dutz (1971)
		Sever and Emanuel (1973)
	Turkey	Cavdar et al. (1980)
	USA	Stewart et al. (1981)
	UK	Soltan and Jenkins (1982)
Spina bifida	West Germany	Bergmann et al. (1980)
Myelomeningocoele	Sweden	Jameson (1976)

enzymes thymidine kinase and DNA polymerase (Dreosti and Hurley 1975; Duncan and Hurley 1978). Recent data (Record et al. 1985) have linked an initiating episode of cell death with zinc-related teratogenesis, which suggests that lack of viability of the zinc-deficient embryonic cells may lie in their inability to meet the requirements of rapid cell division.

Zinc and Brain Function

In animals, zinc restriction during late prenatal or early postnatal life is not overtly teratogenic (Buell et al. 1977) but does lead to changes in behaviour which may persist even after repletion with zinc (Lokken et al. 1973; Sandstead et al. 1975). Affected animals typically display reduced learning capacity (Halas 1983), diminished emotional control and less exploratory activity (Strobel and Sandstead 1984). In humans, some evidence exists of impaired concentration, jitteriness, lethargy, depression and mood lability, as well as disturbed taste and smell perception in individuals suffering from suspected zinc deficiency (Prasad et al. 1961; Henkin et al. 1975; Sivasubramanian and Henkin 1978; Ohlsson 1981; Hansen et al. 1983). Table 15.2 presents the zinc status in the blood and/or brains of patients suffering from a variety of psychological disorders. However the evidence to date is not definitive and a causal nexus remains to be firmly established between zinc deficiency and aberrant behaviour in humans.

Microanatomical Dysmorphology

To date, not much is known with certainty concerning the neurophysiological basis underlying zinc-related behavioural defects, and the involvement of some measure of cytoarchitectural dysmorphology of the neuropil cannot be excluded. Certainly, cellularity is reduced in rats exposed to early postnatal zinc impoverishment and the formation of dendrites and interconnections between

Table 15.2. Zinc status associated with psychiatric disorders in humans

Disorder	Zinc status	Reference
Senile dementia	Decreased	Burnet (1981)
		Srinivasan (1984)
Depression	Decreased	McLardy (1975)
		Moynahan (1976)
Fifth-day fits	Decreased	Goldberg and Sheehy (1982)
Mental retardation	Decreased	Pihl and Parkes (1977)
		Kritscher (1978)
Schizophrenia	Decreased	Pfeiffer and Iliev (1972)
Epilepsy	Increased	Barbeau and Donaldson (1974)
Pick's disease	Increased	Constantinidis and Tissot (1981)

neurons is disturbed (Sandstead et al. 1975). More specifically, electron microscopic studies have revealed that the number of cerebellar granular cells and parallel fibres is depressed, as also is the arborization and synaptogenesis of Purkinje cell dendrites. Similar dendritic abnormalities occur in stellate and basket cells (Dvergsten et al. 1984; Sandstead 1985). Recently, evidence has emerged of a role for zinc as a constituent of a neurotrophic molecule involved in axon sprouting (Frederickson et al. 1984a), which may relate to a zinc–glutathione complex isolated from hippocampal mossy boutons (Sato et al. 1984).

Biochemically, depressed synthesis of nucleic acids and proteins probably underlies many of these developmental defects (Duerre et al. 1977; Sandstead 1985), although disturbed myelination (Sandstead et al. 1972; Dreosti et al. 1981) and diminished polymerization of microtubular tubulin (Hesketh 1984) in zinc-deficient brains could also be involved, both structurally and functionally (Dreosti 1984; Kasarskis 1984).

Cerebral Zinc-Dependent Enzymes

Several zinc-related enzymes which have particular relevance to brain development and brain function are listed in Table 15.3.

The enzymes 2'3'-cyclic nucleotide 3'-phosphohydrolase and alkaline phosphatase relate to the process of myelination (Cohn and Richter 1956; Cohen 1970) and are therefore important during brain maturation. Dopamine-β-hydroxylase and phenylethanolamine-N-methyl transferase which convert dopamine to norepinephrine and norepinephrine to epinephrine respectively, possibly also affect brain catecholamine levels, which have been reported to be higher in zinc-deficient animals (Wallwork et al. 1982), and have been suggested to be related to some of the accompanying neuropsychological symptoms. The effect of zinc deficiency on the activity of brain glutamate dehydrogenase has evoked considerable interest, as the enzyme is directly involved in the metabolism of glutamic acid, a putative amino acid neuro-

Table 15.3. Enzymes exhibiting reduced activity in the brains of zinc-deficient rats

Enzyme	Brain region	Reference
2'3'-Cyclic nucleotide	Cerebellum	Prohaska et al. (1974)
3'-Phosphohydrolase	Cerebellum, hippocampus	Dreosti et al. (1981)
Alkaline phosphatase	Foetal brain	Dreosti et al. (1980)
Glutamate dehydrogenase	Cerebellum, hippocampus, neocortex	Dreosti et al. (1981) Wolf and Schmidt (1982)
Dopamine-β-hydroxylase	Cortex, cerebellum, hippocampus	Wenk and Stemmer (1982)
Phenylethanolamine N-methyl transferase	Cortex, cerebellum, hippocampus	Wenk and Stemmer (1982)
Thymidine kinase	Foetal brain	Record and Dreosti (1979)

transmitter, in several brain regions including the hippocampus (Wolf and Schmidt 1982) and the cerebellum (Tran and Snyder 1979).

Although, in the context of trace-element-related metabolic responses the cuprozinc enzyme superoxide dismutase has only been studied in the brain in relation to copper deficiency (Morgan and O'Dell 1977), the enzyme nevertheless deserves mention as it probably serves to protect against superoxide damage at all stages of brain development. The influence of zinc deficiency on activity would be of particular interest since the enzyme occurs abundantly in synaptosomes and in catecholamine-rich areas of the brain (Ledig et al. 1982), where it may specifically protect catecholamines against superoxide-mediated oxidation.

Glutamate decarboxylase, the enzyme involved in the synthesis of the inhibitory neurotransmitter γ-aminobutyric acid, has received attention in relation to excess cerebral zinc, rather than zinc deficiency. Intraventricular injection of zinc into the brains of rats has been reported by several workers to be epileptogenic (Ebadi and Pfeiffer 1984). Work reported recently by Ebadi et al. (1984) has linked zinc to the regulation of glutamate decarboxylase in several ways. At physiological concentrations, zinc was found to stimulate the enzyme pyridoxal kinase and thus formation of the co-factor pyridoxal phosphate, thereby enhancing activity of glutamate decarboxylase. At supra-physiological levels, zinc inhibits glutamate decarboxylase directly, and also appears to reduce the binding capacity of γ-aminobutyric acid receptor sites. The relevance of these non-physiological studies to epilepsy in humans is unclear, but the relationship receives some support from the work of Mody and Miller (1985) who have demonstrated increased levels of hippocampal zinc following epileptic fits induced in rats by commissural kindling.

Zinc and Enkephalin

The possibility that zinc might interact with the enkephalins was raised by Stengaard-Pedersen et al. (1981) when their studies on guinea pigs revealed that the immunocytochemical staining pattern for enkephalin in the hippocampal mossy fibres matched precisely that for zinc obtained with a silver-sulphide stain. Since it was known that opioid peptides exert a profound effect on hippocampal electrophysiology and that metal ions inhibit binding to opiate receptor sites, the suggestion was made that zinc may act neurochemically by reducing the binding of enkephalin to opioid receptors (Stengaard-Pedersen et al. 1981). Alternative sites for the zinc–enkephalin interaction have since been proposed which include the possibility that zinc may function in the mossy fibres as a co-factor for the metalloenzyme enkephalinase, which degrades enkephalin, and possibly also for the neurotransmitter cholecystokinin-octapeptide (Crawford 1983; Stengaard-Pedersen et al. 1984). Other interactions have been suggested which involve the formation of zinc–enkephalin complexes or the co-localization of opioid peptides and zinc with an excitatory amino acid (McGinity et al. 1984). However, most discussion at this stage is perforce speculative as the available data are insufficient to allow definitive conclusions to be drawn.

Zinc and Receptor Binding

Evidence that zinc might act to down-regulate binding of enkephalins to opioid receptors has triggered further studies into the role of zinc in the binding of other neuroactive substances, including γ-aminobutyric acid, acetylcholine and benzodiazepine (Baraldi et al. 1984; Ebadi et al. 1984; Slevin and Kasarskis 1985). Together the findings are highly provocative and present an emerging role for zinc as a modulator of postsynaptic receptor affinities, acting in part as an inhibitor of excitatory synapses at physiological levels (Slevin and Kasarskis 1985), and of inhibitory synapses at supranormal levels (Ebadi et al. 1984).

Zinc and Synaptic Activity

Because of the high concentration of zinc in the hippocampus, most studies relating to the role of zinc in synaptic transmission have focused on this brain region, where the metal appears to accumulate in synaptic vesicles in the terminal boutons of the mossy fibre axons (Haug 1967; Crawford and Connor 1972; Danscher and Zimmer 1978; Frederickson et al. 1983). Thus, the suggestion has been made that zinc is involved in synaptic signalling (Crawford and Connor 1975; Hesse 1979; Assaf and Chung 1984; Howell et al. 1984) either directly or in a neuromodulatory role, possibly relating to the neurotransmitter receptor regulation discussed in the previous section. Recent evidence has shown that zinc is released into the extracellular space (Assaf and Chung 1984; Howell et al. 1984) and taken up again following electrical stimulation of hippocampal slices and that prolonged stimulation of the hippocampus results in a loss of silver-sulphide staining in the mossy fibre pathway and irreversible neuronal damage (Charlton et al. 1985; Sloviter 1985). However, whether zinc participates in synaptic signalling directly, for example by control of the amount of neurotransmitter glutamic acid released from glutamatozinc dihydrate in presynaptic vesicles (Sloviter 1985), or whether it acts as a neuromodulator, possibly through regulation of neuroreceptor affinity or by influencing neuronal sodium channels (Wright 1984), is not clear.

Zinc Metabolism in the Brain

Although estimates of the amount of zinc in the brain vary substantially (Crawford 1983), the average concentration in most species appears to be little over 25 μg/g, dry weight (Crawford and Connor 1972). However, the distribution of the metal between various brain regions is uneven, with the highest levels (> 65 μg/g dry weight) occurring in the hippocampus the lowest (< 25 μg/g dry weight) in the thalamus–hypothalamus and medulla oblongata and intermediate amounts in the olfactory bulbs, cerebellum, striatum, cortex and mid-brain (Crawford and Connor 1972; Donaldson et al. 1973). Unlike most other brain regions which at birth generally have similar concentrations of zinc to the adult, in the rat, hippocampal levels increase substantially during the first 3 weeks postnatally, especially around the 2nd week (Szerdahelyi and Kasa 1983). At this time the metal appears to shift out of the perikarya by axoplasmic transport and into the terminal field (Wolf et al. 1984). Proportionately (two- to threefold) higher concentrations of zinc are found in the terminal boutons of the mossy fibres than in other hippocampal areas (Crawford 1983; Klitenick et al. 1983), which confirms earlier histological studies (Haug 1967) and supports the view that the concurrent accumulation of zinc in the mossy fibres and the electrophysiological maturation of this brain pathway point to a role for zinc in mossy fibre synaptic function (Crawford 1983).

The mechanism by which zinc is taken up into the central nervous system is largely unknown. Also, whether plasma zinc levels are related to the concentration of the metal in brain cells is not clear (Crawford 1983). Generally, the passage of a nutrient across the cerebral capillaries (blood–brain barrier) depends upon a specific transport mechanism, or is a function of lipid solubility (Pardridge 1986). Passive diffusion of zinc into the brain would be expected to be minimal, unless enhanced by lipophylic chelating agents which could possibly even lead to neurotoxicological effects (Aaseth et al. 1979). Certainly the elevated concentration of zinc in hippocampal mossy fibres, which is 20-fold higher than in the plasma and 100 times greater than in the cerebrospinal fluid, suggests a high-affinity active process in some brain regions (Crawford 1983). A degree of doubt also exists at present concerning whether cerebral zinc is derived mainly from the cerebrospinal fluid via the choroid plexus, or whether it enters directly across the blood–brain barrier. Indications are that radioactive zinc continues to accumulate in the brain long after it can no longer be detected in cerebrospinal fluid. This suggests that the choroid plexus does not play a significant role in general intracranial zinc metabolism, although it may act to protect against excess zinc by transporting the metal out of the intraventricular space (Kasarskis 1984).

Studies concerning the uptake and turnover of zinc in rodent brains (Bergman and Soremark 1968; Kasarskis 1984) indicate very slow uptake of the metal following a single dose of radioisotope, with maximal accumulation being only 0.5% of the administered dose. Turnover of zinc in brain tissue is also slow, with an average biological half-life of 12.3 d, although significant regional differences are evident (Kasarskis 1984). Uptake of radioisotopic zinc is greatly increased in severely zinc-deficient animals, and turnover is reduced, resulting in greater zinc retention. Thus it appears that the central nervous system conserves its zinc with great avidity, with the result that zinc levels in the brain do not fall even under conditions of near-lethal zinc restriction (O'Neal et al. 1970; Wallwork et al. 1983; Kasarskis 1984). Interestingly, this highly effective homeostatic mechanism seems to operate most actively in the hippocampus – the area of naturally high zinc levels (Kasarskis 1984).

Onset of Zinc Deficiency in the Brain

The rate of onset of zinc deficiency with respect to brain development in the embryo is very rapid and in rodents, exposure of the pregnant dam to short periods (2–3 d) of dietary zinc restriction during the period of embryonic organogenesis is highly teratogenic. Underlying this effect no doubt is the rapidity with which plasma zinc levels decline following dietary restriction (Dreosti et al. 1985) and the lack in early embryos of an effective blood–brain barrier (Pardridge 1986) which would protect to some extent against fluctuations in the humoral supply of zinc. In the late prenatal and early postnatal periods, when brain zinc levels are more highly conserved and brain growth has slowed, nutritional zinc deficits are not overtly teratogenic but do induce substantial behavioural abnormalities. Some level of neuronal dysmorphology is possibly still involved at this stage, but zinc-related cerebral dysfunction probably contributes most to the psychological disturbances.

Generally, the times of maximum vulnerability of the brain to zinc impoverishment coincide with periods of active neuronal proliferation and differentiation. However, some studies on rats exposed to zinc deficiency after 35 d of age, when brain development is completed, have resulted in significant learning and behavioural deficits (Halas 1983). Certainly, several reports have shown that induced zinc deficits in adult humans have led to psychological disorders (Table 15.2). Again, little is known with certainty concerning the metabolic lesions responsible for zinc-deficiency-related neuropathy in the mature brain, but it is most likely that the involvement of zinc in the metabolism of neurotransmitters and in neuroreceptor function is of primary importance.

In this connection it is worth noting that although the level of zinc in the postembryonic brain seems to be relatively unaffected by dietary zinc restriction, effects on certain brain enzymes occur comparatively quickly. Thus most of the changes in enzyme activity listed in Table 15.3 were observed within 1–2 months of dietary zinc restriction, while defects in hippocampal electrophysiology have been demonstrated in adult animals depleted of zinc for about 2 months (Hesse 1979). Behavioural deficits occur in young adult rats after 1 month (Hesse et al. 1979) and after 3 months in mature animals (Gordon 1984). In humans, a histidine-induced zinc deficiency has been reported to elicit detectable neuropsychological disorders after only 1–3 weeks (Henkin et al. 1975), although similar effects were not seen in volunteers receiving an experimental zinc-deficient diet for 3–4 months (Tucker and Sandstead 1984).

In summary, zinc is critically needed during brain development and for brain function. In utero, zinc restriction is both physically and behaviourally teratogenic, but severe cerebral dysmorphology occurs only if the deficiency is imposed around the time of embryonic organogenesis. Early postnatal zinc impoverishment mainly affects brain function, which no doubt reflects a measure of neuronal dysmorphogenesis as well as neurophysiological dysfunction. The accompanying behavioural abnormalities generally respond well to zinc repletion and are, to some extent, reversible. Zinc depletion in adult animals elicits milder behavioural aberrations, which appear to be fully reversible. Biochemically, the effect of zinc deficiency on brain growth probably operates largely through the requirement for the metal during cell division. Neurochemically, the role of zinc is more complex as the element is almost certainly involved in several neurological processes where it functions at various levels of importance. Of particular interest must be the emerging role of zinc in synaptic transmission and its proposed involvement in neuroreceptor modulation, a research area which promises to hold the focus on zinc in the brain into the twenty-first century.

References

Aaseth J, Soli NE, Forre O (1979) Increased brain uptake of copper and zinc in mice caused by diethyldithiocarbamate. Acta Pharmacol Toxicol 45: 41–44

Apgar J (1968) Comparisons of the effects of copper, manganese and zinc deficiencies on parturition in the rat. Am J Physiol 215: 428–432

Assaf SY, Chung SH (1984) Release of endogenous Zn^{2+} from brain tissue during activity. Nature 308: 734–736

Baraldi M, Caselgrandi E, Santi M (1984) Effect of zinc on specific binding of GABA to rat brain membranes. In: Frederickson CJ, Howell GA, Kasarskis EJ (eds) The neurobiology of zinc, part A. Alan R. Liss, New York, pp 59–72

Barbeau A, Donaldson J (1974) Zinc, taurine and epilepsy. Arch Neurol 30: 52–54

Bergman B, Soremark R (1968) Autoradiographic studies on the distributions of zinc-65 in mice. J Nutr 94: 6–12

Bergmann KE, Makosch E, Tews KH (1980) Abnormalities of hair zinc concentration in mothers of newborn infants with spina bifida. Am J Clin Nutr 33: 2145–2150

Buell SJ, Fosmine GJ, Ollerich DA et al. (1977) Effects of postnatal zinc deficiency on cerebellar and hippocampal development in the rat. Exp Neurol 54: 199–210

Burnet FM (1981) A possible role of zinc in the pathology of dementia. Lancet I: 186

Caldwell DF, Oberleas D, Clancy JJ et al. (1970) Behavioral impairment in adult rats following acute zinc deficiency. Proc Soc Exp Biol Med 133: 1417–1421

Cavdar AO, Arcasoy A, Baycu T et al. (1980) Zinc deficiency and anencephaly in Turkey. Teratology 22: 141

Charlton G, Rovira C, Ben-Ari Y et al. (1985) Spontaneous and evoked release of endogenous Zn^{2+} in the hippocampal mossy fibre zone of the rat in situ. Exp Brain Res 58: 202–205

Cohen SR (1970) Phosphatases. In: Lathja A (ed) Handbook of neurochemistry, vol 3. Plenum Press, New York, pp 87–137

Cohn P, Richter D (1956) Enzymic development and maturation of the hypothalamus. J Neurochem 1: 166–172

Constantinidis J, Tissot R (1981) Role of glutamate and zinc in hippocampal lesions of Pick's disease. In: Di Chiara G, Gessa GL (eds) Glutamate as a neurotransmitter. Raven Press, New York, pp 413–424

Crawford IL (1983) Zinc and the hippocampus. In: Dreosti IE, Smith RM (eds) Neurobiology of the trace elements, vol 1. Humana Press, Clifton, New Jersey, pp 169–211

Crawford IL, Connor JD (1972) Zinc in the maturing rat brain: hippocampal concentration and localization. J Neurochem 19: 1451–1458

Crawford IL, Connor JD (1975) Zinc and hippocampal function. J Orthomol Psychiatr 4: 39–52

Damyanov I, Dutz W (1971) Anencephaly in Shiraz, Iran. Lancet I: 82

Danscher G, Zimmer J (1978) An improved Timm sulphide silver method for light and electron microscopic localization of heavy metals in biological tissues. Histochemistry 55: 27–40

Donaldson J, St Pierre T, Minnich JL et al. (1973) Determination of Na^+, K^+, Mg^{2+}, $Cu^{2+} \cdot Zn^{2+}$ and Mn^{2+} in rat brain regions. Can J Biochem 51: 87–92

Dreosti IE (1982) Zinc in prenatal development. In: Prasad AS, Dreosti IE, Hetzel BS (eds) Clinical applications of recent advances in zinc metabolism. Alan R. Liss, New York, pp 19–38

Dreosti IE (1983) Zinc and the central nervous system. In: Dreosti IE, Smith RM (eds) Neurobiology of the trace elements, vol 1. Humana Press, Clifton, New Jersey, pp 135–162

Dreosti IE (1984) Zinc in the central nervous system: the emerging interactions. In: Frederickson CJ, Howell GA, Kasarskis EJ (eds) The neurobiology of zinc, part A. Alan R. Liss, New York, pp 1–26

Dreosti IE, Hurley LS (1975) Depressed thymidine kinase activity in zinc-deficient rat embryos. Proc Soc Exp Biol Med 150: 161–165

Dreosti IE, Grey PC, Wilkins PJ (1972) Deoxyribonucleic acid synthesis, protein synthesis and teratogenesis in zinc-deficient rats. S Afr Med J 46: 1585–1588

Dreosti IE, Record IR, Manuel SJ (1980) Incorporation of ^{3}H-thymidine into DNA and the activity of alkaline phosphatase in zinc-deficient fetal rat brains. Biol Trace Element Res 2: 21–29

Dreosti IE, Manuel SJ, Buckley RA et al. (1981) The effect of late prenatal and/or early postnatal zinc deficiency on the development and some biochemical aspects of the cerebellum and hippocampus in rats. Life Sci 28: 2133–2141

Dreosti IE, Manuel SJ, Record IR (1985) Zinc deficiency and the developing embryo. Biol Trace Element Res 7: 103–122

Duerre JA, Ford KM, Sandstead HH (1977) Effect of zinc deficiency on protein synthesis in brain and liver of suckling rats. J Nutr 107: 1082–1093

Duncan JR, Hurley LS (1978) Thymidine kinase and DNA polymerase activity in normal and zinc-deficient, developing rat embryos. Proc Soc Exp Biol Med 159: 39–43

Dvergsten CL, Johnson LA, Sandstead HH (1984) Alterations in the postnatal development of the cerebellar cortex due to zinc deficiency, part III. Dev Brain Res 16: 21–26

Ebadi M, Pfeiffer RF (1984) Zinc in neurological disorders and in experimentally induced epileptiform seizures. In: Frederickson CJ, Howell GA, Kasarskis EJ (eds) The neurobiology of zinc, part B. Alan R. Liss, New York, pp 307–324

Ebadi M, Wilt S, Ramaley R et al. (1984) The role of zinc and zinc-binding proteins in regulation of glutamic acid decarboxylase in brain. In: Evangelopoulos AE (ed) Chemical and biological aspects of vitamin B6 catalysis. Alan R. Liss, New York, pp 255–275

Frederickson CJ, Klitenick MA, Manton WI et al. (1983) Cytoarchitectonic distribution of zinc in the hippocampus of man and the rat. Brain Res 273: 335–339

Frederickson CJ, Gage FH, Howell GA et al. (1984a) A possible role of mossy fibre zinc in sympathetic sprouting. In: Frederickson CJ, Howell GA, Kasarskis EJ (eds) The neurobiology of zinc, part A. Alan R. Liss, New York, pp 173–188

Frederickson CJ, Howell GA, Kasarskis EJ (eds) (1984b) The neurobiology of zinc, parts A and B. Alan R. Liss, New York

Goldberg HJ, Sheehy EM (1982) Fifth day fits: an acute zinc deficiency syndrome. Arch Dis Child 57: 633–634

Gordon EF (1984) Behavioral correlates of experimental zinc deficiency. In: Frederickson CJ, Howell GA, Kasarskis EJ (eds) The neurobiology of zinc, part B. Alan R. Liss, New York, pp 77–90

Halas ES (1983) Behavioral changes accompanying zinc deficiency in animals. In: Dreosti IE, Smith RM (eds) Neurobiology of the trace elements, vol 1. Humana Press, New York, pp 213–243

Hambidge KM, Neldner KH, Walravens PA (1975) Zinc, acrodermatitis enteropathica and congenital malformations. Lancet I: 577–578

Hansen CR, Malecha M, McKenzie T et al. (1983) Copper and zinc deficiencies in association with depression and neurological findings. Biol Psychiatry 18: 395–401

Haug FMS (1967) Electron microscopical localization of zinc in hippocampal mossy fibre synapses by a modified sulfide silver procedure. Histochemie 8: 355–368

Henkin RI, Patten BM, Re PK (1975) A syndrome of acute zinc loss. Arch Neurol 32: 745–752

Hesketh JE (1984) Microtubule assembly in rat brain extracts. Int J Biochem 16: 1331–1339

Hesse GW (1979) Chronic zinc deficiency alters neuronal function of hippocampal mossy fibers. Science 205: 1005–1007

Hesse GW, Frank-Hesse KA, Catalanotto FA (1979) Behavioral characteristics of rats experiencing chronic zinc deficiency. Physiol Behav 22: 211–215

Howell GA, Welch MG, Frederickson CJ (1984) Stimulation-induced uptake and release of zinc in hippocampal slices. Nature 308: 736–738

Hurley LS (1968) The consequences of fetal impoverishment. Nutrition Today 3: 3–10

Hurley LS (1981) Teratogenic aspects of manganese, zinc and copper nutrition. Physiol Rev 61: 249–295

Hurley LS, Swenerton H (1966) Congenital malformations resulting from zinc deficiency in rats. Proc Soc Exp Biol Med 123: 692–697

Hurley LS, Shrader RE (1972) Congenital malformations of the nervous system in zinc-deficient rats. In: Pfeiffer CC (ed) Neurobiology of the trace metals zinc and copper. Academic Press, New York, pp 7–51

Hurley LS, Gowan J, Swenerton H (1971) Teratogenic effects of short-term and transitory zinc deficiency in rats. Teratology 4: 199–204

Jameson S (1976) Effects of zinc deficiency on human reproduction. Acta Med Scand [Suppl] 593: 5–89

Kasarskis EJ (1984) Zinc metabolism in normal and zinc-deficient rat brain. Exp Neurol 85: 114–127

Klitenick MA, Frederickson CJ, Manton WI (1983) Acid-vapor decomposition for determination of zinc in brain tissue by isotope dilution mass spectrometry. Anal Chem 55: 921–923

Kritscher KN (1978) Copper and zinc in childhood behavior. Psychopharmacol Bull 14: 58–59

Ledig M, Rainer F, Ziessel M et al. (1982) Regional distribution of superoxide dismutase in rat brain during post natal development. Dev Brain Res 4: 333–337

Lokken PM, Halas ES, Sandstead HH (1973) Influence of zinc deficiency on behavior. Proc Soc Exp Biol Med 144: 680–682

McGinity JF, Henriksen SJ, Chavkin C (1984) Is there an interaction between zinc and opioid peptides in hippocampal neurons? In: Frederickson CJ, Howell GA, Kasarskis EJ (eds) The neurobiology of zinc, part A. Alan R. Liss, New York, pp 73–90

McLardy T (1975) Hippocampal zinc and structural deficit in brains from chronic alcoholics and some schizophrenics. Orthomol Psychiatr 4: 32–36

Mody I, Miller JJ (1985) Levels of hippocampal calcium and zinc following kindling-induced epilepsy. Can J Physiol Pharmacol 63: 159–161

Morgan RF, O'Dell BL (1977) Effects of copper deficiency on the concentrations of catecholamines and related enzyme activities in the rat brain. J Neurochem 28: 207–213

Moynahan EJ (1976) Zinc deficiency and disturbances of mood and visual behaviors. Lancet I: 91

Ohlsson A (1981) Acrodermatitis enteropathica. Acta Paediatr Scand 70: 269–273

O'Neal RM, Pla GW, Fox MRS et al. (1970) Effect of zinc deficiency and restricted feeding on protein and ribonucleic acid metabolism of rat brain. J Nutr 100: 491–497

Pardridge WM (1986) Blood–brain transport of nutrients. Fed Proc 45: 2047–2049

Pfeiffer CC, Iliev V (1972) A study of zinc deficiency and copper excess in schizophrenia. In: Pfeiffer CC (ed) Neurobiology of the trace metals zinc and copper. Academic Press, New York, pp 141–165

Pihl RO, Parkes M (1977) Hair element content in learning disabled children. Science 198: 204–206

Prasad AS (1979) Clinical biochemical and pharmacological role of zinc. Ann Rev Toxicol 20: 393–420

Prasad AS, Halsted JA, Nadimi M (1961) Syndrome of iron deficiency anemia, hepatosplenomegaly, hypogonadism, dwarfism and geophagia. Am J Med 31: 532–539

Prohaska JR, Luecke RW, Jasinski R (1974) Effect of zinc deficiency from day 18 of gestation and/or during lactation on the development of some rat brain enzymes. J Nutr 104: 1525–1531

Record IR, Dreosti IE (1979) Effects of zinc deficiency on the liver and brain thymidine kinase activity in the fetal rat. Nutr Rep Int 20: 749–755

Record IR, Tulsi RS, Dreosti IE, Fraser FJ (1985) Cellular necrosis in zinc-deficient rat embryos. Teratology 32: 397–405

Record IR, Dreosti IE, Tulsi AS et al. (1986) Maternal metabolism and teratogenesis in zinc-deficient rats. Teratology 33: 311–318

Sandstead HH (1985) Zinc: essentiality for brain development and function. Nutr Rev 43: 129–137

Sandstead HH, Gillespie DD, Brady RN (1972) Zinc deficiency: effect on brain of the suckling rat. Pediatr Res 6: 119–125

Sandstead HH, Fosmire GJ, McKenzie JM et al. (1975) Zinc deficiency and brain development in the rat. Fed Proc 34: 86–88

Sato CM, Frazier JM, Goldberg AM (1984) A kinetic study of the in vivo incorporation of ^{65}Zn into the rat hippocampus. J Neurosci 4: 1671–1675

Sever LE, Emanuel I (1973) Is there a connection between maternal zinc deficiency and congenital malformations of the central nervous system in man? Teratology 7: 117–119

Sivasubramanian KN, Henkin RI (1978) Behavioral and dermatological changes and low serum zinc and copper concentrations in two premature infants after parenteral alimentation. J Pediatr 93: 847–851

Slevin JT, Kasarskis EJ (1985) Effects of zinc on markers of glutamate and aspartate neurotransmission in rat hippocampus. Brain Res 334: 281–286

Sloviter RS (1985) A selective loss of hippocampal mossy fiber Timm stain accompanies granule cell seizure activity induced by perforant path stimulation. Brain Res 330: 150–153

Soltan MH, Jenkins DM (1982) Maternal and fetal plasma zinc concentrations and fetal abnormality. Br J Obstet Gynaecol 89: 56–58

Srinivasan DP (1984) Trace elements in psychiatric illness. Br J Hosp Med 32: 77–79

Stengaard-Pedersen K, Fredens K, Larson LI (1981) Enkephalin and zinc in the mossy fiber system. Brain Res 212: 230–233

Stengaard-Pedersen K, Larson LI, Fredens K et al. (1984) Modulation of cholecystokinin concentrations in the rat hippocampus by chelation of heavy metals. Proc Natl Acad Sci USA 81: 5876–5880

Stewart C, Katchen B, Collipp PJ et al. (1981) Zinc and birth defects. Pediatr Res 15: 515

Strobel DA, Sandstead HH (1984) Social and learning changes following prenatal or postnatal zinc deprivation in rhesus monkeys. In: Frederickson CJ, Howell GA, Kasarskis EJ (eds) The neurobiology of zinc, part B. Alan R. Liss, New York, pp 121–138

Swenerton H, Shrader RE, Hurley LS (1969) Zinc-deficient embryos: reduced thymidine incorporation. Science 166: 1014–1015

Szerdahelyi P, Kasa P (1983) Variations in trace metal levels in rat hippocampus during ontogenic development. Anat Embryol 167: 141–149

Tran VT, Snyder SH (1979) Amino acid neurotransmitter candidates in rat cerebellum: selective effects of kainic acid lesions. Brain Res 167: 345–353

Tucker DM, Sandstead HH (1984) Neuropsychological function in experimental zinc deficiency in humans. In: Frederickson CJ, Howell GA, Kasarskis EJ (eds) The neurobiology of zinc, part B. Alan R. Liss, New York, pp 139–152

Wallwork JC, Botnen JH, Sandstead HH (1982) Influence of dietary zinc on rat brain catecholamines. J Nutr 112: 514–519

Wallwork JC, Milne DB, Sims RL et al. (1983) Severe zinc deficiency: effects on the distribution of nine elements in regions of the rat brain. J Nutr 113: 1895–1905

Warkany J, Petering HG (1972) Congenital malformations of the central nervous system in rats produced by maternal zinc deficiency. Teratology 5: 319–334

Wenk GL, Stemmer KL (1982) Activity of enzymes dopamine-beta-hydroxylase and phenylethanol-amine-N-methyltransferase in discrete brain regions of the copper–zinc deficient rat following aluminum ingestion. Neurotoxicology 3: 93–99

Wolf G, Schmidt W (1982) Zinc (II) as a putative regulatory factor of glutamate dehydrogenase activity in glutamatergic systems. In: Marsan A, Mathies H (eds) Neuronal plasticity and memory formation. Raven Press, New York, pp 437–440

Wolf G, Schutte M, Romhild W (1984) Uptake and subcellular distribution of ^{65}zinc in brain structures during the postnatal development of the rat. Neurosci Lett 51: 277–280

Wright DM (1984) Zinc: effect and interactions with other cations in the cortex of the rat. Brain Res 311: 343–347

Chapter 16

Zinc in Endocrine Function

G.E. Bunce

Introduction

Endocrinology is the study of the structure and function of the endocrine glands and their secretory products (hormones), including the consequences of excessive or deficient production of the latter. Hormones function as primary chemical messengers that deliver their information or signal to a selected site or target cell by virtue of the presence in the latter of a highly specific receptor to which the hormone will bind. As a result of the hormone–receptor interactions, a sequence of events is initiated, often mediated by second and third messengers, the end result of which is determined both by the chemical nature of the hormone and the type of cell in which the receptor resides. Phenomena ranging from ion transport across the plasma membrane to modification of genome transcription may be affected by hormone action. Hormone production and release and receptor synthesis and metabolism are subject to elaborate positive and negative feedback controls. Thus, under normal circumstances, the entire hormonal network operates in a highly integrated and harmonious fashion to insure growth, development and maintenance of a dynamic steady state.

Chemically, one may group the hormones into two broad categories:

1. hormones based upon the steroid molecule as the parent structure, and
2. hormones that are proteins or peptides or are metabolically derived from amino acids.

In this chapter, the role of zinc in examples from both categories will be considered.

Gonadotropins, Sex Hormones and Zinc

Hypogonadism is a major manifestation of severe zinc deficiency in the male rat (Follis et al. 1941; Millar et al. 1958, 1960). In zinc-deficient rats compared with pair-fed controls, the testes are significantly smaller and there is atrophy of the seminiferous epithelium. The consequent testicular hypofunction affects both spermatogenesis and the output of testosterone by the Leydig cells. Oligospermia is present in mild to moderate deficiency states.

It has been suggested that the hypogonadism of zinc deficiency is secondary to an inhibition of pituitary gonadotropin output brought about by inanition and growth failure. A recent paper by McClain et al. (1984) has addressed this question effectively. These authors fed a semipurified diet containing 0.7 mg zinc/kg with no added phytate to male weanling Sprague–Dawley rats for 36 d and employed ad libitum-fed, pair-fed and weight-restricted control animals supplied with 30 mg zinc/l in their drinking water. After this time, basal levels of serum testosterone were reduced in both zinc-deficient rats (0.64 ± 0.21 ng/ml) and weight-restricted rats (0.62 ± 0.08 ng/ml) compared with either pair-fed controls (1.19 ± 0.23 ng/ml) or ad libitum-fed controls (1.60 ± 0.21 ng/ml). Similarly the weight of the testicles was equivalent in the zinc-deficient and weight-restricted groups and about one-half the weight recorded in the ad libitum-fed controls. The reduction in testicular mass was equivalent to the relative sizes of the animals. The weights of the prostate gland, epididymis and seminal vesicle were significantly lower, however, in the zinc-deficient rats than in the weight-restricted group ($P < 0.001$), suggesting that target cell response to available testosterone was also diminished. The basal serum levels of follicle-stimulating hormone (FSH) and luteinizing hormone (LH) were unaffected by restriction of either food in general or zinc in particular. Injection of gonadotropin-releasing hormone (GnRH) elicited an equivalent elevation of serum FSH and LH in the three control groups and an enhanced response in the zinc-deficient animals, but the capacity to produce testosterone after exposure to human chorionic gonadotropin (HCG) was diminished by more than 50% by zinc deficiency in comparison with weight restriction alone.

These careful studies reinforce and extend previous work by Lei et al. (1976) and Root et al. (1979) and establish that zinc-deficient rats maintain adequate basal levels of gonadotropins and show a heightened response to GnRH stimulation. Thus the growth failure and inanition characteristic of zinc deficiency do not disable the hypothalamic–pituitary axis. Rather, it would appear that zinc restriction leads to Leydig cell decline, a result that is also consistent with the reduction and disorganization of smooth endoplasmic reticulum seen in the Leydig cells of the zinc-deficient boar (Hesketh 1982).

Zinc deficiency in human males also provokes a hypogonadal reaction. The classic papers in this field were the studies conducted in Egypt and Iran in 1961–1963 (Prasad et al. 1961, 1963a, 1963b) and in Iran 10 years later (Halsted et al.1972), which are discussed in greater detail in Chap. 18. Zinc-responsive hypogonadism, testicular atrophy and impotence also have been recorded as secondary complications of sickle cell anaemia (Abbasi et al. 1976) and of chronic haemodialysis (Antoniou et al. 1977; Mahajan et al. 1982).

In a further study, Abbasi et al. (1980) restricted dietary zinc intake in five adult male volunteers (51–65 years of age) to 2.7 to 5.0 mg daily for 24 to 40 weeks. The semipurified diet employed texturized soya protein and supplied 1865–2352 kcal/d, 53–58 g protein/d and all essential vitamins and minerals except zinc at the level of the recommended dietary allowances. Oligospermia appeared in four of the five subjects and persisted for between 6 to 14 months. Return to normal occurred only 6 to 12 months after the initiation of zinc repletion. Serum testosterone did not decline during the zinc-restricted phase, fell significantly during the first 12 weeks of repletion (30 mg of zinc as zinc acetate orally) and returned to normal in the subsequent 8 weeks. There was a slight decline in the maximal rise of serum testosterone after GnRH stimulation during zinc restriction and a more significant decline during the early phase of zinc repletion with recovery to normal during the late phase of zinc repletion. Plasma LH and FSH did not change significantly at any time during the study. The authors interpreted these results as being indicative of a decline in Leydig cell function and concluded that a dietary restriction of zinc has an adverse effect upon testicular function in humans as well as experimental animal species.

Because the reports on zinc-deficient humans in Egypt and Iran in 1961–1963 involved only male subjects and because the testes but not the ovaries contain large quantities of zinc, it was thought at first that the syndrome of dwarfism, anaemia and retarded sexual maturation was restricted to males. A subsequent study in Iran, however, found two female subjects with a similar syndrome (Ronaghy and Halsted 1975). During treatment these young women, 19 and 20 years of age, were fed a well-balanced natural diet with ample animal protein providing about 12 mg zinc/d. This regimen corrected the anaemia, but growth and expression of secondary sexual characteristics were only minimal until a zinc supplement was added.

The most dramatic evidence of altered gonadal function in the female may be the marked distress and dystocia which accompanies delivery of a litter by the rat (Apgar 1968a). Female rats fed a diet containing 1 mg zinc/kg or less beginning on the 1st day after mating experienced delayed, prolonged and difficult labour accompanied by excessive bleeding. The neonates were abandoned and placentae left unconsumed. Numerous additional studies by Apgar (1968b, 1970, 1972, 1973, 1975, 1976, 1977a, b; Gombe et al. 1973) showed that these effects were relatively specific to zinc deficiency and were not induced by restriction either of total food intake or of dietary protein, thiamin, copper or manganese. The typical syndrome still emerged even when the zinc-deficient diet was begun as late as day 18 of gestation and could be prevented by zinc repletion beginning on day 19 in rats maintained up to that time on a zinc-depleted ration. Kalinowski and Chavez (1984) have observed prolonged delivery in swine maintained on a low-zinc diet for the last 4 weeks of pregnancy. The existence of similar behaviour in humans has been suggested by several studies but is much more difficult to establish with confidence. Jameson (1976) reported that women with abnormal deliveries, inefficient labour and atonic bleeding showed lower serum zinc concentrations during early pregnancy than did women with normal deliveries. Durá-Travé et al. (1984) also found a significant ($P < 0.05$) correlation between low plasma zinc and prolonged duration of the active phase of delivery in 92 full-term human pregnancies in Spain. More studies will be necessary to determine if the observed clinical disorders of labour in humans are genuine consequences of inadequate zinc nutrition.

What do we know about the biochemical role(s) of zinc in the phenomena described above? In the male animal, the evidence suggests that gonadotropin output is satisfactory but that it fails to elicit the appropriate response of the target organ, the Leydig cells, the consequence being low production and release of testosterone. Since zinc is known to be either a structural or catalytic factor for more than 80 enzymes representing all of the major classification categories, the pathology of zinc deficiency has often been ascribed to depletion of crucial zinc-requiring enzymes. It must be noted, however, that most enzymes are not rate-limiting in their metabolic pathways, that the fractional rates of synthesis and degradation of enzymes differ widely and that zinc enzymes show a wide range in their affinities for zinc (Chesters 1978). If there are, in fact, zinc metalloenzymes which put net testosterone production at risk, they have not yet been identified (Prasad 1985).

It would seem more likely that zinc ions are necessary for the proper function of the gonadotropins. Kellokumpu and Rajaniemi (1981, 1982) injected 100 μl of 5 mM $ZnCl_2$ into rat testes and immediately afterwards administered [^{125}I]-iodo-HCG into the tail vein. Testicular uptake of HCG was significantly higher in zinc-treated testes, and there was evidence of increased receptor–hormone complex formation and stability, but the primary testosterone response was diminished rather than enhanced in both plasma and testicular tissue. The specific biochemical cause(s) of Leydig cell failure induced by zinc deficiency remain unclear.

There is reason to suspect that the parturient difficulties of the zinc-deficient female rat may result from poor compliance to oestrogen "instructions". Lytton and Bunce (1986) recorded uterine pressure cycles during oxytocin-induced labour and observed poor synchronization and low amplitude of contractions in those rats maintained on a diet containing 3 mg zinc/kg. Coordination and propagation of uterine peristaltic contractions during labour require the efficient flow of electrical information through a gap junction network. Garfield et al. (1977, 1978, 1980) have shown that gap junctions are absent from the uterine myometrium during gestation but develop during the last 48 h prepartum, coincident with progesterone withdrawal and oestrogen stimulation. Bunce et al. (1983) found a normal preparturient decline in plasma progesterone and increase in plasma oestrogen in zinc-deficient rats but delayed induction of the ovarian luteolytic marker enzyme, 20α-hydroxysteroid dehydrogenase. Dylewski et al. (1986) found fewer than 50% as many uterine myometrial gap junctions in spontaneously delivering rats fed a diet containing 3 mg zinc/kg as compared with ad libitum-fed controls (40 mg zinc/kg).

The essentiality of zinc for in vitro steroid receptor binding to chromatin has been proposed for mouse mammary gland oestrogen receptor (Shymala and Yeh 1975), for chick oviduct progesterone receptor (Lohmar and Toft 1975) and for prostate dihydrotestosterone receptor (Colvard and Wilson 1984). In vivo support of this theory comes from a paper by Bunce and Vessal (1987), who observed changes in subcellular distribution of uterine oestrogen receptors and in retention of injected [^{3}H] oestrogen in zinc-deficient rats. Matusik et al. (1986) have identified a gene from the lateral lobe of the rat prostate that codes for a prostate fluid protein and is inducible by either androgen or zinc. Of special interest is the recognition of repetitive zinc-binding domains in the protein transcription factor IIIA from *Xenopus* oocytes (Miller et al. 1985). Zinc ions tetrahedrally coordinated to cysteine and histidine stabilize a protein structure with repeated protein loops or "fingers" that are rich in DNA-binding amino acids. A variety of other regulatory and nucleic-acid-binding proteins contain sequences that are suggestive of metal-binding domains, among them the human glucocorticoid receptor (Weinberger et al. 1985) and the chicken oviduct oestrogen receptor (Krust et al. 1986). Considered together, this array of data provides a circumstantial base of evidence to support the hypothesis that zinc ions participate in sex-steroid-directed gene expression and may also be required for maximum effectiveness of progesterone and corticosteroids.

Insulin and Zinc

The discoveries of the essentiality of dietary zinc for rats (Todd et al. 1934) and of the presence of considerable quantities of zinc in crystalline insulin (Scott 1934) have stimulated many investigations into the physiological role of zinc in insulin synthesis, storage, secretion and peripheral metabolism. Grant et al. (1972), Blundell et al. (1972) and Emdin et al. (1980) have performed detailed and extensive studies on the structures of insulin and proinsulin crystals, and readers are urged to consult the original papers for further insights and information. Insulin is synthesized at the ribosome as a single-chain polypeptide, preproinsulin. Removal of the N-terminal leader sequence yields proinsulin, which aggregates into hexamers with two zinc atoms per hexamer coordinated to the six B_{10} histidines. The non-polar regions are buried, and the surface is largely covered by the connecting peptide. Proinsulin is a highly soluble molecule itself, and its solubility is further enhanced by the weak binding of up to 30 atoms of zinc per hexamer, primarily in ionic attraction to numerous carboxylate groups. This prohormone is transported rapidly to the Golgi apparatus where the connecting peptide is removed. The loss of hydrophilic groups and exposure of non-polar residues drastically reduce both zinc binding and solubility. As a consequence, insulin precipitates and is packaged into the B-cell storage vesicles, probably as the two zinc-coordinated hexamers, whence it can be released into the circulation upon the appropriate signal.

While it seems likely that zinc facilitates the processing and packaging of insulin, there are reasons to doubt that it is obligatory for this operation. In studies with islet cells maintained in zinc-depleted culture medium for 9 d, no effect was observed on proinsulin biosynthesis, conversion of proinsulin to insulin or ability to store insulin in granules (Howell et al. 1978; Hoftiezer et al. 1985). Moreover, while the insulin molecule moves into the blood as the crystalline hexamer, it probably circulates as the zinc-free monomer, and this is considered to be its active form (Blundell et al. 1972; Goldman and Carpenter 1974). The ability of certain zinc chelators to strip pancreatic granules of zinc and induce diabetes has been explained quite satisfactorily by Epand et al. (1984), who showed that diabetogenicity correlates most directly with the ability to acidify the vesicle interior by release of protons in exchange for zinc. Acidification promoted insulin solubilization and consequently a marked increase in internal osmotic pressure followed by vesicular rupture or exocytosis.

Another approach to an inquiry into the role of zinc in insulin metabolism is to evaluate circulating insulin and glucose tolerance in intact rats. This seemingly straightforward plan is in fact complicated by factors such as food intake, body weight and the route and size of the glucose dose, with the consequence of confusing and sometimes contradictory results that have been amply documented elsewhere (Chesters 1978; Roth and Kirchgessner 1981; Hambidge et al. 1986). A recent paper by Reeves and O'Dell (1983), however, demonstrates that under carefully controlled conditions one may observe a reduction in glucose utilization by zinc-deficient rats. When control rats are pair-fed to the level of voluntary food intake of zinc-deficient rats, they tend to become meal-eaters, consuming their entire 24-h allotment within 2 h; a pattern that will enhance glucose utilization. Reeves and O'Dell therefore designed an experiment with four groups of male Wistar rats fed as follows: group 1, meal-fed, low zinc; group 2, meal-fed, adequate zinc; group 3, ad libitum-fed, low zinc; and group 4, ad libitum-fed, adequate zinc.

After 3 weeks, all animals were fasted for 22 h, given an intragastric dose of 20 mmol of glucose containing 20 μCi [U-^{14}C] glucose per kg body weight and killed 3 h later. Rats meal-fed an adequate level of zinc incorporated 120 times more glucose carbon into the fatty acids of the epididymal fat pads per g tissue than did those given the same diet ad libitum. Zinc deficiency showed no effect in those rats fed ad libitum but caused a 75% reduction of incorporation in the meal-fed groups. Further experiments showed that the differences in fat cell accumulation of glucose carbon as fatty acids were probably occurring because of reduced entry rather than accelerated catabolism. Finally, the effect of zinc deficiency on glucose absorption was determined by measuring glucose content of the gut 90 min after incubation. The 13% greater absorption displayed by the controls was considered to be too little to explain the large differences in label incorporation.

Fat cells require insulin for the uptake of glucose. The considerable decline in glucose utilization by fat cells in zinc-deficient meal-fed rats is most readily explained by one of two mechanisms: either circulating insulin levels are low, or the response to insulin is reduced. Reeves and O'Dell correctly noted that previous studies on the effect of dietary zinc on plasma insulin have yielded conflicting and inconsistent results; they concluded that diminished insulin synthesis and secretion do not adequately explain the observed effects of zinc deficiency on glucose metabolism. Further support for this viewpoint comes from a recent paper by Park et al. (1986). This group tested a model using force-feeding of zinc-deficient and control rats. The loss of appetite of zinc-deficient rats slows zinc depletion by preventing zinc deposition in growing tissues and by stimulating catabolic release of tissue zinc into the plasma. Park et al. consider that the force-fed rat exhibits a pure zinc deficiency.

Weanling Sprague–Dawley rats were accustomed to an intragastric tube-feeding procedure in which they received three daily feedings of a liquid diet prepared by blending the conventional semipurified ration with water. Ad libitum food intake of the powdered zinc-adequate diet was determined in a group of seven rats. A group of 20 animals was offered the powdered zinc-deficient (< 1 ppm zinc) diet ad libitum. Two additional groups of 20 animals each received by intragastric tube for 8 d either a zinc-deficient or zinc-adequate (25 ppm zinc) diet in the amount consumed by the seven control rats. The rats with a pure zinc deficiency (i.e. force-fed) displayed a considerable glucose intolerance (approx. 400 mg glucose/dl vs 200–250 mg glucose/dl at either 30, 60 or 90 min after an intraperitoneal injection of 2 g glucose/kg body weight) despite normal serum glucagon and slightly elevated serum insulin. Histological studies revealed no differences in insulin, glucagon or somatostatin-producing pancreatic islet cells.

What roles may zinc be playing that are necessary to the peripheral function of insulin? Arquilla et al. (1978) found that pretreatment of mice with zinc resulted in an accelerated and increased magnitude of $[^{125}I]$-iodoinsulin binding to liver plasma membranes and that zinc added to isolated liver plasma membranes slowed the rate of insulin degradation. Micromolar quantities of zinc ions have also been implicated in protein phosphorylation–dephosphorylation reactions, but the effects are highly complex. Brautigan et al. (1981), for example, reported that the zinc ion inhibited a phosphotyrosyl-protein phosphatase for the epidermal growth factor (EGF) receptor, but Pang and Shafer (1985) found that zinc inhibited both the autophosphorylation and dephosphorylation of the insulin receptor. Alitalo et al. (1983) also observed inhibition of the phosphorylation of EGF by zinc but found other polypeptides to have increased phosphorylation mainly at serine residues. Ansorge et al. (1984) have shown that the insulin- and glucagon-degrading proteinase of rat liver cytosol is a metalloenzyme and that either zinc, cobalt or manganese is the probable physiological ligand. Coulston and Dandona (1980) found that ionic zinc alone stimulated rat adipocyte lipogenesis, but May and Contoreggi (1982) concluded that a substantial portion of that effect was mediated by H_2O_2 produced during in vitro incubation of the isolated adipocytes under conditions that would not occur in vivo. It seems plausible to conclude that the pancreas functions normally in the zinc-deficient rat but that the peripheral response to insulin is muted by mechanisms that must yet be defined. The possibility that zinc ions are required in normal amounts for binding and uptake of insulin or for subsequent phosphorylation–dephosphorylation steps is provocative and deserving of additional study.

Summary

In this brief review, I have examined critical references related to the malfunction of two hormone systems, sex steroids and insulin, in the zinc-deprived animal. Readers seeking a wider scope of references on these and other endocrine systems will find them in Roth and Kirchgessner (1981), Prasad (1985) and Hambidge et al. (1986).

Chesters (1978) has proposed "that the critical process for growth requires not, as has been supposed, a zinc metalloenzyme but participation of freely exchangeable zinc in a process required for alteration in the expression of the genetic information stored in the cell's chromatin". Chesters was specifically referring to a role for zinc in gene masking and unmasking, but a recent paper by Grummt et al. (1986) suggests that zinc may also be involved at the level of signalling. They have been investigating diadenosine tetraphosphate ($A_{p4}A$) as the putative intracellular agent for mitogenic induction by PDGF and EGF. They have found that mitogen exposure provokes an increase in zinc ion uptake by quiescent mammalian cells and that the increase in intracellular zinc is essential for the 1000-fold expansion of the $A_{p4}A$ pool which follows. An analogy to the cAMP second-messenger system characteristic of several hormones is obvious. It would appear that several chemical properties of zinc

– its presence in solution as a divalent cation without a tendency for oxidation–reduction and its ability to form readily reversible complexes with proteins and nucleic acids – make it especially valuable to the facilitation of certain types of hormonal information transfer and to the prompt execution of metabolic instructions.

References

Abbasi AA, Prasad AS, Ortega J, Congco E, Oberleas D (1976) Gonadal function abnormalities in sickle cell anemia: studies in adult male patients. Ann Intern Med 85: 601–605

Abbasi AA, Prasad AS, Rabbani P, DuMouchelle E (1980) Experimental zinc deficiency in man. Effect on testicular function. J Lab Clin Med 96: 544–550

Alitalo K, Keski-Oja J, Bornstein P (1983) Effects of Zn^{2+} ions on protein phosphorylation in epithelial cell membranes. J Cell Physiol 115: 305–312

Ansorge S, Bohley P, Kirschke H, Langner J, Wiederanders B (1984) The insulin and glucagon degrading proteinase of rat liver: a metal-dependent enzyme. Biomed Biochim Acta 43: 39–46

Antoniou LD, Shalhoub RJ, Sudhakar T, Smith JC Jr (1977) Reversal of uraemic impotence by zinc. Lancet II: 895–898

Apgar J (1968a) Effect of zinc deficiency on parturition in the rat. Am J Physiol 215: 160–163

Apgar J (1968b) Comparison of the effect of copper, manganese and zinc deficiencies on parturition in the rat. Am J Physiol 45: 1478–1481

Apgar J (1970) Effect of zinc deficiency on maintenance of pregnancy in the rat. J Nutr 100: 470–476

Apgar J (1972) Effect of zinc deprivation from day 12, 15 or 18 of gestation on parturition in the rat. J Nutr 102: 343–348

Apgar J (1973) Effect of zinc repletion late in gestation on parturition in the zinc-deficient rat. J Nutr 103: 973–981

Apgar J (1975) Effects of some nutritional deficiencies on parturition in the rat. J Nutr 105: 1553–1561

Apgar J (1976) Zinc requirement for normal parturition in rats. Nutr Rep Int 13: 281–286

Apgar J (1977a) Use of EDTA to produce zinc deficiency in the pregnant rat. J Nutr 107: 539–545

Apgar J (1977b) Effect of zinc deficiency and zinc repletion during pregnancy on parturition in two strains of rats. J Nutr 107: 1399–1403

Arquilla ER, Packer S, Tarmas W, Miyamoto S (1978) The effect of zinc on insulin metabolism. Endocrinology 103: 1440–1449

Blundell T, Dodson G, Hodgkin D, Mercola D (1972) Insulin: the structure in the crystal and its reflection in chemistry and biology. Adv Protein Chem 26: 279–402

Brautigan DL, Bornstein P, Gallis B (1981) Phosphotyrosyl-protein phosphatase. Specific inhibition by Zn^{2+}. J Biol Chem 256: 6519–6522

Bunce GE, Vessal M (1987) Effect of zinc and/or pyridoxine deficiency upon oestrogen retention and oestrogen receptor distribution in the rat uterus. J Steroid Biochem 26: 303–308

Bunce GE, Wilson GR, Mills CF, Klopper A (1983) Studies on the role of zinc in parturition in the rat. Biochem J 210: 761–767

Chesters JK (1978) Biochemical functions of zinc in animals. World Rev Nutr Diet 32: 135–164

Colvard DS, Wilson EM (1984) Zinc potentiation of androgen receptor binding to nuclei in vitro. Biochemistry 23: 3471–3478

Coulston L, Dandona T (1980) Insulin-like effect of zinc on adipocytes. Diabetes 29: 665–667

Durá-Travé T, Puig-Abuli M, Monreal I, Villa-Elizaga I (1984) Relation between maternal plasmatic zinc levels and uterine contractility. Gynecol Obstet Invest 17: 247–251

Dylewski DP, Lytton FDC, Bunce GE (1986) Dietary zinc and parturition in the rat. II. Myometrial gap junctions. Biol Trace Element Res 9: 165–175

Emdin SO, Dodson GG, Cutfield JM, Cutfield SM (1980) Role of zinc in insulin biosynthesis. Diabetologia 19: 174–182

Epand RM, Stafford AR, Tyers M, Nieboer E (1984) Mechanism of action of diabetogenic zinc-chelating agents. Mol Pharmacol 27: 366–374

Follis RH, Day HG, McCollum EV (1941) Histological studies of the tissues of rats fed a diet extremely low in zinc. J Nutr 22: 223–237

Garfield RE, Sims S, Daniel EE (1977) Gap junctions: their presence and necessity in myometrium during parturition. Science 198: 958–959

Garfield RE, Sims S, Kannan MS, Daniel EE (1978) Possible role of gap junctions in activation of myometrium during parturition. Am J Physiol 235: C168–C179

Garfield RE, Kannan MS, Daniel EE (1980) Gap junction formation in myometrium: control by estrogens, progesterone and prostaglandins. Am J Physiol 238: C81–C89

Goldman J, Carpenter FH (1974) Zinc binding, circular dichroism and equilibrium sedimentation studies on insulin (bovine) and several of its derivatives. Biochemistry 13: 4566–4574

Gombe S, Apgar J, Hansel W (1973) Effect of zinc deficiency and restricted food intake on plasma and pituitary LH and hypothalamic LRF in female rats. Biol Reprod 9: 415–419

Grant PT, Coombs TL, Frank BH (1972) Differences in the nature of the interactions of insulin and proinsulin with zinc. Biochem J 126: 433–440

Grummt F, Weinman-Dorsch C, Schneider-Schanlies J, Lux A (1986) Zinc as a second messenger of mitogenic induction. Effects on diadenosine tetraphosphate ($A_{p4}A$) and DNA synthesis. Exp Cell Res 163: 191–200

Halsted JA, Ronaghy HA, Abadi P et al. (1972) Zinc deficiency in man: the Shiraz experiment. Am J Med 53: 277–284

Hambidge KM, Casey CE, Krebs NF (1986) Zinc. In: Mertz W (ed) Trace elements in human and animal nutrition, vol 2. Academic Press, Orlando, Florida, pp 1–137

Hesketh JE (1982) Effects of dietary zinc deficiency on Leydig cell ultrastructure in the boar. J Comp Pathol 92: 239–247

Hoftiezer V, Berggren P-O, Hellman B (1985) Effects of zinc during culture of an insulin-producing rat cell line (RINm5F). Cancer Lett 29: 15–22

Howell SL, Tyhurst M, Durefelt H, Anderson A, Hellerstrom C (1978) Role of zinc and calcium in the formation and storage of insulin in the pancreatic β-cell. Cell Tissue Res 188: 107–118

Jameson S (1976) Effects of zinc deficiency in human reproduction. Acta Med Scand [Suppl] 593: 3–89

Kalinowski J, Chavez BR (1984) Effect of low dietary zinc during late gestation and early lactation in the sow and neonatal piglets. Can J Anim Sci 64: 749–758

Kellokumpu S, Rajaniemi H (1981) Effect of zinc on the uptake of the human chorionic gonadotropin in rat testis and testosterone response in vivo. Biol Reprod 24: 298–305

Kellokumpu S, Rajaniemi H (1982) Dissociation of receptor-bound human chorionic gonadotropin from rat testicular membranes in vitro as a high molecular weight complex and its inhibition by heavy metals and alkylating agents. Biochim Biophys Acta 718: 26–34

Krust A, Green S, Argos P et al. (1986) The chicken oestrogen receptor sequence: homology and v-erbA and the human oestrogen and glucocorticoid receptors. EMBO J 5: 891–897

Lei KY, Abbasi A, Prasad AS (1976) Function of the pituitary–gonadal axis in zinc-deficient rats. Am J Physiol 230: 1730–1732

Lohmar PH, Toft DO (1975) Inhibition of the binding of progesterone receptor to nuclei: effects of o-phenanthroline and rifamycin AF/013. Biochem Biophys Res Comm 67: 8–15

Lytton FDC, Bunce GE (1986) Dietary zinc and parturition in the rat. I. Uterine pressure cycles. Biol Trace Element Res 9: 151–163

McClain CJ, Gavaler JS, VanThiel DH (1984) Hypogonadism in the zinc-deficient rat: localization of the functional abnormalities. J Lab Clin Med 104: 1007–1015

Mahajan SK, Abbasi AA, Prasad AS, Rabbani P, Briggs WA, McDonald FD (1982) Effect of oral zinc therapy on gonadal function in hemodialysis patients. Ann Intern Med 97: 357–361

Matusik RJ, Kreis C, McNicol P et al. (1986) Regulation of prostatic genes: role of androgens and zinc in gene expression. Biochem Cell Biol 64: 601–607

May JM, Contoreggi CS (1982) The mechanism of the insulin-like effects of ionic zinc. J Biol Chem 257: 4362–4368

Millar MJ, Fischer MI, Elcoate PV, Mawson CA (1958) The effects of dietary zinc deficiency on the reproductive system of male rats. Can J Biochem Physiol 36: 557–569

Millar MJ, Elcoate PV, Fischer MI, Mawson CA (1960) Effect of testosterone and gonadotropin injections on the sex organ development of zinc-deficient male rats. Can J Biochem Physiol 38: 1457–1466

Miller J, McLachlan AD, Klug A (1985) Repetitive zinc-binding domains in the protein transcription factor IIIA from *Xenopus* oocytes. EMBO J 4: 1609–1614

Pang DT, Shafer JA (1985) Inhibition of the activation and catalytic activity of insulin receptor kinase by zinc and other divalent metal ions. J Biol Chem 260: 5126–5130

Park JHY, Grandjean CJ, Hart MH, Erdman JH, Pour P, Vanderhoof JA (1986) Effect of pure zinc deficiency on glucose tolerance and insulin and glucagon levels. Am J Physiol 251: E273–E278

Prasad AS (1985) Clinical, endocrinological and biochemical effects of zinc deficiency. In: Cohen MP, Foa PP (eds) Special topics in endocrinology and metabolism, vol 7. Alan R. Liss, New York, pp 45–76

Prasad AS, Halstead JA, Nadimi M (1961) Syndrome of iron deficiency anemia, hepatosplenomegaly, hypogonadism, dwarfism and geophagia. Am J Med 31: 532–546

Prasad AS, Miale A, Farid Z, Sandstead HH, Schulert AR, Darby WJ (1963a) Biochemical studies on dwarfism, hypogonadism and anemia. AMA Arch Int Med 111: 407–428

Prasad AS, Miale A, Farid Z, Schulert A, Sandstead HH (1963b) Zinc metabolism in patients with the syndrome of iron deficiency anemia, hepatosplenomegaly, dwarfism and hypogonadism. J Lab Clin Med 61: 537–547

Reeves PG, O'Dell BL (1983) The effect of zinc deficiency on glucose metabolism in meal-fed rats. Br J Nutr 49: 441–452

Ronaghy HA, Halsted JA (1975) Zinc deficiency occurring in females. Report of two cases. Am J Clin Nutr 28: 831–836

Root AW, Duckett G, Sweetland M, Reiter RO (1979) Effects of zinc deficiency upon pituitary function in sexually mature and immature male rats. J Nutr 109: 958–964

Roth HP, Kirchgessner M (1981) Zinc and insulin metabolism. Biol Trace Element Res 3: 13–32

Scott DA (1934) Crystalline insulin. Biochem J 28: 1592–1602

Shymala G, Yeh Y-F (1975) Is the estrogen receptor of mammary glands a metallo-protein? Biochem Biophys Res Commun 64: 408–415

Todd WR, Elvehjem CA, Hart EG (1934) Zinc in the nutrition of the rat. Am J Physiol 107: 146–156

Weinberger C, Hollenberg SM, Rosenfeld MG, Evans RM (1985) Domain structure of human glucocorticoid receptor and its relationship to the v-erbA oncogene product. Nature 318: 670–672

Chapter 17

Severe Zinc Deficiency

P.J. Aggett

The realization that acrodermatitis enteropathica (AE) was a zinc-responsive syndrome (Moynahan and Barnes 1973) provided the archetype of severe human zinc deficiency which aided the recognition of sporadic severe zinc deficiency accompanying total parenteral nutrition (TPN) in adults (Kay et al. 1976) and children (Arakawa et al. 1976) and in a variety of other conditions. This chapter outlines the features of severe zinc deficiency and the various conditions in which this has been observed.

Clinical Features

The individual clinical features of zinc deficiency are non-specific and protean. However, they occur in a syndrome complex which should invariably alert one to the possible diagnosis of zinc deficiency, the classic tetrad of which comprises neuropsychiatric changes, circumorificial and acral dermatitis, diarrhoea and alopecia.

Neuropsychiatric Manifestations

The earliest features comprising a change in mood, loss of affect and emotional lability have been noted in patients on TPN (Kay et al. 1976) and in AE (Moynahan 1976) in which these features are reliable warnings to parents that the affected children are becoming zinc deficient.

In a prospective study of these changes Henkin et al. (1975) gave patients with systemic sclerosis large doses of histidine to induce zincuria. They noted a reproducible sequence of effects. The patients rapidly became anorexic and 2–3 d thereafter developed dysfunction of smell and of taste. Bitter taste was lost before that of sweetness; this was attributed to there being fewer receptors for the former function. The subjects then became irritable, depressed and easy to anger; coincidentally they became lethargic and sleepy. The ability to perform simple mental arithmetic or to interpret proverbs was impaired; one subject lost short-term memory, another developed receptive aphasia and a third developed visual agnosia. Signs of cerebellar disturbances then emerged manifested by a fine tremor, an ataxic and increasingly wide-based gait and slurred speech. Supplementation with 50 mg of elemental zinc, even with continued administration of histidine, reversed these features.

It is noteworthy that these phenomena occurred with an apparently small depletion of whole-body zinc. The patients lost 5.3 ± 0.8 mg of zinc per day in their urine which represents less than 0.5% of estimated whole-body zinc. Alternatively, since histidine can easily cross the blood–brain barrier, it may have selectively depleted the limbic, cortical and cerebellar pools of zinc. Nonetheless the neuropsychiatric features are those seen in clinical zinc depletion, and some adults with undiagnosed AE had histories of treatment for neurological, "schizoid" and "depressive" psychiatric illnesses (Olholm-Larsen 1978; Graves et al. 1980). Gaze aversion and avoidance of eye-to-eye contact is a striking feature in children, but it is not clear whether this is purely behavioural or partially due to retinal dysfunction (Moynahan 1976).

Zinc-deficient babies cry excessively and are irritable and inconsolable (Sivasubrumanian and Henkin 1978; Aggett et al. 1980). Jitteriness (Sivasubrumanian and Henkin 1978) and possibly fits (Entwistle 1965) secondary to zinc deficiency have been recorded, as has an ataxic gait in an older child with AE (Aquilera-Diaz 1971).

Impaired central vision, photophobia (Moynahan 1976) and a zinc-responsive impairment of dark adaptation have been described in patients with sickle cell anaemia (Warth et al. 1981), Crohn's disease (McClain et al. 1983) and alcoholic cirrhosis (Keeling et al. 1982).

Eye Abnormalities

A mild dry conjunctivitis may progress to a more marked bilateral xerosis and keratomalacia. Increased keratinization of the eyelids can lead to blepharitis, stenosis of the lacrimal puncta and esotropia. Corneal oedema can progress to corneal clouding and to punctate and radial intra- and subepithelial opacities, and 1 mm broad band opacities concentric with the upper limbus have also been noted (Wirsching 1962; Mata et al. 1975; Feldberg et al. 1981). Vitamin A therapy is ineffective on such lesions which can deteriorate to corneal ulceration and to perforation (Feldberg et al. 1981). Whereas healing of these more advanced lesions may leave residual corneal scarring, minor lesions may resolve completely. Reversible changes in visual acuity were noted in the study of Henkin and his colleagues (1975), and in severe deficiency bilateral cataracts have developed (Racz et al. 1978).

Some patients with AE had optic atrophy and, whereas this may be secondary to their treatment with hydroxyquinoline derivatives, it has been suggested that chronic zinc deficiency may also be responsible (Leopold 1978; Sturtevant 1980).

Gastrointestinal Features

An increased awareness and earlier diagnosis of zinc deficiency has meant that the classic bulky, frothy, watery stools of zinc deficiency are being observed less often. Now this feature is more apparent retrospectively when after zinc supplementation, bowel frequency reduces and faeces become more formed (e.g. Aggett et al. 1980). In earlier cases, diarrhoea had been a prominent, but not invariable, feature of AE (Portnoy and Marsden 1961; Tompkins and Livingood 1969) in which abnormal fat and protein absorption and lactose intolerance had been described (Milla and Moynahan 1972; Moynahan and Barnes 1973). A degree of small intestinal villus atrophy with a mixed inflammatory cellular infiltrate (Kelly et al. 1976; Bohane et al. 1977) can occur but in some symptomatic patients the mucosa may be normal, in others dilated intercellular spaces and cytoplasmic vacuoles are evident by light microscopy (Moynahan et al. 1963).

On electron microscopy, cytoplasmic spherical and rod-shaped granules and lysosomal inclusion bodies are evident in the enterocytes and Paneth cells (Lombeck et al. 1974; Bohane et al. 1977; Braun et al. 1977; Jones et al. 1983). The number and size of these inclusions reduced after zinc therapy in patients with AE, and were not found in a biopsy taken from a patient who had been on prolonged zinc treatment (Bohane et al. 1977; Jones et al. 1983). Similar lesions have been noted in the intestinal mucosae of zinc-deficient rats (Otto and Weitz 1972; Koo and Turk 1977) but they are probably a non-specific change, since similar abnormalities have been noted in the small intestinal mucosae of patients with other enteropathies, Crohn's disease and beta-lipoprotein deficiency (Jones et al. 1983).

A diffuse abnormality of the rectal mucosa with crypt abscesses, disorganized rectal glands and giant cells has been described in one brief case report. This patient also had evidence of impaired ileal absorption of vitamin B_{12} (Ament and Broviac 1973).

The pathogenesis of the loose stools of zinc deficiency is probably complex and it has not been studied extensively. Increased faecal loss of sodium and water may occur in zinc-deprived children (Golden and Golden 1985), but changes in mucosal disaccharidase activity have not been correlated with the diarrhoea.

Dermatological Features

The earliest cutaneous manifestations of severe zinc deficiency are the development, in the naso labial and retro-auricular folds, of an erythematous scaling eruption (see Figs. 17.1 and 17.2). This later involves skinfolds in the neck, the groins, the axillae and the perineum where it has a moist eczematoid character. Simultaneously an angular cheilitis, stomatitis and glossitis may develop causing dysphagia and drooling. Areas exposed to friction and trauma also become involved; these include the knees, elbows, heels and the occiput. On the trunk and limbs small papular and vesicular satellite lesions may be apparent. With continuing deficiency the dermatitis extends to the trunk and becomes exudative. Normally the rash is symmetrical and it often has an orange-brown discolouration and crusting. Affected areas may bleed easily. Lesions develop early on the tips of the digits; in these areas the vesicular and bullous features predominate. Secondary infection of the rash with *Candida* and staphylococci is common. A postinflammatory hyperpigmentation may be evident after skin lesions resolve.

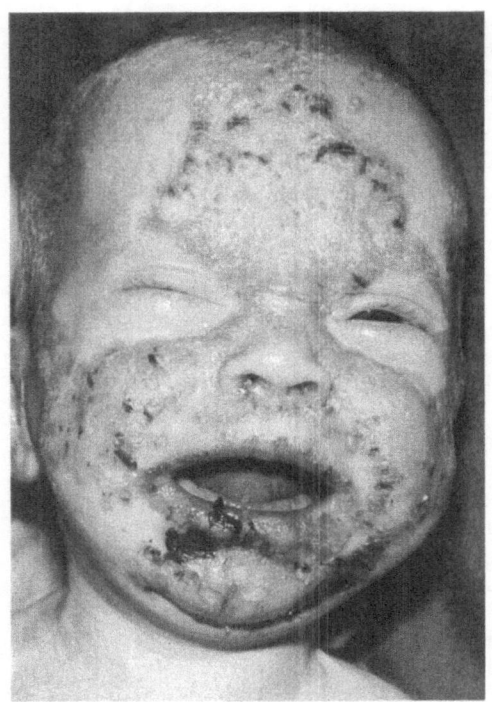

Fig. 17.1. Typical symmetrical facial dermatitis of severe zinc deficiency. Note the bleeding areas on the chin.

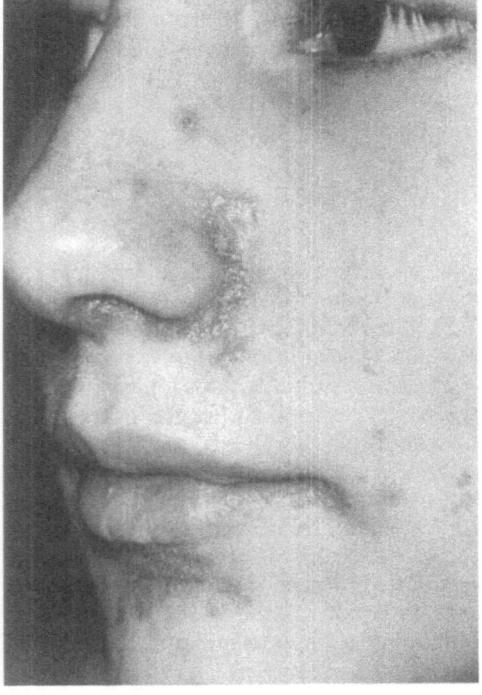

Fig. 17.2. Early skin rash of zinc deficiency involving the nostrils, corner of the mouth and chin. Separate, i.e. satellite, lesions can be seen on the nose and cheek.

Some patients complain of an uncomfortable tingling in their fingers, soles and toes before skin lesions appear at these sites. They may find it difficult to stand; an arthralgia has been described in AE.

Paronychiae may develop and nail dystrophy with white ungual flecks appears; the nails may become flattened and shed. After adequate treatment the regrowing nails have Beau's lines as evidence of their earlier growth arrest (Weismann 1977).

With prolonged zinc deficiency the skin develops hyperkeratotic lesions. These are common on the extensor surfaces, over bony prominences and pressure areas such as the ischial tuberosities, the back, knuckles and feet and they can develop in inadequately healed lesions elsewhere.

In some cases a papular acneiform rash with a haemorrhagic brown discolouration develops (see Figs. 17.3 and 17.4).

The histological appearance of affected skin comprises a number of essentially non-specific features which occur in a combination which is almost pathognomonic of zinc deficiency (Gonzalez et al. 1982). There is acanthosis, with elongation of the rete pegs and loss of the granular layer and a perivascular monocytic infiltration with persistence or preservation of cellular nuclei in the stratum corneum. In chronic lesions hyperkeratosis and parakeratosis with irregular acanthoses are evident in the epidermis. An associated epidermal pallor has been attributed to intracellular oedema in the spinous layer with accumulation of lipid droplets indicative of deranged intracellular metabolism (Ginsburg et al. 1976).

The differential diagnosis of the skin rash includes psoriasis, congenital icthyosiform dermatitis, moniliasis, epidermolysis bullosa, familial benign chronic pemphigus and necrolytic migratory erythema as has been encountered in glucagonoma syndrome (Amon et al. 1976).

Discernible changes in hair usually occur after the dermatitis. The hair may be hypopigmented (Dupre et al. 1979) or acquire a reddish hue (Ginsburg et al. 1976), and it can be fine and brittle. Patchy loss of hair is a common feature and it may progress to extensive loss of scalp hair, eyebrows, eyelashes and body hair. These changes persist after zinc supplementation and hair may still be excessively shed for up to 6 months following institution of treatment.

Microscopic examination of hair shows mild trichonodosis in which the shaft has a variable calibre, transverse striae and short longitudinal splits. The tip of the hair may have a spindly appearance whilst the adjacent shaft may be bent and may acquire a swan-neck deformity (Dupre et al. 1979; Weismann and Hagdrup 1981). The varying calibre of the shaft has been attributed to fluctuations of zinc supply and the concomitant changes in hair protein formation resulting in alternating periods of rapid (narrow shaft) and slow (broad shaft) growth rates. This phenomenon and the striae are therefore non-specific indicators of growth arrest of the hair and as such both features are analogous to Beau's lines in the nails.

Growth Retardation and Reproduction

Failure to thrive is a prominent feature of children with AE (Brandt 1936; Danbolt and Closs 1942; Neldner and Hambidge 1975). Similarly, impaired weight gain has been noted in association with other causes of zinc deficiency

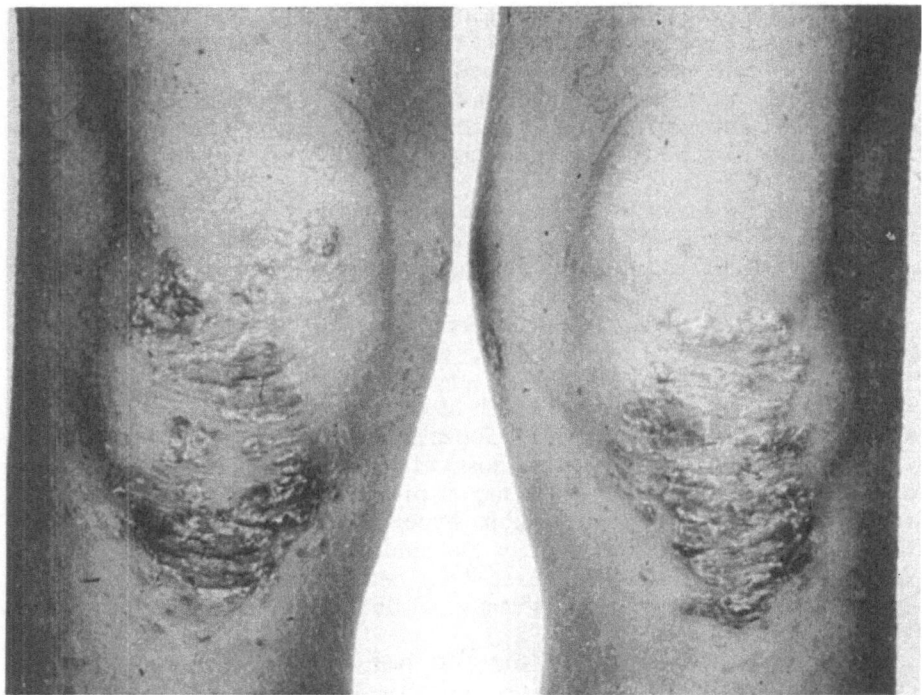

Fig. 17.3. Parakeratotic thickened skin lesions associated with protracted and mild severe symptomatic zinc deficiency.

in which zinc supplements induce catch-up growth and an anabolic response (Kay et al. 1976; Wolman et al. 1979; Golden and Golden 1981).

There are no clear data on the occurrence of hypogonadism in severe zinc deficiency; one 18-year-old youth with AE was described as being affected by this (Gartside and Allen 1975), and an untreated female did not experience menarche until 19 years of age (Graves et al. 1980). Further evidence that at least some untreated patients reach sexual maturity is provided by the pregnancies which have been reported in women with AE.

Immunological Features

The immune defects which accompany zinc deficiency correspond to the role of zinc in cell-mediated and humoral immunity. Thymic hypoplasia (Rodin and Goldman 1969; Julius et al. 1973) has been noted in acrodermatitis enteropathica and in association with protein–energy malnutrition (Golden et al. 1976). Impaired cutaneous responses to mitogens (Golden et al. 1978; Chandra 1980), and in vitro lymphoblast responses have been noted in patients (Chandra 1980) as have zinc-responsive defects in monocyte and neutrophil chemotaxis (Weston et al. 1977).

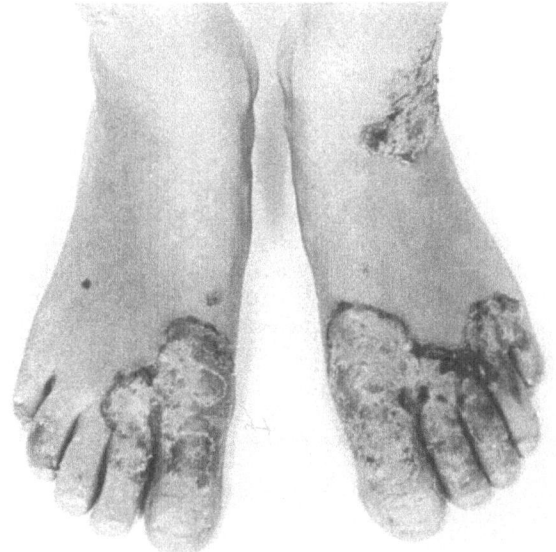

a

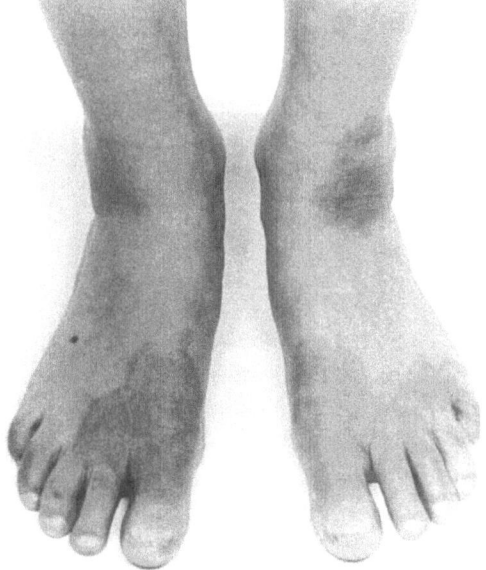

b

Fig. 17.4a, b. Parakeratotic skin lesions of both feet and the left ankle before (a) and after (b) zinc therapy in a patient with acrodermatitis enteropathica. The dermatitis (b) has healed leaving some residual pigmentation of he skin.

Clinical Situations in which Severe Zinc Deficiency has been Encountered

Table 17.1 lists conditions which induce a risk of zinc deficiency. Although there are numerous reports linking these conditions with zinc deficiency, the bases for diagnosing the deficiency state have normally been reduced plasma or serum zinc concentrations. The difficulty of interpreting these and related criteria is discussed elsewhere in this volume, and it is noteworthy that in comparison with the quantity of these reports, severe symptomatic zinc deficiency is a relatively infrequent occurrence.

Table 17.1.   Conditions which could predispose to zinc deficiency in man

1. *Inadequate dietary intake and absorbability of zinc*
   Protein–calorie malnutrition
   Vegetarianism
   Restricted protein diets
   Synthetic diets, e.g. for management of inborn errors of metabolism or
       malabsorption states
   Commercial preparations to replace meat
   Low socioeconomic groups (elderly)

2. *Maldigestion and malabsorption of zinc*
   Immaturity of absorptive mechanisms
   Inborn errors of absorption: acrodermatitis enteropathica
   Coeliac disease and other enteropathies
   Chronic inflammatory bowel disease
   Intestinal resection
   Exocrine pancreatic insufficiency
   Hepatic disease

3. *Increased zinc losses*
   Starvation, burns, diabetes mellitus
   Diuretic therapy, proteinuria, hepatic disease
   Intravascular haemolysis, porphyria
   Chelating agent therapy
   Chronic blood loss, dialysis
   Exfoliative dermatitis, excessive sweating
   Protein-losing enteropathies

4. *Intravenous feeding*

Acrodermatitis Enteropathica

This rare condition is inherited in an autosomal recessive manner. Parents are consanguineous in about 10% of reported cases and the sexes are affected equally; on the basis of the more accurate case reports, a sibship segregation ratio of approximately 3 : 1 can be calculated. The condition was probably first described by Wende (1902). In 1936 Brandt described four cases whom he could not diagnose confidently. Danbolt and Closs (1942) deduced that these four cases and two similar ones of their own constituted a distinct entity to which they gave the descriptive name of acrodermatitis enteropathica. The untreated disease usually followed a progressive course with an unpredictable pattern of remissions and relapses in which, however, relapses became more prolonged and normally culminated during childhood with cachexia and death from an overwhelming infection such as bronchopneumonia. However some patients survive untreated to adulthood (Olholm-Larsen 1978; Graves et al. 1980; Bronson et al. 1983).

Brandt (1936) noted that the features of the disease did not appear in infants until they started solids or discontinued breast feeding completely. Thus onset is earlier in formula-fed infants than in breast-fed babies. Furthermore he noted that reintroducing breast milk induced a clinical remission. The breast milk need not be fresh; boiled, frozen and thawed bank breast milk may be as effective (Entwistle 1965). In 1953 Dillaha et al. confirmed the unpublished observation of Elias H. Schlomovitz that 5,7-diiodo-8-hydroxyquinoline (DIH) could induce a complete clinical remission in these patients. Many similar 8-hydroxyquinoline-derivative chelating compounds have been similarly efficacious (Neldner and Hambidge 1975; Aggett et al. 1979a), and even penicillamine has been reported to induce a mild remission in AE (Javett 1963).

The role of zinc in AE was appreciated when Moynahan and Barnes (1973) described a girl being fed a lactose-free milk preparation who was unresponsive to DIH therapy. She made a remarkable recovery after she had been given zinc supplements to correct a suspected dietary zinc deficiency. This fortuitous observation was rapidly confirmed by Moynahan (1974) who also realized the similarity of the clinical features of AE to those of zinc-deficient animal models and of Friesian cattle with the probably homologous A46 trait (Brummerstedt et al. 1977).

The basic defect in AE has not been identified but it is probably related to impaired intestinal uptake and transfer of zinc. In young patients with AE, intestinal absorption as assessed by whole-body retention of orally administered ^{65}Zn is reduced (Lombeck et al. 1975; Weismann et al. 1979). Surprisingly, the latter study showed a comparable uptake of the radioisotope in healthy adults and adults with AE. This finding is at variance with metabolic studies of adults and children (Jackson 1977; Aggett et al. 1978) in which net intestinal loss of zinc occurs in untreated patients. However, the finding of Weismann et al. (1979) could be attributable to a relatively higher specific activity of the intestinal luminal zinc pool in patients with AE compared with that of healthy individuals; one could speculate that such a difference may arise from a homeostatic reduction of endogenous zinc secretion into their intestine by the zinc-deprived patients.

Hambidge et al. (1978) found a reduced zinc content in a small intestinal biopsy from an untreated patient and suggested that intestinal uptake of zinc in this condition may be abnormal. Studies in vitro have shown impaired accumulation of ^{65}Zn by proximal small intestinal biopsies from treated and untreated patients (Atherton et al. 1979).

Inferential evidence for a defect in the intestinal uptake of zinc can be derived from the therapeutic efficacy of chelating compounds and of a polyene antibiotic, amphotericin, which induced a resolution in a child who was on an inadequate oral zinc supplement (Aggett et al. 1981). This latter compound forms non-specific pores in membranes. These pores are permeable to divalent cations including zinc (Aggett et al. 1982). On the other hand DIH enables the transfer of zinc across lipid bilayers by acting as an ionophore (Aggett et al. 1979a). If these compounds are acting similarly in vivo then it could be construed that they are bypassing a defective mucosal mechanism for the uptake and transfer of zinc at customary luminal concentrations. The therapeutic effect of large oral doses of zinc in AE may be attributable to the creation of a sufficient intraluminal concentration to effect net intestinal transfer of zinc by a less specific high-capacity mechanism analogous to the system proposed for the effectiveness of magnesium supplements in the management of primary hypomagnesaemia (Milla et al. 1979).

The diagnosis of AE is essentially clinical. Monitoring plasma zinc concentrations and biochemical changes are adjuncts to observing the clinical response following zinc supplementation. In AE, plasma zinc concentrations may need to be interpreted cautiously. Gross symptomatic zinc deficiency can be present with normal plasma or serum zinc concentrations (Garretts and Molokhia 1977; Olholm-Larsen 1978). Additionally in AE, as has been seen in patients on TPN or with malnutrition (Golden and Golden 1981a), rapid growth, even if it is induced by the provision of zinc supplements, can result in marked depressions of plasma and serum zinc concentrations.

The delicate balance in which adult patients may be in respect of zinc deficiency is illustrated by the effect of pregnancy on the metabolism of zinc in adult women with AE. There have been 14 reported pregnancies in five women who had not been treated with zinc. One woman who was not treated until DIH was introduced at the end of the first trimester (i.e. after organogenesis) had an achondroplastic dwarf; she remained on treatment and her subsequent three pregnancies were normal (Epstein and Vedder 1960). Another patient who was already receiving DIH needed her dose increased because her symptoms were exacerbated during pregnancy; she gave birth to an infant weighing 2.52 kg. Although this could be regarded as light, the mother herself was small being only 1.51 m tall and weighing only 40 kg. Possibly her growth had been influenced by her disease and, in fact, her infant needed to be delivered by caesarean section because of cephalo-pelvic disproportion (Verburg et al. 1974).

Another woman being treated with DIH had a spontaneous abortion followed by an anencephalic stillbirth (Neldner and Hambidge 1975). Four other pregnancies occurred in a woman who was not diagnosed until 40 years of age and who had had no specific treatment at all. Her first two pregnancies proceeded to term with delivery of a normal boy on each occasion. However during each pregnancy her skin disease was grossly exacerbated and improved only after delivery. During her next two pregnancies her dermatitis was so severe that she had therapeutic abortions, each of which was followed by a prompt remission. This woman was eventually diagnosed when her longstanding depressive illness progressed to a parkinsonian syndrome (Olholm-Larsen 1978). The remaining three pregnancies occurred in a black woman who was not diagnosed until the end of her third pregnancy. She too had suffered a severe exacerbation of her dermatitis during her first pregnancy and this resolved promptly in the puerperium after she had given birth to a healthy boy. Her next pregnancy was terminated during the first trimester. In her subsequent pregnancy her disease deteriorated considerably until she was diagnosed in the 9th month. Zinc supplements were provided during the last 10 d of the pregnancy with immediate effect. She gave birth to a healthy boy weighing 3.06 kg (Bronson et al. 1983).

Some of these anecdotal reports illustrate that zinc deficiency may be teratogenic in human pregnancy, and that the metabolism of zinc is altered in pregnancy, but most fascinating is the clear demonstration that despite the disturbed zinc metabolism evidenced by the clinical deterioration during pregnancy, some of these pregnancies proceeded to a normal outcome.

Intravenous Feeding

Most patients on TPN who have become zinc deficient had conditions which predisposed them to zinc deficiency; predominant amongst these are protracted diarrhoea in infants and children, chronic inflammatory bowel disease and short bowel syndromes. Patients on TPN were often found to have a greatly increased zinc loss in the urine such as has been attributed to the adventitious release of zinc from catabolized tissues (Lindeman et al. 1972; Fell et al. 1973). Kay et al. (1976) noted that, while most patients lost in their urine 4–7 mg of zinc/d, one lost 22.6 mg of zinc daily. It is intriguing that zinc deficiency with TPN had not been noted earlier. This is probably because until the early 1970s protein (casein) hydrolysates which had a variable but high zinc content (about 4 mg/l) had been used, and could provide 4–5 mg of zinc daily. These products were replaced by crystalline amino acid compounds which have a lower zinc content (0.67–1.02 mg/l) (van Caillie et al. 1978). The native zinc content of one particular protein hydrolysate probably accounted for its therapeutic efficacy in treating a child with acrodermatitis enteropathica; it was retrospectively calculated that the product, which contained 3.7–4.2 mg of zinc/l, would have supplied the patient with approximately 4 mg of zinc/d (Frier and Jungreis 1976).

Two- to threefold variations have been noted in the native trace element content of many intravenous feeding products. Van Caillie et al. (1978) found that the zinc content of 5% dextrose/0.45% saline solutions varied between 4 and 56 μg/dl, whilst that for a glucose amino acid preparation varied between 67 and 102 μg/dl. Much of this variability may arise from contamination of the solution with zinc leached from the rubber stoppers of the bottles (van Caillie et al. 1978).

The urinary losses of zinc and plasma zinc concentrations are reduced during an anabolic response to TPN (Kay et al. 1976); similar changes in plasma zinc concentrations occur in patients transferring to oral feeding from TPN (Fleming et al. 1976), and in malnourished children during treatment (Golden and Golden 1981a). Although it would seem that in these circumstancs other metabolic events and nutrients (e.g. energy and protein) would be the primary determinants of the changes in zinc metabolism, Wolman et al. (1979) noted that increasing zinc supplementation to a level that achieved positive zinc balance induced an increased and positive nitrogen retention. Simultaneously, increased zinc supply was associated with increased serum insulin concentrations and glucose tolerance. Wolman et al. (1979) found that, although significant urinary losses of zinc occurred (about 2.0 mg/d), these were not as large as those observed by Kay et al. (1976), and that the major routes of zinc loss in people on TPN were via intestinal fluids, fistulas and diarrhoea. They determined that if patients were not losing more than 300 g of stool or small intestinal drainage daily, net zinc retention and optimization of nitrogen balance and glucose tolerance could be achieved by a daily zinc supplement of 3 mg. For other patients in whom there was significant loss of intestinal fluids they derived a formula in which the required zinc replacement (mg/d) is 2.0 + 17.1A + 12.2B, where A is the mass (g) of stool or ileostomy loss in patients with intact small bowel, B is the mass of small bowel fluid lost by a fistula stoma or enterostomies and 2.0 is an allowance for urinary loss.

The zinc requirements for preterm infants on TPN have been calculated on the basis of regression analysis of balance data to be 7.5 μmol (490 μg)/kg body weight/d (James et al. 1979).

Another possible cause of zincuria in TPN patients arises from the use of combined sugar and amino acid compounds. Freeman et al. (1975) reported that heat-sterilized mixtures of crystalline amino acids or casein hydrolysates with dextrose generated sugar amine compounds from an interaction between the free aldehyde group of the reducing sugar and the amino moiety of the amino acid. These compounds were detected in the infusate, serum and urine of patients and volunteers, and in the latter group were associated with up to a fourfold increase of urinary zinc loss [0.8 (± 0.24) increasing to 4.5 (± 0.34) mg of zinc daily]. The formation of these compounds is not entirely dependent on heat sterilization since they develop in glucose–amino acid mixtures kept at 4°C (Fry and Stegink 1982).

In contrast to parenteral nutrition, zinc deficiency in patients receiving enteral nutrition has not been extensively noted. Pekarek et al. (1979) recorded such an instance in a decerebrate youth who had been on long-term enteral feeding. Symptomatic zinc deficiency has also developed in Japanese children with lactose intolerance being treated with a lactose-free milk (Morishima et al. 1980). It is possible that a child with phenylketonuria who developed AE (Ermacora and Benelli 1968) may have had a deficiency secondary to an inadequate quality or quantity of zinc in its synthetic diet.

Breast-Fed Infants

Severe zinc deficiency has been noted both in preterm and term infants. The gestations of the preterm infants varied between 26 and 34 weeks, and their birthweights between 0.71 and 2.2 kg. Amongst reported cases, boys predominate and most infants presented at about 3 months of age. One report describes a similar occurrence in infants fed a cows' milk-based formula (Bonifazi et al. 1980), but the account is not extensive enough to enable all the cases to be assessed reliably.

The pathogenesis of zinc deficiency in these infants may arise from one or more of a number of factors including a preceding period of parenteral nutrition, impaired or immature intestinal absorption and homeostasis of zinc, increased requirement of zinc imposed by rapid growth and an inadequate intake of zinc from their mother's milk. The likelihood of the last possibility is emphasized by the uneventful histories of several children before presentation (Aggett et al. 1980; Blom et al. 1980; Zimmerman et al. 1982) in whom zinc intakes and the contents of zinc in maternal milk were low. Zimmerman et al.

(1982) re-studied one mother's milk when she was breast feeding a subsequent child born at term. They considered the milk to have a low zinc content and surmised that the mother had defective mammary secretion of zinc. They and others have found that zinc supplements given to such mothers do not increase the zinc content of the milk. However, in contrast, Murphy et al. (1985) reported a low zinc content in the milk of one woman with a zinc-deficient, breast-fed, preterm infant, and higher (normal) levels during a later lactation after a full-term pregnancy. Thus the low zinc content may in some women be a phenomenon of preterm milk.

Severe zinc deficiency has now been described in five breast-fed term infants (Bye et al. 1985; Kuramato et al. 1986; Roberts et al. 1987). These infants had no antecedent predisposing factors and they presented 3 to 5 months postnatally. Again the zinc content of the maternal breast milk was found to be low and zinc supplementation of the mother failed to increase the zinc content of the milk.

The onset of symptoms while being breast-fed counts against the diagnosis of AE in these children; furthermore they respond to smaller doses of zinc (5–10 mg elemental zinc/d) than would be expected in a child with AE. These children continue to thrive after the zinc supplements are withdrawn at about 10 months to 1 year of age when the child has become established on solid feeding. Occasionally this test may need to be repeated. In one child who developed zinc deficiency after prolonged intravenous feeding features of zinc deficiency recurred after the first attempt to withdraw supplements (Sivasubramanian and Henkin 1978).

"Protein–Energy Malnutrition"

Probably the most common cause of severe zinc deficiency in the world is "protein–energy malnutrition". The most extensive work on this has been performed in Jamaica (Golden and Golden 1979, 1981a). These workers have shown that zinc supplementation in malnourished children will reverse thymic atrophy as assessed by the size of the thymic silhouette on chest x-ray (Golden et al. 1976) and topical zinc improves the hypersensitivity reaction to some mitogens (Golden et al. 1978). Further studies by the Goldens have shown that similar application of zinc will also induce the healing of cutaneous ulcers in children recovering from protein–energy malnutrition, and zinc has been found to be a limiting nutrient affecting the rate of weight gain in children being renourished (Golden and Golden 1981a, b). This topic is discussed in more detail in Chaps. 8, 9, 18 and 20.

Gastrointestinal and Hepatic Disease

McClain et al. (1980) described two young men with longstanding active Crohn's disease who developed symptomatic severe zinc deficiency. Impaired intestinal absorption of zinc in other similar patients has been inferred from data from an oral zinc tolerance test (McClain et al. 1980) and from measurement of the gastrointestinal transfer of the radioisotope ^{69}Zn (Sturniolo et al. 1980). One patient with Crohn's disease was reported to have had abnormal dark adaptation which responded to zinc supplements (McClain et al. 1983).

In patients who have had extensive resections of the small intestine the absorption of zinc may be expected to be compromised (Andersson et al. 1976). Hessov et al. (1983) have shown that such patients have a reduced zinc content in muscle, but the clinical significance of such findings is not yet apparent. Gross zinc deficiency in such patients has only been demonstrated clearly in one case (Weismann et al. 1978), and less conclusively in one other (Smith 1977).

The risk that patients with alcoholic pancreatitis on TPN may develop zinc deficiency has been highlighted (Williams et al. 1979). Although abnormal data from "zinc tolerance tests" of absorption in adults (Boosalis et al. 1983) and negative zinc balances in children with exocrine pancreatic insufficiency (Aggett et al. 1979b) suggest the existence of defects in zinc absorption, severe zinc deficiency is rare in this disorder. Dodge and Yassa (1978) have described an adolescent male with cystic fibrosis and growth retardation which was responsive to zinc, and Hansen et al. (1983) have reported a patient who developed an AE-like syndrome.

In the light of the pivotal role that the liver plays in the metabolism of zinc it is unsurprising that metabolism of zinc is disturbed extensively and variably by hepatic disease. Dermatological manifestations of severe zinc deficiency have been reported in patients with alcoholic cirrhosis (Weismann et al. 1980; Gaveau et al. 1987). Gaveau et al. (1987) investigated 33 patients with alcoholic cirrhosis. Of these, 21 had skin lesions which in nine were cured by oral zinc replacement therapy alone. Only did the symptoms in one of these nine patients resemble classic AE. The other patients had dry eczematous lesions on the extensor surfaces and in the perianal and genital regions, with accompanying glossitis, stomatitis, chelitis and, in some, alopecia. All these lesions and accompanying behavioural disturbances responsed to zinc supplementation. The plasma zinc concentrations were depressed in all the patients but were most markedly reduced in those with the zinc-responsive dermatitis. However, zinc concentrations in erythrocytes and hair from all the patients were similar to those of a healthy reference population.

The mechanisms for loss of zinc in liver disease are unclear and since cirrhosis is such a heterogeneous entity it is not surprising that several possible aetiologies have emerged. It has been proposed that some patients have a reduced dietary intake of zinc (McClain et al. 1979). Information on zinc absorption by cirrhotics varies. Valberg et al. (1985), using a faecal monitoring technique, found normal intestinal absorption of zinc in a group of cirrhotics, whereas Mills et al. (1983), using a whole-body monitoring technique to detect the retention of ^{65}Zn 7 d after its ingestion, found evidence of increased zinc absorption in patients. Similarly Milman et al. (1983) noted increased zinc absorption in alcoholics with compensated cirrhosis, some of whom were on diuretics which increase zincuria (Wester 1980), and who were still consuming at least 50 g of alcohol per day. The sizes of the test doses used in the previous two studies were more physiological than that used in a study by Dinsmore et al. (1985), who found reduced plasma zinc tolerance curves in uncompensated cirrhotics.

The liver content of zinc is reduced in alcoholic cirrhosis even after allowance has been made for the fibrosis which occurs in alcoholic liver disease (Keeling et al. 1980; Kiilerich et al. 1980; Valberg et al. 1985). Mills et al. (1983), by relating the amount of zinc in liver tissue in proportion to that of magnesium as an intracellular marker, confirmed a reduced hepatocytic zinc content. The reason for this is unclear. It is not known if the hepatocyte in alcoholic liver disease loses zinc excessively, but Keeling et al. (1981) have shown impaired hepatointestinal clearance of intravenously administered zinc in cirrhotics, thus suggesting that hepatocytic uptake of zinc may be defective.

Functional manifestations of zinc deficiency in chronic liver disease include zinc-responsive abnormalities in dark adaptation (Morrison et al. 1978; Keeling et al. 1980). Sullivan et al. (1979) noted hyperammonaemia, hyperbilerumin-aemia and raised alkaline phosphatase activities which normalized following zinc supplementation, but were justifiably cautious in interpreting their data because they were not derived from a randomized or blind controlled study. Recently oral zinc supplementation has been reported to improve features of hepatic encephalopathy (Reding et al. 1984). These patients had no other features of zinc depletion and it is uncertain whether or not this therapeutic effect is indicative of a preceding zinc deficiency.

Anorexia Nervosa

There is no evidence that zinc deficiency has any prominent role in anorexia nervosa (Ainley et al. 1986). However, it is possible that some patients with severe self-induced malnutrition may develop some zinc-responsive features, and two case reports describe the possible benefits of zinc therapy in anoretics who had Kwashiorkor-like syndromes with a skin rash akin to that of zinc deficiency (Esca et al. 1979; Grillet and Harms 1980).

Chelating Agents and Drugs

The clinical use of chelating agents such as penicillamine in the treatment of Wilson's disease (Klingberg et al. 1976) and of di-ethylene triamine penta-acetate (DTPA) in the treatment of the iron overload of thalassaemia major (Ridley 1982) has precipitated severe zinc deficiency. Anti-convulsant drugs, in particular sodium valproate, have also been implicated in causing zinc deficiency (Lewis-Jones et al. 1985).

Deranged Zinc Metabolism in Renal Disease

Severe symptomatic zinc deficiency has not been reported in renal disease but there is evidence of disturbed metabolism of zinc in renal failure. Most patients have hypozincaemia irrespective of their treatment, but there are conflicting reports of raised, normal or reduced erythrocytic zinc content in patients with chronic renal failure, both with and without dialysis treatment.

Prolonged zinc therapy has caused an increase in zinc content of leucocytes, hair and plasma. The presence of altered zinc metabolism was inferred more strongly by alterations in functional parameters such as a decrease in plasma ammonium concentrations and a fall in plasma ribonuclease activity (Mahajan et al. 1979, 1982a). This same group of workers has observed some evidence of improved gonadal function in uraemic men who have been given zinc supplements (Mahajan 1982b), but this is not an invariable response (Brook et al. 1980).

In the light of some evidence of zinc-responsive features in patients with renal failure it is paradoxical that an autopsy study found increased zinc concentrations in some tissues of patients with uraemia (Smythe et al. 1982). In comparison with reference data, a small reduction of muscle zinc in patients was accompanied by increased zinc content in the heart, liver and spleen irrespective of whether they had been dialysed or not. The zinc content of bone was only measured in patients who had been dialysed, but this too was increased. Thus there would appear to occur in chronic renal failure a systemic recompartmentalization of zinc. The occurrence of this phenomenon needs to be confirmed but it raises the intriguing possibility that there may occur in conditions of chronic stress in humans a situation in which zinc-responsive symptoms may develop in the presence of a normal or increased body content of the element.

Treatment of Zinc Deficiency

Both oral and intravenous preparations of zinc salts are available. Zinc sulphate heptahydrate [50 mg (0.77 mmol) of elemental zinc in 220 mg] is used commonly but other salts such as the gluconate [13.4 mg (205 μmol) of zinc in 100 mg] and zinc acetate [30 mg (460 μmol) of zinc in 100 mg] can be used. Some effervescent preparations and the latter salts are more palatable than zinc sulphate. Another means of improving the acceptability of zinc solutions and disguising their metallic taste is to prepare them in ice-cold fruit-flavoured solutions or, usually for children, as lollipops.

The dose should be tailored to match the patient's clinical response. Daily doses of between 35 mg and 100 mg of elemental zinc are usually adequate for the management of AE. Smaller doses are usually effective in the treatment of severe symptomatic zinc deficiency secondary to the other causes outlined above. Relatively larger doses are needed during accelerated growth and immediately following initiation of treatment. The use of zinc therapy in other conditions is discussed elsewhere in this volume.

References

Aggett PJ, Atherton DJ, Delves HT et al. (1978) Studies in acrodermatitis enteropathica. In: Kirschgessner M (ed) Trace element metabolism in man and animals. Arbeitskreis für Tierernährungsforschung, Weihenstephan, pp 418–422
Aggett PJ, Thorn JM, Delves HT, Clayton BE, Harries JT (1979a) Trace element malabsorption in exocrine pancreatic insufficiency. Monogr Paediatr 19: 8–11

Aggett PJ, Delves HT, Harries JT, Bangham AD (1979b) The possible role of diodoquin as a zinc ionophore in the treatment of acrodermatitis enteropathica. Biochem Biophys Res Commun 87: 513–517

Aggett PJ, Atherton DJ, More J, Davey J, Delves HT, Harries JT (1980) Symptomatic zinc deficiency in a breast fed preterm infant. Arch Dis Child 58: 547–550

Aggett PJ, Delves HT, Thorn JM, Atherton DJ, Harries JT, Bangham AD (1981) The therapeutic effect of amphotericin in acrodermatitis enteropathica. Eur J Pediatr 137: 23–25

Aggett PJ, Fenwick PK, Kirk H (1982) The effect of amphotericin B on the permeability of lipid bilayers to divalent trace metals. Acta Biochim Biophys 684: 291–294

Ainley CC, Casson J, Carlsson L, Thompson RPH, Slavin BM, Naughton KR (1986) Zinc state in anorexia nervosa. Br Med J 293: 992–993

Ament ME, Broviac J (1973) Acrodermatitis enteropathica: demonstration of small and large intestinal mucosal lesions. Gastroenterology 64: A-9/692

Amon RB, Swenson KH, Hanifen JM, Hambidge KM (1976) The glucogonoma syndrome (necrolytic migratory erythema) and zinc. N Engl J Med 295: 962

Andersson KE, Bratt L, Denker H, Lanner E (1976) Some aspects of the intestinal absorption of zinc in man. Eur J Clin Pharmacol 9: 423–428

Aquilera-Diaz LF (1971) Un nouveau symptome dans l'acrodermatite enteropathique: la démarche ataxique. Bull Soc Fr Dermatol Syphilgr 78: 259–260

Arakawa T, Tamura T, Igarashi Y, Suzuki H, Sandstead HH (1976) Zinc deficiency in two infants during parenteral alimentation. Am J Clin Nutr 29: 197–204

Atherton DJ, Muller DPR, Aggett PJ, Harries JT (1979) A defect in zinc uptake by jejunal biopsies in acrodermatitis enteropathica. Clin Sci 56: 505–507

Blom I, Jameson S, Crook F, Larsson-stymne B, Wranne L (1980) Zinc deficiency with transitory acrodermatitis enteropathica in a boy of low birthweight. Br J Dermatol 104: 459–464

Bohane TD, Cutz E, Hamilton JR, Gall DG (1977) Acrodermatitis enteropathica, zinc, and a paneth cell. Gastroenterology 73: 587–592

Bonifazi E, Rigillo N, De Simone B, Meneghini CL (1980) Acquired dermatitis due to zinc deficiency in a premature infant. Acta Derm Venereol 60: 449–451

Boosalis MG, Evans GW, McClain CJ (1983) Impaired handling of orally administered zinc in pancreatic insufficiency. Am J Clin Nutr 37: 268–271

Brandt T (1936) Dermatitis in children with disturbances of the general condition and the absorption of food elements. Acta Derm Venereol 17: 513–546

Braun OH, Heilmann K, Rossner JA, Pauli W, Bergman KE (1977) Acrodermatitis enteropathica, zinc deficiency and ultrastructural findings. Eur J Pediatr 125: 153–161

Bronson DM, Barksky R, Barksky S (1983) Acrodermatitis enteropathica: recognition at long last during a pregnancy. J Am Acad Dermatol 9: 140–144

Brook AC, Johnston DG, Ward MK, Watson MJ, Cook DB, Kerr DNS (1980) Absence of a therapeutic effect of zinc in the sexual dysfunction of haemodialysis patients. Lancet II: 618–619

Brummerstedt E, Bass EA, Flagstead T, Andresen E (1977) Acrodermatitis enteropathica, zinc metabolism: animal model: lethal trait A46 in cattle. Am J Pathol 87: 725–728

Bye AM, Goodfellow A, Atherton DJ (1985) Transient zinc deficiency in a full term breast fed infant of normal birthweight. Pediatr Dermatol 2: 308–311

Chandra RK (1980) Acrodermatitis enteropathica: zinc levels and cell-mediated immunity. Pediatrics 66: 789–791

Danbolt N, Closs K (1942) Acrodermatitis enteropathica. Acta Derm Venereol 23: 127–169

Dillaha CJ, Larencz AL, Aavic OR (1953) Acrodermatitis enteropathica: review of the literature and a report of a case successfully treated with diodoquin. J Am Med Ass 152: 509–512

Dinsmore W, Callander ME, McMaster D, Todd SJ, Love AHG (1985) Zinc absorption in alcoholics using zinc-65. Digestion 32: 238–242

Dodge JA, Yassa JG (1978) Zinc deficiency syndrome in the British youth with cystic fibrosis. Br Med J 1: 411

Dupre A, Bonafe JL, Carriere JP (1979) The hair in acrodermatitis enteropathica – a disease indicator. Acta Derm Venereol 59: 177–178

Entwistle BR (1965) Acrodermatitis enteropathica. Aust J Dermatol 8: 13–21

Epstein S, Vedder JS (1960) Acrodermatitis enteropathica persisting into adulthood. Arch Dermatol Syph 82: 189–190

Ermacora E, Benelli MG (1968) Acrodermatite enteropathica in bambino fenil-chetonurico. Min Derm 41: 523–524

Esca SA, Brenner W, Mach K, Gschnait F (1979) Kwashiorkor like zinc deficiency syndrome in anorexia nervosa. Acta Derm Venereol 59: 361–364

Feldberg F, Yassur Y, Ben-Sera I, Varsano I, Zelikovitz I (1981) Keratomalacia in acrodermatitis enteropathica. Metab Pediatr Syst Ophthalmol 5: 207–211

Fell GS, Cuthbertson DP, Morrison C et al. (1973) Urinary zinc levels as an indication of muscle catabolism. Lancet I: 280–282

Fleming CR, Hodges RE, Hurley LS (1976) A prospective study of serum copper and zinc levels in patients receiving total parenteral nutrition. Am J Clin Nutr 29: 70–77

Freeman JB, Stegink LD, May PD, Fry LK, Denbesten L (1975) Excessive urinary zinc losses during parenteral alimentation. J Surg Res 18: 463–469

Frier S, Jungreis E (1976) Zinc and acrodermatitis enteropathica. Lancet I: 914–915

Fry LK, Stegink LD (1982) Formation of Maillard reaction products in parenteral alimentation solutions. J Nutr 112: 1631–1637

Garretts M, Molokhia M (1977) Acrodermatitis enteropathica without hypo-zincaemia. J Pediatr 91: 492–494

Gartside JM, Allen BR (1975) Treatment of acrodermatitis enteropathica with zinc sulphate. Br Med J 2: 521–522

Gaveau D, Piette F, Cortot A, Dumur V, Vergoend H (1987) Manifestations cutanées du déficit en zinc dans la cirrhose ethylique. Ann Dermatol Venereol 114: 39–53

Ginsburg R, Robertson A, Michel B (1976) Acrodermatitis enteropathica; abnormalities of fat metabolism and integumental ultrastructures in infants. Arch Dermatol 112: 653–660

Golden BE, Golden MHN (1979) Plasma zinc and the clinical features of malnutrition. Am J Clin Nutr 32: 2490–2494

Golden BE, Golden MHN (1981a) Plasma zinc, rate of weight gain, and the energy cost of tissue deposition in children recovering from severe malnutrition in a cow's milk or soya protein based diet. Am J Clin Nutr 34: 892–899

Golden BE, Golden MHN (1981b) Dietary zinc and the output and composition of faeces. In: Howell JMcC, Gawthorne JM, White CL (eds) Trace element metabolism in man and animals – TEMA-4. Australian Academy of Sciences, Canberra, pp 73–76

Golden BE, Golden MHN (1985) Zinc, sodium and potassium loss in the diarrhoeas of malnutrition and zinc deficiency. In: Mills CF, Bremner I, Chesters JK (eds) Trace elements in man and animals – TEMA-5. CAB International, Farnham Royal, UK, pp 228–231

Golden MHN, Golden BE (1981) Effect of zinc supplementation on the dietary intake, rate of weight gain, and energy cost of tissue deposition in children recovering from severe malnutrition. Am J Clin Med 34: 900–908

Golden MHN, Jackson AA, Golden BE (1976) Effect of zinc on thymus of recently malnourished children. Lancet II: 1057–1059

Golden MHN, Golden BE, Harland PSEG, Jackson AA (1978) Zinc and immuno-competence in protein-energy malnutrition. Lancet I: 1226–1228

Gonzalez JR, Botet MV, Sanchez JL (1982) The histopathology of acrodermatitis enteropathica. Am J Dermatol Pathol 4: 303–311

Graves K, Kestenbaum T, Kalivas J (1980) Hereditary acrodermatitis enteropathica in an adult. Arch Dermatol 116: 562–564

Grillet JP, Harms M (1980) Kwashiorkor et déficit en zinc chez une anorexique mentale adulte. Ann Dermatol Venereol 107: 1187–1191

Hambidge KM, Neldner KH, Walravens PA et al. (1978) Zinc and acrodermatitis enteropathica. In: Zinc and copper in clinical medicine. Spectrum Publications, Holliswood, New York, pp 81–98

Hansen RC, Lennen R, Revsin R (1983) Cystic fibrosis manifesting with acrodermatitis enteropathica-like eruption. Arch Dermatol 119: 51–55

Henkin RI, Patten BM, Re PK, Bronzert DA (1975) A syndrome of acute zinc loss: cerebellar dysfunction, mental changes, anorexia, and taste and smell dysfunction. Arch Neurol 32: 745–751

Hessov I, Hasselblad C, Fasth C, Hulten N (1983) Zinc depletion after small bowel resections or Crohn's disease. Hum Nutr Clin Nutr 37C: 353–359

Jackson MJ (1977) Zinc and di-iodohydroxyquinoline therapy in acrodermatitis enteropathica. J Clin Pathol 30: 284–287

James BE, Hendry PG, MacMahon RA (1979) Total parenteral nutrition of premature infants. II. Requirement for micronutrient elements. Aust Paediatr J 15: 67–71

Javett SN (1963) Regrowth of hair during penicillamine treatment of acrodermatitis enteropathica. Lancet I: 504

Jones JG, Elmes ME, Aggett PJ, Harries JT (1983) The effect of zinc therapy on lysosomal inclusion bodies in intestinal epithelial cells in acrodermatitis enteropathica. Pediatr Res 17: 354–357

Julius R, Schulking M, Sprinkle T, Rennert O (1973) Acrodermatitis enteropathica with immune deficiency. J Pediatr 83: 1007–1011

Kay RG, Tasman-Jones C, Pybus J, Whiting R, Black H (1976) A syndrome of acute zinc deficiency during total parenteral alimentation in man. Ann Surg 183: 331–340

Keeling PWN, Jones RB, Hilton PJ, Thompson RPH (1980) Reduced leucocyte zinc in liver disease. Gut 21: 561–564

Keeling PWN, Rus W, Boll A, Hannigan B, Thompson RPH (1981) Direct measurement of hepatointestinal extraction of zinc in cirrhosis and hepatitis. Clin Sci 61: 441–444

Keeling PWN, O'Day J, Rus W, Thompson RPH (1982) Zinc deficiency and photo receptor dysfunction in chronic liver disease. Clin Sci 62: 109–111

Kelly R, Davidson GP, Rugely R, Townley W, Campbell PE (1976) Reversible intestinal mucosal abnormalities in acrodermatitis enteropathica. Arch Dis Child 51: 219–222

Kiilerich S, Dietrichson O, Loud FB et al. (1980) Zinc depletion in alcoholic liver diseases. Scand J Gastroenterol 15: 363–367

Klingberg WG, Prasad AS, Oberleas D (1976) Zinc deficiency following penicillamine therapy. In: Prasad AS, Oberleas D (eds) Trace elements in human health and disease, vol 1. Academic Press, New York, pp 51–65

Koo SI, Turk DE (1977) Effect of zinc deficiency on the ultrastructure of the pancreatic acinar cell and intestinal epithelium in the rat. J Nutr 107: 896–908

Kuramato Y, Igarashi Y, Kato S, Tagami H (1986) Acquired zinc deficiency in two breast fed mature infants. Acta Derm Venereol 56: 359–361

Leopold IH (1978) Zinc deficiency and visual impairment? Am J Ophthalmol 85: 871–873

Lewis-Jones MS, Evans S, Culshaw MA (1985) Cutaneous manifestations of zinc deficiency during treatment with anticonvulsants. Br Med J 290: 603–604

Lindeman RD, Bottomley RG, Cornelison L, Jacobs LA (1972) Influence of acute tissue injury on zinc metabolism in man. J Lab Clin Med 79: 452–460

Lombeck I, von Bassenitz DB, Becker K, Tinschmann P, Kastner H (1974) Ultra-structural findings in acrodermatitis enteropathica. Pediatr Res 8: 82–88

Lombeck I, Schnippering HG, Kasperek K et al. (1975) Akrodermatitis enteropathica – Eine Zinkstoffwechselstörung mit Zinkmalabsorption. Z Kinderheilk 120: 181–189

McClain CJ, Van Thiel DH, Parker S, Badzin LK, Gilbert H (1979) Alterations in zinc, vitamin A, and retinal-binding protein in chronic alcoholics: a possible mechanism for night blindness and hypogonadism. Alcoholism 3: 135–141

McClain CJ, Soutar C, Zieve L (1980) Zinc deficiency: a complication of Crohn's disease. Gastroenterology 78: 272–279

McClain CJ, Le-Chu S, Gilbert H, Cameron D (1983) Zinc deficiency induced retinal dysfunction in Crohn's disease. Dig Dis Sci 28: 85–87

Mahajan SK, Prasad AS, Rabbani P, Briggs WA, McDonald RD (1979) Zinc metabolism in uraemia. J Lab Clin Med 94: 693–698

Mahajan SK, Prasad AS, Rabbani P, Briggs WA, McDonald FD (1982a) Zinc deficiency: a reversible complication of uremia. Am J Clin Nutr 36: 1177–1183

Mahajan SK, Abbasi AA, Prasad AS, Rabbani P, Briggs WA, McDonald FD (1982b) Effect of oral zinc therapy on gonadal function in hemodialysis patients: a double blind study. Ann Intern Med 97: 357–361

Mata CS, Felker GV, Ide CH (1975) Eye manifestations in acrodermatitis enteropathica. Arch Ophthalmol 93: 140–142

Milla PJ, Moynahan EJ (1972) Acrodermatitis enteropathica with lactose intolerance. Proc R Soc Med 65: 600–601

Milla PJ, Aggett PJ, Wolff OH, Harries JT (1979) Studies in primary hypomagnesaemia: evidence for defective carrier mediated small intestinal transport of magnesium. Gut 20: 1028–1033

Mills PR, Fell GS, Bessant RG, Nelson OM, Russell RI (1983) A study of zinc metabolism in alcoholic cirrhosis. Clin Sci 64: 527–535

Milman N, Hvid-Jacobsen K, Hegnhoj J, Sorenson SS (1983) Zinc absorption in patients with compensated alcoholic cirrhosis. Scand J Gastroenterol 18: 871–875

Morishima T, Yagi S, Kuwabara A, Endo M, Takemura T (1980) An acquired form of acrodermatitis enteropathica due to long term lactose free milk alimentation. J Dermatol 7: 121–125

Morrison SA, Russell RM, Carney EA, Oaks EV (1978) Zinc deficiency: a case of abnormal dark adaptation in cirrhosis. Am J Clin Nutr 31: 276–281

Moynahan EJ (1974) Acrodermatitis enteropathica: a lethal inherited human zinc deficiency disorder. Lancet II: 399–400

Moynahan EJ (1976) Zinc deficiency and disturbances of mood and behaviour. Lancet I: 91

Moynahan EJ, Barnes PM (1973) Zinc deficiency and a synthetic diet for lactose intolerance. Lancet I: 676–677

Moynahan EJ, Johnson FR, McMinn RMH (1963) Acrodermatitis enteropathica: demonstration of possible enzyme defect. Proc R Soc Med 56: 300

Murphy JF, Gray OP, Randall JR, Hann S (1985) Zinc deficiency: a problem with preterm breast milk. Early Hum Dev 10: 303–307

Neldner KH, Hambidge KM (1975) Zinc therapy of acrodermatitis enteropathica. N Engl J Med 292: 879–882

Olholm-Larsen P (1978) Untreated acrodermatitis enteropathica in adults. Dermatologica 156: 155–166

Otto HF, Weitz H (1972) Elektronenmikroskopische Untersuchungen an Paneth-Zellen der Ratte unter zinkaner Diät. Beitr Pathol 145: 336–342

Pekarek RS, Sandstead HH, Jacob RA, Barcome DF (1979) Abnormal cellular immune responses during acquired zinc deficiency. Am J Clin Nutr 32: 1466–1471

Portnoy B, Marsden CW (1961) Acrodermatitis enteropathica without diarrhoea. Arch Dermatol 83: 420–424

Racz P, Kovacs B, Varga L, Ujlaki E, Zombai E, Karbuczky S (1978) Bilateral cataract in acrodermatitis enteropathica. J Pediatr Ophthalmol Strabismus 16: 180–182

Reding P, Duchateau J, Bataille C (1984) Oral zinc supplementation improves hepatic encephalopathy. Lancet II: 493–495

Ridley CM (1982) Zinc deficiency developing in treatment for thalassaemia. J R Soc Med 75: 38–39

Roberts LJ, Shadwick CF, Bergstresser PR (1987) Zinc deficiency in two full term breast fed infants. J Am Acad Dermatol 16: 301–304

Rodin AE, Goldman AS (1969) Autopsy findings in acrodermatitis enteropathica. Am J Clin Pathol 51: 315–322

Sivasubramanian KN, Henkin RI (1978) Behavioural and dermatologic changes and a low serum zinc and copper concentrations in two premature infants after parenteral alimentation. J Pediatr 93: 847–851

Smith SZ (1977) Skin changes in short-bowel syndrome. Kwashiorkor-like syndrome. Arch Dermatol 113: 657–659

Smythe WR, Alfrey AC, Craswell PW et al. (1982) Trace elements abnormalities in chronic uraemia. Ann Intern Med 96: 302–310

Sturniolo GC, Molokhia MM, Shields R, Turnberg LA (1980) Zinc absorption in Crohn's disease. Gut 33: 387–391

Sturtevant FM (1980) Zinc deficiency, acrodermatitis enteropathica, optic atrophy, subacute myelo-optic neuropathy, and 5,7-Dihalo-8-quinolols. Pediatrics 65: 610–613

Sullivan JF, Williams RV, Burch RE (1979) The metabolism of zinc and selenium in cirrhotic patients during six weeks of zinc ingestion. Alcoholism 3: 235–239

Tompkins RR, Livingood CS (1969) Acrodermatitis enteropathica persisting into adulthood. Arch Dermatol 99: 190–195

Valberg LS, Flannagan PR, Jhant CN, Chamberlain MJ (1985) Zinc absorption and leukocyte zinc in alcoholic and non-alcoholic cirrhosis. Dig Dis Sci 30: 329–333

van Caillie M, Degenhart H, Luigendijk I, Fernandes J (1978) Zinc content of intravenous solutions. Lancet II: 200–201

Verburg DJ, Burd LI, Hoxtell EO, Merrill LK (1974) Acrodermatitis enteropathica and pregnancy. Obstet Gynecol 44: 223–227

Warth JA, Prasad AS, Zwas F, Frank RN (1981) Abnormal dark adaptation in sickle cell anaemia. J Lab Clin Med 98: 189–194

Weismann K (1977) Lines of Beau: possible markers of zinc deficiency. Acta Derm Venereol 57: 88–89

Weismann K, Hagdrup HK (1981) Hair changes due to zinc deficiency in a case of sucrose malabsorption. Acta Derm Venereol 61: 444–447

Weismann K, Wadskov S, Nikkelsen HI, Knudsen L, Khristensen KC, Sgorgaard L (1978) Acquired zinc deficiency dermatosis in man. Arch Dermatol 114: 1509–1511

Weismann K, Hoe S, Knudsen L, Sorensen SS (1979) ^{65}Zn absorption in patients suffering from acrodermatitis enteropathica and in normal adults assessed by whole-body counting technique. Br J Dermatol 101: 573–579

Weismann K, Hoyer H, Christensen E (1980) Acquired zinc deficiency in alcoholic liver cirrhosis: report of two cases. Acta Derm Venereol 60: 447–449

Wende GW (1902) Epidermolysis bullosa hereditaria. J Cutan Dis 20: 532–547

Wester PO (1980) Urinary zinc excretion during treatment with different diuretics. Acta Med Scand 280: 209–212

Weston WL, Clark-Huff J, Humbert JR, Hambidge KM, Neldner KH, Walravens TA (1977) Zinc correction of a defective chemotaxis in acrodermatitis enteropathica. Arch Dermatol 113: 422–425

Williams RB, Russell RM, Dutta SK (1979) Alcoholic pancreatitis; patients at high risk of acute zinc deficiency. Am J Med 66: 889

Wirsching L (1962) Eye symptoms in acrodermatitis enteropathica. Acta Ophthalmol 40: 567–574

Wolman SL, Anderson GH, Marliss EB, Jeejeebhoy KN (1979) Zinc in total parenteral nutrition: requirements and metabolic effects. Gastroenterology 76: 458–467

Zimmerman AW, Hambidge KM, Lapow ML, Greenberg RD, Stover ML, Casey CE (1982) Acrodermatitis in breast-fed premature infants: evidence for a defect of mammary zinc secretion. Pediatrics 59: 176–183

Chapter 18

Mild Zinc Deficiency in Human Subjects

K.M. Hambidge

Introduction

The controversies and uncertainties associated with "mild" zinc deficiency in human subjects are as evident today as at any time in the 30-year interval since the possibility that alcoholic cirrhosis may be associated with a secondary zinc deficiency state was first suggested (Vallee et al. 1957). Incidentally, continuing uncertainties about the interrelationship between alcoholic cirrhosis and zinc nutritional status can serve very effectively to illustrate the extraordinary difficulties still encountered by investigators in this field. That profound alterations in zinc metabolism occur in alcoholic liver disease and other hepatic disorders (Hambidge et al. 1987) is not in question. However, whether or not these changes in metabolism are associated with, or lead to, impaired zinc nutriture remains controversial. On the one hand, it has been reported that the characteristic hyperzincuria is not associated with negative zinc balance because of a corresponding increase in zinc absorption from the gastrointestinal tract (Milman et al. 1983). On the other hand, there have been case reports of zinc-responsive "acrodermatitis enteropathica syndromes" with severe hypozincaemia in association with alcoholic cirrhosis (Ilchyshyn and Mendelsohn 1982); these extreme examples of acute, severe zinc deficiency occur only in very special circumstances and are considered to represent indisputable proof of clinically significant zinc deficiency.

Current uncertainties extend even to semantics of this subject. It is necessary to draw attention to this, as it has been the custom of this author in previous review articles to draw an arbitrary distinction between "mild" and "moderate" human zinc deficiency (Hambidge et al. 1984; Hambidge et al. 1986). In this chapter, both of these entities will be discussed under the single heading of "mild" zinc deficiency. In order to address this topic clearly, it will be necessary to risk overlap with certain other chapters in this monograph. Thus, for example, it will be necessary to include reference to diagnostic difficulties in order to provide the basis for a critical review of our present state of knowledge of mild zinc deficiency. As another example, it is not possible to consider mild zinc deficiency without some discussion of the reproductive cycle.

This review will focus primarily on the fundamental question "does mild zinc deficiency exist in humans?" Given a positive, albeit incomplete, answer to this central question, the circumstances in which mild zinc deficiency may occur and its health-related significance will be considered. Finally, the challenge of how to obtain more definitive answers to these questions will be addressed.

Theoretical Considerations

It is inherently unlikely that human zinc deficiency is "an all-or-none" phenomenon, i.e. that an individual's zinc nutritional status is either normal for all practical, clinical purposes or that he has a life-threatening specific nutrient deficiency disorder. The alternative concept of a "graded response" to progressive degrees of deficiency is supported by studies using animal models. For example, when the zinc content of the diet of weanling rats was severely restricted (less than 1 mg zinc/kg), growth arrest occurred after 4–5 d (Williams and Mills 1970). Under the conditions of that study, a diet containing 12 mg zinc/kg supported normal growth. Between 12 and 1 mg zinc/kg there was a graded response with increasingly severe impairment of growth velocity, the more severe the dietary restriction of zinc. Thus, with 9 mg zinc/kg in the diet, weight was reduced by about 20%. It could be argued that the reduction in growth is a component of the adaptation process to relatively low dietary zinc intakes. However, current concepts of paediatric care do not yet include ready acceptance of any factors that inhibit maximal physiological growth potential.

While studies of laboratory animals such as that cited above provide convincing evidence of the occurrence of mild growth-limiting zinc deficiency states in the young animal, they do not provide any indication of the complexity of this subject. For example, there is relatively little change in total body zinc even in severe zinc deficiency states. Jackson et al. (1982) have calculated that the deficit is of the order of 20% in severely zinc-depleted rats and is probably similar in man; however, a high percentage of this zinc loss is from bone where its metabolic importance, except in the epiphyseal region, is uncertain. Most tissues, including skeletal muscle, have no measurable decline in zinc concentration. This kind of observation is partly responsible for the concept that when diets are zinc restricted, certain unidentified, small, physiologically important body pools of zinc may be depleted. Chesters (1982) has postulated that the effects of zinc on gene expression in vivo are dependent on a small intracellular pool of readily exchangeable zinc and that this pool is rapidly depleted when dietary zinc is restricted. Membrane-bound zinc may also be readily depleted (Bettger and O'Dell 1981), but too little is known about these possibly vulnerable, metabolically important pools of zinc to allow much more than conjecture at this stage. More extensive knowledge of the biochemistry and metabolism of zinc will be necessary before it is possible to understand the sequence of events at different stages of the life cycle when available dietary zinc is progressively decreased from the lower end of the optimal range.

Evidence for the Occurrence of Mild Deficiency of Zinc in Man

Two outstanding factors responsible for difficulties in the detection or confirmation of mild human zinc deficiency have been the lack of a reliable sensitive laboratory index of zinc deficiency and the lack of any specific clinical features. In these circumstances, it would be of little value, and may be counterproductive, to consider the numerous reports that have relied entirely on laboratory assays to assess zinc nutriture. Mention will be made, however, of selected reports in which laboratory data have been linked to clinical features that could be explained by mild zinc deficiency. Such papers included one from Denver (Hambidge et al. 1972) which provided the incentive for subsequent studies in the University of Colorado School of Medicine programme. More recently, similar reports have emerged from Beijing, China (Chen et al. 1985) and Guelph, Canada (Smit-Vanderkooy and Gibson 1987). None of these studies included adequately designed supplementation protocols and they were not definitive. An impressive feature of the Beijing study was the large number of subjects included. A well-designed supplementation study with an equally large population would represent an invaluable contribution to this field. Similar associations between low plasma and hair zinc concentrations and low growth percentile were reported for pre-adolescent boys in Yugoslavia (Buzina et al. 1980), but again, more definitive studies are required (Dorea and Paine 1985).

More definitive confirmation of mild zinc deficiency remains dependent on the results of adequately designed, randomized, controlled studies of dietary supplementation with physiological quantities of zinc. Studies of this nature have been limited in number and in scope. Most of those studies that have been undertaken have been based on measurements of physical growth rates, which are affected progressively by increasingly severe dietary restrictions of zinc in young animals but are not affected by pharmacological quantities of zinc. For example, modest, but statistically significant, increases in growth rates have been associated with small zinc supplements in double-blind controlled studies in Denver, Colorado. The first of these focused on the effects of a zinc-supplemented cows' milk-based formula vs the same formula without a zinc supplement given to normal term infants (Walravens and Hambidge 1976). The unsupplemented formula had a zinc concentration approximately 60% of that of cows' milk, the result of diluting the protein content in the "humanizing" process. Male, but not female, infants who were given the zinc-supplemented formula had a mean weight increment over the first 6 months that was significantly greater than the corresponding increment for control infants.

Subsequently, zinc-supplemented, low-income, male, Hispanic preschool children, selected on the basis of low height-for-age percentiles, were found to have a significantly greater mean height increment and height-for-age percentile increment than their matched controls (Hambidge and Walravens 1978). Because of the small number of subjects, a second similar study was undertaken (Walravens et al. 1983). The results in the second study were also significant and very similar to those of the previous pilot study. Zinc supplementation in the second of these studies was also associated with a significant increase in calculated energy and protein intake compared with the placebo-treated children (Krebs et al. 1984). Concurrently with these studies, zinc supplementation of middle-income Anglo children who were unselected with respect to growth percentiles was not associated with any increase in growth velocity (Hambidge et al. 1979); in that study, the zinc was incorporated into a breakfast cereal.

The growth percentiles of the Hispanic children, who appeared to respond to zinc supplementation with increased growth velocity, started to decline in infancy. Hence, if zinc deficiency was an aetiological factor contributing to the poor growth of these children, it was concluded that this deficiency must have commenced in infancy (Hambidge 1986). This hypothesis provided the rationale for a further study that has recently been completed (Walravens et al. 1986). Again, significant treatment-related effects were observed on growth velocity. In this study the effect was observed in girls as well as in boys.

The small numbers of participants in these studies underline the need for caution in trying to reach any final conclusions and especially in attempting to extrapolate beyond the immediate groups that have been studied. However, the data derived from them do provide strong evidence that mild zinc deficiency states can occur in sections of the free-living population. Our experience in Denver has been used in this section to illustrate the type of evidence that has led to and supported the concept of mild human zinc deficiency. Other evidence and other circumstances will be considered in the next section.

Circumstances in which Mild Human Zinc Deficiency has been Reported

Mild human zinc deficiency has been reported as an isolated dietary deficiency in certain sections of the free-living population, in association with multiple nutrient deficiencies and in various disease states.

An isolated dietary deficiency of zinc has been documented primarily at stages of the life cycle when requirements are relatively high. This applies especially to periods of rapid growth in infancy and childhood, some of the evidence for which has been considered in the previous section. This is not entirely surprising in view of the uniquely rapid growth rates of young infants and the lack of substantial neonatal zinc stores. However, even in infancy, calculated dietary zinc requirements are quite modest provided reasonable absorption is achieved (Krebs and Hambidge 1986). Thus, in general, the zinc status of the breast-fed infant appears to be favourable despite a progressive decline in zinc intake to levels that are relatively low after the 1st month or two of age (Krebs et al. 1985). There is an abundance of evidence that absorption of zinc from cows' milk infant formulas (Casey et al. 1981; Sandström et al. 1983; Ehrenkranz et al. 1986) as well as from mixed diets is considerably less than from human milk.

Even taking these differences into account, however, the baseline calculated dietary zinc intakes to which the addition of a supplement has been associated with a growth response have been surprisingly high. For example, calculations of zinc intake from dietary sources in the most recent of the Denver studies, which involved older infants and toddlers, indicated an average of about 450 μg zinc/kg body weight/d (Walravens and Hambidge, unpublished work). With a calculated requirement for net absorption of only about 50 μg zinc/kg body weight/d at this age (Krebs and Hambidge 1986) and a metabolic rqeuirement (i.e. requirement for true absorption) probably of the order of 100 μg zinc/kg body weight/d, it appears that it should have required considerably less than 20% net absorption and less than 25% true absorption to meet requirements. Dietary histories could be obtained on only a minority of participants in this study and the validity of the records in this population is uncertain. Nevertheless, discordant data such as these remain puzzling. They underline the need for additional, careful studies of dietary zinc supplementation to confirm the occurrence of this growth-limiting zinc deficiency syndrome. Meanwhile, subject to further confirmation, these data suggest that zinc deficiency may occur or persist at levels of zinc intake that are surprisingly high in terms of requirements calculated on a factorial basis. If these intakes are evaluated in terms of the current recommended dietary allowances (RDA) (Committee on Dietary Allowances 1980), they do appear low. But this approach merely sidesteps the very real and challenging questions related to the aetiology of mild chronic zinc deficiency states in otherwise apparently normal infants and children.

As there has not been any notable documented reduction of dietary zinc intake in the subjects included in these and other studies, these findings appear to suggest that the zinc intake of the general population lacks a comfortable margin of safety so that relatively minor deviations from a "usual" intake can be growth limiting. This is especially puzzling in the context of recent reports that have documented an impressive ability of normal adult humans to adapt to quite severe dietary zinc restrictions (Wada et al. 1985; Milne et al. 1987). Better understanding of the epidemiology of mild human zinc deficiency will depend on improved knowledge of several interrelated factors, including dietary zinc requirements, how these differ according to the type of diet and the extent to which adaptation can occur and be regarded as acceptable at different stages of the life cycle.

There is evidence that zinc deficiency in infants suffering from protein–energy malnutrition can be of considerable clinical significance. The most extensive data have been derived from studies in Jamaica. Zinc supplementation during the recovery stage was associated with an immediate increase in the rate of weight gain, a growth of the thymus and activation of the sodium pump (Golden and Golden 1981). There was also a decrease in the calculated energy cost of new tissue deposition, indicating improved synthesis of lean body tissue. The increment in the rate of weight gain and the energy cost of tissue deposition were both correlated significantly with plasma zinc. Again, the baseline dietary zinc intake of these infants was quite high (approximately 400 μg zinc/kg body weight/d), but poor absorption, especially from the soya-protein formula fed to some of the subjects, and rapid catch-up growth are both likely to have contributed to unusually high dietary zinc requirements. This study was neither randomized nor blind controlled. However, the same investigators have reported two small randomized controlled studies of topical zinc which showed beneficial effects on open skin sores (Golden et al. 1980) and delayed-type cutaneous hypersensitivity (Golden et al. 1978). In a randomized, double-blind, controlled study of dietary zinc supplementation of 32 malnourished infants in Chile (Castillo-Duran et al. 1987), the zinc supplement was associated with a greater increment in weight-for-length over the first 60 d, a decrease in the incidence of skin infections, a decrease in the incidence of anergic infants and an increase in serum IgA levels. In non-breastfed Amazonian infants, zinc supplementation was associated with higher salivary secretory IgA levels (Lehti et al. 1983).

Golden and Golden (1985) have de-emphasized the practical significance of mild zinc deficiency. Though no reference was made to their own research in their review, this has served as a timely reminder that reliable data documenting the clinical significance of mild zinc deficiency remain quite tenuous.

Though the premature infant appears to be more susceptible to severe acute zinc deficiency states than is the infant delivered at term, evidence for milder deficiency states is surprisingly limited (Kumar and Anday 1984; Hambidge 1985). Indeed, the results of the one supplementation study to be reported (Hascke et al. 1985) suggest that growth rates of premature infants are *less* susceptible to low dietary zinc intakes than are those of term infants. The unsupplemented intake of the male subjects in this study, in which zinc supplementation was not associated with accelerated growth, was lower than that for normal term infants in a study in which zinc supplementation was associated with increased growth velocity (Walravens and Hambidge 1976). The approximate figures were 200 μg zinc/kg/d for the preterm infants and 270 μg zinc/kg/d for the infants born at term. These intakes extended over intervals of 4 and 6 months respectively. It has been suggested (Solomons 1986) that the high iron concentration in the formula administered to the term infants may have impaired zinc absorption. However, calculated zinc requirements of the premature infant (Shaw 1979) are considerably greater than for infants born at term and these would not be met even with 100% absorption of the dietary zinc provided to the control group in the above study (Hascke et al. 1985). Hence, in striking contrast with the Denver study, a growth response was a reasonable expectation in view of the low zinc intake from the formula, but surprisingly was not observed. The major sources of concern about the zinc status of premature infants are theoretical considerations of calculated needs versus intake and the results of most, but not all, zinc balance studies which have indicated that adequate zinc retention is not achieved generally prior to 34–36 weeks postconceptional age (Hambidge 1987). Major unexplained differences between centres have been observed for traditional balance data and also have been reported recently for measurements of true absorption using stable isotope techniques (Ehrenkranz et al. 1984; Peirce et al. 1988).

Evidence that dietary zinc deficiency can occur in otherwise apparently normal young children has been discussed in the previous section. To some extent, it is possible that the impression that particular sections of the population are at risk from dietary zinc deficiency is an artefact reflecting the choice of subjects included in previously reported studies. Subject selection in the Denver studies could also be responsible for the impression that this entity is more common in low-income Hispanic children than in middle-income Anglo children. In those studies, it was only the Hispanic children who were preselected on the basis of low and/or declining growth percentiles. There is, indeed, evidence that this syndrome is not limited to any one socioeconomic group (Hambidge et al. 1972). There is also suggestive evidence that school-aged children as well as infants and younger children can be affected (Hambidge et al. 1972) and that this may occur in various cultures and in various geographical locations (Buzina et al. 1980; Chen et al. 1985; Smit-Vanderkooy and Gibson 1987).

The syndrome of adolescent nutritional dwarfism provided the focus for the first studies of human zinc deficiency related to inadequate dietary zinc intake. These studies followed the original, historically important, hypothesis of Prasad and co-workers (1961), that a nutritional deficiency of zinc can occur in the human and that, among other consequences, growth can be affected adversely. The studies in the Middle East extended through the 1960s and early 1970s. The early studies in Egypt explored zinc status with laboratory investigations and case studies of zinc supplementation (Prasad et al. 1963a, b, c, d; Sandstead et al. 1967). An effect of zinc supplementation on growth was not confirmed by an independent study of Carter et al. (1969) in Cairo. Subsequently a series of supplementation studies in Iran gave mixed results (Ronaghy et al. 1969; Halsted et al. 1972; Ronaghy et al. 1974; Mahloudji et al. 1975). These studies have been critically reviewed recently (Golden and Golden 1985). Significant questions about the role of zinc deficiency in this syndrome of adolescent nutritional dwarfism include the documented occurrence of other nutrient deficiencies in some of these subjects, e.g. iodide, protein and iron deficiencies.

Not all studies were "double-blind" and only one that demonstrated a positive effect of zinc supplementation on growth was randomized. Despite these concerns, there appears to be a weight of evidence in support of zinc deficiency being one factor that contributed to the profound growth retardation of these adolescents and young adults. It should be noted, however, that pre-adolescent aboriginal children in Northwest Australia with a similar syndrome and with low plasma zinc concentration have not shown a growth response to zinc in a double-blind study (Smith et al. 1985). The full details of this study have not been published and cannot be evaluated adequately at this time. Tentatively, however, these findings suggest that zinc deficiency is not an essential or universal aetiological factor in syndromes of reversible growth retardation in older children. Remarkably, data on the zinc status of free-living adolescents in North America and Europe are virtually non-existent.

The occurrence of isolated zinc deficiency in normal free-living adults has not been, and is unlikely to be, documented. The one exception to this statement concerns women during the reproductive cycle. The increase in metabolic requirements for zinc during pregnancy is quite modest, especially early in gestation. Despite this, zinc nutrition during pregnancy has attracted more attention than that at any other stage of the life cycle, with the possible exception of the infant. In general, though, the design of the reported studies has either not been intended or has not been adequate to determine whether a clinically significant degree of zinc depletion occurs in some pregnancies.

It has been established that there is a substantial longitudinal decline in plasma zinc across gestation, which starts early in the first trimester (Breskin et al. 1983; Hambidge et al. 1983). Thus, levels which are markedly depressed by non-pregnant standards are not necessarily indicative of zinc deficiency. Plasma zinc concentrations that are lower than can be explained by physiological changes have been documented in selected populations of pregnant women and in some individual women. For example, unusually low plasma zinc concentrations occur in vegetarian Hindu women resident in London, England (Campbell-Brown et al. 1985). The dietary zinc intake of these subjects was remarkably low and the combination of these findings leaves little doubt that these women were zinc-depleted to some extent. However, the clinical significance of this finding could not be determined from the study performed.

Several studies have found associations between low levels of plasma or tissue zinc and complications of pregnancy and delivery or impairment of foetal development (Jameson 1976; Cavdar et al. 1980; Cherry et al. 1981; Meadows et al. 1981; McMichael et al. 1982; Soltan and Jenkins 1982; Baumah et al. 1984; Mukherjee et al. 1984; Zimmerman et al. 1984; Simmer and Thompson 1985; Simmer et al. 1985). A detailed review of these findings is beyond the scope of this chapter. In general, there has tended to be a lack of consistency in findings between different studies. Once again, it is not justifiable to reach any definitive conclusions about cause–effect relationships from this type of study. Few studies of zinc supplementation during pregnancy have been "double-blind" with subjects randomly allocated a zinc supplement or a placebo. In those instances where such studies have been undertaken in populations of pregnant women considered to be at relatively high risk of zinc deficiency, the effects of the supplement have been quite limited and have not been consistent. For example, in a study involving mature low-income women of Mexican descent, Hunt et al. (1984) found a significantly lower incidence of pregnancy-induced hypertension in the zinc-supplemented than in the placebo-treated group. In a subsequent study of teenage pregnancies in the same population, this was not discerned (Hunt et al. 1985). Nor was it observed in a study of black adolescent teenagers in New Orleans (Cherry et al. 1987) despite an association between low plasma zinc concentrations and pregnancy-induced hypertension in an earlier study of that population (Cherry et al. 1981). Despite the lack of clear-cut consistent clinical benefits from zinc supplementation in these studies, the results provided some laboratory evidence and some limited clinical evidence that improved zinc status did result from the supplement. This should encourage further endeavours in select populations, preferably commencing the supplement prior to conception and enrolling a large number of subjects. Furthermore, the parameters to be evaluated merit careful consideration. For example, it is possible that premature delivery (Apgar 1987; Cherry et al. 1987) may justify special attention.

In comparison with pregnancy, the increase in metabolic requirement for zinc during lactation is relatively large, especially in the early weeks. No definitive studies of maternal zinc status or of the effects of maternal zinc status on the zinc intake of the fully breastfed infant have been reported. While some investigators had concluded on the basis of limited data that milk zinc concentrations are not affected by maternal zinc intake and maternal zinc status, the results of one study suggest that the rate of decline in zinc concentrations of human milk across lactation is dependent on maternal zinc intake (Krebs et al. 1985). With a dietary zinc intake of approximately 11 mg zinc/d an abnormally steep rate of decline in milk zinc concentration occurred. Zinc supplementation of urban Amazonian mothers during lactation was associated with significantly higher retinol levels in maternal serum and in milk (Shrimpton et al. 1983).

Plasma zinc concentrations and calculated dietary zinc intakes have been found to be relatively low in the elderly. However, there is evidence that zinc requirements are lower in the elderly (Bunker et al. 1987) and no adverse clinical consequences of the lower plasma zinc concentrations have yet been confirmed. Treatment of elderly subjects with zinc in quantities considerably greater than dietary intakes has been found to improve immune status (Duchateau et al. 1981). These effects are likely to be pharmacological and do not provide evidence of a zinc deficiency state.

Though the effect of phytate and of fibre on the absorption of dietary zinc remains a controversial topic, the results of some studies have provided strong evidence for an inhibitory effect of these dietary factors, especially phytate (Turnlund et al. 1984). High levels of dietary phytate (Reinhold et al. 1973) and fibre (Ismail-Beigi et al. 1977) have been indicted as a major aetiological factor in the zinc deficiency of adolescents in Iran. Other studies have failed to demonstrate an adverse effect of high-phytate diets on zinc status (e.g. Anderson et al. 1981).

Evidence of mild zinc deficiency has been reported in association with numerous disease states. These include diseases such as inflammatory bowel disease (Solomons et al. 1977; McClain et al. 1980), coeliac disease (Naveh et al. 1983) and cystic fibrosis (Caillie-Bertrand et al. 1982) in which zinc absorption may be compromised. In other instances, e.g. sickle cell disease (Prasad et al. 1976) and insulin-dependent diabetes (Canfield et al. 1984), excessive losses of endogenous zinc can occur. In yet other instances, e.g. growth hormone deficiency during replacement therapy (Ghavami-Maibodi et al. 1983), a deficiency state secondary to increased requirements has been postulated. In none of these circumstances have reports of zinc deficiency been consistent (Palin et al. 1979; Fleming et al. 1981; Richards and Marshall 1983; Abshire et al. 1988) and both the frequency and clinical significance of zinc deficiency require further study.

Clinical Significance

Impaired growth velocity is the clinical feature of mild zinc deficiency that has received most recognition in the infant, child and adolescent. In part, this may reflect the extent to which physical growth rates have been studied in comparison with other parameters. Demonstrable effects of zinc supplementation in controlled studies had been documented for weight gain in infants (Walravens and Hambidge 1976; Walravens et al. 1986) and for linear growth in children (Hambidge et al. 1985) and adolescents (Halsted et al. 1972). In some instances, the extent of growth retardation has been severe and some subjects appear to have responded to zinc supplementation with major gains in height (Halsted et al. 1972). In other studies, in which growth retardation has been much less severe, the response to zinc supplementation has been quite modest, i.e. averaging about 10% greater than placebo-treated controls (Hambidge et al.

1985). It should be emphasized that not all studies have demonstrated a difference in growth velocity between zinc-supplemented and control groups. This applies to some studies of adolescents in the Middle East (Carter et al. 1969) and to premature infants in North America (Hascke et al. 1985).

Mild zinc deficiency may also affect the quality of growth. Thus, zinc supplementation of malnourished infants has been associated with a decrease in the energy cost of growth, which, it was concluded, was probably related to improved synthesis of lean body tissue (Golden and Golden 1981). Confirmation of this observation may be of greater practical importance than the documentation of increased growth velocity, as this may have broad implications for the quality of "catch-up" growth and for the extent of final recovery from malnutrition in infancy and early childhood.

A reduction in food consumption is, together with growth deceleration, the earliest reported feature of dietary zinc restriction in animal models. Quantitative assessment of appetite in free-living humans is difficult. In one of the Denver studies, zinc supplementation, under randomized double-blind controlled conditions, was associated with a significantly greater intake of energy and protein than occurred in the placebo-treated children over the study period (Krebs et al. 1984). The calculated mean energy intake of the placebo-treated children remained low, while that for the zinc-supplemented children increased to a more acceptable level. In a recent study from this centre, involving a younger age group of infants and toddlers, no effect of zinc supplementation on food consumption was demonstrated (Walravens and Hambidge, unpublished work). A clear-cut negative conclusion was, however, not possible because of the low compliance rate with records of dietary intake. Appetite was not impaired in malnourished Jamaican infants who were considered to be zinc deficient, but this was attributed to the concurrent deficiency of dietary protein (Golden and Golden 1981).

It has been suggested that zinc supplementation may be beneficial in severe appetite disorders. Unfortunately, in one randomized controlled study of zinc therapy in anorexia nervosa (Katz et al. 1987), pertinent measures were not reported. This study included only six zinc-supplemented subjects. Despite this small number, zinc therapy (50 mg Zn^{2+} per day) was associated with a significant improvement in the Zung Depression Scale and the State–Trait Anxiety Inventory. The placebo group demonstrated no change in either test over the 6-month study period. Prior to commencing the supplement, serum and urine zinc were negatively correlated with depression. Urine zinc excretion rates, but not serum zinc concentrations, were significantly lower in anorexia nervosa patients than in controls. Behavioural changes, especially loss of hedonic tone (Walravens et al. 1978), are a notable feature of severe zinc deficiency states, but little work has been done on behavioural changes in milder zinc deficiency states.

It has been proposed that zinc has an important physiological role in the special sense of taste (Henkin 1984). Improvement in idiopathic dysguesia was reported in a single-blind trial of zinc therapy but a subsequent double-blind trial failed to confirm this. It seemed reasonable to hypothesize that a positive result would depend on appropriate selection of patients with detectable abnormalities of zinc metabolism. As yet, however, there is no published report of such a study. Several reports have linked hypoguesia with evidence of impaired zinc nutriture (Hambidge et al. 1986) but definitive reports are lacking.

Of all nutrient deficiencies, the effects of zinc deficiency on the immune system have been most clearly documented (Fraker et al. 1986). Zinc deficiency in experimental animals has been shown to depress delayed-type hypersensitivity, T lymphocyte numbers, T cell mitogen responses, T-helper function, natural killer function and cytotoxic killer cell activity. There is involution of the thymus and thymocyte depletion. Little is yet known about the possible effects of mild human zinc deficiency on T cell function and on other components of the immune system. Delayed-type hypersensitivity (DTH) responses have been reported to improve with zinc supplementation in a number of circumstances in which there is no evidence of severe zinc deficiency, as well as in severe zinc deficiency states. The improved DTH responses with zinc supplementation in the absence of severe zinc deficiency have been seen in obese subjects (Chandra and Kutty 1980), those with Down's syndrome (Bjorksten et al. 1980), infants with protein–energy malnutrition and the elderly (Bogden et al. 1987). Unlike physical growth, there is good evidence that the immune system is responsive to pharmacological quantities of zinc and a response to zinc supplementation does not necessarily indicate a zinc deficiency state.

In contrast with severe zinc deficiency, epithelial tissues are not obviously affected in milder forms of zinc deficiency. In infants with kwashiorkor, skin lesions have been reported to be more evident in association with hypozincaemia (Golden and Golden 1979). Roughening of the skin has been attributed to zinc deficiency in Egyptian adolescents. An association between mild human zinc deficiency and impaired healing of wounds and ulcers appears quite likely, but has not been convincingly or uniformly demonstrated (Hambidge et al. 1986). Quantities of zinc used in most studies have been higher than those likely to be needed to correct a deficiency state.

Hypogonadism, with retarded development of secondary sexual characteristics, was a conspicuous feature of "adolescent nutritional dwarfism" in Egypt and Iran. This has been attributed to zinc deficiency. The hormonal changes reported in these subjects (Sandstead et al. 1967) are, however, more consistent with protein–energy deficiency or a co-existing protein and zinc deficiency than with a pure zinc deficiency (Kulin et al. 1984). The latter results in local effects on the gonads with decreased testosterone levels or a decreased testosterone response to LHRH, rather than effects at the level of the hypothalamus or pituitary (Hambidge et al. 1986). Zinc deficiency has also been reported, inconsistently, to be a reversible cause of secondary gonadal dysfunction, especially in uraemia (Mahajan et al. 1982) and sickle cell disease (Abbasi et al. 1976). Reversible oligospermia has been reported in mild experimental zinc deficiency (Abbasi et al. 1980).

Because of the conclusively demonstrated importance of adequate zinc nutriture for normal foetal development and obstetric course in animal models, the potential importance of adequate maternal zinc nutrition during the human reproductive cycle is very substantial. A variety of complications of pregnancy and delivery together with foetal morbidity have been attributed to maternal zinc deficiency during human pregnancy. These complications include congenital malformations (e.g. neural tube defects), intrauterine growth retardation, pregnancy-induced hypertension, prematurity, prolonged labour and intrapartum haemorrhage. The extent of supportive evidence for these observations has been discussed in a previous section.

In addition to changes in zinc concentrations and in the activity of zinc-dependent enzymes, a number of other laboratory changes have been documented in careful studies of the effects of experimental dietary zinc restriction in normal adult subjects. These studies have, for example, provided further evidence for the role of zinc in host defence mechanisms with the demonstration of decreased total lymphocyte counts and of impaired polymorphonuclear leucocyte chemotaxis (Baer et al. 1985). Thyroid hormone levels have been found to be decreased, the basal metabolic rate has fallen significantly, protein utilization was impaired and fasting blood glucose levels have been elevated (Wada and King 1986). Alterations in protein metabolism include decreased circulating levels of hepatic proteins, i.e. albumin, pre-albumin, retinol-binding protein and transferrin (Wada and King 1986). There is some evidence that these abnormalities can be sufficiently severe to cause oedema in the premature infant (Kumar and Anday 1984).

Conclusions and Future Research Needs

There is a substantial body of evidence that mild zinc deficiency does occur in humans and that this can be found in sections of the free-living population as well as in more specialized circumstances, e.g. in various disease states or under conditions of experimental zinc deprivation. However, there have been few studies that can be regarded as reasonably definitive and little is yet known about the incidence, causes or clinical consequences.

In order to understand the causes of mild zinc deficiency in man, it is essential to achieve better understanding of dietary requirements and factors that affect these requirements including other dietary components. We also need to determine the limits of adaptation to relatively low zinc intakes, and how this may vary depending on such factors as the stage of the life cycle and the type of diet.

It is difficult to foresee achieving a good estimate of the incidence of mild zinc deficiency until a reliable laboratory assay is available that can be used in population surveys. Meanwhile, an increased commitment to well-designed supplementation studies on a more global basis offers the most promise. For example, if such studies were to be undertaken for infants and children in Beijing or for pregnant Hindu women in London, it would be possible to ascertain whether the pilot data that have been reported for these population groups did indicate a mild zinc deficiency state. By this laborious but necessary approach it would also be possible to achieve a more reasonable understanding of the clinical importance of zinc deficiency. For example, studies in these particular populations could help to answer the question of whether maternal zinc deficiency during pregnancy affects prematurity rates or the rate of intrauterine growth of the foetus. Investigations of children in Beijing could help to determine the effects of mild human zinc deficiency on appetite and taste perception. The immune system and possibly certain aspects of brain function are other examples of areas that merit careful evaluation if we are to understand the extent and the limits of the clinical relevance of these mild zinc deficiency states.

Acknowledgements. The original work undertaken by the author to which reference is made in this chapter was supported in part by a grant from the National Institutes of Health, NIADDKD, 5R22 AMI2432, and grant RR-69 from National Institutes of Health, General Clinical Research Center.

References

Abbasi AA, Prasad AS, Ortega J, Congco E, Oberleas D (1976) Gonadal function abnormalities in sickle cell anemia: studies in adult male patients. Ann Intern Med 85: 601–605

Abbasi AA, Prasad AS, Rabbani P, DuMouchelle E (1980) Experimental zinc deficiency in man: effect on testicular function. J Lab Clin Med 96: 544

Abshire TC, English JL, Githens JH, Hambidge KM (1988) Zinc (Zn) status in children and young adults with sickle cell disease. Clin Res 36:209A

Anderson BM, Gibson RS, Sabry JH (1981) The iron and zinc status of long-term vegetarian women. Am J Clin Nutr 34: 1042–1048

Apgar J (1987) Effect on the guinea pig of low zinc intake during pregnancy. Fed Proc 46: 2515

Baer MT, King JC, Tamura T et al. (1985) Nitrogen utilization, enzyme activity, glucose intolerance and leukocyte chemotaxis in human experimental zinc depletion. Am J Clin Nutr 41: 1220–1235

Baumah PK, Russell M, Bates G, Milford WA, Skillen AW (1984) Maternal zinc status: a determination of central nervous system malformation. Br J Obstet Gynaecol 91: 788–790

Bettger WJ, O'Dell BL (1981) A critical physiological role of zinc in the structure and function of biomembranes. Life Sci 28: 1425–1438

Bjorksten B, Back O, Gustavson KH, Hallmans G, Hagglof B, Tarvnik A (1980) Zinc and immune function in Down's syndrome. Acta Paediatr Scand 69: 183–187

Bogden JD, Oleske JM, Munves EM et al. (1987) Zinc and immunocompetence in the elderly: baseline data on zinc nutriture and immunity in unsupplemented subjects. Am J Clin Nutr 45: 101–109

Breskin MW, Worthington-Roberts BS, Knopp RH et al. (1983) First trimester serum zinc concentrations in human pregnancy. Am J Clin Nutr 38: 943–953

Bunker VW, Hinks LJ, Stansfield MF, Lawson MS, Clayton BE (1987) Metabolic balance studies for zinc and copper in housebound elderly people and the relationship between zinc balance and leukocyte zinc concentrations. Am J Clin Nutr 46: 353–359

Buzina R, Jusic M, Sapunar J, Milanovic N (1980) Zinc nutrition and taste acuity in school children with impaired growth. Am J Clin Nutr 33: 2262–2267

Caillie-Bertrand MV, De Bieville F, Neijens H, Kerrebijn K, Fernandes J, Degenhart H (1982) Trace metals in cystic fibrosis. Acta Paediatr Scand 71: 203–207

Campbell-Brown M, Ward RJ, Haines AP, North WRS, Abraham R, McFadyen IR (1985) Zinc and copper in Asian pregnancies – is there evidence for a nutritional deficiency? Br J Obstet Gynaecol 92: 875–885

Canfield WK, Hambidge KM, Johnson LK (1984) Zinc nutriture in type I diabetes mellitus: relationship to growth measures and metabolic control. J Pediatr Gastro Nutr 3: 577–584

Carter JP, Grivetti LE, Davies JT et al. (1969) Growth and sexual development of adolescent Egyptian village boys: effects of zinc, iron and placebo supplementation. Am J Clin Nutr 22: 59–78

Casey CE, Walravens PA, Hambidge KM (1981) Availability of zinc: loading tests with human milk, cow's milk and infant formulas. Pediatrics 68: 394–396

Castillo-Duran C, Heresi G, Fisberg M, Uauy R (1987) Controlled trial of zinc supplementation during recovery from malnutrition: effects on growth and immune function. Am J Clin Nutr 45: 602–608

Cavdar AO, Arcasoy A, Baycu T, Himmetoglu O (1980) Zinc deficiency and anencephaly in Turkey. Teratology 22: 141

Chandra RK, Kutty KM (1980) Immunocompetence in obesity. Acta Paediatr Scand 69: 25–30

Chen Xue-Cun, Yin Tai-An, He Jin-Sheng, Ma Qiu-Yan, Han Zhi-Min, Li Li-Xiang (1985) Low levels of zinc in hair and blood, pica, anorexia, and poor growth in Chinese preschool children. Am J Clin Nutr 42: 694–700

Cherry FF, Bennett EA, Bazzano GS et al. (1981) Plasma zinc in hypertension/toxemia and other reproductive variables in adolescent pregnancy. Am J Clin Nutr 34: 2367–2375

Cherry F, Sandstead H, Bazzano G et al. (1987) Zinc nutriture in adolescent pregnancy: response to zinc supplementation. Fed Proc 46: 2519

Chesters JK (1982) Metabolism and biochemistry of zinc. In: Prasad AS (ed) Clinical, biochemical, and nutritional aspects of trace elements. Alan R. Liss, New York, pp 221–238

Committee on Dietary Allowances Food and Nutrition Board (1980) Recommended dietary allowances, 9th edn. National Research Council, Washington DC

Dorea JG, Paine PA (1985) Hair zinc in children: its uses, limitations and relationship to plasma zinc and anthropometry. Hum Nutr Clin Nutr 39C(6): 389–398

Duchateau J, Delepesse G, Vrijens R, Collet H (1981) Beneficial effects of oral zinc supplementation on the immune response of old people. Am J Med 70: 1001–1004

Ehrenkranz RA, Ackerman BA, Nelli CM, Janghorbani M (1984) Determination with stable isotopes of the dietary bioavailability of zinc in premature infants. Am J Clin Nutr 40: 72–81

Ehrenkranz RA, Nelli CM, Gettner PA et al. (1986) The influence of food on zinc (Zn) absorption in premature infants. Pediatr Res 20: 409A

Fleming CR, Huizenga KA, McCall JT, Gildea J, Dennis R (1981) Zinc nutrition in Crohn's disease. Dig Dis Sci 26: 865–870

Fraker PJ, Gershwin ME, Good RA, Prasad AS (1986) Interrelationships between zinc and immune function. Fed Proc 45: 1474–1479

Ghavami-Maibodi SZ, Collipp PJ, Castro-Magana M, Stewart C, Chen SY (1983) Effect of oral zinc supplements on growth, hormonal levels, and zinc in healthy short children. Ann Nutr Metab 27: 214–219

Golden BE, Golden MHN (1979) Plasma zinc and the clinical features of malnutrition. Am J Clin Nutr 32: 2490–2494

Golden MHN, Golden BE (1981) Effect of zinc supplementation on the dietary intake, rate of weight gain, and energy cost of tissue deposition in children recovering from severe malnutrition. Am J Clin Nutr 34: 900–908

Golden MHN, Golden BE (1985) Problems with the recognition of human zinc-responsive conditions. In: Mills CF, Bremner I, Chesters JK (eds) Trace elements in man and animals – TEMA-5, Commonwealth Agricultural Bureaux, London, pp 933–938

Golden MHN, Golden BE, Harland PSEG, Jackson AA (1978) Zinc and immunocompetence in protein-energy malnutrition. Lancet I: 1226–1227

Golden MHN, Golden BE, Jackson AA (1980) Skin breakdown in kwashiorkor responds to zinc. Lancet I: 1256

Halsted JA, Ronaghy HA, Abadi P et al. (1972) Zinc deficiency in man. Am J Med 53: 277–284

Hambidge KM (1985) Zinc deficiency in the premature infant. Pediatr Rev 6: 209–216

Hambidge KM (1986) Zinc deficiency: how important? Acta Paediatr Scand 323: 52–58

Hambidge KM (1987) Trace element requirements of premature infants. In: Roy CC (ed) Nutritional requirements of the low-birth-weight neonate. Excerpta Medica, Princeton, pp 9-21

Hambidge KM, Walravens PA (1978) Zinc supplementation of low income preschool children. In: Kirchgessner M (ed) Trace element metabolism in man and animals, vol 3. Arbeitskreis für Tierernährungs-forschung. Weihenstephan, pp 296–299

Hambidge KM, Hambidge C, Jacobs MA, Baum JD (1972) Low levels of zinc in hair, anorexia, poor growth and hypoguesia in children. Pediatr Res 6: 868–874

Hambidge KM, Chavez MN, Brown RM, Walravens PA (1979) Zinc nutritional status of young middle-income children and the effects of consuming zinc-fortified breakfast cereals. Am J Clin Nutr 32: 2532–2539

Hambidge KM, Krebs NF, Jacobs MA, Favier A, Guyette L, Ikle DN (1983) Zinc nutritional status during pregnancy: a longitudinal study. Am J Clin Nutr 37: 429–442

Hambidge KM, Krebs NF, Walravens PA (1984) Growth – the importance of zinc. In: Zinc in human medicine. Til Publications, Isleworth, pp 81–94

Hambidge KM, Krebs NF, Walravens PA (1985) Growth velocity of young children receiving a dietary zinc supplement. Nutr Res [Suppl] 1: 306–316

Hambidge KM, Casey CE, Krebs NF (1986) Zinc. In: Mertz W (ed) Trace elements in human and animal nutrition, vol 2, 5th edn. Academic Press, Florida, pp 1–137

Hambidge KM, Krebs NF, Lilly JR (1987) Plasma and urine zinc in infants and children with extra-hepatic biliary atresia. J Pediatr Gastro Nutr 6: 872–877

Hascke F, Singer P, Baumgartner D, Steffan I, Schilling R, Lothaller H (1985) Growth, zinc and copper nutritional status of male premature infants with different zinc intake. Ann Nutr Metab 29: 95–102

Henkin RI (1984) Zinc in taste function: a critical review. Biol Trace Element Res 6: 263–280

Hunt IF, Murphy NJ, Cleaver AE et al. (1984) Zinc supplementation during pregnancy: effects on selected blood constituents and on progress and outcome of pregnancy in low-income women of Mexican descent. Am J Clin Nutr 40: 508–521

Hunt IF, Murphy NJ, Cleaver AE et al. (1985) Zinc supplementation during pregnancy in low-income teenagers of Mexican descent: effects on selected blood constituents and on progress and outcome of pregnancy. Am J Clin Nutr 42: 815–828

Ilchyshyn A, Mendelsohn S (1982) Zinc deficiency due to alcoholic cirrhosis mimicking acrodermatitis enteropathica. Br Med J 284: 1676

Ismail-Beigi F, Reinhold JG, Faraji B, Abadi P (1977) Effects of cellulose added to diets of low and high fiber content upon the metabolism of calcium, magnesium, zinc and phosphorus by man. J Nutr 107: 510–518

Jackson MJ, Jones DA, Edwards RHT (1982) Tissue zinc levels as an index of body zinc status. Clin Physiol 2: 333–343

Jameson S (1976) Effects of zinc deficiency in human reproduction. Acta Med Scand 593: 1–89

Katz RL, Keen CL, Litt IF, Hurley LS, Kellams-Harrison KM, Glader LJ (1987) Zinc deficiency in anorexia nervosa. J Adolesc Health Care 8 (5): 400–406

Krebs NF, Hambidge KM (1986) Zinc requirements and zinc intakes of breast fed infants. Am J Clin Nutr 43: 288–292

Krebs NF, Hambidge KM, Walravens PA (1984) Increased food intake of young children receiving a zinc supplement. Am J Dis Child 138: 270–273

Krebs NF, Hambidge KM, Jacobs MA, Oliva-Rasbach J (1985) The effects of a dietary zinc supplement during lactation on longitudinal changes in maternal zinc status and milk zinc concentrations. Am J Clin Nutr 41: 560–570

Kulin HE, Bwibo N, Mutie D, Santner SJ (1984) Gonadotropin excretion during puberty in malnourished children. J Pediatr 105: 325–328

Kumar SP, Anday EK (1984) Edema, hypoproteinemia, and zinc deficiency in low-birth weight infants. Pediatrics 73: 327–329

Lehti R, Shrimpton R, Alencar FH, Waterlow JC (1983) Zinc supplementation and secretory immunoglobulin production in non-breast-fed Amazonian infants. Proc Nutr Soc 42: 123A

McLain C, Soutor C, Zieve L (1980) Zinc deficiency: a complication of Crohn's disease. Gastroenterology 78: 272–279

McMichael AJ, Dreosti IE, Gibson GT, Hartshorne JM, Buckley RA, Colley DP (1982) A prospective study of serial maternal serum zinc levels and pregnancy outcome. Early Hum Dev 7: 59–69

Mahajan SK, Prasad AS, Rabbani P, Briggs WA, McDonald FD (1982) Zinc deficiency: a reversible complication of uremia. Am J Clin Nutr 36: 1177–1183

Mahloudji M, Reinhold JG, Haghsehnass M, Ronaghy KA, Fox MRS, Halsted JA (1975) Combined zinc and iron compared with iron supplementation of diets of 6- to 12-year old village schoolchildren in southern Iran. Am J Clin Nutr 28: 721–725

Meadows NJ, Smith MF, Keeling PWN et al. (1981) Zinc and small babies. Lancet II: 1135–1137

Milman N, Jacobsen-Hvid K, Hegnhoj J, Sorensen SS (1983) Zinc absorption in patients with compensated alcoholic cirrhosis. Scand J Gastroenterol 18: 871–875

Milne DB, Canfield WK, Gallagher SK, Hunt JR, Klevay LM (1987) Ethanol metabolism in postmenopausal women fed a diet marginal in zinc. Am J Clin Nutr 46: 688–693

Mukherjee MD, Sandstead HH, Ratnaparkhi MV, Johnson LK, Milne DB, Stelling HP (1984) Maternal zinc, iron, folic acid, and protein nutriture and outcome of human pregnancy. Am J Clin Nutr 40: 496–507

Naveh Y, Lightman A, Zinder O (1983) A prospective study of serum zinc concentration in children with celiac disease. J Pediatr 5 (102): 734–736

Palin D, Underwood BA, Denning CR (1979) The effect of oral zinc supplementation on plasma levels of vitamin A and retinol-binding protein in cystic fibrosis. Am J Clin Nutr 32: 1253–1259

Peirce PL, Hambidge KM, Fennessey PV, Miller L, Goss CH (1988) Zinc absorption in premature infants. In: Trace elements in man and animals – TEMA 6. Plenum, New York (in press)

Prasad AS, Halsted JA, Nadimi N (1961) Syndrome of iron deficiency anemia, hepatosplenomegaly, hypogonadism, dwarfism and geophagia. Am J Med 31: 532–546

Prasad AS, Miale A, Farid Z, Sandstead HH, Schulert AR (1963a) Zinc metabolism in patients with the syndrome of iron deficiency anemia, hepatosplenomegaly, dwarfism and hypogonadism. J Lab Clin Med 61: 537–548

Prasad AS, Miale A, Farid Z, Sandstead HH, Schulert AR, Darby WJ (1963b) Biochemical studies on dwarfism, hypogonadism, and anemia. Arch Intern Med 3: 407–427

Prasad AS, Sandstead HH, Schulert AR, Rooby AS (1963c) Zinc and iron deficiencies in male subjects with dwarfism and hypogonadism but without ancylostomiasis, schistosomiasis or severe anemia. Am J Clin Nutr 12: 437–444

Prasad AS, Sandstead HH, Schulert AR et al. (1963d) Urinary excretion of zinc in patients with the syndrome of anemia, hepatosplenomegaly, dwarfism and hypogonadism. J Lab Clin Med 62: 591–599

Prasad AS, Ortega J, Brewer GJ, Oberleas D, Schoomaker EB (1976) Trace elements in sickle cell disease. J Am Med Ass 235: 2396–2398

Reinhold JG, Nasr K, Lahimgarzadeh A, Hedayati H (1973) Effects of purified phytate and phytate-rich bread upon metabolism of zinc, calcium, phosphorus, and nitrogen in man. Lancet II: 283–288

Richards GA, Marshall RN (1983) The effect of growth hormone treatment alone or growth hormone with supplemental zinc on growth rate, serum, and urine zinc and copper concentrations and hair zinc concentration in patients with growth hormone deficiency. J Am Coll Nutr 2: 133–140

Ronaghy H, Fox MRS, Garn SM et al. (1969) Controlled zinc supplementation for malnourished school boys: a pilot experiment. Am J Clin Nutr 22: 1279–1289

Ronaghy H, Reinhold JG, Mahloudji M, Ghavami P, Fox MRS, Halsted JA (1974) Zinc supplementation of malnourished schoolboys in Iran: increased growth and other effects. Am J Clin Nutr 27: 112–121

Sandstead HH, Prasad AS, Schulert AR et al. (1967) Human zinc deficiency, endocrine manifestations and response to treatment. Am J Clin Nutr 20: 422–442

Sandström B, Cederblad A, Lönnerdal B (1983) Zinc absorption from human milk, cow's milk and infant formulas. Am J Dis Child 137: 726–729

Shaw J (1979) Trace elements in the fetus and young infant. Am J Dis Child 133: 1260–1268

Shrimpton R, Marinho HA, Rocha YS, Alencar FH (1983) Zinc supplementation in urban Amazonian mothers: concentrations of Zn and retinol in maternal serum and milk. Proc Nutr Soc 42: 122A

Simmer K, Thompson RPH (1985) Maternal zinc and intrauterine growth retardation. Clin Sci 68: 395–399

Simmer K, Punchard NA, Murphy G, Thompson RPH (1985) Prostaglandin production and zinc depletion in human pregnancy. Pediatr Res 19: 697–700

Smit-Vanderkooy PD, Gibson RS (1987) Food consumption patterns of Canadian preschool children in relation to zinc and growth status. Am J Clin Nutr 45: 609–616

Smith RM, King RA, Spargo RM, Cheek DB, Field JB, Veitch LG (1985) Growth retarded aboriginal children with low plasma zinc levels do not show a growth response to supplementary zinc. Lancet I: 923–924

Solomons N (1986) Competitive interaction of iron and zinc in the diet: consequences for human nutrition. J Nutr 116: 927–935

Solomons NW, Rosenberg IH, Sandstead HH, Vo-Khactu KP (1977) Zinc deficiency in Crohn's disease. Digestion 16: 87–95

Soltan MH, Jenkins DM (1982) Maternal and fetal plasma zinc concentration and fetal abnormality. Br J Obstet Gynaecol 89: 56–58

Turnlund JR, King JC, Keyes WR, Gong B, Michel MC (1984) A stable isotope study of zinc absorption in young men: effects of phytate and α-cellulose. Am J Clin Nutr 40: 1071–1077

Vallee BL, Wacker Warren EC, Bartholomay AF, Hoch FL (1957) Zinc metabolism in hepatic dysfunction. N Engl J Med 257: 1055–1065

Wada L, King JC (1986) Effect of low zinc intakes on basal metabolic rate, thyroid hormones and protein utilization in adult men. J Nutr 116: 1045–1053

Wada L, Turnlund JR, King JC (1985) Zinc utilization in young men fed adequate and low zinc intakes. J Nutr 115: 1345–1354

Walravens PA, Hambidge KM (1976) Growth of infants fed a zinc supplemented formula. Am J Clin Nutr 29: 1114–1121

Walravens PA, Van Doorninck WJ, Hambidge KM (1978) Metals and mental function. J Pediatr 93: 535

Walravens PA, Krebs NF, Hambidge KM (1983) Linear growth of low income preschool children receiving a zinc supplement. Am J Clin Nutr 38: 195–201

Walravens PA, Hambidge KM, Koepfer D (1988) Zinc supplements in infants with failure to thrive: effects on weight gains. Pediatrics (in press)

Williams RB, Mills CF (1970) The experimental production of zinc deficiency in the rat. Br J Nutr 24: 989

Zimmerman AW, Dunham BS, Nochimson DJ, Kaplan BM, Clive JM, Kunkel SL (1984) Zinc transport in pregnancy. Am J Obstet Gynecol 149: 523–528

Chapter 19

Putative Therapeutic Roles for Zinc

N.W. Solomons, M. Ruz and C. Castillo-Duran

Introduction

The trace element, zinc, is both a nutrient and a drug. As a nutrient, its role in correcting zinc deficiency states is incontrovertible. That topic has been reviewed in Chaps. 17 and 18 of this volume. However, a variety of therapeutic roles for zinc, apart from those associated with the reversal of nutritional depletion or recovery from zinc-deficiency-related symptoms, have been suggested. This domain of zinc as a drug will be the focus of the present chapter.

Zinc is widely available for oral administration to humans in the form of sulphate, acetate, gluconate and orotate salts, and in a host of chelated complexes (Solomons and Cousins 1984). As with any therapeutic agent, the validity of using zinc in the treatment of specific diseases and health conditions must be confirmed before its use can be recommended or accepted. It is our present purpose to review the various putative roles for zinc in therapy, and to evaluate the strength of the evidence supporting its specific efficacy for each condition.

Methodology of Inquiry and Review

For our present inquiry, we have set out some specific procedures. We shall restrict discussion to oral administration of zinc; topical and parenteral zinc administration will not be considered. Moreover, we shall attempt to avoid discussion of situations related to zinc deficiency. We recognize that establishing a diagnosis of human zinc depletion is difficult (Solomons 1979), but we have tried to exclude from consideration in the present review those populations (or patients within sample populations) who present with low circulating zinc levels or other clinical evidence of zinc deficiency. When this cannot be avoided entirely, we shall qualify our judgments based on considerations of zinc deficiency per se.

We shall not discuss the prophylactic use of zinc to avoid the development of deficiency in situations such as thermal burns or systemic chelation therapy, in which the individual is at risk of zinc depletion. Moreover, we shall restrict our review to somatic and organic illness. Cognitive and affective (psychological/ psychiatric) conditions such as learning impairment, hyperactivity and attention disorders, psychosis and depression – each often addressed with zinc treatment by orthomolecular therapists – have been considered beyond the scope of this inquiry.

Classification System

A hierarchy of judgment can be applied to the scientific rigour of the observations presented as evidence for a therapeutic role of any agent. This hierarchy is based on the nature of clinical observations and the design and execution of the clinical trials. Shown in Table 19.1 is the classification scheme we have adopted in our review and dissection of the evidence. The first level

Table 19.1. Classification scheme for reports of putative effects of oral zinc by level of experimental rigour

Level one (I)

A simple, anecdotal observation(s) of an apparent beneficial or non-beneficial effect of oral zinc on the manifestations or course of a given disease or condition, published in the form of a case report

Level two (II)

A report on the therapeutic response of a series of patients with a common diagnosis or condition treated with oral zinc as a "phase-one" trial

Level three (III)

A report on the therapeutic response of a series of patients with a common diagnosis or condition, treated with oral zinc, including a comparison with the results in another group of patients with the same diagnosis of condition left untreated, or treated with an alternative, traditional therapy. This is a non-controlled trial, as the comparison group is neither contemporaneous nor strictly matched and randomly assigned to the alternative treatment

Level four (IV)

A report on the therapeutic response of a series of patients with a common diagnosis or condition, in which some were assigned by random allocation (or some other distribution scheme) to receive oral zinc as the intervention and others were assigned in a similar fashion to receive a placebo drug or an alternative therapy. This represents a controlled clinical trial consists of a simple case report or clinical note, an anecdotal observation that the administration of zinc was coincident with an apparent improvement in a given illness or clinical condition. The second level includes what are commonly called "phase-one" studies in the parlance of pharmacological trials; that is, the administration of the agent (zinc) to a series of patients with a given condition in a non-controlled fashion and without a comparison reference group. The third level consists of a comparative study in which zinc is compared to another treatment (or to a placebo, or to a situation of no treatment), but in which no attempt to randomize or control the design has been made. The fourth level includes the prospective, simultaneous, randomized, double-blind, clinical trial (Chalmers 1981), or some reasonable facsimile thereof. The variants of level four include oral zinc treatment compared to a placebo or to another therapy or therapies; the study can be in a single phase, or in a cross-over design in which all subjects get both treatments. It is our basic tenet that as one goes up the hierarchical scheme of scientific rigour from levels one to four, the increasing objectivity allows greater confidence to be placed in the claims about the therapeutic efficacy of oral zinc as a reasonable therapeutic option.

Survey of Putative Roles for Oral Zinc Therapy

After a search and review of the literature, we have identified more than 25 illnesses or conditions in which oral zinc has been used therapeutically for a purpose other than to restore zinc status to normal. These represent situations in which some specific clinical sign(s) or symptom(s) of an organic, somatic disease entity or clinical manifestation was addressed. As shown in Table 19.2, 19 specific disease entities (i.e. diagnoses) and six less specific disorders, which can be primary or secondary conditions (i.e. manifestations), which respond to zinc therapy have been identified. To facilitate discussion and analysis further, we have organized the index conditions into categories related to the putative pathophysiological mechanisms (Table 19.3). The relevant literature associated with each condition identified is cited in the Appendix, which also summarizes the reported responses to zinc. In the subsequent sections, we shall classify and discuss the evidence for a therapeutic role for oral zinc treatment in each of these conditions.

Zinc and Red Cell Clogging of the Vasculature

Sickle Cell Disease

Sickle cell disease (SCD) is an inborn error of haemoglobin metabolism that results in conformational changes in the red cells (sickling). These sickled cells

Table 19.2. List of illnesses (diagnoses) and conditions (manifestations) to be considered in relation to their responses to oral zinc therapy

Disease entities
Acne
Acute lymphoblastic leukaemia
Alopecia areata (and related hair-loss conditions)
Coryza (the common cold)
Ehlers–Danlos syndrome
Oesophageal dysplasia
Furunculosis
Hepatic encephalopathy
Herpes simplex II (genital herpes)
Leprosy
Peptic disease (gastric and duodenal ulcers)
Perifolliculutis capitis abscedens et suffodiens (Hoffman)
Porphyria
Primary biliary cirrhosis
Prostatitis
Rheumatoid arthritis
Sickle cell anaemia
Toxic optic neuropathy (tobacco–alcohol amblyopia)
Wilson's disease

Conditions/manifestations
Cutaneous ulcers
Hypogeusia (impaired taste acuity)
Hyposmia (impaired smell acuity)
Impotence
Infertility
Surgical wound healing

Table 19.3. Classification by putative pathophysiological mechanisms of the diseases and conditions for which therapeutic responses to zinc have been investigated

Red cell clogging of vasculature
    Sickle cell disease

Connective tissue disorders
    Ehlers–Danlos syndrome

Copper storage diseases
    Wilson's disease
    Primary biliary cirrhosis

Epithelial reactions
    Peptic disease
    Alopecia areata
    Psoriasis
    Cutaneous ulcers
    Surgical wound healing

Gonadal dysfunction
    Infertility
    Impotence

Inflammations and infections
   Acne
   Furunculosis
   Perifolliculitis capitis abscedens et suffodiens (Hoffman)
   Herpes simplex II
   Leprosy
   Prostatitis
   Rheumatoid arthritis
   Coryza

Metabolic disorders
   Hepatic encephalopathy
   Porphyria

Neoplasia
   Acute lymphoblastic leukaemia
   Oesophageal dysplasia

Sensory disorders
   Toxic optic neuropathy
   Hypogeusia
   Hyposmia

---

clog the microvascular circulation and produce local ischaemia and necrosis. This explains the pathophysiology of the episodic, acute "sickle crises". It is known that patients with SCD are susceptible to zinc deficiency (Prasad et al. 1975, 1976, 1977). The issues at hand, however, are direct, pharmacological effects of zinc on the erythrocyte.

Both in vitro and in vivo demonstrations of zinc effects on red cells have been presented. It is known that virtually any agent that maintains maximal oxygen tension in cells containing haemoglobin-S helps them to resist sickling. Oelshlegel et al. (1973, 1974) showed in vitro binding of zinc to the haemoglobin molecule which increased O_2 affinity. This occurs, however, at zinc concentrations well in excess of those encountered in the human circulation. In 1979, Brewer and Krukeberg showed that zinc, even at low concentrations, directly influenced the membrane of red cells in SCD patients, and increased their in vitro filterability through small nucleopore filters. Even more elegant was the finding by Brewer and Bereza (1982) that zinc can expand red cell membranes and inhibit calmodulin, both effects beneficial in the treatment of SCD.

Several clinical trials of oral zinc as an anti-sickling agent have been reported. Brewer et al. (1977a) administered a dose of 25 mg of zinc as zinc acetate every 4 h to 12 SCD patients. The mean percentage of irreversibly sickled cells fell from 28% to 19% during treatment. When this treatment regimen was provided to 13 subjects in an uncontrolled (level-two) supplementation trial for between 2 and 30 months, a significant reduction in the frequency of painful crises from 6.1 per year before therapy to 2.3 per year during zinc treatment was observed (Brewer et al. 1977b). A controlled, long-duration formal clinical trial of zinc therapy in SCD also was initiated by the Michigan group (Brewer 1980). Although never published in full, the results of the prospective, placebo-controlled trial (level four) did not confirm the finding of an improved clinical response with oral zinc therapy (G.J. Brewer, personal communication). Thus, we conclude that the use of zinc to restore adequate zinc nutriture in zinc-depleted patients with SCD is justifiable, but its use to abort or reduce sickling crises has not been shown to be a reliable therapy.

Zinc and Connective Tissue Disorders

Ehlers–Danlos Syndrome

Ehlers–Danlos syndrome (EDS) is a connective tissue disease characterized by abnormal laxity of skin, joints and other connective tissues due to defective formation of collagen (McKusick 1972). Emser (1978a, b) in Germany provided a case report (level one) of an 8-year-old child with type I Ehlers–Danlos syndrome treated with 50 mg of zinc as zinc sulphate daily for 20 d and 25 mg of zinc per day for 10 d. A remarkable improvement in the clinical condition was observed, including a reduction of ecchymoses, a decrease in joint hyperelasticity and improved muscular tone. Initial serum zinc was not measured, but the concentration after 20 d of therapy was 1.12 mg/l, i.e. in the normal range. We assume that the child was not initially deficient. A year later, also from Germany, came another case report (level one) of zinc sulphate therapy in two additional type I, EDS patients, one aged 14 years and the other aged 4 years (Lenard and Lombeck 1979). They had normal circulating zinc levels at the beginning of zinc therapy. For 16 weeks, they received, respectively, 11 and 22 mg of elemental zinc daily; no beneficial effects were noted.

It has been suggested that deranged copper metabolism, reducing the activity of the cuproenzyme, lysyl oxidase, may be involved in the aetiology of EDS (Kuivaniemi et al. 1982). This leads to the intriguing possibility of mineral–mineral interactions. Even with the designation of EDS by clinical types, some pathophysiological heterogeneity may still exist. The children from the two reports may not have had an identical disease. In any event, the data on both sides of the zinc therapy issue in EDS are level one in nature; given the rarity of the disorder, a large controlled trial at a multicentre level will be needed to provide more conclusive information on the value of oral zinc.

Zinc and Copper Storage Disease

There is an antagonistic interaction between zinc and copper at the level of intestinal epithelium (Fischer et al. 1981; Menard et al. 1981). High doses of oral zinc produce copper deficiency in normal individuals, and recently have been used to block copper absorption in certain copper storage conditions.

Wilson's Disease

Wilson's disease (hepatolenticular degeneration) is an autosomal, recessive, inborn error of metabolism leading to toxic accumulation of copper in tissues such as liver and the basal ganglia of the brain. The conventional treatment is systemic chelation therapy with penicillamine (β β' dimethyl-cysteine) (Walshe 1984). However, a host of side-effects often emerge with chronic administration of penicillamine. Schouwink (1961) was the first to report the use of oral zinc to control the progression of Wilson's disease in two patients. There was amelioration and then stabilization of the condition in both. One of these individuals continued on 135 mg of zinc as the sulphate daily for 14 years with highly satisfactory control and no side-effects (Hoogenraad et al. 1979). Other case reports, each involving a small series (level one) of Wilson's disease patients treated with oral zinc as the copper-blocking agent, have appeared (Hoogenraad et al. 1978, 1984; Walshe 1984; Alexiou et al. 1985; Ramadori et al. 1985; Van Caillie-Bertrand et al. 1985). In all reports but that by Walshe (1984), evidence of acceptable control of the clinical course, biochemical variables and/or hepatic copper stores have been demonstrated.

A prospective study of a larger series of Wilson's patients was conducted at Utrecht by Hoogenraad et al. (1983). They followed four male patients aged 17 to 46 years for 3 years of close clinical monitoring. Control of neurological symptoms was good and biochemical indices remained stable.

Two metabolic balance studies from Ann Arbor, Michigan, USA, provide strong objective proof that the effect of oral zinc in patients with Wilson's disease is indeed to increase the amount of copper excreted in the faeces and to produce either zero, or even negative, balance with regard to total-body copper reserves (Brewer et al. 1983; Hill et al. 1988). They have shown that a constant period of oral zinc treatment is needed to induce the blockage of copper absorption and reabsorption by the intestine. They have also shown that the dosages may have to be highly individualized from patient to patient. This modality of oral zinc in Wilson's disease points now to its use as the initial, primary, therapy. Clearly, D-penicillamine is a more toxic drug than oral zinc. If the latter proves to be equally as effective in both "decoppering" and clinical control, a new era in the treatment of this inborn error of copper metabolism will have begun.

Primary Biliary Cirrhosis

Primary biliary cirrhosis (PBC) is an idiopathic disease common in middle-aged women, characterized by periportal hepatitis and destructive cholangitis leading to cirrhosis and portal hypertension (Sherlock and Scheuer 1973; James et al. 1981). As the hepatic deposition of copper in PBC was thought to be part of the destructive process, clinical trials with chelation therapy (penicillamine) have been carried out, most with negative therapeutic results. By analogy to Wilson's disease, others felt that copper levels could be controlled by blocking copper uptake at the gastrointestinal level with oral zinc. Hoogenraad and Sindram (1982) reported the reduction of copper levels in hepatic tissue in one PBC patient after a year of oral zinc therapy with 135 mg of zinc as zinc sulphate daily (level one); biochemical abnormalities remained unchanged. Olsson (1982) reported data from a series of four PBC sufferers treated for 1 year using 135 mg of elemental zinc daily for 9 months and 202 mg daily for the last 3 months (level two). No change in the clinical outcome or in the variable, hepatic copper content, was observed. Experience is limited, and no large-scale, well-designed trials have been executed. However, neither with penicillamine nor oral zinc as decoppering agents is there a sufficiently promising effect to think that removing the copper irritant in the hepatic tissues will mitigate the pathological features of PBC.

Zinc and Epithelial Reactions

Zinc is found in high concentration in the skin. It is felt to have some role in the keratinization of tissues (Lindelöf 1984). A host of diseases of epithelial tissue, both integumentary and gastrointestinal, have been treated with oral zinc. The evidence for and against a valuable role for zinc as a drug in five epithelial conditions is reviewed below. Topical zinc in the form of ointments and salves has long been used in dermatology; this form of administration, however, is beyond the scope of this chapter.

Peptic Disease

The therapeutic value of zinc for the treatment of both gastric and duodenal forms of peptic ulcer disease in humans has been investigated. Much recent experimental animal work supports a protective role for zinc against peptic ulceration.

Gastric Ulcer. A suggestion of an adjunctive effect was seen by Fraser et al. (1972) in London when zinc was given to, or withheld from, 12 patients respectively with gastric ulcers, each receiving 300 or 600 mg daily of an analogue of carbenoxolone in a level three-type study. The dose of zinc was 150 mg (as 660 mg of zinc sulphate). The degree of healing over 4 weeks averaged over 50% with zinc added to the regimen, and 22% without the metal. A case history (level one) of a patient with a persistent bleeding gastric ulcer which failed repeatedly to respond to conservative medical therapy showed that it finally stopped haemorrhaging when oral zinc (150 mg) was added to the regimen (Orr 1976). Frommer (1975) in Australia performed a controlled, double-blind study (level four) on 18 patients with gastric ulcers who received 150 mg of zinc as zinc sulphate daily. Over 3 weeks, the zinc group showed a statistically significant, threefold more rapid rate of ulcer healing than the control group. The data from all levels are convergent in showing a beneficial effect of zinc in gastric ulcer disease.

Duodenal Ulcer. In a study from Madrid, Alcala-Santaella et al. (1985) reported a 1-month, controlled, randomized, double-blind trial of zinc acexamate versus a placebo in the treatment of a duodenal ulcer. Eighty-six per cent of the 14 zinc-treated patients versus 44% of the placebo-treated patients were endoscopically healed at the end of the 4-week trial ($P < 0.05$). An interpretative issue in this study is which of the components had the therapeutic activity, as the acexamic acid (N-acetyl-amino-6-hexanoic acid) itself has activity as a topical wound-healing agent.

Alopecia Areata and Related Hair-Loss Conditions

Hair loss has been observed in situations of acquired zinc deficiency associated with diabetes mellitus (Nishiyama 1976), leprosy (Mathur et al. 1983a, b), Down's syndrome (DuVivier and Munro 1975) and penicillamine therapy (Klingberg et al. 1976). Alopecia areata (AA) is an idiopathic condition of partial patchy loss of scalp hair. Two closely allied, but more severe, conditions are alopecia totalis capillitii (full loss from the scalp) and alopecia universalis (or maligna) in which eyebrows, eyelashes, axillary and pubic hair are also affected. These entities will be treated collectively. In a comparative study of patients with AA versus controls, no difference in fasting serum zinc was seen, but the avidity of intestinal zinc uptake was greater in the AA patients (Sonnichsen et al. 1984). Numerous case reports (level one) and a cumulative phase-one study (level two) have come from a dermatology clinic in Warsaw, Poland (Wolowa and Jablonska 1976; Wolowa and Stachow 1978, 1980; Jablonska and Wolowa 1981; Wolowa 1982). Dosages of zinc from 135 to 170 mg daily were given for up to several years in all forms of idiopathic alopecia. In the most recent tally (Wolowa 1982) of 42 patients with AA treated for at least 3 months, 60% had regrowth of hair. Hair loss commenced again when zinc therapy was suspended.

Occasional success was reported with oral zinc treatment of alopecia totalis and universalis, but less consistently. No comparative or controlled studies have come from the Polish group. In the one controlled clinical trial of oral zinc therapy (level four) in AA from Manchester, England, 42 patients with AA or the more severe variants were randomized to receive 100 mg of zinc daily or a placebo for 3 months (Ead 1981). No differences in the activity of the disease were observed between treatment groups. Given the length of time before a response was seen in the uncontrolled trials in Poland, one could question whether the interval in the controlled Manchester trial was of sufficient duration. Clearly, further (longer) level-four studies of oral zinc supplementation in alopecia areata are warranted.

Psoriasis

A possible therapeutic effect of oral zinc on the course of psoriasis has been examined in two double-blind (level four) clinical trials, each with a series of 19 adult patients. Voorhees et al. (1969) found no improvement in the clinical disease with 2 months of daily treatment with 150 mg of oral zinc given as the sulphate. Similarly, Greaves and Dawber (1970) found no effect with 6 weeks of therapy with the same dosage schedule. Psoriasis is not among the inflammatory disorders for which oral zinc therapy holds promise.

Surgical Wound Healing

Zinc has specific functions in RNA and DNA metabolism, and specifically in the proliferation of collagenous connective tissue. Experimental zinc deficiency in humans reduces collagen formation in vivo (Prasad et al. 1978b). Thus, when a person is zinc deficient, it would be logical to expect problems in wound healing. Controversy reigns, however, as to whether oral zinc is of benefit in the healing of surgical wounds in individuals with only marginal (or no) evidence of impaired zinc nutriture. This topic has been reviewed in detail (Wacker 1976; Hallbook 1984). The classic study was by Pories et al. (1967a, b) and involved young, healthy adult men (airmen at a United States Air Force hospital) in a double-blind, randomized trial of 150 mg of oral zinc (as 220 mg of zinc sulphate) three times daily, or a placebo, for up to 60 d. They observed an increased rate of granulation and decreased time to complete healing in pilonidal cyst excision tracts in those treated with zinc. The zinc group ($n = 10$) healed at a rate of 1.25 ml per day, and showed complete healing in 46 days, whereas the placebo-treated group ($n = 10$) healed at a rate of 0.44 ml per day, and complete healing took 80 ± 14 d. These rates and intervals were statistically different.

Barcia (1970) performed an almost identical trial. His also involved 20 subjects undergoing pilonidal cyst excisions, who were randomized in the same way to zinc treatment or control groups, and who were supplied with 150 mg of zinc as the sulphate or with a placebo in the postoperative period. These subjects were US army personnel on an army base. The respective healing rates were 0.72 ml/d (with zinc) and 0.72 ml/d (without zinc), and the average number of days to complete healing were, respectively, 59.6 ± 12 and 63.7 ± 20 d. These differences were not statistically different.

Initial wound volumes and the mean ages of the subjects were comparable across studies. No immediate reason for the drastic contrast in results is apparent. Moreover, no subsequent replications of this study have been published. The value of oral zinc therapy in wound healing is a topic still meriting investigation, both with pilonidal cyst excision and in other standard, elective surgical procedures.

Cutaneous Leg Ulcers

Ulceration of epithelial surfaces can occur in almost any region of the body. It can result from pressure necrosis, from compromised blood supply and as a consequence of inflammation. Oral zinc was first suggested to be of therapeutic benefit in cutaneous ulcers in the form of decubitus ulcers (bedsores) (Cohen 1968). Since that time, oral zinc has been used in the treatment of cutaneous ulcerations resulting from venous stasis (Carruthers 1969; Husain 1969; Greaves and Skillen 1970; Myers and Cherry 1970; Clayton 1972; Greaves and Ive 1972a, b; Hallbook and Lanner 1972; Phillips et al. 1977; Floersheim and Lais 1980), and of ischaemic ulcers (Haeger and Lanner 1974) and trophic ulcers related to sickle cell disease (Sergeant et al. 1970), leprosy (Mathur and Bumb 1983) and systemic lupus erythematosus (Wang and Adams 1982), of tropical ulcers (Watkinson et al. 1985) and even of recurrent oral ulcers (Merchant et al. 1977).

The most available series of studies for evaluation are those for venous stasis ulcers. Six of these were randomized, controlled trials. Of the controlled studies, two (Husain 1969; Hallbook and Lanner 1972) showed a positive effect of oral zinc on healing, and four (Myers and Cherry 1970; Clayton 1972; Greaves and Ive 1972a, b; Phillips et al. 1977) showed no benefit from zinc administration.

In the assorted individual reports of the effects of zinc therapy on the healing of other types of ulcers, for example tropical ulcers and those found in sickle cell anaemia, leprosy, SLE etc., results are again variable. In general, however, the evidence for beneficial effects on ulcer healing of oral zinc is not overwhelming. We should reserve its use until such a time as additional studies clearly overturn the present preponderance of negative findings for the effects of oral zinc treatment on cutaneous ulcers.

Zinc and Gonadal Dysfunction

From the early observations of Prasad et al. (1963), it was known that zinc deficiency could produce hypogonadism and delayed sexual maturation. Experimental human zinc deprivation produces gonadal failure characterized by depressed testosterone levels (Abbasi et al. 1980) and oligospermia (Abbasi et al. 1979). Thus, it was logical to suppose that certain forms of gonadal dysfunction might be related to deranged zinc metabolism.

Infertility

Zinc plays an important role in spermatogenesis and sperm physiology. Studies by Kvist et al. (1985) and Bjorndahl et al. (1986) have demonstrated that an appreciable amount of prostatic zinc is taken up by sperm during ejaculation, and that this zinc seems to be necessary to inhibit the condensation of the spermatic chromatin. A further uptake of zinc within the female genital tract is necessary to allow the sperm to penetrate the ovum. It has been noted that a sizeable number of sperms of patients with prostatic dysfunction show premature chromatin condensation (Kvist and Eliasson 1980).

Marmar et al. (1975) in an open, uncontrolled study (level two), gave 11 patients with oligospermia and low seminal zinc concentrations 54 mg of zinc daily (as 240 mg of zinc sulphate) for 6 months. They noted improvement in sperm counts, and pregnancies resulted in three of the couples. Hartoma et al. (1977) in Finland, in a similarly uncontrolled trial (level two), supplemented 10 oligospermic, infertile men with 150 mg of zinc daily (as zinc sulphate) for 4 to 8 weeks; sperm counts and circulating testosterone levels both increased. In yet another level-two, uncontrolled, non-comparative study, this time from Rochester, New York, 20 infertile men were selected who had low seminal zinc concentrations and hypomotile sperm. After 2 months of supplementation with 100 mg of zinc daily as zinc sulphate, sperm motility improved, and three pregnancies were confirmed (Caldamone et al. 1979). A more complex study design from the same Rochester fertility clinic compared different treatment regimens in men with poor sperm motility and low seminal zinc concentrations. These treatments were as follows: (a) a synthetic androgen (fluoxymesterone); (b) zinc at 100 mg per day as the sulphate; and (c) a combined therapy. Combination therapy proved to be the most effective regimen for improving sperm motility. The results are promising, yet inconclusive. Even better designed, controlled studies are warranted in the field of male infertility and oral zinc therapy.

Impotence

In uraemic zinc deficiency, oral zinc therapy has been shown to reverse impotence (Antoniou et al. 1977; Mahajan et al. 1980). It is thus legitimate to inquire if zinc has any influence on erectile dysfunction in zinc-adequate subjects. Despite such mythology concerning oysters (a zinc-rich food) and sexual prowess, and the inclusion of zinc in aphrodisiac and potency-restoring preparations, we found no reports of the use of oral zinc in the treatment of idiopathic impotence in the scientific literature.

Zinc and Inflammatory Conditions and Infections

A series of inflammatory conditions and infections have been studied with respect to the beneficial effects of oral zinc.

Acne

Acne vulgaris is a common affliction, specifically related to changes in sebum secretion during the adolescent transition. In some instances, however, it becomes so persistent and disfiguring that it requires medical intervention. Strain and Pories (1966) first suggested, on theoretical grounds, that zinc treatment would be beneficial for acne. However, in a subsequent, open, non-comparative trial of oral zinc therapy in eight acne patients, no improvement was found (Burton and Goolamali 1973). Experience since that time has been extensive. We have been able to review nine clinical trials of oral zinc treatment in acne with various dosages and durations (Hillstrom et al. 1977; Michaelsson et al. 1977a, b; Weissman et al. 1977; Goransson et al. 1978; Orris et al. 1978; Weimar et al. 1978; Cunliffe et al. 1979; and Verma et al. 1980). Weissman et al. (1977) and Orris et al. (1978) found no beneficial effects of zinc therapy. The studies of Weimar et al. (1978) and Cunliffe et al. (1979) had equivocal results, and the remaining studies, four of them double-blind controlled trials, had positive results with oral zinc therapy. Michaelsson (1984) reviewed a total of 14 studies of oral zinc therapy in acne, 12 of which involved comparisons with a placebo ($n = 8$), with vitamin A ($n = 2$), with vitamin E ($n = 1$) or with tetracyclines ($n = 3$). He notes a differentially beneficial effect of oral zinc over the comparative options in 12 of the reports. Thus, it is reasonable to conclude that zinc is more effective than some other popular remedies in the resolution of severe acne.

Furunculosis

In a comparative, uncontrolled (level-three) study from Sweden, 15 patients with recurrent furunculosis were followed (Brody 1977). In seven, conventional excision and drainage together with antibiotic therapy was instituted, and the boils returned. In the remaining eight subjects, 135 mg of zinc was taken daily for 3 months, and the lesions regressed. Complicating the interpretation of this study as to its pharmacological or nutritional relevance is the finding of low serum zinc levels (0.45 to 0.60 mg/l) in the patients. The value of zinc therapy in clearly zinc-adequate furunculosis patients should be subjected to scrutiny in prospective, controlled clinical trials.

Perifolliculitis Capitis Abscedens et Suffodiens (Hoffman)

Perifolliculitis capitis abscedens et suffodiens (Hoffman) is a rare, suppurative, dissecting cellulitis of the scalp. In a 24-year-old man with normal serum zinc levels and a 1-year history of Hoffman's perifolliculitis capitis resistant to antibiotic therapy, oral zinc therapy was instituted (Berne et al. 1985). The initial dose was 270 mg of zinc as 1200 mg of zinc sulphate daily for the first 10 weeks, followed by 135 mg daily thereafter. The cellulitis resolved with complete regrowth of hair after 12 weeks. Neutrophil function tests, initially impaired, were also normalized during the course of the zinc treatment. This is a rare condition and it would be hard to amass a series of cases. The promising and dramatic results in this Swedish patient call for additional experience with oral zinc as a therapeutic intervention.

Herpes Simplex II (Genital Herpes)

Zinc is capable of inhibiting replication of herpes simplex virus in vitro (Zimmer et al. 1974; Gordon et al. 1975). An open trial of zinc supplementation (level two) was carried out on 10 Australian patients with recurrent genital herpes (Jones 1979); they received a daily regimen of 50 mg of zinc as zinc sulphate (along with 50 mg of magnesium sulphate, 5 mg of thiamine and 5 mg of riboflavin) for 3 months. Both frequency of attacks and number of days of herpes recurrence were reduced by 50% as compared with the preceding period. These results with humans are at variance with those of Tennican et al. (1979) suggesting that topical, but not systemic, zinc administration prevented the in vivo replication of the herpes simplex-2 virus in mice. The human trial dealt with recurrent infection (Jones 1979), whereas the murine model dealt with primary infection (Tennican et al. 1979). Clearly the possible contribution of other nutrients obscures the interpretation of the role of zinc, per se, in this trial. The "phase-one" nature of the design limits our confidence in the therapeutic efficacy of the regimen. A truly double-blind and cross-over design would be needed to evaluate the role of oral zinc as a therapeutic agent in recurrent human genital herpes.

Leprosy

Mathur and colleagues (1983b, 1984) have explored oral zinc as an adjunctive therapy in patients with lepromatous leprosy at a leprosarium near Jaipur, India. The bactericidal effects of antibiotics against *Mycobacterium leprae* are not curative until the depressed immune system removes the dead, but antigenically intact, bacteria. In a comparison of standard antibiotic (dapsone) therapy in 10 lepromatous leprosy patients with dapsone therapy plus 50 mg daily of zinc as 220 mg of zinc sulphate in 15 similar subjects (level three) (Mathur et al. 1984), the zinc-treated lepers showed more rapid clinical improvement, a greater fall in bacterial index, more abundant regrowth of eyebrows and a stronger cellular immune response. In eight additional patients with recurrent, leprosy-induced erythema nodosum, in a level-two study, Mathur et al. (1983b) administered 50 mg daily of elemental zinc. The need for suppressive medication and the incidence and severity of the erythema nodosum were reduced over a period of 3–4 months. Circulating zinc levels are mildly depressed in leprosy patients, and some confusion remains as to whether a nutritional or a pharmacological phenomenon is occurring. Appraisal of the value of oral zinc in leprosy awaits a truly controlled study to produce definitive conclusions about its therapeutic role.

Prostatitis

Chronic bacterial prostatitis (CBP) is not an uncommon affliction in men past middle life. Studies from Washington University in St. Louis, Missouri (Fair and Wehrer 1973; Fair et al. 1973; Fair et al. 1976) suggest that zinc, in a low-molecular-weight complex, may be a bacteriostatic/bactericidal agent ("prostatic antibacterial factor"). In CBP, prostatic fluid zinc levels are greatly depressed (Marmar et al. 1975; Fair and Heston 1977). Treatment of 42 prostatitis patients aged 27 to 70 years who had initially normal circulating zinc levels, with 50 or 100 mg of oral zinc daily as zinc gluconate resulted in a rise in serum zinc, but not prostatic fluid zinc, levels (Fair and Heston 1977). More importantly, no improvement in the clinical course of prostatic infection was observed. Zinc metabolism is obviously deranged in CBP, but most likely as a secondary phenomenon. These level-two studies provide little encouragement for suggestions that better-designed, comparative studies involving oral zinc might show it to be a useful therapeutic agent.

Rheumatoid Arthritis

At least some studies (Grennan et al. 1980; Kennedy et al. 1980) suggest that rheumatoid arthritis (RA) patients maintain lower plasma zinc levels than controls. This, of course, could be due to a redistribution of circulating zinc caused by the inflammatory stimulus rather than a true sign of depletion. Such observations, however, raise questions about a potential role for zinc as an anti-arthritic agent in RA. Experimental animal work with Freund's adjuvant-induced arthritis, a rodent model of human RA, has shown varied results. In France, Job et al. (1980) found no effect of an unspecified amount of zinc sulphate given to male rats with adjuvant arthritis. Swerdel and Cousins (1984) in Gainesville, Florida, however, found 0.4 mg of zinc given intraperitoneally daily reduced inflammation, possibly through a metallothionein-mediated anti-inflammatory mechanism.

Human experience with oral zinc treatment of RA is extensive and conflicting. In two open, uncontrolled (level-two) trials, both positive results (decreased inflammation and lowered sedimentation rates) in a group of 15 French RA patients (Menkes et al. 1978; Job et al. 1980) and negative results (no improvement or clinical deterioration) in 22 Dutch RA patients (Rasker and Kardaun 1982) have been reported.

Three double-blind, controlled (level-four) trials have been reported. Simkin (1976) in Seattle, Washington, randomly selected 24 RA patients to add either 150 mg of zinc daily as zinc sulphate or a placebo to their prescribed, individual medications. After 12 weeks, morning stiffness, walking time and joint swelling were all improved in the zinc group. The placebo group was switched to oral zinc therapy for another 12 weeks, in an open trial. During this phase, both groups improved. Unfortunately, neither Menkes et al. (1981) in France with 35 RA subjects given zinc or a placebo for 4 months, nor Mattingly and Mowat (1982) in England with 27 RA patients given zinc or a placebo for 6 months showed any objective anti-rheumatic effects from oral zinc. The balance of the evidence from the well-designed clinical trials in rheumatoid arthritis patients, therefore, is inclined against oral zinc as a useful agent in this disease.

Common Cold (Coryza)

The effect of sucking zinc gluconate lozenges during a typical cold was studied in a randomized, double-blind clinical trial format (level four) on the university campus in Austin, Texas (Eby et al. 1984). Treatment was initiated with 46 mg of zinc given as two zinc gluconate lozenges, dissolved in the mouth, and one lozenge (23 mg of zinc) was taken every 2 waking hours thereafter. After 7 d, 86% of the 37 zinc-treated cold victims and only 46% of the 28 placebo-treated subjects were asymptomatic. It was estimated that the oral zinc shortened the duration of the common cold by 1 week. It is not clear whether a local effect on the pharyngeal mucosa or a systemic effect is responsible. Surprisingly, this study has not been repeated. Based on the one experience, well designed and executed, we conclude that a palliative role exists for zinc lozenges in coryza.

Zinc and Metabolic Disorders

Hepatic Encephalopathy

Hepatic encephalopathy is often the result of acute hepatic necrosis or end-stage chronic cirrhosis. Hepatologists in Belgium noted serendipitously that administering 600 mg of zinc acetate (222 mg of zinc) in an effort to raise serum zinc levels also improved intellectual performance (on the "trial-making test") in five, abstinent, outpatient, alcoholic cirrhotics with stable, chronic hepatic encephalopathy (level two) (Reding et al. 1984). Patients were already receiving lactulose and a protein-restricted diet. Prospectively, a randomized, controlled trial was then initiated involving 22 similar patients who were given a placebo ($n = 12$) or the same dose of zinc acetate ($n = 10$). Mean serum zinc levels were low at the beginning of the trial in both groups, but had risen by 60% in the zinc-treated group by the end of the trial. Most importantly, intellectual function improved by 20% in the zinc acetate group, and remained unchanged in the controls. Again, a potential interaction of zinc deficiency is possible, rather than a pure drug effect. Since hepatic encephalopathy only occurs in situations that also predispose to zinc depletion, however, the question is rather academic. A good rationale for oral zinc therapy as an adjunct to other chronic measures in stable encephalopathic cirrhotics is provided by this report.

Porphyria

Roman et al. (1967, 1969) describe in case-report format (level one) the history of a 32-year-old Australian woman with recent onset of acute intermittent porphyria, who received apparent relief from the abdominal pain crises with the administration of 50 mg of zinc as zinc sulphate every 8 h over 72 h. The zinc did not reverse the biochemical feature of urinary porphobilogen excretion. This form of porphyria is rare. Further observations on the analgesic benefits to be derived from oral zinc administration during attacks of acute intermittent porphyria are merited.

Zinc and Neoplasia

Acute Lymphoblastic Leukaemia

Oral zinc therapy has been instituted in patients with acute lymphoblastic leukaemia (ALL) in an attempt to boost immunocompetence, and hence to extend remissions. Wazewska-Czyzewska et al. (1978), in Poland, performed a non-randomized, comparative (level-three) trial in 27 ALL patients, aged 4 to 21 years, on maintenance chemotherapy after induction of remission. Fifteen subjects received 0.02 mg of oral zinc/kg body weight daily and 12 received a placebo during the last 8 d of a 22-d chemotherapy cycle. A statistically significant increase in T lymphocytes was observed in the zinc group. Its significance for the long-term outcome in ALL cannot be interpreted since firm clinical variables such as tolerance to maintenance chemotherapy, relapse rate and ultimate mortality were not monitored. Studies including the aforementioned indicators would be of greater transcendence to the field of haematological oncology.

Oesophageal Dysplasia

The search for risk factors associated with precancerous oesophageal lesions has been directed towards nutritional and dietary issues. Low intakes of riboflavin, vitamin A and zinc have been encountered in regions of high risk of oesophageal cancer (Thurnham et al. 1982). Huixian in Henan Province of China is an endemic high-risk area for oesophageal carcinoma. A supplementation trial of 610 subjects, aged 35 to 64 years, living in this area, was carried out. During 13.5 months they were randomized to receive a weekly treatment of a placebo or an "active" capsule with 50 mg of zinc, 200 mg of riboflavin and 50 000 IU of retinol (Munoz et al. 1985). A final, single endoscopic examination and oesophageal biopsy was performed. Cancers were diagnosed in four subjects from each group. The prevalence of oesophagitis, with or without atrophy or dysplasia, was 45% in the placebo group and 49% in the treatment cohort; these results are not statistically different. Once again, the combination of nutrients would have confounded interpretation as to the effective agent. If we assume that the groups were comparable at baseline (only one endoscopic examination was logistically acceptable), we must conclude that oral nutrient therapy including zinc was ineffective.

Sensory Disorders

Optic Neuropathy (Alcohol–Tobacco Amblyopia)

Bechetoille et al. (1983) followed a large number of patients with optic nerve impairment due to tobacco and alcohol abuse. In an open trial, they had noted improvement in visual function with oral zinc administration. They then conducted a randomized double-blind trial in 17 male patients with visual impairment characterized by decreased acuity, central visual field loss (scotoma) and abnormalities of colour discrimination. Eight were given 90 mg of oral zinc daily as the sulphate salt, and nine received a placebo. Over the course of the trial, circulating zinc levels rose only in the patients given oral zinc, but there was an improvement in visual function in all parameters in both groups. The only differential improvement that could be ascribed to oral zinc was a reduction in central scotoma. It is likely that the nutritional attention, per se, of participating in the study benefited all subjects. The utility of zinc beyond this may be marginal. It is unlikely to represent a specific therapeutic agent for tobacco–alcohol optic neuropathy.

Hypogeusia (Impaired Taste Acuity)

Hypogeusia is a loss of acuity in taste perception both for the detection and recognition of flavours. Dysgeusia is the perception of unpleasant tastes, and often occurs along with hypogeusia. Loss of taste can occur as the result of tumours, surgical interventions and neurological lesions. Experimental and acquired zinc deficiency produces taste impairments (Russell et al. 1983). It was felt, therefore, that idiopathic hypogeusia (or dysgeusia), with or without evidence of altered zinc metabolism, might be benefited by oral zinc treatment. An initial observation of a potential ameliorating effect on taste disorders for oral zinc was documented by Henkin and Bradley (1970) in a case report (level one). In a subsequent open, staged, single-blind, comparative trial (level three), Schecter et al. (1972) reported that oral zinc was more effective than a placebo for improving hypogeusia. Finally, in a prospective study with a very exacting, double-blind, partial cross-over design, Henkin et al. (1976) studied 106 consecutive patients attending the taste and smell clinic of the National Institutes of Health. Patients were randomized to one of four sequences: (a) oral zinc for both of two consecutive 3-month periods; (b) a placebo on both occasions; (c) zinc followed by placebos; and (d) a placebo followed by zinc. The dose of zinc was 100 mg daily as 220 mg of zinc sulphate. Among the patients with idiopathic taste disorders (hypogeusia with or without dysgeusia), no objective effects of taste-test performance could be demonstrated.

Hyposmia (Impaired Olfactory Acuity)

The analogous "smell" disorder to hypogeusia is hyposmia, in which the threshold for the detection and recognition of odours is raised. It is often associated with dysosmia, the perception of unpleasantness in odours. The Henkin and Bradley (1970) level-one study also had shown some potential for divalent cations to improve smell function. However, in the same elaborate, cross-over supplementation study with oral zinc reported from the NIH by Henkin et al. (1976), no remedial effect of zinc over placebo was noted among the patients with idiopathic olfactory acuity disorders. Thus, we must conclude that, in the absence of nutritional zinc deficiency, no reliable or predictable therapeutic effect of oral zinc therapy can be expected for patients with sensory disorders of taste or smell.

Conclusions and Recommendations

It is a fundamental goal of medical science to provide the most reliable, safe and effective therapeutic modalities for the treatment of disease and disability. It is also a fundamental aim of the branch of nutritional sciences concerned with the trace elements, and specifically with zinc, to maintain outstanding scientific credibility. These two imperatives made this inquiry into the putative therapeutic use of oral zinc both timely and important. Zinc is promoted as a useful agent for the therapy of numerous conditions and diseases, both psychic and somatic. When the manifestation is indeed due to impaired nutriture or diminished body stores of zinc, oral zinc therapy is appropriate and indispensible. When the manifestation is not due to zinc deficiency, whether the person is adequate or subadequate in his or her zinc nutriture, it will only respond to the pharmacological (drug) action of zinc if zinc has an inherent beneficial influence on the pathophysiology of the disorder.

Two reasons necessitating that one be reasonably certain that zinc is better than a placebo before instituting therapy are obvious: firstly, to prevent wasting the patients' time and effort, and perhaps delaying the institution of a truly effective remedy; and secondly, to prevent exposing the person to the side-effects of oral zinc which are not inconsiderable. Oral zinc produces a metallic taste, nausea and epigastric irritation (Walshe 1984), and can cause mucosal erosions (Moore 1978) and overt copper deficiency anaemia (Pfeiffer and Jenney 1978; Porter et al. 1977; Prasad et al. 1978a, b; Patterson et al. 1985).

Our present dissection of the vast literature on oral zinc in somatic diseases unrelated to zinc deficiency per se has revealed some firm data from well-designed studies suggesting a therapeutic role in some instances. Most important of these are a role for zinc in Wilson's disease to produce decoppering, in acne and in peptic ulcer disease. In such common diseases as genital herpes and the common cold, confirmatory, well-designed studies should be carried out to follow up on promising results seen in isolated cases. This is also the case with regard to less common conditions such as infertility, leprosy, chronic hepatic encephalopathy, alopecia areata and other dermatological conditions reviewed. It is probably reasonable to conclude on the other side, based on the preponderance of evidence reviewed, that for primary biliary cirrhosis, rheumatoid arthritis, cutaneous leg ulcers, impaired wound healing, chronic bacterial prostatitis and idiopathic losses of taste and smell acuity unassociated with concomitant systemic zinc deficiency, little benefit of zinc over placebos is to be expected, and zinc is not a reliably effective therapeutic agent.

Sincere therapeutic decision making requires a familiarity with clinical trial results. A large number of the trials were well designed with appropriate controls and placebo therapy. It is hoped that physicians and nurse practitioners will make use of the more rigorous evidence concerning oral zinc in attempting to devise therapeutic regimens that will have the highest benefit/risk ratios for their patients.

Appendix

Citation	Type of study	Therapeutic effect of zinc
A: Red cell clogging of the vasculature		
Sickle cell disease		
G. Brewer (personal communication)	Controlled (IV)	−
Brewer et al. (1977a)	Phase one (II)	+
Brewer et al. (1977b)	Phase one (II)	+
B: Connective tissue disorders		
Ehlers–Danlos syndrome		
Emser (1978a, b)	Case report (I)	+
Lenard and Lombeck (1979)	Case report (I)	−
C: Copper storage disorders		
Primary biliary cirrhosis		
Hoogenraad and Sindram (1982)	Case report (I)	±
Olsson (1982)	Phase one (II)	−
Wilson's disease		
Alexiou et al. (1985)	Case report (I)	+
Brewer et al. (1983)	Phase one (II)	+
Hill et al. (1987)	Phase one (II)	+
Hoogenraad et al. (1983)	Phase one (II)	+
Hoogenraad et al. (1978)	Case report (I)	+
Hoogenraad et al. (1979)	Case report (I)	+
Hoogenraad et al. (1984)	Case report (I)	+
Ramadori et al. (1985)	Case report (I)	+
Schouwink (1961)	Case report (I)	+
Van Caillie-Bertrand et al. (1985)	Case report (I)	+
Walshe (1984)	Case report (I)	−
D. Epithelial reactions		
Peptic disease		
1. Gastric ulcer		
Fraser et al. (1972)	Controlled (IV)	±
Frommer (1975)	Controlled (IV)	+
Orr (1976)	Case report (I)	+
2. Duodenal ulcer		
Alcala-Santaella et al. (1985)	Controlled (IV)	+
Alopecia areata		
Ead (1981)	Controlled (IV)	−
Sonnichsen et al. (1984)	Phase one (II)	+
Wolowa and Jablonska (1976)	Phase one (II)	+
Wolowa and Stachow (1980)	Phase one (II)	+
Psoriasis		
Greaves and Dawber (1970)	Controlled (IV)	−
Voorhees et al. (1969)	Controlled (IV)	−
Cutaneous leg ulcer		
Clayton (1972)	Controlled (IV)	−
Cohen (1968)	Case report (I)	+
Floersheim and Lais (1980)	Non-controlled (III)	−
Greaves and Skillen (1970)	Phase one (II)	+
Greaves and Ive (1972a, b)	Controlled (IV)	−

Hallbook and Lanner (1972)	Controlled (IV)	+
Husain (1969)	Controlled (IV)	+
Myers and Cherry (1970)	Controlled (IV)	−
Phillips et al. (1977)	Controlled (IV)	−

Surgical wound healing

Barcia (1970)	Controlled (IV)	−
Pories et al. (1967a, b)	Controlled (IV)	+

E. Gonadal dysfunction

Infertility

Caldamone et al. (1979)	Phase one (II)	+
Hartoma et al. (1977)	Phase one (II)	+
Marmar et al. (1975)	Non-controlled (III)	+

F. Inflammatory conditions and infections
Acne

Burton and Goolamali (1973)	Phase one (II)	−
Cunliffe et al. (1979)	Controlled (IV)	±
Goransson et al. (1978)	Controlled (IV)	+
Hillstrom et al. (1977)	Controlled (IV)	+
Michaelsson et al. (1977a)	Controlled (IV)	+
Michaelsson et al. (1977b)	Controlled (IV)	+
Orris et al. (1978)	Controlled (IV)	−
Verma et al. (1980)	Controlled (IV)	+
Weimar et al. (1978)	Non-controlled (III)	±
Weissman et al. (1977)	Non-controlled (III)	

Furunculosis

Brody (1977)	Non-controlled (III)	+

Perifolliculitis capitis abscedens et suffodiens (Hoffman)

Berne et al. (1985)	Case report (I)	+

Herpes simplex, type II

Jones (1979)	Phase one (II)	+

Leprosy

Mathur et al. (1983b)	Phase one (II)	+
Mathur et al. (1984)	Non-controlled (III)	+

Prostatitis

Fair and Heston (1977)	Case report (I)	−

Rheumatoid arthritis

Job et al. (1980)	Controlled (IV)	−
Mattingly and Mowat (1982)	Controlled (IV)	−
Menkes et al. (1978)	Phase one (II)	+
Menkes et al. (1981)	Controlled (IV)	−
Rasker and Kardaun (1982)	Phase one (II)	−
Simkin (1976)	Controlled (IV)	+

Common cold

Eby et al. (1984)	Controlled (IV)	+

G. Metabolic disorders

Hepatic encephalopathy

Reding et al. (1984)	Controlled (IV)	+

Porphyria

Roman et al. (1967)	Case report (I)	+
Roman et al. (1969)	Case report (I)	+

H. Neoplasia

Acute lymphoblastic leukaemia		
Wazewska-Czyzewska et al. (1978)	Controlled (IV)	+
Oesophageal dysplasia		
Munoz et al. (1985)	Controlled (IV)	−

I. Sensory disorders

Optic neuropathy (tobacco–alcohol)		
Bechetoille et al. (1983)	Controlled (IV)	±
Hypogeusia		
Henkin and Bradley (1970)	Case report (IV)	+
Henkin et al. (1976)	Controlled (IV)	−
Schechter et al. (1972)	Phase one (II)	+
Hyposmia		
Henkin et al. (1976)	Controlled (IV)	−

References

Abbasi AA, Prasad AS, Rabbani PR (1979) Experimental zinc deficiency in man. Effect on spermatogenesis. Trans Assoc Am Physicians 92: 292–302

Abbasi AA, Prasad AS, Rabbani PR, DuMouchelle E (1980) Experimental zinc deficiency in man. Effect on testicular function. J Lab Clin Med 96: 544–550

Alcala-Santaella R, Castellanos D, Velo JL, Gonzalez-Lara V (1985) Zinc acexamate in treatment of duodenal ulcer. Lancet II: 157

Alexiou D, Hatzis T, Koutselinis A (1985) Traitement d'entretien de la maladie de Wilson par le zinc per os. A propos d'un enfant traite pendant 4 ans. Arch Fr Pediatr 42: 447–449

Antoniou LD, Shalhoub RJ, Sudhakar T, Smith JC Jr (1977) Reversal of uraemic impotence by zinc. Lancet II: 895–898

Barcia PJ (1970) Lack of acceleration of healing with zinc sulfate. Ann Surg 172: 1048–1050

Bechetoille A, Ebran JM, Allain P, Mauras Y (1983) Effet therapeutique du sulfate de zinc sur le scotome central des neuropathies optiques alcoolo-tabagiques. J Fr Ophtalmol 6: 237–242

Berne B, Venge P, Ohman S (1985) Perifolliculitis capitis abscedens et suffodiens (Hoffman): complete healing associated with oral zinc therapy. Arch Dermatol 121: 1028–1030

Bjorndahl L, Kjellberg S, Roomans GM, Kvist U (1986) The human sperm nucleus takes up zinc at ejaculation. Int J Androl 9: 77–80

Brewer GJ (1980) Zinc and copper in hematology. In: Karcioglu ZA, Sarper RM (eds) Zinc and copper in medicine. Charles C. Thomas, Springfield, pp 347–375

Brewer GJ, Bereza U (1982) Therapy of sickle cell anemia with membrane expander/calmodulin inhibitor classes of drugs. In: Prasad AS (ed) Clinical, biochemical and nutritional aspects of trace elements. Alan R. Liss, New York, pp 211–220

Brewer GJ, Krukeberg WC (1979) The anticalcium and erythrocyte membrane effects of zinc, and their potential value in the treatment of sickle cell anemia. In: Rosa J. Beuzard Y, Hercules J (eds) Development of therapeutic agents for sickle cell disease. North Holland Biomedical Press, New York, p 195–204

Brewer GJ, Brewer LF, Prasad AS (1977a) Suppression of irreversibly sickled erythrocytes by zinc therapy in sickle cell anemia. J Lab Clin Med 90: 549–554

Brewer GJ, Schoomaker EB, Leichtmen DA, Kruckeberg WC, Brewer LF, Meyers N (1977b) The use of pharmacological doses of zinc in the treatment of sickle cell anemia. In: Brewer GJ, Prasad AS (eds) Zinc metabolism; current aspects in health and disease. Alan R. Liss, New York, pp 241–254

Brewer GJ, Hill GM, Prasad AS, Cossack ZT, Rabbani R (1983) Oral zinc therapy for Wilson's disease. Ann Intern Med 99: 314–320

Brody I (1977) Treatment of recurrent furunculosis with oral zinc. Lancet II: 1358

Burton JL, Goolamali SK (1973) Zinc and sebum excretion. Lancet I: 1448

Caldamone AA, Freytag MK, Cockett ATK (1979) Seminal zinc and male infertility. Urology 13: 280–281

Carruthers R (1969) Oral zinc sulphate in leg ulcers. Lancet I: 1264

Chalmers TC (1981) The clinical trial. Milbank Mem Fund Q 59: 324–339

Clayton RJ (1972) Double-blind trial of oral zinc sulphate in patients with leg ulcers. Br J Clin Pract 26: 368–370

Cohen C (1968) Zinc sulphate and bed sores. Br Med J 2: 561

Cunliffe WJ, Burke B, Dodman B, Gould DJ (1979) A double-blind trial of a zinc sulfate–citrate complex and tetracycline in the treatment of acne vulgaris. Br J Dermatol 101: 321–325

DuVivier A, Munro DD (1975) Alopecia areata autoimmunity and Down's syndrome. Br Med J 1: 191–192

Ead RD (1981) Oral zinc sulphate in alopecia areata – a double blind trial. Br J Dermatol 104: 483–484

Eby GA, Davis DR, Halcomb WW (1984) Reduction in duration of common colds by zinc gluconate lozenges in a double-blind study. Antimicrob Agents Chemother 25: 20–24

Emser W (1978a) Zink-Therapie beim Ehlers–Danlos–Syndrom. Mschr Kinderheilk 126: 347–348

Emser W (1978b) Fall eines Ehlers–Danlos–Syndrom und seine Behandlung mit Zink. Klin Padiatr 190: 397–402

Fair WR, Heston WDW (1977) The relationship of bacterial prostatitis and zinc. In: Brewer GJ, Prasad AS (eds) Zinc metabolism: current aspects in health and disease. Alan R. Liss, New York, pp 129–140

Fair WR, Wehrer N (1973) The antibacterial action of canine prostatic fluid and human seminal plasma in an agar diffusion assay system. Invest Urol 10: 262–265

Fair WR, Couch J, Wehrer N (1973) The purification and assay of the prostatic antibacterial factor (PAF). Biochem Med 8: 329–339

Fair WR, Couch J, Wehrer N (1976) The prostatic antibacterial factor: identity and significance. Urology 7: 169–177

Fischer PW, Giroux A, L'Abbe MR (1981) The effect of zinc on intestinal copper absorption. Am J Clin Nutr 34: 1670–1675

Floersheim G, Lais E (1980) Fehlender Einfluss von oralem Zinksulfat auf die Wundheilung bei Ulcus cruris. Schweiz Med Wochenschr 110: 1138–1145

Fraser PM, Doll R, Langman MJS, Misiewicz JJ, Shawdon HH (1972) Clinical trial of a new carbenoxolone analogue (BX24), zinc sulphate, and vitamin A in the treatment of gastric ulcer. Gut 13: 459–463

Frommer D (1975) The healing of gastric ulcers by zinc sulphate. Med J Aust 2: 793–796

Goransson K, Liden S, Odsell L (1978) Oral zinc and acne vulgaris. A clinical and methodological study. Acta Derm Venereol (Stockh) 58: 443–448

Gordon YJ, Asher Y, Becker Y (1975) Irreversible inhibition of Herpes simplex virus replication in BSC-1 cells by zinc ions. Antimicrob Agents Chemother 8: 377–380

Greaves MW, Dawber R (1970) Zinc in psoriasis. Lancet I: 1295

Greaves MW, Ive FA (1972a) Double-blind trial of zinc sulphate in the treatment of chronic venous leg ulceration. Br J Dermatol 87: 632–634

Greaves MW, Ive FA (1972b) Serum-zinc and healing of venous leg ulcers. Lancet II: 1261

Greaves MW, Skillen AW (1970) Effects of long-continued ingestion of zinc sulphate in patients with venous leg ulceration. Lancet II: 889–892

Grennan DM, Knudson JML, Dunckley J, Mackinnon MJ, Myers DB, Palmer DG (1980) Serum copper and zinc in rheumatoid arthritis and osteoarthritis. NZ Med J 91: 47–50

Haeger K, Lanner E (1974) Oral zinc sulphate and ischaemic leg ulcers. Vasa 3: 77–81

Hallbook T (1984) Zinc and wound healing. In: Zinc in human medicine. TIL Publications, Isleworth, Toronto, pp 51–59

Hallbook T, Lanner E (1972) Serum zinc and healing of venous leg ulcers. Lancet II: 780–782

Hartoma TR, Nahoul K, Netter A (1977) Zinc, plasma androgen and male sterility. Lancet II: 1125–1126

Henkin R, Bradley D (1970) Hypogeusia corrected by Ni^{++} and Zn^{++}. Life Sci 9: 701–709

Henkin R, Schecter PJ, Friedewald WT, Demets DL, Raff M (1976) A double blind study of the effects of zinc sulfate on taste and smell dysfunction. Am J Med Sci 272: 285–299

Hill GM, Brewer GJ, Prasad AS, Hydrik CR, Hartmann DE (1987) Treatment of Wilson's disease with zinc. I. Oral zinc therapy regimens. Hepatology 7: 522–528

Hillstrom L, Pettersson L, Hellbe L, Kjellin A, Leczinsky C-G, Nordwall C (1977) Comparison of oral treatment with zinc sulphate and placebo in acne vulgaris. Br J Dermatol 97: 681–684

Hoogenraad T, Koevoet R, de Ruyter Korver E (1979) Oral zinc sulphate as long-term treatment in Wilson's disease (hepatolenticular degeneration). Eur Neurol 18: 205–211

Hoogenraad T, Sindram J (1982) Oral zinc in primary biliary cirrhosis. N Engl J Med 307: 122–123

Hoogenraad T, Van den Hamer C, Koevoet R, de Ruyter Korver E (1978) Oral zinc in Wilson's disease. Lancet II: 1262

Hoogenraad T, Van den Hamer C, de Ruyter Korver E (1983) Three years of continuous oral zinc therapy in 4 patients with Wilson's disease. Acta Neurol Scand 67: 356–364

Hoogenraad T, Van den Hamer C, van Hattum J (1984) Effective treatment of Wilson's disease with oral zinc sulphate: two case reports. Br Med J 289: 273–276

Husain SL (1969) Oral zinc sulphate in leg ulcers. Lancet I: 1069–1071

Jablonska S, Wolowa F (1981) Zinc sulphate and alopecia totalis. Br J Dermatol 105: 485–488

James O, Macklon A, Watson A (1981) Primary biliary cirrhosis – a revised clinical spectrum. Lancet I: 1278–1281

Job C, Menkes CJ, Delbarre F (1980) Zinc sulfate in the treatment of rheumatoid arthritis. Arthritis Rheum 23: 1408–1409

Jones R (1979) Genital herpes and zinc. Med J Aust 1: 236

Kennedy A, Bessent RG, Reynolds PM (1980) Effect of oral zinc sulfate and penicillamine on zinc metabolism in patients with rheumatoid arthritis. J Rheumatol 7: 639–644

Klingberg WG, Prasad AS, Oberleas D (1976) Zinc deficiency following penicillamine therapy. In: Prasad AS (ed) Trace elements in human health and disease, vol 1. Zinc and copper. Academic Press, New York, pp 51–65

Kuivaniemi H, Peltonen L, Palotie A, Kaitila I, Kivirikko KI (1982) Abnormal copper metabolism and deficient lysyl oxidase activity in a heritable connective tissue disorder. J Clin Invest 69: 730–733

Kvist U, Eliasson R (1980) Influence of seminal plasma on the chromatin stability of ejaculated human spermatozoa. Int J Androl 3: 130–142

Kvist U, Bjorndahl L, Kjellberg S, Lindholmer C, Roomans GM (1985) Sperm nuclear zinc: its prostatic origin and its significance for human fertility. Acta Physiol Scand [Suppl 542] 124: 389

Lenard HG, Lombeck I (1979) Die Behandlung des Ehlers–Danlos Syndrom mit Zink. Klin Padiatr 191: 578–579

Lindelöf B (1984) Zinc and hair. In: Zinc in human medicine. TIL Publications. Isleworth, Toronto, pp 103–123

McKusick VA (1972) Heritable disorders of connective tissue, 4th edn. Mosby, St Louis, pp 292–371

Mahajan SK, Prasad AS, Briggs WA, McDonald FD (1980) Effect of zinc therapy on sexual dysfunction in hemodialysis patients. Trans Am Soc Artif Intern Organs 26: 139–141

Marmar J, Katz S, Praiss DE, DeBenedictis TJ (1975) Semen zinc levels in infertile and postvasectomy patients and patients with prostatitis. Fertil Steril 26: 1057–1063

Mathur NK, Bumb RA (1983) Oral zinc in trophic ulcers of leprosy. Int J Lepr 51: 410–411

Mathur NK, Bumb RA, Mangal HN (1983a) Zinc restores hair growth in lepramatous leprosy. Br J Dermatol 107: 240

Mathur NK, Bumb RA, Mangal HN (1983b) Oral zinc in recurrent erythema nodosum leprosum reaction. Lepr India 55: 547–552

Mathur NK, Bumb RA, Mangal HN, Sharma ML (1984) Oral zinc as an adjunct to Dapsone in lepromatous leprosy. Int J Lepr 52: 331–338

Mattingly PC, Mowat A (1982) Zinc sulfate in rheumatoid arthritis. Ann Rheum Dis 41: 456–457

Menard MP, McCormick CC, Cousins RJ (1981) Regulation of intestinal metallothionein biosynthesis by dietary zinc. J Nutr 112: 1352–1361

Menkes CJ, Job C, Delbarre F (1978) Traitement de la polyarthrite rhumatoide par le sulfate de zinc per os. Nouv Presse Med 7: 760

Menkes CJ, Job C, Bruneaux F, Delbarre F (1981) Traitement de la polyarthrite rhumatoide par le sulfate de zinc: resultat d'un essai en double aveugle. Rev Rhum Mal Osteartic 48: 223–227

Merchant HW, Gangarosa LP, Glassman AB et al. (1977) Zinc sulfate supplementation for treatment of recurring oral ulcers. South Med J 70: 559–561

Michaelsson G (1984) Zinc in relation to some skin diseases. In: Zinc and human medicine. TIL Publications, Isleworth, Toronto, pp 125–134

Michaelsson G, Juhlin L, Vahlquist A (1977a) Effects of oral zinc and vitamin A in acne. Arch Dermatol 113: 31–36

Michaelsson G, Juhlin L, Ljunghall K (1977b) A double-blind study of the effect of zinc and oxytetracycline in acne vulgaris. Br J Dermatol 97: 561–566

Moore R (1978) Bleeding gastric erosion after oral zinc sulphate. Br Med J 1: 754

Muñoz N, Wahrendorf J, Bang LS et al. (1985) No effect of riboflavine, retinol, and zinc on prevalence of precancerous lesions of esophagus. Randomized double-blind intervention study in high-risk population of China. Lancet II: 111–114

Myers MB, Cherry G (1970) Zinc and the healing of chronic leg ulcers. Am J Surg 120: 77–81

Nishiyama S (1976) Hair abnormality in systemic disease. In: Kobori T, Montagna W (eds) Biology and disease of the hair. University of Tokyo Press, Tokyo, pp 329–336

Oelshlegel FJ Jr, Brewer GJ, Prasad AS, Knutsen C, Schoomaker EB (1973) Effect of zinc on increasing oxygen affinity of sickle and normal red blood cells. Biochem Biophys Res Commun 53: 560–566

Oelshlegel FJ Jr, Brewer GJ, Knutsen C, Prasad AS, Schoomaker EB (1974) Studies on the interaction of zinc with human hemoglobin. Arch Biochem Biophys 163: 742–748

Olsson R (1982) Oral zinc treatment in primary biliary cirrhosis. Acta Med Scand 212: 191–192

Orr KB (1976) Healing of gastric ulcers by zinc sulphate. Med J Aust 1: 244

Orris L, Shalita AR, Sibulkin D, London SJ, Gans EH (1978) Oral zinc therapy of acne: absorption and clinical effect. Arch Dermatol 114: 1018–1020

Patterson WP, Winkelman M, Perry MC (1985) Zinc-induced copper deficiency: megamineral sideroblastic anemia. Ann Intern Med 103: 385–386

Pfeiffer CC, Jenney EH (1978) Excess oral zinc in man lowers copper levels. Fed Proc 37: 324

Phillips A, Davison M, Greaves MW (1977) Venous leg ulceration: evaluation of zinc treatment, serum zinc and rate of healing. Clin Exp Dermatol 2: 395–399

Pories WJ, Henzel JH, Rob CG, Strain WH (1967a) Acceleration of wound healing in man with zinc sulphate given by mouth. Lancet I: 121–124

Pories WJ, Henzel JH, Rob CG, Strain WH (1967b) Acceleration of healing with zinc sulphate. Ann Surg 165: 432–436

Porter KG, McMaster D, Elmes ME, Love AHG (1977) Anaemia and low serum-copper during zinc therapy. Lancet II: 774

Prasad AS, Miale A, Farid Z, Schulert A, Sandstead HH (1963) Zinc metabolism in patients with the syndrome of iron deficiency anemia, hypogonadism and dwarfism. J Lab Clin Med 61: 537–549

Prasad AS, Schoomaker EB, Ortega J, Brewer GJ, Oberleas D, Oelshlegel FJ Jr (1975) Zinc deficiency in sickle cell disease. Clin Chem 21: 582–587

Prasad AS, Ortega J, Brewer GJ, Oberleas D, Schoomaker EB (1976) Trace elements in sickle cell disease. J Am Med Ass 235: 2396–2398

Prasad AS, Abbasi A, Ortega J (1977) Zinc deficiency in man: studies in sickle cell disease. In: Brewer GJ, Prasad AS (eds) Zinc metabolism: current aspects in health and disease. Alan R. Liss, New York, pp 211–236

Prasad AS, Brewer GJ, Schoomaker EB, Rabbani P (1978a) Hypocupremia induced by zinc therapy in adults. J Am Med Ass 240: 2166–2168

Prasad AS, Rabbani P, Abbasi A, Bowersox E, Fox MRS (1978b) Experimental zinc deficiency in humans. Ann Intern Med 89: 483–490

Ramadori G, Keidl G, Hutteroth T et al. (1985) Orale Zink-Therapie bei Morbus Wilson – eine Alternative zu D-Penicillamin. Z Gastroenterol 23: 25–29

Rasker JJ, Kardaun SH (1982) Lack of beneficial effect of zinc sulfate in rheumatoid arthritis. Scand J Rheumatol 11: 168–170

Reding P, Duchateau J, Bataille C (1984) Oral zinc supplementation improves hepatic encephalopathy: results of a randomised controlled trial. Lancet II: 493–495

Roman W, Oon R, West RF, Reid DP (1967) Zinc sulphate in acute porphyria. Lancet II: 716

Roman W, Oon R, West RF, Reid DP (1969) A case of acute intermittent porphyria treated successfully with zinc sulphate and chelation. Med J Aust 1: 633–638

Russell RM, Cox ME, Solomons NW (1983) Zinc and the special senses. Ann Intern Med 99: 227–239

Schecter PJ, Friedewald WJ, Bronzert DA, Raff MS, Henkin RI (1972) Idiopathic hypogeusia: a description of the syndrome and a single blind study with zinc sulfate. In: Pfeiffer CC (ed) International review of neurobiology. Supplement 1. Academic Press, New York, pp 125–140

Schouwink G (1961) De hepato-cerebrale degeneratie (met een onderzoek nar de zinkstowisseling). Thesis, University of Amsterdam

Sergeant GR, Galloway RE, Gueri MC (1970) Oral zinc sulphate in sickle-cell ulcers. Lancet II: 891–893

Sherlock S, Scheuer PJ (1973) The presentation and diagnosis of 100 patients with primary biliary cirrhosis. N Engl J Med 289: 674–678

Simkin PA (1976) Oral zinc sulphate in rheumatoid arthritis. Lancet II: 539–542

Solomons NW (1979) On the assessment of zinc and copper status in man. Am J Clin Nutr 32: 856–871

Solomons NW, Cousins RJ (1984) Zinc. In: Solomons NW, Rosenberg IH (eds) Absorption and malabsorption of mineral nutrients. Alan R. Liss, New York, pp 125–197

Sonnichsen VN, Reinicke C, Herrman C, Glatzel E (1984) Zinktherapie der Alopecia Areata. Dermatol Monatsschr 170: 437–442

Strain WH, Pories WJ (1966) Zinc levels of hair as tools in zinc metabolism. In: Prasad AS (ed) Zinc metabolism. Charles C. Thomas, Springfield, pp 378–394

Swerdel MR, Cousins RJ (1984) Reduction of adjuvant induced arthritis and concomitant increases in ceruloplasmin and metallothionein synthesis in rats by zinc and 13-cis-retinoic acid. Fed Proc 43: 686

Tennican PO, Carl GZ, Chvapil M (1979) The diverse effects of topical and systemic administration of zinc on the virulence of herpes simplex genitalis. Life Sci 24: 1877–1883

Thurnham DI, Rathakette P, Hambidge KM, Munoz N, Crespi M (1982) Riboflavin, vitamin A and zinc status in Chinese subjects in a high-risk area for oesophageal cancer in China. Hum Nutr Clin Nutr 36c: 337–349

Van Caillie-Bertrand M, Degenhart HJ, Visser HKA, Sainaasappel M, Bouquet J (1985) Oral zinc sulphate for Wilson's disease. Arch Dis Child 60: 656–659

Verma KC, Saini AS, Dhamija SK (1980) Oral zinc sulphate therapy in acne vulgaris: a double-blind trial. Acta Derm Venereol (Stockh) 60: 337–340

Voorhees JJ, Chakrabarti SG, Botero F, Miedler L, Harrell BR (1969) Zinc therapy and distribution in psoriasis. Arch Dermatol 100: 669–673

Wacker WEC (1976) Role of zinc in wound healing: a critical review. In: Prasad AS (ed) Trace elements in human health and disease, vol 1. Zinc and copper. Academic Press, New York, pp 107–113

Walshe J (1984) Treatment of Wilson's disease with zinc sulphate. Br Med J 289: 558–559

Wang F, Adams BA (1982) Zinc in the treatment of chronic leg ulcers of systemic lupus erythematosus. Singapore Med J 23: 171–173

Watkinson M, Aggett PJ, Cole TJ (1985) Zinc and acute tropical ulcers in Gambian children and adolescents. Am J Clin Nutr 41: 43–51

Wazewska-Czyzewska M, Wesierska-Gadek J, Legutko L (1978) Immunostimulatory effect of zinc in patients with acute lymphoblastic leukemia. Folia Haematol (Leipz) 105: 727–732

Weimar VM, Puhl SC, Smith WH, tenBroeke JE (1978) Zinc sulphate in acne vulgaris. Arch Dermatol 114: 1776–1778

Weissman K, Wadskov S, Sondergaard J (1977) Oral zinc sulphate for acne vulgaris. Acta Derm Venereol 57: 357–360

Wolowa F (1982) Behandlung of Alopecia areata totalis and maligna with Solvezink. Z Hautkr 57: 393–405

Wolowa F, Jablonska S (1976) Zinc in the treatment of alopecia areata. In: Kobori T, Montagna W (eds) Biology and disease of the hair. University of Tokyo Press. Tokyo, pp 305–308

Wolowa F, Stachow A (1978) Zinc in the treatment of alopecia areata. Przegl Dermatol 65: 687–696

Wolowa F, Stachow A (1980) Behandlung der Alopecia areata mit Zinksulfat. Z Hautkr 55: 1125–1134

Zimmer C, Luck G, Triebel H (1974) Conformation and reactivity of DNA. IV. Base binding ability of tranairion metal ions to native DNA and effect on helix conformation with special reference to DNA-Zn (II). Biopolymers 13: 425–453

Chapter 20

The Diagnosis of Zinc Deficiency

M.H.N. Golden

Introduction

The very large number of indices that have, from time to time, been evaluated, promulgated and, with experience, discarded as methods of assessing the zinc status of a subject is true testimony to the difficulties involved (Solomons 1979; Kirchgessner and Roth 1981; Golden and Golden 1985). None of the methods currently used can be recommended, with the possible exception of the monitoring of responses to zinc supplementation with plasma zinc estimation in conjunction with clinical evaluation.

The remarkable report by Beach et al. (1982) clearly shows the extraordinary extent of the difficulties involved, if all the features of zinc deficiency are to be properly ascribed. These authors fed pregnant mice a moderately zinc-deficient diet from day 7 of gestation until parturition. After birth, the pups were cross-fostered to normally nourished mice. They showed depressed immune function for at least 6 months. This is at least understandable; however, the second and the third generation mice, all of which were fed absolutely normal diets, continued to manifest reduced immunocompetence. This finding is staggering in its implications. If the same holds true for humans, as for these mice, you may have a compromised immune system because your grandmother took a zinc-deficient diet during her pregnancy! You would thus have a manifestation of zinc deficiency. I cannot conceive, at the moment, of the mechanism involved; I am also quite confident that no test so far devised, or even contemplated, will specifically identify this defect as being due to zinc deficiency. Clearly, our studies and ideas about zinc nutrition are at an early and conceptually naive stage.

Zinc Deficiency – The Diagnostic Problem

> Water, water, everywhere, nor any drop to drink.
> (Coleridge, *The Ancient Mariner*)

The most notable feature of both mild and severe experimental zinc deficiency, which is crucial to any appreciation of tests to assess zinc status, is that there is almost no reduction in the tissue concentration of zinc, even in overt, severe, symptomatic zinc deficiency (Williams and Mills 1970; Kirchgessner et al. 1976; Aggett et al. 1983). Muscle and white cell zinc are stubbornly preserved. There is a small decrease in liver and pancreatic zinc in prolonged, profound deficiency. The only tissue with an unequivocal reduction in zinc is bone. This is not a very useful diagnostic feature for human studies, particularly as the zinc sequestered in bone is neither readily available nor exchangeable, so that zinc deficiency can be present without any change in bone zinc and vice versa. However, even in very severe zinc deficiency, the quantitative reduction in total body zinc is small. There is still plenty of zinc on an absolute basis and the change in the zinc concentration in the tissues of metabolic importance is trivial.

It seems that the paradoxical observation of suffering from overt zinc deficiency and yet having a normal tissue zinc concentration is a particularly difficult feature of zinc deficiency for researchers to come to terms with. For example, Thompson (1985) argued that the pigs studied experimentally by Aggett et al. (1983), which had been put on a zinc-deficient diet, had stopped growing, had developed all the classical features of zinc deficiency and were reported to be preterminal and yet had normal concentrations of zinc in their white cells and muscle, could not, therefore, be zinc deficient (sic). Thompson (1985) stated that "the white cells are showing that the animals were not short of zinc, but that, if the diet were continued, the level would eventually fall and then the animals might be truly deficient".

Certainly, the clinicians who looked after these debilitated animals had no doubt about the diagnosis and would disagree fundamentally with Dr. Thompson's interpretations in the light of their experience that adding zinc alone to the diets of these animals restored their health completely to normal and stopped them from dying. Surely it is not in question that they were in fact suffering from actual zinc deficiency, and *only* from zinc deficiency, despite the normal levels of tissue zinc. It is equally undeniable that there is a major conceptual problem when we try to understand the nature of zinc deficiency. Such an understanding is essential before we can devise tests to determine which individuals are actually suffering from zinc deficiency. We also require techniques capable of differentiating between such zinc-deficient subjects and those showing changes in zinc metabolism due to factors totally unrelated to zinc nutriture. The nature of the response to a deficiency of zinc is at the heart of the problem.

Types of Nutritional Deficiency

An animal kept on a low-selenium diet grows normally. It will develop a low tissue concentration of selenium and thus be vulnerable to noxious stresses. Only after it manifests clinical signs of illness does it sometimes stop growing, as a secondary phenomenon, in association with definite ill-health. In total contrast, an animal at a stage of development at which sensitivity to zinc deficiency is high stops growing almost immediately when given a low-zinc diet. It maintains a normal concentration of zinc in its tissues. Whether additional clinical signs develop is contingent usually upon the extent and nature of externally imposed challenges such as stress or trauma. There is clearly a fundamental and quite distinct difference between these two responses to a single nutritional deficiency. I have suggested (Golden 1987) that these two types of response should be clearly differentiated, one from the other, and nutrients classified according to whether they give rise to a "type I" or a "type II" response.

Type I Nutrient Deficiency

The "type I" nutrient deficiency is characterized by a reduction in the tissue concentration of the nutrient and a defect in one or more specific metabolic pathways. There is a resultant loss of function so that the deficiency first presents with specific clinical signs. There is no primary effect on growth. Deficiencies of such nutrients as iron, iodine, copper, calcium, thiamine, riboflavine, ascorbic acid, retinol, tocopherol, cobalamin and selenium fall into this category.

When one conceives of a specific nutritional deficiency, one automatically considers a type I deficiency, as exemplified by Thompson's (1985) remarks. Such a deficiency results in a conceptually pleasing and easily envisaged chain of events. The diet has a specifically low content of the limiting nutrient; this results in a reduced tissue concentration and hence leads to an identifiable major defect in a metabolic pathway. This, in turn, gives rise to characteristic clinical signs and symptoms. The diagnosis is relatively straightforward; all one has to do is measure the concentration of the nutrient in a convenient tissue, measure the vulnerable pool of the nutrient, test the metabolic pathway where the defect lies, demonstrate an in vitro effect of adding the nutrient to some functional system or even simply recognize the specific clinical sign, which ever is most specific and/or convenient.

The point is that with any of these type I nutrient deficiencies, there is no doubt about the mechanisms involved, the way in which the disease arises and how to set about designing and evaluating a diagnostic test that everyone can understand. The arguments surround the relative merits of the various approaches and their specificity, sensitivity, convenience and cost. This is in marked contrast to the type II nutrient deficiencies, as exemplified by zinc.

Type II Nutrient Deficiency

The type II deficiencies present with a primary diminution or cessation of growth, without a reduction in tissue concentration of the limiting nutrient and are not normally associated with any diagnostically specific signs and symptoms. After cessation of growth, with a very profound deficiency, relatively non-specific signs may occur due to a generalized dysfunction, particularly affecting the most metabolically active tissues. Deficiencies of zinc, nitrogen, essential amino acids, phosphorus, magnesium, potassium and possibly sulphur and sodium all fall into this category.

The position with respect to type II nutrients, such as zinc, is thus quite different from that of the type I nutrients. The type II nutrient deficiencies are those that have always given trouble in diagnosis; for none are there satisfactory procedures for diagnosis. None of the methods that can be used to diagnose a type I deficiency can be used, unequivocally, to diagnose deficiency of a type II nutrient. This gives rise to major conceptual and practical difficulties. Since these difficulties are central to the attempts to diagnose zinc deficiency I will consider them in more detail below.

1. The primary response to zinc deficiency, growth failure or loss of whole tissue, is common to a deficiency of each of the type II nutrients, so that, although growth failure is a very sensitive response, it is totally non-specific.

2. There does not seem to be a functional body store of any of the type II nutrients which will serve the needs of the body for more than a day or two. Thus, if an excess of the nutrient is made available, it is metabolized and excreted, rather than saved for future use.

3. There is a common repertoire of metabolic changes and reductive adaptations that take place in response to a deficiency of any one of the type II nutrients.

Each of these nutrients is locked in tissue and is fundamental to the composition, to the integrity and to the day-to-day functioning of that tissue; they are only released if the whole tissue is catabolized.

Small amounts may be released if there is a metabolic re-adjustment, which is often totally unrelated to a deficiency of the nutrient under consideration. For example, if there is a reduction in protein synthesis rate, for whatever reason, then the complement of enzymes used for protein synthesis will not be maintained. There will, of necessity, be a reduction in the concentration of any component which is an integral part of the enzymes involved. As most of these enzymes are zinc metalloenzymes, a reduction in protein synthesis rate must, of itself, lead to a reduction in the functional tissue zinc concentration. In practice these metabolic re-adjustments are proportionately small. Indeed, a reduction in protein synthesis rate is the probable reason for the small changes in the liver and pancreatic zinc concentrations in zinc deficiency: even this change is therefore totally non-specific.

However, because whole tissue is being broken down during deficiency of any one of these nutrients, those liberated from the tissue in excess relative to the deficient nutrient have to be metabolized and excreted. Thus, in the face of a deficiency of any one of the type II nutrients, we would predict a negative body balance for them all. A negative zinc balance is thus as likely to be due to a dietary protein deficiency or to a metabolic re-adjustment of the tissue as to a zinc deficiency.

For example, Davies (see Golden and Golden 1985) offered 12 male weanling rats a diet deficient in tryptophan and cysteine with a full mineral mix providing adequate dietary zinc (40 mg zinc/kg diet). A control group was given 1%

tryptophan and cysteine. The amino-acid-deficient diet significantly reduced kidney and muscle zinc concentrations and thus achieved what a zinc-deficient diet had failed to achieve – a reduction in tissue zinc concentration. There are many pools of zinc in the body, each affected by disease processes and metabolic states irrespective of the subject's "zinc status". Gross tissue concentrations of zinc do not reflect zinc status. Furthermore, isotopic turnover and measurement of the metabolism of zinc are equally unlikely to faithfully represent zinc status, although they may say something about the general metabolic state of the subject.

4. When a diet deficient in any one of the type II nutrients is given, the body begins to conserve that particular nutrient avidly. For this reason, it is not easy to produce an overt deficiency of any of these nutrients in the non-growing animal by dietary means. Thus zinc losses can be reduced to very low levels in the face of a dietary deficiency (Golden and Golden 1981; Baer and King 1984) and, as with other type II nutrients, the appearance of overt signs of zinc deficiency is often contingent upon a pathological failure of the ability to regulate these losses.

The corollary of this is important for identification of the type II nutrients likely to be the most deficient. If a negative balance develops because of a general catabolism and is accompanied by an excess excretion of an essential nutrient, then the deficiency initiating this response is likely to be of another nutrient altogether. Because type II tissue nutrients liberated by tissue metabolism are excreted if the dietary supply is not limiting, it is a suboptimal intake of the type II nutrient with the *least* negative balance which we should suspect as the cause of a pathological response. Balance studies are unlikely to be helpful in defining zinc status. If the growth failure or weight loss is due to zinc deficiency, per se, then the urinary and faecal zinc excretion should be very low (Baer and King 1984).

5. The rate of growth is the major determinant of the dietary requirement for type II nutrients. For example, children given a diet which supplied just enough energy to maintain their body weight, without growing, were able to remain in zinc balance with a normal plasma zinc concentration, with an intake of only 1.3 µmol/kg/d. When the same children were subsequently given enough energy to gain weight rapidly, their plasma zinc fell to very low concentrations. This occurred despite a tenfold increase in the amount of zinc the children consumed (14 µmol/kg/d). Clearly, any diagnosis of zinc "deficiency" should be seen in the context of the metabolic state of the subject and what the demands are for zinc.

6. As growth is the major determinant of requirement, when we supply the missing nutrient, there should be a "catch-up" growth response. While circumstances exist in which this response has diagnostic value as demonstrated by Hambidge's group (Walravens and Hambidge 1976; Walravens et al. 1983) and by Golden and Golden (1981) this is not always so. Thus, if deficiency was developing and whole tissue was being catabolized there would be a deficit of *all* the components of that tissue, irrespective of the nutrient originally deficient which caused the loss of tissue. There will consequently be a greatly increased dietary requirement for *all* the nutrients during "catch-up" growth. The "catch-up" response to giving the originally deficient nutrient may thus be very short-lived and we should not necessarily expect the weight or growth deficit to be restored to normal. It may well go unobserved under certain

circumstances, such as when a subject is admitted to hospital for investigation and given a non-habitual diet. The rate of "catch-up" may now be limited by another nutrient which, although provided by the initial diet in a quantity capable of sustaining *normal* growth, is now being provided at a level insufficient to support *accelerated* growth. The important variable is the ratio of the requirement for "normal" and for "accelerated" growth for each type II nutrient compared with their ratio in the particular diet being used. In this case a secondary "catch-up" response may be observed when the nutrient that was present in the original diet in adequate amounts for normal growth rates, is supplemented. Thus, although suggestive of deficiency, a response to supplementation cannot be taken as unequivocal evidence of an original deficiency.

Fortunately, the response to supplying zinc is extraordinarily rapid. Often within a matter of hours the clinical picture changes, enzymes are resynthesized (Williams and Mills 1970) and even a growth response in children can be observed (Golden and Golden 1981). This is a "fast motion" deficiency. The onset is very abrupt when a deficient diet is given and the response is equally rapid to supplementation. This is probably a consequence of having no body store and requiring a continual dietary supply of each of the type II nutrients to sustain growth. There is a consequence of this immediacy that is not obvious. Clinical signs of zinc deficiency should be reversed with very small amounts of zinc, even relative to a normal intake. Because there is almost no reduction in the tissue zinc concentration we do not have a large depleted pool which has to be repleted before we see a response. Instead we need to give just sufficient zinc to replete the very small vulnerable pool that is giving rise to the specific sign. This is what is indeed observed in clinical practice.

We have treated a child recovering from malnutrition on a low-zinc, soya-based diet. He developed diarrhoea, skin lesions and had a plasma zinc of 4.1 μmol/1 (0.27 mg/l). A single oral dose of 1 mg zinc per kg body weight stopped the diarrhoea and skin lesions healed within 24 h, although plasma zinc did not change. We were able to continue his diet for a further 6 d and do a full metabolic balance study without any recurrence of signs or symptoms of zinc deficiency.

It may even be possible to "treat" clinical zinc deficiency with brief starvation so that endogenous supplies are made available to the vulnerable points. This very small "requirement" to redress the tissue type II nutrient balance and "cure" signs of clinical zinc deficiency is a necessary consequence of the lack of change in tissue concentration with deficiency; there will of course not be a growth response to these low levels of supplementation. This phenomenon may underlie the cyclical food intake pattern seen under some experimental conditions and the rapid waxing and waning of clinical signs in profound zinc deficiency. Further, it is likely to underlie much of the controversy that surrounds the presence or absence of clinically manifest zinc deficiency in such conditions as Crohn's disease (see Golden and Golden 1985).

Once the tissue balance is redressed, it is no longer a problem of zinc deficiency (although zinc deficiency was the aetiological agent), it is now a problem of whole tissue deficiency and all the nutrients will need to be given to bring about recovery.

7. One response which seems to be a common, but not essential (Golden and Golden 1981), feature of a type II deficiency is anorexia. This is corrected

if the nutrient is supplied, and is particularly pronounced if there is a gross imbalance of type II nutrients in the diet. As anorexia is common, if a child with zinc deficiency is supplemented with zinc, he will regain his appetite and thus have an increased intake of protein, potassium and even of intrinsic dietary zinc. Clearly, with this type of response it is extremely difficult to interpret dietary intake data. One might conclude, from a dietary survey, that zinc is very low in a diet and thus the subjects must be suffering from zinc deficiency. But what if, for example, the low dietary zinc intake was actually caused by anorexia secondary to a protein-deficient diet? Clearly dietary data must be viewed with great caution and not used to draw aetiological conclusions for the type II nutrients, or even for the type I nutrients in the face of a real type II nutrient deficiency.

As growth will be limited by the most deficient nutrient, it is only possible to have a deficiency, in a classical sense, of one type II nutrient at a time – the limiting one. Thus, even if a diet contains very reduced quantities of zinc, no response to supplementation and no specific consequences of the low zinc intake are to be expected if another type II nutrient is even more deficient. There may, however, be a low faecal output of zinc and, possibly, urine output as well.

Diagnostic Relevance of Zinc Pools in Tissues

Free Pools

One of the characteristics of type II nutrients is that, although they have large fixed pools in the tissues, they each have a small, rapidly turning over, free pool. This free pool is particularly sensitive to a deficiency of the individual nutrient (not potassium or sodium). Thus, plasma or serum zinc has most frequently been used to diagnose zinc deficiency. It is a convenient sample to take and the measurement is relatively straightforward. And yet plasma zinc has been denigrated as a measure of zinc "status". To illustrate, I quote from the article by Solomons (1979),

statements such as "Serum zinc (known to reflect body stores) was measured . . ." are often seen. The assumption that circulating zinc levels reflect total body stores of the nutrient is one that must come under serious reevaluation in light of mounting evidence to the contrary.

Clearly, the author quoted by Solomons was rightly criticized because serum does not relate to tissue zinc concentration or to stores. However, Solomons makes the assumption that it is invalid as a measure because it does not reflect total body zinc. But, as I have argued total body zinc itself probably bears little relationship to zinc "status" or the likelihood of an individual suffering from zinc deficiency. That is not the nature of zinc deficiency; Solomons has assumed that zinc behaves as a type I nutrient. Indeed the very use of the word "status" with respect to zinc nutriture requires careful definition, as it probably subsumes a presumption of the nutrient in question giving rise to a type I deficiency.

Hypoalbuminaemia is frequently accompanied by hypozincaemia. It is often argued that plasma zinc may be low in such situations because albumin is low and there is thus a reduced plasma concentration of zinc-binding sites. This argument does not hold up to scrutiny, despite the statistical correlation between plasma zinc and albumin concentration (Giroux et al. 1976). With a molecular weight of 69 000 and a concentration of 35 g/l, the molality of albumin is about 500 μM. The molality of albumin-bound zinc is about 10 μM. Thus, even when plasma albumin is at the lower limit of normal, only about one of every 50 albumin molecules is associated with a zinc atom. It is very difficult to see how a reduction in albumin can lead to a significant loss of available binding sites for zinc under almost any circumstances. However, the reverse is not unlikely; a marginal zinc status may well affect albumin synthesis and excretion. Indeed, in one study in an old peoples' home, the major effect of zinc supplementation was an increase in plasma albumin concentration.

And yet, it is quite true to say that plasma zinc is *not* a very specific measure of zinc deficiency, although it appears to be a sensitive measure. This is because plasma zinc is subject to metabolic alteration as well as nutritional limitation. It is well recognized that low plasma zinc levels are seen in endotoxaemia, infection, carcinoma, steroid administration, etc. (see Solomons 1979 for references), and that these changes are a reflection of metabolic redistribution within the free pool and not to a change in zinc status. It is for this reason that plasma zinc has been regarded as an unreliable indicator of zinc deficiency; in the same way, plasma iron, by itself, is regarded as an unreliable index of iron status, because it undergoes similar changes in response to stress. These changes are part of the acute phase response; if we had an independent measure of the acute phase response we may be able to decide which fraction of plasma zinc was reduced because of zinc deficiency and which decrement to ignore because it originated from a metabolic redistribution of zinc.

Metallothionein

It is not entirely true to say there is no zinc "store", only that there is no functional zinc "store" in the conventional sense. There seem to be small "stores", which I conceive of as typical of the "metabolic buffer pools" of the type II nutrients. When we take a large protein meal, for example, the calculated rise in the total body free amino acid pool is much less than would be expected if the amino acids derived from the protein simply expanded the free pool. So, although no "storage protein" has been identified, there must be some, as yet unidentified, way of temporarily "parking" amino acids at meal times for postprandial use, perhaps as small peptides. Similarly, the pool of pyrophosphate probably prevents a catastrophic reduction of inorganic phosphate whenever glucose is ingested and has to be metabolized.

With zinc we have identified the metabolic buffer pool: it is zinc metallo-thionein-I. Herein lies the key to making the diagnosis of zinc deficiency, for it is this same metabolic buffer pool which is also used to sequester free zinc in response to each of the confounding factors which cause a reduction in plasma zinc concentration secondary to metabolic re-adjustment. Indeed, the different transcriptional control sequences for metal induction, steroid regulation, interleukin (endotoxin) regulation and interferon regulation of the metallothionein-I gene have been identified. Very elegant recombinant gene-splicing experiments have demonstrated the importance of these sequences for the induction of metallothionein-I synthesis, and hence for the control of the withdrawal of zinc from the free pool (Hamer 1986). In future we may be able to differentiate the types of hypozincaemia on the basis of the measurement of metallothionein-I-specific RNA.

If there is zinc bound to metallothionein-I, it is prima facie evidence that there is sufficient zinc immediately available to satisfy metabolic demand and hence there is very unlikely to be a zinc deficiency. It would strongly argue against zinc deficiency if we were able to demonstrate metallothionein-I-bound zinc, irrespective of the clinical, nutritional, biochemical or physiological state of the subject. Further, because increased hepatic metallothionein-I synthesis is the mechanism whereby zinc is redistributed in the body in response to all those confounding conditions which lower plasma zinc, metallothionein-I serves as a mechanism whereby the cause of a low plasma zinc concentration can be clearly determined. If it is due to a zinc deficiency, zinc-bound metallothionein-I should be reduced or absent. If it is an inflammatory, or steroid hormone-induced reduction there should be an increase in metallothionein-I. Herein lies the resolution of the problem of the diagnosis of zinc deficiency. Measurement of plasma zinc concentration and metallothionein-I, together, should clearly indicate if an individual is, or is not, zinc deficient.

However, as metallothionein is a hepatic protein, there are several practical problems to be addressed before determinations of metallothionein-I and plasma zinc concentrations can be used in combination to unequivocally make or refute a diagnosis of zinc deficiency. These problems are being addressed through the pioneering work of Dr. I. Bremner and his collaborators (Bremner and Morrison 1986; Bremner et al. 1986; Bremner et al. 1987; Morrison and Bremner 1987). They have demonstrated, in the following series of important investigations, the potential of the metallothionein-I assay in the diagnosis of zinc deficiency:

1. Metallothionein-I is detectable by radioimmunoassay in plasma where its concentration is closely correlated with that in liver. Thus there is no need to do a liver biopsy to obtain a useful proxy measure of hepatic metallothionein-I.

2. The concentration of metallothionein-I increases in plasma in response to endotoxin and other stresses as well as in response to cadmium and copper toxicity.

3. There is no reduction in plasma metallothionein-I in a deficiency of copper, or if the level of exposure to cadmium and mercury is very low.

4. The concentration of metallothionein-I decreases in plasma in response to a zinc-deficient diet in a dose-dependent manner.

5. Most importantly, when there is an infection, endotoxaemia or other stress, the rise in the metallothionein-I concentration in plasma seen in the zinc-replete animal does not occur if the subject is zinc deficient.

However, perhaps the most exciting finding has been the observation that metallothionein-I is present in red blood cells in concentrations similar to those in plasma and that the red cell metallothionein-I concentration is sensitive to dietary zinc intake. However, unlike plasma metallothionein-I, injection of endotoxin and other stresses have no effect on the red cell metallothionein-I concentration. Copper deficiency had no effect whatsoever on the red cell metallothionein-I concentration.

These results clearly indicate that the assay of plasma zinc and red blood cell metallothionein-I can be used to make an unequivocal diagnosis of zinc deficiency. Preliminary experiments (Dr. Ian Bremner, personal communication) with an ELISA assay in man give similar results to those found in the rat. It would seem that after much controversy the place of red cell metallothionein-I assay will become established in the near future. The test is likely to become more refined when monoclonal antibodies are raised to specifically recognize the protein only when it is carrying bound zinc.

Conclusion

Because we have been dissatisfied with all the methods of measuring zinc status and have invested considerable research effort into investigating metallothionein-I, the "metabolic buffer pool" of zinc, we can at last see the way forward to a definite test of zinc deficiency and thus to a much clearer understanding of the nature of zinc deficiency. This is unlike the situation which pertains for any of the other type II nutrients where, in general, unsatisfactory methods of making a diagnosis have been pragmatically accepted.

Acknowledgement. I thank Dr. Ian Bremner and Dr. N.T. Davis for making their unpublished results available and for permission to quote them. The concepts put forward in this contribution were generated out of results of experiments fully supported by the Wellcome Trust.

References

Aggett PJ, Crofton RW, Chapham M, Humphries WR, Mills CF (1983) Plasma leucocyte and tissue zinc concentrations in young zinc deficient pigs. Pediatr Res 17: 433

Baer MT, King JC (1984) Tissue zinc levels and zinc excretion during experimental zinc depletion in young men. Am J Clin Nutr 39: 556–570

Beach RS, Gershwin ME, Hurley LS (1982) Gestational zinc deprivation in mice: persistence of immunodeficiency for three generations. Science 218: 469–471

Bremner I, Morrison JN (1986) Assessment of zinc, copper and cadmium status in animals by assay of extracellular metallothionein-I. Acta Pharmacol Toxicol 59 Suppl 7: 502–509

Bremner I, Mehra RK, Morrison JN, Wood AM (1986) Effects of dietary copper supplementation of rats on the occurrence of metallothionein-I in liver and its secretion into blood, bile and urine. Biochem J 235: 735–739

Bremner I, Mehra RK, Sato M (1987) Metallothionein. In: Kagi JHR, Kojima Y (eds) Proc 2nd international meeting on metallothionein and the low molecular weight metal binding proteins. Birkhauser Verlag, Basel, pp 507–518

Giroux EL, Durieux M, Schechter PJ (1976) A study of zinc distribution in human serum. Bioinorg Chem 5: 211–219

Golden BE, Golden MHN (1981) Dietary zinc and the output and composition of faeces. In: Howell JM'C, Gawthorn JM, White CL (eds) Trace element metabolism in man and animals, TEMA-4. Australian Academy of Science, Canberra, pp 73–76

Golden MHN (1987) The role of individual nutrient deficiencies in growth retardation of children as exemplified by zinc and protein. In: Waterlow JC, Goyens P (eds) Linear growth retardation in Third World children. Raven Press, New York (in press)

Golden MHN, Golden BE (1981) Effects of zinc supplementation on the dietary intake, rate of weight gain and energy cost of tissue deposition in children recovering from severe malnutrition. Am J Clin Nutr 34: 900–908

Golden MHN, Golden BE (1985) Problems with the recognition of human zinc responsive conditions. In: Mills CF, Bremner I, Chesters JK (eds) Trace elements in man and animals, TEMA-5. Commonwealth Agricultural Bureau, Farnham Royal, pp 933–938

Hamer DH (1986) Metallothionein. Annu Rev Biochem 55: 913–951

Kirchgessner M, Roth HP (1981) Problems and possibilities in diagnosing zinc deficiency. In: Howell JM, Gawthorn JM, White CL (eds) Trace element metabolism in man and animals, TEMA-4. Australian Academy of Science, Canberra, pp 327–330

Kirchgessner M, Roth MP, Weigard E (1976) Biochemical changes in zinc deficiency. In: Prasad AS (ed) Trace elements in human health and disease, vol 1. Academic Press, New York, pp 189–225

Morrison JN, Bremner I (1987) Effects of maternal zinc supply on blood and tissue metallothionein I concentrations in suckling rats. J Nutr (in press)

Solomons NW (1979) On the assessment of zinc and copper nutriture in man. Am J Clin Nutr 32: 856–871

Thompson RPH (1985) Discussion following Prasad (1985) "Diagnostic approaches to trace element deficiencies". In: Chandra RK (ed) Trace elements in nutrition of children. Raven Press, New York, pp 34–39

Walravens PA, Hambidge KM (1976) Growth of infants fed a zinc supplemented formula. Am J Clin Nutr 29: 1114–1121

Walravens PA, Krebs NF, Hambidge KM (1983) Linear growth of low income pre-school children receiving a zinc supplement. Am J Clin Nutr 38: 195–201

Williams RB, Mills CF (1970) The experimental production of zinc deficiency in the rat. Br J Nutr 24: 989–1003

Chapter 21

Human Zinc Requirements

Janet C. King and Judith R. Turnlund

Introduction

The dietary requirement for zinc is literally the amount required in the diet to maintain optimally the various metabolic and physiological functions of life (Smith et al. 1983). The dietary zinc requirement for a population is not a single value. Instead it varies over a wide range depending on the age and physiological state of the individuals and on composition of the diet, particularly with respect to the amount and proportion of organic and inorganic components of the diet which influence zinc absorption and utilization (Hambidge et al. 1986).

The true requirement for zinc differs from the dietary requirement. It is the amount which must be absorbed to replace all sources of endogenous loss and to provide for tissue synthesis during growth or for milk secretion. The bioavailability of the form(s) of ingested zinc determines the total dietary zinc which must be supplied to meet the true requirement. If the true requirement and bioavailability of dietary zinc are known, the dietary requirement can be calculated by dividing the metabolic requirement by the fractional absorption of zinc from the diet.

Factors influencing the bioavailability of zinc in the diet were discussed in Chap. 4. The focus of this chapter, therefore, will be the true requirement for zinc and factors which influence this requirement.

Basic Concepts

The two major components of the true requirement for zinc are replacement of endogenous losses and provision for growth or for milk secretion. The endogenous losses of zinc include faecal, urinary, integumental (desquamated skin cells and sweat) and other, small miscellaneous losses from the body. Examples of the latter are zinc losses in hair and nail growth, vaginal secretions, menstrual fluid and semen.

Endogenous losses of zinc may be quantitated by measuring its excretion when virtually no zinc is fed in the diet or by estimating the turnover of an isotopic zinc tracer. If an isotopic zinc tracer is not used to separate endogenous from unabsorbed dietary zinc in the faeces, faecal endogenous losses can only be quantitated when the unabsorbed dietary zinc is eliminated by feeding a zinc-free diet. By extrapolating the relationship of total endogenous losses with time back to zero time, it is possible to calculate minimal endogenous losses, presumably before the calculations are influenced by the development of a zinc deficiency (Hess et al. 1977). This method underestimates total endogenous losses of zinc in subjects at equilibrium with an adequate zinc status because urinary and faecal losses drop quickly when a zinc-free diet is fed (Baer and King 1984). This fall in urinary and faecal losses lowers the estimation of the total loss at time zero by linear regression.

The advantage of isotopic turnover studies is that this approach allows measurement of endogenous zinc loss without the use of a controlled zinc-free diet and metabolic collections for 2–3 weeks. The endogenous losses measured in this way are probably more representative of those of the general population consuming typical zinc-adequate diets than are those estimated by regression analysis of data obtained with a zinc-free diet. The risks associated with the use of radioactive zinc isotopes and the analytical problems associated with the use of stable zinc isotopes have limited the number of studies of zinc turnover in humans, however.

The needs for zinc for growth or for milk secretion are the other components of the metabolic requirement. The amount of zinc required for growth or for milk secretion can be calculated from factorial data, i.e. the quantity of tissue gained and the concentration of zinc in the tissue, or the volume of milk secreted and the concentration of zinc in the milk. Since it is difficult to estimate the composition of tissue gained during growth, the balance method is often used to quantitate zinc growth requirements. With the balance method the difference between the total intake and total loss is the amount retained for growth. Balance studies require a controlled diet with metabolic collections and are difficult to perform, especially in young, growing children. Furthermore, an adaptation period with a controlled diet over at least 4 weeks prior to the study period is probably required (Schwartz et al. 1986). Balance study results may be altered by previous low intakes or health problems. Long-term (about 6–8 week) balance studies in healthy infants with no recent illness have not been done. Thus, factorial data must be used to estimate the needs for growth.

An alternative method for estimating zinc requirements during growth is to measure the amount of zinc absorbed using a stable isotope of zinc as a tracer. During growth, the amount of zinc absorbed would equal the amount required to replace endogenous losses plus the growth need. Quantitation of total zinc intake and faecal zinc output is required. This method has been used extensively to measure zinc requirements in adults, but zinc absorption has been studied in only one group of healthy, growing infants (Ziegler et al. 1987). Others have measured the absorption in preterm infants using stable isotopes (Ehrenkranz et al. 1984; Peirce et al. 1987).

Adaptation of Zinc Requirements

The amount of dietary zinc needed to replace endogenous zinc losses has been used to estimate the true zinc requirement for healthy adults. Endogenous zinc losses do not appear to be constant in an individual. Instead, they vary with the amount of zinc consumed, making the true requirement difficult to ascertain. This is because healthy individuals adapt to changes in dietary zinc levels by increasing or decreasing zinc losses. Data from metabolic studies of individuals fed extreme levels of dietary zinc illustrate this point.

Jackson and co-workers (1984) measured the changes in urinary and faecal zinc losses, zinc absorption and gastrointestinal (GI) zinc secretion in a 29-year-old male fed either 7, 15 or 30 mg zinc/d. Faecal zinc excretion increased progressively with increases in dietary zinc; no changes in urinary zinc excretion occurred (Table 21.1). Using a stable isotope of zinc, GI zinc secretion and zinc absorption were quantitated. When dietary zinc increased from 7 to 30 mg/d, the amount of zinc absorbed doubled, but the quantity of zinc secreted into the GI tract also increased to attain overall balance. Changes in GI secretion were seen before changes in absorption. The authors concluded that the control of body zinc is achieved in two ways: small daily variations in intake are dealt with by rapid responses in the GI secretion; larger changes in intake can only be dealt with by slower adjustments in absorption.

The response to *reductions* in dietary zinc was studied in two metabolic balance studies at the University of California, Berkeley. In these studies the dietary zinc of six healthy men was reduced from about 16 mg/d to either 5.5 mg/d (Wada et al. 1985) or 0.3 mg/d (Baer and King 1984) (Table 21.2). When it was reduced from 16 to 5.5 mg/d, all but one of the six men achieved zinc balance within 9 d. Balance was accomplished by a 70% reduction in faecal zinc; urinary zinc losses were unchanged. The decrease in faecal zinc probably is due primarily to a decrease in unabsorbed dietary zinc, but a decrease in endogenous gastrointestinal zinc secretion could also cause a fall in total faecal zinc losses. Zinc absorption, as measured by a stable isotope tracer, decreased from 4.1 mg/d to 2.9 mg/d. Although the amount of zinc absorbed fell by about 1 mg/d, all but one of the men maintained zinc balance suggesting that gastrointestinal zinc secretion also was reduced in order to achieve balance.

Table 21.1. Changes in zinc utilization with increases in dietary zinc (mg/d)

Dietary zinc	7	15	30
Urinary zinc	0.63	0.60	0.70
Faecal zinc	6.83	14.43	30.16
GI zinc secretion	3.02	4.60	5.82
Zinc absorption:			
total	3.30	4.72	6.34
% of intake	47	32	21

From Jackson et al. (1984).

In the second study, the dietary zinc of six healthy men was reduced from 16 to 0.3 mg/d (Baer and King 1984). A semi-purified, liquid diet was required in order to supply a virtually zinc-free diet adequate in all other nutrients. None of the men in this study achieved zinc balance, but it is surprising how close they came. At the end of the study, which varied in length from 4 to 9 weeks depending on the onset of the symptoms of zinc deficiency, urinary and faecal zinc losses totalled only 0.67 mg/d (Table 21.2). These losses may be considered to be an estimate of the minimal obligatory zinc loss, i.e. the amount of zinc lost after the body has adapted to the maximum extent possible to the extremely low zinc intake.

The results of these two studies show that healthy adults can adapt and achieve balance when the diet supplies only 5.5 mg zinc/d, but not when it supplies 0.3 mg zinc/d. A systematic study has not been done to determine the lowest zinc intake to which healthy adults can successfully adapt and achieve balance. However, a group of seven young men fed only 1.7 mg zinc/d at the University of Missouri attained zinc balance in 3 weeks by significant reductions in both urinary and faecel zinc losses (H.L. Anderson and B.L. O'Dell 1987, personal communication). Plasma zinc levels also fell significantly during the 3-week period. Milne and co-workers (1983) fed 3.4 mg zinc/d to healthy men for 18 weeks. Zinc balance was achieved for the first 5–6 weeks, but then it became negative and remained negative for the remainder of the study. These results suggest that early adaptation to a low zinc intake may not be maintained over the long term. Also, the data of Wada and King (1986) show that attainment of balance does not assure maintenance of physiological and metabolic functions. Although balance was achieved by five of the six men fed 5.5 mg/zinc/d, levels of serum albumin, pre-albumin, retinol-binding protein, thyroid-stimulating hormone, thyroxine (T_4), free thyroxine and basal metabolism declined significantly. The changes were small, however, and all values remained within the normal range. Such functional changes may become more important when combined with a health problem such as trauma, infection or a metabolic abnormality, e.g. diabetes.

These data show that zinc intake and losses are related, i.e. high intakes are associated with high losses and vice versa. An association between intake

Table 21.2. Faecal and urinary zinc changes with reductions in dietary zinc (mg/d)

	Beginning of study	End of study
Dietary Zinc: From 16.5 to 5.5[a]		
Urine	0.50±0.17	0.53±0.15
Faeces	14.5 ±1.9	4.5 ±0.9
Dietary Zinc: From 15.7 to 0.3[b]		
Urine	0.45±0.27	0.17±0.21
Faeces	10.5 ±2.9	0.5 ±0.2

[a] From Wada et al. (1985).
[b] From Baer and King (1984).

and endogenous losses suggests that the amount of zinc required for an individual may be related to his/her current zinc status. For example, an individual with a good zinc status may have higher endogenous losses so that more zinc would be required to maintain the high status than would be required to maintain poor status. In previous studies of human zinc requirements (King 1986), it was found that endogenous zinc loss was significantly correlated with prestudy serum zinc levels ($r = 0.574$; $P < 0.05$). It has been questioned whether circulating zinc is a good indicator of tissue zinc status. However, if it is a marker of tissue zinc status, these data suggest that zinc intake and status are related. It seems logical that zinc losses would be reduced in an individual with poor zinc status in order to conserve whole-body zinc.

Because zinc losses are influenced by dietary zinc intake and, possibly, by zinc status, it is inappropriate to use data on the zinc needed to replace endogenous losses as the only criterion for the true requirement. An alternative approach may be to determine the lowest zinc intake which prevents any changes in physiological or metabolic functions. This approach is difficult for zinc, however, because a sensitive, specific functional test for zinc status is lacking. Such considerations are discussed in Chap. 24 in relation to problems associated with the terms currently used to describe requirement. It has also been suggested that erythrocyte metallothionein may be a marker for zinc stores (Bremner and Morrison 1986). If this is true, erythrocyte metallothionein levels may provide a standard for the requirement. Further research is needed before either of these alternative approaches can be used to assess zinc requirements.

True "Physiological" Requirement for Zinc

Healthy Adults

Although the amount of zinc required to replace that lost endogenously is not an ideal criterion for the adult true requirement, it is the only approach for which sufficient data are available to derive an estimate. Because there is an adaptation of endogenous losses to zinc intake, this estimate is influenced by the amount of zinc consumed. At the present time there seems to be no way to avoid this circular relationship in establishing a requirement, i.e. the more you eat, the more you appear to require.

The true zinc requirement for adults has been estimated by three methods:

1. from total endogenous zinc losses of individuals fed zinc-free diets extrapolated back to time zero using linear regression;
2. from total zinc turnover of adults given radioactive tracers of zinc; and
3. from total zinc absorption measured with stable zinc isotopes. With this approach, it is assumed that in a person in equilibrium the amount of zinc absorbed equals the amount required to replace total endogenous loss. If the amount absorbed exceeds endogenous losses in a person previously in an equilibrium state, the intake exceeds the requirement and zinc is deposited in tissues. If the amount absorbed is not sufficient to replace endogenous losses, the requirement has not been met.

Total endogenous zinc losses in six men and ten women fed zinc-free diets have been measured using linear regression analysis to estimate the losses at time zero (Hess et al. 1977; Baer and King 1984). The endogenous faecal and urinary zinc losses of the men (1.36 ± 0.59 mg/d) were significantly higher than those of the women (0.84 ± 0.37 mg/d) (King 1986). This difference may be related to the greater amount of lean tissue in men than in women. Lean body mass was not measured in these studies, but there was no correlation with other estimates of body mass such as total body weight or body weight to the 3/4th power. These endogenous losses are an underestimate of the true requirement for men and women because of adaptations in urinary and faecal zinc output with the zinc-free diet.

Whole-body zinc turnover is another estimate of the true zinc requirement. In a 5-d turnover study using [69m]Zn, the zinc turnover of a group of men and women between 21 and 73 years of age averaged 2.7 mg/d (Foster et al. 1979). Gender differences were not reported.

We have used stable isotopes of zinc as tracers for between one and six absorption measurements in 68 individuals studied with a variety of protocols (Turnlund et al. 1981, 1982, 1984, 1985, 1986, 1987; Swanson et al. 1983; Wada et al. 1985). In all of the studies, the stable zinc isotopes were divided among all of the meals fed during a 24- or 48-h period. Polyethylene glycol was given with the stable zinc isotope as a faecal marker, and total stool output was collected for the next 15 to 21 d. Both conventional and formula diets were used; there was no effect of type of diet on zinc absorption. Metabolic collections were done and total urinary and faecal zinc losses were determined. Endogenous faecal zinc excretion was calculated from the stable isotope absorption data and the balance data. Thus, we were able to determine the total amount of endogenous zinc lost in urine and faeces as well as the amount absorbed to replace those losses and the integumental losses.

Zinc absorption was measured in adult men and women between 21 and 33 years of age in six different studies. The dietary zinc intake ranged from 4.6 to 19 mg/d. Zinc absorption increased with an increase in dietary zinc (Table 21.3). When the diet provided less than 12 mg/d, zinc absorption averaged about 2.7 mg/d; when more than 13 mg zinc was consumed per day, zinc absorption averaged about 4.4 mg/d. The amount of zinc absorbed was not greater when 19 mg was fed than it was when 13–16 mg was fed. Endogenous faecal and urinary zinc losses were less than the amount absorbed at all levels

Table 21.3. Effect of dietary zinc level on zinc absorption and endogenous loss (mg/d)

Dietary zinc	4.6–4.5	7.4–11.5	13.0–16.8	19
(Number of measurements)	(24)	(20)	(41)	(6)
Zinc absorption, total	2.9 ±0.2	2.6 ±0.2	4.4 ±0.7	4.4
Endogenous faecal zinc	1.7 ±0.2	1.7 ±0.5	3.1 ±0.9	3.5
Urinary zinc	0.59±0.01	0.52±0.23	0.62±0.21	0.68
Total endogenous loss[a]	2.3 ±0.2	2.2 ±0.3	3.8 ±0.9	4.1

Compiled from data published by Turnlund et al. (1982); Swanson et al. (1983); Turnlund et al. (1984); Wada et al. (1985); Turnlund et al. (1986); Turnlund et al. (1987).
[a] Total endogenous loss includes only faecal and urinary losses; integumental and other losses are not included.

of intake studied. The difference between the amount absorbed and the endogenous losses in the faeces and urine would include the integumental and miscellaneous losses and zinc retained in the body.

Only five of the young adults in these six studies were women. The zinc absorption of those women did not appear to differ from that of men fed comparable amounts of zinc. For example, six young men fed 16.4 mg zinc/d absorbed 4.1 mg/d; the five young women fed 16.8 mg zinc/d absorbed 4.0 mg/d. The data are limited, but when zinc intake was high there did not appear to be a gender difference in zinc absorption among the young adults studied. The gender differences observed in the previous studies were with zinc-free diets (Hess et al. 1977; Baer and King 1984) and, therefore, reflect obligatory losses, not endogenous losses.

The amount of zinc absorbed appears to decline with ageing (Turnlund et al. 1986). A group of five healthy men between 65 and 74 years of age absorbed 2.7 mg zinc/d when 15.4 mg zinc/d was fed as zinc sulphate in a formula diet (Table 21.4). The fractional absorption was only about 17%. A group of six young men between 22 and 30 years of age fed the same diet absorbed 5.1 mg zinc/d or 33% of the amount fed. The endogenous faecal and urinary losses of the elderly men were less than those of the young men, 2.2 vs 4.7 mg/d. The zinc absorption by the elderly men was sufficient to replace their endogenous losses. These studies were done with diets providing 15 mg zinc daily. With a more typical intake of about 10 mg/d, the total amount of zinc absorbed by the elderly probably is lower. In the above studies the elderly absorbed about 55% of the zinc absorbed by the young adults. If the same difference exists between elderly and young individuals consuming 10 mg/d, an elderly person may absorb about 1.5 mg/d.

Dietary phytate also influences the amount of zinc absorbed and the endogenous faecal and urinary zinc losses (Turnlund et al. 1984). When 2.3 g sodium phytate was added to a formula diet containing 15 mg zinc, the fractional zinc absorption was reduced from 33% to 17%. Total zinc absorption averaged 2.6 mg/d with the high-phytate diet (Table 21.4). Endogenous losses were greater than the amount absorbed resulting in a net whole-body zinc loss.

Table 21.4. Effect of age, dietary phytate and nitrogen balance on zinc absorption and endogenous loss (mg/d)

	Age (years)		Dietary phytate		Nitrogen balance	
	>65	<30	High	Zero	Negative	Equilibrium
Dietary zinc	15.4	15.5	15.0	15.0	15.5	15.5
Zinc absorption	2.7	5.1	2.6	5.1	4.6	5.1
Endogenous faecal zinc	1.5	3.8	3.2	2.7	4.5	3.8
Urinary zinc	0.7	0.9	0.4	0.5	0.9	0.9
Total endogenous loss	2.2	4.7	3.6	3.2	5.4	4.7

Compiled from data published by Turnlund et al. (1982); Turnlund et al. (1984); Turnlund et al. (1986).

Since this experimental diet was a formula fortified with a source of soluble phytate, it is difficult to conclude from those data what the true zinc requirement is when vegetable diets intrinsically rich in a variety of phytate sources are consumed. It appears likely that the requirement is higher, however, because the reabsorption of endogenous GI zinc is less efficient. After allowance for integumental and miscellaneous losses, the total endogenous loss in this study was about 4.0 mg/d.

The overall utilization of zinc is influenced by the intake of other nutrients, particularly other 'trace elements and protein. In one of our studies, the participants were in negative nitrogen balance due to a low protein intake for one of the metabolic periods (Turnlund et al. 1986). During the period of negative nitrogen balance, endogenous faecal and urinary zinc losses increased from 4.7 to 5.4 mg/d. Total zinc absorption did not change, and a net loss of whole-body zinc occurred. Tissue catabolism in association with a negative nitrogen balance resulted in an increase in GI zinc· secretion.

In sum, these data show that age, dietary zinc intake, diet composition and nutritional state all influence the true zinc requirement. Healthy adults between 20 and 35 years of age who are consuming a typical industrialized diet providing about 10–12 mg zinc daily from highly refined animal and plant food sources seem to require approximately 2.5 mg zinc/d to replace their endogenous losses. Healthy elderly adults over 65 years of age consuming a similar diet require about 1.5 mg/d. The true requirement for individuals consuming only plant foods has not been studied as thoroughly as that for individuals consuming mixed food sources, but it could be up to 4.0 mg/d.

Pregnant Women

The additional need for zinc during pregnancy can be calculated from the weight of tissues gained during gestation and the concentration of zinc in those tissues. Widdowson and Dickerson (1964) measured the concentration of zinc in 24 human foetuses and full-term babies ranging in weight from 0.75 to 4373 g. The zinc concentration per kilogram fat-free body tissue did not change consistently as body weight increased from 400 to 4400 g; the concentration averaged 18.4 mg/kg fat-free body. Using those foetal zinc data and measured foetal growth rates, Shaw (1979) calculated the rate of zinc accumulation by a human foetus growing along the 10th, 50th or 90th percentiles. The zinc accumulation rate for a foetus growing along the 50th percentile increased progressively from 0.21 mg/d at the 24th week of gestation to 0.67 mg/d at the 36th week.

In addition to the foetus, placental tissue, amniotic fluid, uterine and mammary tissue and maternal blood are also gained during gestation. The zinc concentration of those tissues has been determined (Tipton and Cook 1963; Mulay et al. 1971; Alexiou et al. 1977; Hambidge et al. 1986; Qvist et al. 1986). Hytten (1980) has calculated the total weight of the pregnancy tissues at term (Table 21.5). Based on the total weight of those tissues and the tissue zinc concentrations, the total zinc requirement for gestation has been calculated to be about 100 mg (Swanson and King 1987). Approximately 60% of the zinc gain is associated with the conceptus.

The additional *daily* need for zinc can be calculated from the rate of tissue gain and the tissue zinc concentrations (Table 21.5). The daily rates of zinc accumulation for the four quarters of pregnancy are 0.08, 0.24, 0.53 and 0.73 mg (Swanson and King 1987). Sandstead (1973) estimated that an additional 0.75 mg zinc/d is gained during the last half of gestation. This estimate was based on the assumption that the zinc concentration of the products of conception exclusive of the foetus are the same as for the foetus. The results of tissue zinc analysis show that uterine tissue zinc concentration is greater than that of the foetus whereas the concentrations in all the other pregnancy tissues are lower (Table 21.5).

The true zinc requirement for pregnancy can be estimated by adding the incremental need for pregnancy to the true requirement for non-pregnant women. This approach assumes that the basal true requirements of pregnant women are the same as non-pregnant women. If endogenous zinc losses are reduced to conserve whole-body zinc, this assumption is invalid. Non-pregnant women consuming about 10–12 mg zinc/d have a true requirement of about 2.5 mg/d. After addition of the incremental need for pregnancy, the true requirements for each of the four quarters of gestation would be 2.6, 2.7, 3.0 and 3.2 mg/d. Women do not seem to increase their zinc intake during pregnancy (Breskin et al. 1983; Hambidge et al. 1983). Average intakes remain constant throughout gestation at near 10 mg/d. To provide the additional zinc required, therefore, zinc absorption must increase or endogenous losses must decrease. In experimental animals, zinc absorption increased significantly in late gestation (Davies and Williams 1977; Schwarz et al. 1981) but a significant increase was not seen in a group of pregnant women (Swanson et al. 1983). The cross-sectional design of the human study may have obviated detection of

Table 21.5. Components of added zinc during pregnancy

	Tissue weight[a] (g)	Zinc concentration (µg/g wet tissue)	Zinc content (mg)
Foetus	3400	17[b]	57.8
Placenta	650	10[c]	6.5
Amniotic fluid	800	0.2[d]	0.5
Uterus	970	25[e]	24.3
Mammary tissue	405	13[f]	5.3
Blood	1259	5[g]	6.3
Total			100.7

[a] Weight of tissues at term (Hytten 1980).
[b] Based on a full-term infant zinc concentration of 19.2 µg/g fat-free tissue (Widdowson and Dickerson 1964) and a body fat of 440 g. Foetal zinc accumulation between 0 and 20 weeks was calculated from foetal weights published by Hytten (1980) and a zinc concentration of 17 µg/g; zinc accumulation between 20 and 40 weeks from Shaw (1979).
[c] Based on zinc analysis of 18 placentae (Alexiou et al. 1977).
[d] Amniotic fluid concentration was increased to 0.6 µg/g during the last 10 weeks (Hambidge et al. 1986).
[e] Uterine tissue analysis (Tipton and Cook 1963).
[f] Mulay et al. (1971).
[g] Qvist et al. (1986).

any effect of pregnancy on zinc absorption. Since the change in fractional absorption required to meet gestational needs is small, i.e. from about 25% to 32%, it may not be detected unless a longitudinal study is done. Urinary zinc losses tend to increase during gestation (Hambidge et al. 1983), but whole-body zinc may be conserved by a reduction in GI zinc secretion. Zinc losses with menstruation cease during gestation, but this saving of only 0.01 mg/d (Greger and Buckley 1977; Umoren and Kies 1982) is quite small relative to the total daily need.

Lactating Women

The additional requirement for zinc varies during lactation with changes in milk volume and milk zinc concentration. Between the 1st and 9th months of lactation the milk zinc concentration drops by about 75%, from 2.6 μg/ml to 0.7 μg/ml (Krebs et al. 1985). Milk zinc concentrations do not appear to be correlated with maternal zinc status or dietary zinc intake (Kirksey et al. 1979; Vaughan et al. 1979; Vuori et al. 1980; Moser and Reynolds 1983). However, these studies were of limited duration. Krebs and co-workers (1985) found that long-term (12 months) administration of a separate 15 mg daily zinc supplement significantly reduced the rate of decline in milk zinc concentration. The authors concluded that the long-term maternal zinc supplementation prevented an abnormally steep rate of decline in milk zinc caused by suboptimal maternal zinc intake. The type of zinc supplement may be important. Self-supplemented lactating women who took 15 mg zinc/d in a multivitamin/mineral preparation did not have any detectable increase in milk zinc concentration above unsupplemented women (Krebs et al. 1985).

Milk volumes also vary during lactation (Butte and Garza 1985). To estimate the additional zinc required for lactation, a standard volume of 750 ml/d was assumed; 850 ml/d was considered to be the likely maximum. Also, to show the potential need for zinc during lactation, milk zinc outputs were calculated for the mean milk zinc concentration with a volume of 750 ml and for the mean + 1 standard deviation in milk zinc concentration with a volume of 850 ml (N.F. Krebs and K.M. Hambidge 1987, personal communication). The mean zinc output of women secreting 750 ml milk/d in the 1st month of lactation is a little over 2 mg/d (Table 21.6). However, the output may be as high as 3.3 mg/d if the milk zinc concentration is 1 standard deviation above the mean and if milk volume is 850 ml/d. By the 3rd and 6th months of lactation the mean zinc output had dropped to about 1 and 0.7 mg/d, respectively. In those women with high milk volumes and high milk zinc concentrations, the total milk zinc outputs are 1.6 and 1 mg/d respectively.

A true zinc requirement for lactating women may be estimated by adding the non-pregnant, non-lactating adult requirement of 2.5 mg/d to the estimated milk zinc output. Using this approach, in the 1st month of lactation the true requirement ranges from 4.6 to 5.8 mg/d; from the 3rd month onward, the true requirement ranges from about 3 to 4 mg/d. There is no evidence that lactating women increase their zinc intake (Krebs et al. 1985). Thus, dietary zinc must be used more efficiently, i.e. there must be increased absorption and/or decreased excretion, to meet the demands for lactation while maintaining maternal homeostasis. We are unaware of any studies of zinc absorption by

Table 21.6. Calculated milk zinc output and true requirements for zinc during lactation

Month of lactation	Milk				True requirement[a]	
	Zinc concentration (μg/ml)	Volume (ml)	Zinc output (mg/d)			
			Mean[b]	Mean+1 SD[c]	Mean	Mean+1 SD
1	2.83 (3.88)[d]	750 (850)[e]	2.12	3.30	4.62	5.80
2	1.25 (1.90)	750 (850)	0.94	1.62	3.44	4.12
3	0.91 (1.23)	750 (850)	0.68	1.04	3.18	3.54

Personal communication from N. F. Krebs and K. M. Hambidge (1987).
[a] True requirement equals the zinc output plus the non-pregnant, non-lactating need of 2.5 mg/d.
[b] Mean zinc output: Mean [Zn] $\times$ 750 ml/1000.
[c] Mean + 1 SD: (Mean [Zn] + 1 SD) $\times$ 850 ml/1000.
[d] Mean concentration (mean concentration + 1 SD).
[e] Assumed standard volume 750 ml (assumed upper end of volume range 850 ml (Butte and Garza 1985)).

lactating women, but an increased zinc absorption has been measured in lactating rats (Davies and Williams 1977). If the endogenous zinc loss is reduced as well by about 50%, for example, to 1.25 mg/d, then the true requirement would range from 3.4 to 4.5 mg/d in the 1st month of lactation and from 2 to 3 mg/d after the 2nd month. Provision of this quantity of absorbable zinc without an increase in total zinc intake would require an increase in the fractional zinc absorption and, possibly, utilization of bone zinc released during early lactation.

Infants

Estimation of zinc requirements during infancy using the factorial approach is difficult because of the lack of data. However, Krebs and Hambidge (1986) have calculated the amount of zinc which must be retained to provide for growth and to replace urinary and integumental losses, i.e. the net zinc absorption. The authors assumed that the fat-free tissue gain contained 30 μg zinc/g and that the combined urinary and integumental losses were 20 μg/kg/d. Since growth velocity declines progressively during the 1st year, the need for net absorbed zinc decreased from a high for male infants of 780 μg/d at 1 month to 480 μg/d in the 5th month and then remained quite constant through the 1st year. Because of slower increases in fat-free tissue and lower body weights, the zinc requirement for females was about 10% lower than for males.

The zinc requirement for growth during the 1st year, interpolated from the figures published by Krebs and Hambidge (1986), is presented in Table 21.7. The estimated zinc requirement for growth declines from about 175 μg/kg/d in the 1st month to about 30 μg/kg/d at 9–12 months. To compute the true zinc requirement, endogenous faecal, urinary and integumental zinc losses need to be added to the requirement for growth. Ziegler and co-workers (1987) recently reported some preliminary findings on the endogenous faecal losses of zinc by

Table 21.7. Estimated true requirements for zinc of infants between 0 and 12 months of age

Age (months)	Growth requirement[a]		Total requirement[b]	
	Male	Female	Male	Female
0–1	175	165	270	260
1–2	130	105	225	200
2–3	85	80	180	175
3–4	55	55	150	150
4–5	45	45	140	140
5–6	45	45	140	140
6–9	40	35	135	130
9–12	30	30	125	125

[a] Interpolated from figures published by Krebs and Hambidge (1986).
[b] Total requirement equals growth need plus endogenous faecal, urinary and integumental losses. Faecal endogenous losses were assumed to be 75 µg/kg/d (Ziegler et al. 1987); urinary and integumental losses were assumed to be 20 µg/kg/d (Krebs and Hambidge 1986).

six normal infants between 41 and 197 d of age. These data suggest that endogenous faecal zinc loss varies with zinc intake; when 237 µg zinc/kg/d was fed, it was about 75 µg/kg/d. Urinary and integumental endogenous losses were assumed to be 20 µg/kg/d (Krebs and Hambidge 1986). Using these values for endogenous losses and the growth estimates published by Krebs and Hambidge (1986), the total requirement can be calculated. As indicated in Table 21.7 it decreases from about 270 µg/kg/d in the 1st month to 125 µg/kg/d at 9–12 months. Since body weight is increasing during this year, the variation in the total requirement is much smaller, from 1.0 to 1.2 mg/d for the males and from 0.9 to 1.05 mg/d for the females.

Data for the true zinc absorption of the six infants studied by Ziegler and co-workers (1987) provide another estimate of the infant true zinc requirement. When these infants were fed 237 µg zinc/kg/d, the zinc absorbed averaged 139 µg/kg/d. It is reassuring that this value is similar to the estimated true zinc requirement for infants between 2 and 7 months of age shown in Table 21.7, 152 µg/kg/d. The infants studied by Ziegler and co-workers (1987) were fed two infant formulas. It would be of interest to measure zinc absorption and endogenous zinc loss in infants fed breast milk since human milk zinc concentrations tend to be lower than those of cows' milk formula and since absorption of zinc from human milk is reported to be especially efficient (Krebs and Hambidge 1986).

Recommended Dietary Intakes of Zinc

Recommendations for dietary zinc intake are published by nine different countries and the World Health Organization (Table 21.8) (WHO 1973; Committee of IUNS 1983). The most widely publicized recommendation is

Table 21.8. Recommended zinc intakes and estimated true requirements

	Adult male	Pregnant woman	Lactating woman	Infant
Recommended intakes (mg/d)				
Australia, 1981	12–16	+4–5	+6	4.5–6
Canada, 1983	9	+0–2	+6	2–3
Czechoslovakia, 1981	8	+8	+10	4
GDR, 1980	12	+1–3	+13	6
Italy, 1978	15	+5	+10	3
New Zealand, 1981	15	+5	+10	3
Spain, 1980	15	+5	+10	3
Uruguay, 1977	15	+5	+10	–
USA, 1980	15	+5	+10	3
WHO, 1973 (20% availability)[a]	11	+1.8–4.0	+16.3	5.5–6.3
Estimated true requirement (mg/d)				
WHO, 1973	2.2	+0.35–0.8	+3.25	1.1–1.25
This chapter	2.5	+0.1–0.7	+1–3.3	1.0–1.2

[a] Recommendations are also made for 10% and 40% dietary zinc availability.

probably that of the United States (NRC 1980). The US recommendation for the adult male is based on the assumption that the turnover of body zinc is 6 mg/d and that 40% of the diet zinc is absorbed. A re-evaluation of the zinc turnover data showed that the 6 mg value was incorrect and was higher than the true rate. Additional allowances of 5 and 10 mg/d were recommended for pregnant and lactating women, respectively. The Committee recognized that these allowances may be generous.

The recommendation by the Canadians is the most recent (CRDSC 1983). Their recommended zinc intake from age 13 onwards is 9 mg/d for males and 8 mg/d for females. This recommendation is based on daily obligatory zinc losses of 90 μg/kg$^{0.75}$, a growth requirement of 0.3 mg and 30% absorption of dietary zinc. The daily obligatory loss figure was derived from the studies of young women fed virtually zinc-free diets (Hess et al 1977) and may, therefore, be an underestimate of the obligatory or endogenous loss of zinc from subjects in which a physiologically adequate zinc status is being maintained. Using this factor of 90 μg/kg$^{0.75}$, the obligatory loss is 2.3 mg/d for a reference male weighing 74 kg and 1.9 mg/d for a reference female weighing 59 kg. The estimated requirement for growth presumably includes zinc for hair and nail loss, menstrual secretions and seminal losses. This may be insufficient to cover the seminal losses depending on the frequency of ejaculations. Baer and King (1984) reported an average loss of 0.63 mg zinc per ejaculum.

The World Health Organization used a factorial approach for making dietary zinc recommendations (1973). The true zinc requirement was calculated to provide for growth, tissue repair and obligatory excretion. Although the approach used to estimate the true requirement differed from that used by us in this chapter, the results are similar (Table 21.8). For example, the two estimates of the true zinc requirement for the adult male were 2.2 and 2.5 mg/d. The WHO made three provisional dietary recommendations accounting for situations in which the fractional absorption of dietary zinc (defined by WHO as "available zinc") was 10%, 20% or 40%. The 10% value may be appropriate for whole-grain cereal-based meals, the 20% value for animal-protein-based meals supplying about 10 mg zinc/d and the 40% value for animal-based meals supplying less zinc (see Chap. 22). The dietary recommendations for an adult male for these three dietary zinc "availabilities" were 22, 11 and 5.5 mg/d, respectively. Our data suggest that the 10% value is low. We observed absorption of 20% or less only in the elderly, when the diet contained high levels of an interfering substance, or when zinc intake exceeded 16 mg/d.

Summary

In the light of the incomplete data and absence of a method without flaws, it is very difficult to delineate human zinc requirements. Furthermore, zinc requirements cannot be defined without considering the dietary zinc source and amount because they both influence whole-body zinc utilization. In general, however, it appears that an intake of approximately 10–12 mg zinc/d from animal-protein-based meals should be sufficient to maintain zinc equilibrium in adults and to provide the zinc needed by pregnant women. It is more difficult to understand how the increased needs of lactating women and infants are met by the intakes typically reported. Further studies of zinc utilization by lactating women and healthy infants are essential before an intake range believed to be sufficient for these two groups can be defined.

Acknowledgements. The authors gratefully acknowledge the helpful comments and suggestions of Dr. K. Michael Hambidge and Dr. Leslie Wada in the preparation of this manuscript.

References

Alexiou D, Grimanis AP, Griman M, Papaevangelou G, Koumantakis E, Papadatos C (1977) Trace elements (zinc, cobalt, selenium, rubidium, bromine, gold) in human placentae and newborn liver at birth. Pediatr Res 11: 646–648
Baer MT, King JC (1984) Tissue zinc levels and zinc excretion during experimental zinc depletion in young men. Am J Clin Nutr 39: 556–570
Bremner I, Morrison JN (1986) Assessment of zinc, copper and cadmium status in animals by assay of extracellular metallothionein. Acta Pharmacol Toxicol (Copenh) [Suppl 7] 59: 502–509
Breskin MW, Worthington-Roberts BS, Knopp RH et al. (1983) First trimester serum zinc concentrations in human pregnancy. Am J Clin Nutr 38: 943–953
Butte NF, Garza C (1985) Energy and protein intake of exclusively breast-fed infants during the first four months of life. In: Garcey M, Falkner F (eds) Nutrition needs and assessment of normal growth. Raven Press, New York
Committee for the Revision of the Dietary Standard for Canada (1983) Recommended nutrient intakes for Canadians. Bureau of Nutritional Sciences, Food Directorate, Health Protection Branch, Department of National Health and Welfare, pp 140–146, 179–180
Committee 1/5 of the IUNS (1983) Recommended dietary intakes around the world, Part 2. Nutr Abst Rev 53: 1076–1119

Davies NT, Williams RB (1977) The effect of pregnancy and lactation on the absorption of zinc and lysine by the rat duodenum in situ. Br J Nutr 38: 417–423

Ehrenkranz RA, Ackerman BA, Nelli CM, Janghorbani M (1984) Determination with stable isotopes of the dietary bioavailability of zinc in premature infants. Am J Clin Nutr 40: 72–81

Foster DM, Aamodt RL, Henkin RI, Berman M (1979) Zinc metabolism in humans: a kinetic model. Am J Physiol 237: R340–R349

Greger JL, Buckley S (1977) Menstrual blood loss of zinc, copper, magnesium and iron by adolescent girls. Nutr Rep Int 16: 639–647

Hambidge KM, Krebs, NF, Jacobs MA, Favier A, Guyette L, Ikle DN (1983) Zinc nutritional status during pregnancy: a longitudinal study. Am J Clin Nutr 37: 429–442

Hambidge KM, Casey CE, Krebs NF (1986) Zinc. In: Mertz W (ed) Trace elements in human and animal nutrition, 5th edn, vol 2. Academic Press, Orlando, Florida, pp 1–138

Hess FM, King JC, Margen S (1977) Zinc excretion in young women on low zinc intakes and oral contraceptive agents. J Nutr 107: 1610–1620

Hytten FE (1980) Weight gain in pregnancy. In: Hytten FE, Chamberlain G (eds) Clinical physiology in obstetrics. Blackwell Scientific Publications, Oxford, pp 193–233

Jackson MJ, Jones DA, Edwards RHT, Swainbank IG, Coleman ML (1984) Zinc homeostasis in man: studies using a new stable isotope-dilution technique. Br J Nutr 51: 199–208

King JC (1986) Assessment of techniques for determining human zinc requirements. J Am Diet Assoc 86: 1523–1528

Kirksey A, Ernst JA, Roepke JL, Tsai TL (1979) Influence of mineral intake and use of oral contraceptives before pregnancy on the mineral content of human colostrum and of more mature milk. Am J Clin Nutr 32: 30–39

Krebs NF, Hambidge KM (1986) Zinc requirements and zinc intakes of breast-fed infants. Am J Clin Nutr 43: 288–292

Krebs NF, Hambidge KM, Jacobs MA, Rasbach JO (1985) The effects of a dietary zinc supplement during lactation on longitudinal changes in maternal zinc status and milk zinc concentrations. Am J Clin Nutr 41: 560–570

Milne DB, Canfield WK, Mahalko JR, Sandstead HH (1983) Effect of dietary zinc on whole body surface loss of zinc: impact on estimation of zinc retention by balance method. Am J Clin Nutr 38: 181–186

Moser PB, Reynolds RD (1983) Dietary zinc intake and zinc concentrations of plasma, erythrocytes, and breast milk in antepartum and postpartum lactating and nonlactating women: a longitudinal study. Am J Clin Nutr 38: 101–108

Mulay IL, Roy R, Knox BE, Suhr NH, Delaney WE (1971) Trace-metal analysis of cancerous and non-cancerous human tissue. J Natl Cancer Inst 47: 1–13

National Research Council (1980) Recommended dietary allowances, 9th edn. National Academy of Sciences, Washington DC, pp 144–147

Peirce P, Hambidge M, Fennessey P, Miller L, Goss C (1987) Evaluation of zinc absorption in the preterm infant using stable isotope technique. Fed Proc 46: 748

Qvist I, Abdulla M, Jagerstad M, Svensson S (1986) Iron, zinc, and folate status during pregnancy and two months after delivery. Acta Obstet Gynecol Scand 65: 15–22

Sandstead HH (1973) Zinc nutrition in the United States. Am J Clin Nutr 26: 1251–1260

Schwartz R, Apgar BJ, Wien EM (1986) Apparent absorption and retention of Ca, Cu, Mg, Mn, and Zn from a diet containing bran. Am J Clin Nutr 43: 444–455

Schwarz FJ, Kirchgessner M, Sherif SY (1981) Zur intestinalen Absorption von Zink während der Gravidität und Laktation. Res Exp Med (Berl) 179: 35–42

Shaw JCL (1979) Trace elements in the fetus and young infant. I. Zinc. Am J Dis Child 133: 1260–1268

Smith JC Jr, Morris ER, Ellis R (1983) Zinc: requirements, bioavailabilities and recommended dietary allowances. In: Prasad AS, Cavdar AO, Brewer GJ, Aggett PJ (eds) Zinc deficiency in human subjects. Alan R. Liss, New York, pp 147–169

Swanson CA, King JC (1987) Zinc and pregnancy outcome. Am J Clin Nutr 46: 763–771

Swanson CA, Turnlund JR, King JC (1983) Effect of dietary zinc sources and pregnancy on zinc utilization in adult women fed controlled diets. J Nutr 113: 2557–2567

Tipton IH, Cook MJ (1963) Trace elements in human tissue. II. Adult subjects from the United States. Health Phys 9: 103–145

Turnlund J, Costa F, Margen S (1981) Zinc, copper, and iron balance in elderly men. Am J Clin Nutr 34: 2641–2647

Turnlund JR, Michel MC, Keyes WR, King JC, Margen S (1982) Use of enriched stable isotopes to determine zinc and iron absorption in elderly men. Am J Clin Nutr 35: 1033–1040

Turnlund JR, King JC, Keyes WR, Gong B, Michel MC (1984) A stable isotope study of zinc absorption in young men: effects of phytate and α-cellulose. Am J Clin Nutr 40: 1071–1077

Turnlund JR, Acord LL, Betschart AA, Kretsch MJ, Sauberlich HE (1985) The effect of vitamin B-6 deficient diets on zinc absorption in young women. Fed Proc 44: 1278

Turnlund JR, Durkin N, Costa F, Margen S (1986) Stable isotope studies of zinc absorption and retention in young and elderly men. J Nutr 116: 1239–1247

Turnlund JR, Betschart AA, Keyes WR, Acord LL (1987) A stable isotope study of zinc bioavailability in young men from diets with white vs whole wheat bread or beef vs soy. Fed Proc 46: 879

Umoren J, Kies C (1982) Menstrual blood losses of iron, zinc, copper and magnesium in adult female subjects. Nutr Rep Int 26: 717–726

Vaughan LA, Weber CW, Kemberling SR (1979) Longitudinal changes in the mineral content of human milk. Am J Clin Nutr 32: 2301–2306

Vuori E, Makinen SM, Kara R, Kuitunen P (1980) The effects of the dietary intakes of copper, iron, manganese, and zinc on the trace element content of human milk. Am J Clin Nutr 33: 227–237

Wada L, King JC (1986) Effect of low zinc intakes on basal metabolic rate, thyroid hormones and protein utilization in adult men. J Nutr 116: 1045–1053

Wada L, Turnlund JR, King JC (1985) Zinc utilization in young men fed adequate and low zinc intakes. J Nutr 115: 1345–1354

Widdowson EM, Dickerson JWT (1964) Chemical composition of the body. In: Comar CL, Bronner F (eds) Mineral metabolism, an advanced treatise, vol 2. Academic Press, New York, p 26

World Health Organization (1973) Trace elements in human nutrition. Report of a WHO Expert Committee. WHO Tech Rept Ser No. 532, pp 9–15

Ziegler EE, Figueroa-Colon R, Serfass RE, Nelson SE (1987) Effect of low dietary zinc on zinc metabolism in infancy: stable isotope studies. Am J Clin Nutr 45: 849

Chapter 22

Dietary Pattern and Zinc Supply

Brittmarie Sandström

Introduction

The range of food items from which diets are constituted and the preparative procedures used in their processing can influence both the total intake of zinc and its biological availability. That such variables can give rise to an inadequate zinc supply is exemplified by the observations of zinc-responsive growth failure in American children with a zinc intake of 5 mg/d (Krebs et al. 1984) and also in Iranian adolescents (Halsted et al. 1972) on diets providing 19–22 mg of zinc daily (Maleki 1973). Sensitive indices of zinc status, suitable for evaluation of availability and adequacy of the zinc supplied by the differing diets of individuals or populations, are still lacking. Present data indicate, however, that dietary attributes influencing the efficiency with which zinc can be absorbed are the main determinants of zinc availability. Thus, experimental studies of zinc absorption from composite meals or diets can provide useful information on the relative values of various food combinations as zinc sources.

Zinc Content of Human Food Sources

Significant quantities of zinc are found in most natural non-processed foodstuffs for human consumption. However, the quantities present in edible animal and plant tissues and organs reflect differences in the functional importance of zinc for each tissue and also the degree of uptake of zinc from the environment. Large variations in zinc content can thus be found between otherwise nutritionally similar food sources. Relationships between the zinc content and the energy and protein contents of some common food sources of human diets are indicated in Table 22.1. Such data illustrate the possible variations in total

Table 22.1. Zinc contents of selected foods expressed on a raw wet weight basis and in relation to their protein and energy contents (from data of Paul and Southgate 1978)

	mg/kg raw wet weight	mg/g protein	mg/MJ
Beef			
lean	43	0.21	8.3
fat	10	0.11	0.4
Pork			
lean	24	0.12	3.9
fat	4	0.06	0.1
Chicken			
light meat	7	0.03	1.4
dark meat	16	0.08	3.0
Fish			
cod	4	0.02	1.2
Milk	3.5	0.11	1.3
Cheese, cheddar type	40	0.15	2.4
Butter	1.5	–	0.05
Lentils	31	0.13	2.4
Wheat			
wholemeal	30	0.23	2.2
white	9	0.08	0.6
Maize (sweetcorn)	12	0.29	2.2
Rice (polished)	13	0.20	0.8
Potatoes	3	0.14	0.8
Yam	4	0.20	0.7
Coconut	5	0.16	0.3

zinc intake that can arise from the selection of foods to cover energy needs. Other primary energy sources, like fats, oils, sugar and alcoholic beverages, have a low zinc content and normally do not contribute significantly to total zinc intake.

Animal Products

The content of zinc in meat depends both on the type of animal and on the tissue that is consumed. Twofold variations in zinc content of individual muscles from the same animal have been reported (Marchello et al. 1985). In general, dark red meat has a higher zinc content than white meat. Fat tissue has a low zinc content and variations in fat content will give large differences in zinc content of various "butcher's cuts" and also influence the zinc content of cured meat products (Kirkpatrick and Coffin 1975). In grazing buffalos an effect of season on zinc content of the meat has been reported, with higher zinc contents after the dry season than during periods of rapid herbage growth (El-Kirdassy et al. 1977). Offal is a good zinc source with a zinc content similar to dark red meat.

The richest food source of zinc is oysters which, depending on water conditions, can contain up to 1 g zinc/kg. Fish meat has a lower zinc content than most meat muscles. Higher zinc levels are reported for some marinated fish products in which the zinc content of the bones probably diffuses into the meat.

Milk and milk products are particularly important sources of zinc for infants and children in most countries and for adults in Northern Europe. Zinc in cows' milk is mainly found in the casein fraction with only minor amounts present in fat (Sandström et al. 1983). The content of zinc is consequently high in cheese and low in cream and butter.

Vegetable Foods

Cereals are the major source of energy and also of zinc in large parts of the world. Large differences in zinc content depending not only on type of cereal but also on variety, class, subclass and growing location have been reported (Koivistoinen et al. 1974; Dikeman et al. 1982; Davis et al. 1984). Davis et al. (1984) found a range in zinc content of wheat on a dry weight basis from 15 to 102 mg zinc/kg depending on its botanical variety or strain and from 29 to 61 mg/kg for the same wheat variety grown in 13 locations and in two crop years.

Limited data on zinc content are available for the large range of legumes that are grown for human consumption. In relation to their protein content, legumes are comparable to lean pork as zinc sources. As with cereals, differences among varieties and strains of legumes have been reported (Meiners et al. 1976).

Because of their high water contents green leafy vegetables and fruits are only modest sources of zinc with a typical range of 1–8 mg/kg. Their contribution to total energy intake from most diets is also relatively low with the exception of starchy roots like potatoes, yam and cassava, which are staple foods in parts of the world. Wide variations in their zinc content reflect differences in their conditions of growth (Warren 1972; Leveille et al. 1974). Locally grown traditionally prepared plant foods, sometimes fortified by the addition of plant ash to dishes, have been found to contribute significantly to the zinc intake of Arizonan Indians (Calloway et al. 1974).

Effect of Processing and Preparation of Food

Values for zinc content in food tables, used to calculate zinc intakes, are usually given on a fresh non-processed basis. Knowledge of losses of nutrients during food preparation and processing is relatively limited and research in this field has mainly been concerned with vitamins. The food process that probably has the most pronounced effect on zinc intake is the refinement of cereals. In grains and seeds the minerals are found mainly in the outer layers of the kernel and with a low extraction rate of the flour large losses of zinc occur (Schroeder 1967). Similar losses occur during the polishing of rice and in the refining of sugar.

Other pretreatments of foods before cooking or consumption and cooking procedures themselves can also influence the zinc content of a diet. Zinc from galvanized cooking utensils was probably in former days an important additional zinc source and may still be so in parts of the world. Increases in the zinc content of canned foods have been observed which probably arise from gradual release of zinc from a zinc-oxide-containing coating of tin plate on cans (Henriksen et al. 1985). Toma and Tabekhia (1979) reported a significant increase in the zinc retention of rice when cooked in domestic tap water. On the other hand, leakage of zinc from food during preparation can, when the cooking water is discarded, reduce the zinc intake (Meiners et al. 1976). Losses of 20%–23% of the zinc from canned products into the canning media have been reported (Schmitt and Weaver 1982), whereas, for frozen food, the losses were 10% or less.

Food preparation can also affect the availability of zinc. Phytic acid in whole grains and legume seeds is a potent antagonist of zinc absorption. Prolonged fermentation of bread reduces the phytic acid content and significantly improves zinc absorption (Nävert et al. 1985). Fermentation is used also in the preparation of soyabean dishes in many Asian diets and for sorghum in some African diets. The extent of phytic acid degradation during these traditional methods of food preparation and its significance for the availability of zinc to man is not known. Extrusion cooking, which is used for example for breakfast cereals, seems to inhibit the degradation of phytic acid in the gut and results in a less efficient apparent absorption of zinc compared with that from similar non-extruded products (Kivistö et al. 1986; Sandberg et al. 1986). Germination has been found to reduce the phytic acid content of beans, while cooking for 3 h had little effect on phytate retention (Tabekhia and Luh 1980).

Zinc Intake and Zinc Status

A number of unsuccessful attempts have been made to correlate zinc intake with indices of zinc status in man. One reason for failure could be the method used for collection of dietary data. Considering the large variations in zinc content of foods, an actual analysis of an identical portion of the food eaten, as used in several studies, would seem to be the most appropriate and precise method. It has, however, been shown that this technique can restrict and reduce normal voluntary food intake sufficiently to give a negative zinc balance (Patterson et al. 1984). Large day-to-day variations in zinc intake have been reported (Record et al. 1985). For protein it has been calculated, from urinary nitrogen excretion, that a minimum of 18 d of food records is needed to obtain a valid estimate of intake within 80% of the subject's mean value (Bingham and Cummings 1985). As the intakes of protein and zinc are closely correlated (Hunt et al. 1979; Anderson et al. 1981) it is likely that sampling over a similar period is needed to obtain a good estimate of zinc intake for an individual on a varied Western diet. In many studies, the uncertainties caused by reliance upon 24-h "recalls" or 3–4 d dietary records to calculate zinc intakes are added to by the frequent failure to derive any estimates of the physiological availability of zinc from the diets in use.

Diet Composition and the Absorption and Utilization of Zinc

Balance Studies

The conventional balance technique, where "apparent" absorption is calculated as the difference between intake and faecal excretion, is a relatively insensitive method for monitoring the absorption of zinc. Its limitations arise partly from the large endogenous intestinal losses of zinc (Turnlund et al. 1982, 1986). For reliable data, an adaptation period of at least 4 weeks on a constant diet and a study period of 2–3 weeks is required according to Schwartz et al. (1986). Few published studies fulfil these criteria. It is also obvious that when intake and/or absorption are low, adaptive mechanisms develop which reduce the excretion of zinc (Wada et al. 1985). Thus it is questionable if a "true" zinc balance ever exists.

Balance studies have shown a high apparent efficiency of zinc absorption and a positive retention of zinc from "refined" American diets and from formula diets based on sources of animal protein and providing at least 15 mg zinc/d (Robinson et al. 1973; Ritchey et al. 1979; Turnlund et al. 1984; Schwartz et al. 1986). In children, lower absorption and retentions of zinc have been observed from a "low-income diet" containing 24 g of mixed animal and plant proteins (Meiners et al. 1977) and from a 100% vegetarian diet based on legumes and nuts (Engel et al. 1966) than from a diet based on sources of animal protein (Ritchey et al. 1979). In contrast, Swanson et al. (1983) observed no difference in the absorption and retention of zinc by pregnant women when a vegetable-based formula was compared with an animal-protein-based formula. However, 30% of the protein was still of animal origin in the vegetable-based formula and the total protein content was fairly high.

Claims that clinical signs of zinc deficiency develop in subjects consuming diets based on wholemeal bread (Halsted et al. 1972) have stimulated a number of studies of cereal-based diets. In early studies by McCance and Widdowson (1942), a positive zinc retention was observed from a mixed diet providing 19.6 mg zinc/d in which 40% of the energy was provided by bread made from 92% extraction rate flour. A similar diet of white bread providing 9.9 mg zinc/d produced a negative zinc retention. Almost identical estimates of the "apparent" efficiency of zinc absorption were made by Andersson et al. (1983) for "normal" UK diets containing either 200 g of wholemeal bread or 200 g of zinc-enriched white bread. Van Dokkum et al. (1982) also found no difference in the absorption and retention of zinc from diets based on animal protein with 240 g of white bread or from the same diet in which white bread was replaced by 240 g of bran-enriched bread.

In contrast, Reinhold et al. (1976) reported negative zinc balancs in a 20-d balance study with two subjects on a diet with 500 g of wholemeal bread providing 18–19 mg/d, whereas the same diet with white bread resulted in a positive zinc balance. A diet in which unleavened wholewheat bread supplied at least half of the food intake of two subjects for 63 d resulted, initially, in a negative (-3.2 mg/d) zinc balance, but during the last 15 d a positive ($+2.7$ mg/d) zinc balance was observed (Campbell et al. 1976).

The absorption and retention of zinc from typical low-income Indian diets has been studied by Rao and Rao (1980). Eleven different diets were served to six subjects, each for 11 d with faecal collections during the last 4 d on each diet. Since no faecal marker was used, the variation in estimates of faecal zinc excretion was large making it impossible to observe any statistically significant differences in absorption or balance. It is however interesting to note that the mean "apparent" absorption of zinc from a diet typical for the West Bengal area based on rice and providing 12.7 mg zinc/d was higher than that from a Maharashtra diet based on Jowar providing 15.4 mg zinc/d but similar to that of an Uttar Pradesh diet based on wheat providing 20 mg zinc/d. Although such studies suggest that the poorer availability of zinc from some diets rich in unrefined cereals can be counteracted by their higher zinc contents, the limitations of many published balance studies make it difficult to identify dietary patterns which are likely to have consistent negative effects on zinc status.

Isotope Studies

The absorption of zinc from single meals can be studied under more standardized conditions and with a higher degree of precision by the use of isotope techniques. A selection of results from absorption studies of composite meals using a radionuclide technique is given in Table 22.2. The fractional absorption of zinc reported in these studies ranged from a mean of 2.4% to 38% for different meals. Although the levels of many nutrients differed between the meals, the balance of evidence suggests that three major factors influence the efficiency and extent of zinc absorption. They are the levels of zinc and of protein and the extent of refinement of the cereal component of diets. From refined, low-zinc, low-phytic-acid, cereal-based meals 35%–40% of the zinc was absorbed, whereas from whole-grain, cereal-based meals only 10%–15% was absorbed and with high levels of bran only 5%–10% may be absorbed.

Inverse relationships between total zinc intake and the fractional efficiency of zinc absorption are evident from studies with diets with relatively high contents of animal protein. Thus the fractional efficiency of zinc absorption from meals providing 3–5 mg zinc was 20%–25% compared with 30%–40% when zinc intake was lower. The few vegetarian meals studied included meals containing soya protein providing 1.2–2.5 mg zinc of which approximately 20% was absorbed.

Zinc absorption studies conducted by using stable isotopes of zinc have usually involved distribution of the isotopic marker between four identical daily meals. Swanson et al. (1983) added 2.5 mg of zinc to a formula providing approximately 4 mg zinc per meal of which 24%–25% was absorbed, a result similar to that of Sandström and Cederblad (1980) for the absorption of 24% of zinc from a beef meal containing 4.6 mg of zinc. Other comparisons indicate reasonable agreement between data derived from stable isotope or radiozinc studies of absorption. Thus the high absorption, 52.6%, observed during a stable isotope study by Wada et al. (1985) using a low-zinc diet containing 1.4 mg per meal is consistent with the relatively high (36%) absorption from a low-zinc chicken meal observed by Sandström and Cederblad (1980) (see

Table 22.2). Also similar are the 25.7% absorption of the 4.1 mg zinc provided by meals prepared by the addition of oysters to a low-zinc diet (Wada et al. 1985) and the 20% absorption from beef-containing meals of similar zinc content found by Sandström and Cederblad (1980). Turnlund et al. (1984) reported that the absorption of 34% of the 3.8 mg zinc provided by meals of a fibre- and phytic-acid-free formula declined to 17.5% when 2.34 g phytate was added. The latter finding is comparable to that from studies with a wholewheat bread and animal-protein-containing meal from which a zinc absorption of 14% was observed (Sandström et al. 1980).

Overall, there is a good consistency among estimates of the fractional absorption of zinc reported from such isotopic studies, particularly when differences in the types of meal and in their zinc contents are taken into account. Thus, it appears feasible to combine data from both "single meal" (radiozinc) studies and from stable isotope studies to derive provisional estimates of the absorbability and potential supply of zinc from differing diets when their zinc contents and gross composition are known.

Dietary Pattern and Zinc Supply

The daily intake of zinc from diets typical of industrialized countries that characteristically are associated with high intakes of fat, refined sugar and protein, is approximately 10–12 mg or 1.2 mg/MJ (Spring et al. 1979; Gibson and Scythes 1982; Welsh and Marston 1982; Murphy and Calloway 1986). Animal protein sources account for about two-thirds of the intake. The phytic acid content of such diets is, on average, low (Murphy and Calloway 1986) and thus a high availability of zinc can be assumed. Assuming a zinc intake of 2–3 mg/meal of which 30% may be absorbed, the daily supply of absorbable zinc from such a diet could amount to 3 mg.

Slightly lower zinc contents have been reported for lacto-vegetarian diets in affluent countries. Abraham et al. (1985) report a daily intake of 8 mg (1.0 mg/MJ) by vegetarian pregnant women in an Asian community in Britain. Animal protein accounted for 41%–47% of the total zinc intake and with this type of diet at least 20% of the zinc is probably absorbed. The type of vegetarian diet taken by Seventh-Day Adventist Canadian women was found to provide 9 mg of zinc mostly from plant products (Anderson et al. 1981). With this type of diet a lower fractional absorption of zinc, approximately 15%, can be expected.

In developing countries, locally produced staple foods determine both the intake and absorbability of zinc. In Table 22.3, food consumptions and reported or calculated zinc intakes are given for five different dietary patterns together with estimates of zinc availability. Data are based on surveys of food intake or of purchases by large numbers of families or from analysis of 24-h duplicate food samples. The completely cereal-based diet described by Cresta et al. (1976) for the Republic of Benin and the Chinese diets based on cereals and starchy roots described by Thurnham et al. (1985) are all likely to contain zinc in forms that are not highly absorbable. Maize and rice, the major cereals in these diets, also have low zinc contents and this further reduces the supply of absorbable zinc.

The Indian diets described by Soman et al. (1969) (Table 22.2) are higher in zinc content and, due to their inclusion of animal protein sources, the efficiency with which this zinc is absorbed is likely to be higher. Although the fish diets taken by low-income families in areas around the Amazonas (Shrimpton 1984) probably provide zinc in a readily absorbable form, their intrinsic content of zinc is low thus reducing total supply of absorbable zinc. It is noteworthy that signs of zinc deficiency have been observed in this population. Similar low zinc intakes associated with low hair zinc levels have been reported by Ross et al. (1986) for a population in Papua New Guinea subsisting on a yam-based diet. Food balance sheet data (FAO 1984) for Egypt are also included in Table 22.3. Assuming that the majority of the wheat consumed is in the form of wholemeal bread, a low utilization of zinc can be expected. Despite the higher zinc content of this diet, the overall supply of potentially utilizable zinc is similar to most of the other diets considered and lower than from a typical "Western" diet. Indeed, the only exceptions to this generalization are the Indian diets described by Soman et al. (1969).

Limited data are available for the metabolic requirements for absorbable zinc needed to replace the urinary and skin losses of zinc and losses through endogenous intestinal secretions. Data from Wada et al. (1985) indicated that 2.9 mg of zinc was absorbed from a refined diet providing 5.5 mg/d and on which zinc balance was achieved. Factors interfering with the absorption of zinc from the diet can also affect the efficiency with which zinc in endogenous secretions is re-utilized within the digestive tract. Thus Turnlund et al. (1984) estimated that for diets with a high phytic acid content at least 1 mg additional zinc needs to be absorbed by healthy Americans to compensate for this effect and maintain "balance".

It is important to emphasize that other variables also influence the interpretation of estimates of the dietary supply of absorbable zinc such as those presented in Table 22.3. Thus integumental losses of zinc may be relatively lower in small-statured populations. Conversely, a hot climate probably increases zinc loss through perspiration. Although adaptation to a low zinc supply or to increased losses may well occur neither the extent to which this is possible nor the metabolic consequences of such adaptation are known.

The estimates and observations presented in Tables 22.1 to 22.3 indicate that the zinc content of the major energy and protein sources of the diet can sometimes be the limiting factor for zinc supply even in situations when the fractional efficiency of zinc absorption is high. From the single-meal studies presented in Table 22.2 it can be seen that the calculated amount of zinc absorbed is less from low-zinc meals than from similar meals with a higher zinc content. This has also been observed by Wada et al. (1985) in studies using stable zinc isotopes. The potential risk for zinc deficiency can therefore be similar on a low-zinc "cafeteria diet" with a high fat and sugar content, on diets based on fish or with starchy roots as staple foods as on a diet based on wholemeal wheat or sorghum with a higher zinc content.

Table 22.2. Zinc absorption from single test meals differing in composition, studied by radionuclide absorption techniques

Meal characteristics	Zinc content mg (added)	Protein content (g)	(n)	Zinc absorption		Reference
Composition (main components)				fractional (%)	total (mg[a])	
White bread, water	0.4	5	8	38.2± 8.1	0.2	Sandström et al. (1980)
White bread, milk	1.6	15	24	37.0±10.8	0.6	Sandström et al. (1987)
Corn flakes, milk, white bread	2.0	16[b]	6	37.7± 7	0.8	Lykken et al. (1986)
Rice crisps, water	15.2 (15.0)	0.1	7	21.5± 6.6	3.2	Farah et al. (1984)
Wheat bran (20 g), water	17.0 (15.0)	0.3	8	2.4± 1.3	0.4	Farah et al. (1984)
Wheat bran (10 g), white bread, milk	1.9	12	8	9.6± 1.4	0.2	Nävert et al. (1985)
Wheat bran bread 16 h fermentation (10 g bran), milk	1.9	12	9	19.8± 4.5	0.4	Nävert et al (1985)
Wheat bran bread 16 h fermentation (30 g bran), milk	3.6	12	8	6.6± 3.1	0.2	Nävert et al. (1985)
Wholemeal bread, milk	2.4	14	16	11.3± 3.8	0.3	Sandström et al. (1987)
Wholemeal bread, milk, cheese	3.2	19	13	14.0± 3.8	0.5	Sandström et al. (1980)
Chicken, potatoes	1.3	25	11	36.2± 6.7	0.5	Sandström and Cederblad (1980)
Turkey	4.0 (1.0)	28	11	28 ± 8	1.1	Valberg et al. (1984)
Beef, potatoes	4.6	25	6	20.4± 6.9	0.9	Sandström and Cederblad (1980)
Soyabeans, potatoes	2.5	25	6	19.6± 6.9	0.5	Sandström and Cederblad (1980)
Cereals, milk, formula	3.6	23	11	18.1± 3.4	0.7	Dawson-Hughes et al. (1986)
Meat sauce, rice	3.1	23	21	25.2± 8.0	0.8	Sandström et al. (1987)
Meat sauce, rice, milk	4.9	31	5	20.7± 3.4	1.0	Sandström et al. (1987)
Soya isolate sauce, rice	1.2	22	8	20.9± 5.9	0.3	Sandström et al. (1987)

[a] Calculated from zinc content of meal and fractional absorption.
[b] Estimates from data in paper.

Table 22.3. Dietary compositions and estimated intakes of total and absorbable zinc from diets typical of some non-Western societies

Country:	Benin	China (Linxian)	India	Brazil (Amazonas)	Egypt
Reference:	Cresta et al. (1976)	Thurnham et al. (1985)	Soman et al. (1969)	Shrimpton (1984)	Food Balance Sheets, FAO (1984)
Ingredient	Intake (g/d)				
Cereals	883[a]	564	450	203	695
Starchy roots, tubers	135	496	–	85	64
Legumes	43	15	67	30	20
Vegetables	36	685	120	38	328
Meat	5	4	40	87	42
Fish	23	–	40	151	14
Eggs	< 1	–	40	12	5
Milk, cheese	< 1	–	86	85	57
Energy intake, MJ	10.1	11.9	(11.5)[b]	8.8	13.3
Zinc intake mg/d	8.0	(12)[b]	16.1	7.3	(20)[b]
mg/MJ	1.0	(1.0)[b]	(1.4)[b]	0.8	(1.5)[b]
Estimated efficiency of zinc absorption, %	15–20	15–20	20	25–30	10–15
Estimated absorbed zinc, mg/10 MJ	1.2–1.6	1.5–2.0	2.8	1.8–2.2	1.5–2.3

[a] Principally maize.
[b] Estimated from data presented.

Needs for Future Research

Our appreciation of differences in the efficiency with which zinc can be utilized from different diets is still very limited especially as regards the various types of mainly vegetarian diets taken by large populations of the world. The effect of "traditional" and of industrial methods of food preparation and processing on zinc availability is also an important research area. To gain a better understanding of the effects of different diets on both the absorption and the overall metabolic turnover of zinc a useful experimental approach would be to combine conventional balance studies over sufficiently long periods of time to achieve equilibration with measurements of the absorption of zinc using its stable or radioactive isotopes. However, the cost of stable zinc isotopes, the labour required to perform the balance studies and the need for specialized analytical equipment for the isotope studies limit the number of diets that can be studied in this way.

Once biochemical criteria have been developed which adequately reflect physiologically relevant changes in zinc status, epidemiological studies will begin to provide much more valuable information on relationships between dietary habits and the risks of deficiency. Such criteria are needed to justify changes in the selection or processing of dietary ingredients to improve the intake or utilization of zinc and for decisions to be made on the advisability of zinc supplementation for populations at risk from deficiency.

However, factors affecting the availability of zinc are also likely to affect the availability of other trace elements. With a few exceptions, there is also a close correlation between zinc content of a diet and the content of other trace elements. Therefore, in any intervention, the overall nutritional status including that of trace elements other than zinc should be followed.

References

Abraham R, Campbell-Brown M, Haines AP, North WRS, Hainsworth V, McFadyen IR (1985) Diet during pregnancy in an Asian community in Britain – energy, protein, zinc, copper, fibre and calcium. Hum Nutr Appl Nutr 39A: 23–35

Anderson BM, Gibson RS, Sabry JH (1981) The iron and zinc status of long-term vegetarian women. Am J Clin Nutr 34: 1042–1048

Andersson H, Nävert B, Bingham SA, Englyst HN, Cummings JH (1983) The effects of breads containing similar amounts of phytate but different amounts of wheat bran on calcium, zinc and iron balance in man. Br J Nutr 50: 503–510

Bingham SH, Cummings JH (1985) Urine nitrogen as an independent validatory measure of dietary intake: a study of nitrogen balance in individuals consuming their normal diet. Am J Clin Nutr 42: 1276–1289

Calloway DH, Giauque RD, Costa FM (1974) The superior mineral content of some American Indian foods in comparison to federally donated counterpart commodities. Ecol Food Nutr 3: 203–211

Campbell BJ, Reinhold JG, Cannell JJ, Nourmand I (1976) The effects of prolonged consumption of wholemeal bread upon metabolism of calcium, magnesium, zinc and phosphorus of two young American adults. Pahlavi Med J 7: 1–17

Cresta M, Allegrini M, Casadei E, Gallorini M, Lanzola E, Panatta GB (1976) Benin: nutritional considerations on trace elements in the diet. Food Nutr 2: 8–18

Davis KR, Peters LJ, Cain RF, LeTourneau D, McGinnis J (1984) Evaluation of the nutrient composition of wheat. III. Minerals. Am Assoc Cereal Chem 29: 246–248

Dawson-Hughes B, Seligson FH, Hughes VA (1986) Effects of calcium carbonate and hydroxyapatite on zinc and iron retention in postmenopausal women. Am J Clin Nutr 44: 83–88

Dikeman E, Pomeranz Y, Lai FS (1982) Minerals and protein contents in hard red winter wheat. Cereal Chem 59: 139–142

El-Kirdassy ZHM, El-Safouri SA, Abdel-Galil AM (1977) Biochemical studies on Egyptian buffalo meat. II. Determination of iron, zinc and copper. Die Nährung 21: 289–295

Engel RW, Miller RF, Price NO (1966) Metabolic patterns in preadolescent children. XIII. Zinc balance. In: Prasad AS (ed) Zinc metabolism. Thomas, Springfield, Illinois, pp 326–338

Farah DA, Hall MJ, Mills PR, Russell RI (1984) Effect of wheat bran on zinc absorption. Hum Nutr Clin Nutr 38C: 433–441

Food and Agriculture Organisation (1984) Food balance sheets 1979–81 average. Rome

Gibson RS, Scythes CA (1982) Trace element intakes of women. Br J Nutr 48: 241–248

Halsted JA, Ronaghy HA, Abadi P et al. (1972) Zinc deficiency in man: the Shiraz experiment. Am J Med 53: 277–284

Henriksen LK, Mahalko JR, Johnson LK (1985) Canned foods: appropriate in trace element studies? J Am Diet Assoc 85: 563–568

Hunt IF, Murphy NJ, Gomez J, Smith JC (1979) Dietary zinc intake of low-income pregnant women of Mexican descent. Am J Clin Nutr 32: 1511–1518

Kirkpatrick DC, Coffin DE (1975) Trace metal content of various cured meats. J Sci Food Agric 26: 43–46

Kivistö B, Andersson H, Cederblad G, Sandberg A-S, Sandström B (1986) Extrusion cooking of a high-fibre cereal product. II. Effects on apparent absorption of zinc, iron, calcium, magnesium and phosphorus in humans. Br J Nutr 55: 255–260

Koivistoinen P, Nissinen H, Varo P, Ahlström A (1974) Mineral element composition of cereal grains from different growing areas in Finland. Acta Agric Scand 24: 327–334

Krebs NF, Hambidge KM, Walravens PA (1984) Increased food intake of young children receiving a zinc supplement. Am J Dis Child 138: 270–273

Leveille GA, Bedford CL, Kraut CW, Lee YC (1974) Nutrient composition of carrots, tomatoes and red tart cherries. Fed Proc 33: 2264–2266

Lykken GI, Mahalko J, Johnson PE et al. (1986) Effect of browned and unbrowned corn products intrinsically labeled with ^{65}Zn on absorption of ^{65}Zn in humans. J Nutr 116: 795–801

McCance RA, Widdowson EM (1942) Mineral metabolism of healthy adults on white and brown bread dietaries. J Physiol 101: 44–85

Maleki M (1973) Food consumption and nutritional status of 13 year old village and city schoolboys in Fars province, Iran. Ecol Food Nutr 2: 39–42

Marchello MJ, Slanger WD, Milne DB (1985) Macro and micro minerals from selected muscles of pork. J Food Sci 50: 1375–1378

Meiners CR, Derise NL, Lau HC, Crews MG, Ritchey SJ, Murphy EW (1976) The content of nine mineral elements in raw and cooked mature dry legumes. J Food Chem 24: 1126–1127

Meiners CR, Taper LJ, Korslund MK, Ritchey SJ (1977) The relationship of zinc to protein utilization in the preadolescent child. Am J Clin Nutr 30: 879–882

Murphy SP, Calloway DH (1986) Nutrient intakes of women in NHANES II, emphasizing trace minerals, fiber, and phytate. J Am Diet Assoc 86: 1366–1371

Nävert B, Sandström B, Cederblad Å (1985) Reduction of the phytate content of bran by leavening in bread and its effect on absorption of zinc in man. Br J Nutr 53: 47–53

Patterson KY, Holbrook JK, Bodner JE, Kelsay JL, Smith JC Jr, Veillon C (1984) Zinc, copper, and manganese intake and balance for adults consuming self-selected diets. Am J Clin Nutr 40: 1397–1403

Paul AA, Southgate DAT (1978) McCance and Widdowson's The composition of foods. 4th edn. Elsevier/North Holland, Biomedical Press, Amsterdam New York Oxford

Rao CN, Rao BSN (1980) Absorption and retention of magnesium and some trace elements by man from typical Indian diets. Nutr Metab 24: 244–254

Record IR, Record SJ, Dreosti IE, Rohan TE (1985) Dietary zinc intake of pre-menopausal women. Hum Nutr Appl Nutr 39A: 363–369

Reinhold JG, Faradji B, Abadi P, Ismail-Beigi F (1976) Decreased absorption of calcium, magnesium, zinc and phosphorus by humans due to increased fiber and phosphorus consumptions as wheat bread. J Nutr 106: 493–503

Ritchey SJ, Korslund MK, Gilbert LM, Fay DC, Robinson MF (1979) Zinc retention and losses of zinc in sweat by preadolescent girls. Am J Clin Nutr 32: 799–803

Robinson MF, McKenzie JM, Thomson CD, van Rij A (1973) Metabolic balance of zinc, cadmium, iron, molybdenum and selenium in young New Zealand women. Br J Nutr 30: 195–212

Ross J, Gibson RS, Sabry JH (1986) A study of seasonal trace element intakes and hair trace element concentrations in selected households from the Wosera, Papua New Guinea. Trop Geogr Med 38: 246–254

Sandberg A-S, Andersson H, Kivistö B, Sandström B (1986) Extrusion cooking of a high-fibre cereal product. I. Effects on digestibility and absorption of protein, fat, starch, dietary fibre and phytate in the small intestine. Br J Nutr 55: 255–260

Sandström B, Cederblad Å (1980) Zinc absorption from composite meals. II. Influence of the main protein source. Am J Clin Nutr 33: 1778–1783

Sandström B, Arvidsson B, Cederblad Å, Björn-Rasmussen E (1980) Zinc absorption from composite meals. I. The significance of wheat extraction rate, zinc, calcium and protein content in meals based on bread. Am J Clin Nutr 33: 739–745

Sandström B, Keen CL, Lönnerdal B (1983) An experimental model for studies of zinc bioavailability from milk and infant formulas using extrinsic labeling. Am J Clin Nutr 38: 420–428

Sandström B, Kivistö B, Cederblad Å (1987) Absorption of zinc from soy protein meals in humans. J Nutr 117: 321–327

Schmitt HA, Weaver CM (1982) Effects of laboratory scale processing on chromium and zinc in vegetables. J Food Sci 47: 1693–1695

Schroeder HA, Nason AP, Tipton IH, Balassa JJ (1967) Essential trace metals in man: zinc. Relation to environmental cadmium. J Chronic Dis 20: 179–210

Schwartz R, Apgar BJ, Wien EM (1986) Apparent absorption and retention of Ca, Cu, Mg, Mn, and Zn from a diet containing bran. Am J Clin Nutr 43: 444–455

Shrimpton R (1984) Food consumption and dietary adequacy according to income in 1,200 families, Manaus, Amazonas, Brazil. Arch Latinoam Nutr 34: 615–629

Soman SD, Panday VK, Joseph KT, Raut SJ (1969) Daily intake of some major and trace elements. Health Phys 17: 35–40

Spring JA, Robertson J, Buss DH (1979) Trace nutrients. 3. Magnesium, copper, zinc, vitamin B_6, vitamin B_{12} and folic acid in the British household food supply. Br J Nutr 41: 487–493

Swanson CA, Turnlund JR, King JC (1983) Effect of dietary zinc sources and pregnancy on zinc utilization in adult women fed controlled diets. J Nutr 113: 2557–2567

Tabekhia MM, Luh BS (1980) Effect of germination, cooking, and canning on phosphorus and phytate retention in dry beans. J Food Sci 45: 406–408

Thurnham DI, Zheng S-F, Munoz N et al. (1985) Comparison of riboflavin, vitamin A and zinc status of Chinese populations at high and low risk for esophageal cancer. Nutr Cancer 7: 131–143

Toma RB, Tabekhia MM (1979) Changes in mineral elements and phytic acid contents during cooking of three California rice varieties. J Food Sci 44: 619–621

Turnlund JR, Michel MC, Keyes WR, King JC, Margen S (1982) Use of enriched stable isotopes to determine zinc and iron absorption in elderly men. Am J Clin Nutr 35: 1033–1040

Turnlund JR, King JC, Keyes WR, Gong B, Michel MC (1984) A stable isotope study of zinc absorption in young men: effects of phytate and α-cellulose. Am J Clin Nutr 40: 1071–1077

Turnlund JR, Durkin N, Costa F, Margen S (1986) Stable isotope studies of zinc absorption and retention in young and elderly men. J Nutr 116: 1239–1247

Valberg LS, Flanagan PR, Chamberlain MJ (1984) Effects of iron, tin and copper on zinc absorption in humans. Am J Clin Nutr 40: 536–541

Van Dokkum W, Wesstra A, Schippers FA (1982) Physiological effects of fibre-rich types of bread. Br J Nutr 47: 451–460

Wada L, Turnlund JR, King JC (1985) Zinc utilization in young men fed adequate and low zinc intakes. J Nutr 115: 1345–1354

Warren HV (1972) Variations in the trace element contents of some vegetables. J R Coll Gen Pract 22: 56–60

Welsh SO, Marston RM (1982) Zinc levels of the US Food supply – 1909–1980. Food Technol 36: 70–76

Chapter 23

Zinc Excess

M.R.S. Fox

Background

In early animal experiments, zinc was shown to have a low order of toxicity. At least some of the results arose from the use of diets containing natural ingredients (sources of phytate and fibre) and high levels of essential nutrients capable of interacting with zinc. In humans, zinc toxicity involved infrequent acute poisoning episodes.

The important functions and therapeutic uses of zinc have been reported in the press, leading to widespread use of supplemental zinc by humans. With the greater understanding of the metabolism of zinc and its interactions with other nutrients that now exists, there is reason to be concerned about the safety of consuming zinc supplements at levels greater than requirement.

Effects of Excess Zinc in Animals

The effects of excess zinc, studied in many domestic and experimental animals, have been reviewed (National Research Council 1979, 1980a; Fox and Jacobs 1986; Hambidge et al.1986). A detailed examination of that extensive literature is beyond the scope of this review.

The types and severity of adverse effects are related to zinc exposure level and duration; animal age, sex, species and nutritional status; and composition of the diet. Food consumption usually declines quickly with high zinc, probably due to unpalatability of the diet. A reduction in growth and/or weight loss follow. Gross abnormalities include rough hair, achromotrichia, pulmonary emphysema, diarrhoea, arthritis, leg paralysis, abortions, non-viable newborn, convulsions and death. Cattle may exhibit pica. A microcytic hypochromic anaemia is common, sometimes with increased erythrocyte fragility and a severe haemolytic anaemia. Other pathology includes fibrosis of the renal cortex and pancreas, fatty liver and necrosis of the liver and areas of the gastrointestinal tract. Total bone minerals are frequently below normal. Cartilage erosion, weak cancellous bone ends and enlargement of the epiphyseal regions of the long bones have been observed.

The earliest and most sensitive effects of excess zinc are changes in tissue minerals, which are frequently useful indices of exposure. Zinc concentration usually increases in the liver, intestinal tissue, pancreas, kidney, testes, heart, bone, milk and hair. Typically there are decreases of copper and iron in the liver and kidney cortex, although there are a few reports of increased iron in the liver and of copper in the spleen and kidney cortex. Manganese decreases in the liver. Serum copper, ceruloplasmin and amine oxidase decline. Hypercholesterolaemia in animals fed excess zinc has been attributed to lowered copper status. Decreased retention of calcium and phosphorus has also been shown. In general, as dietary levels of the elements antagonized by zinc (copper, iron, manganese and calcium) decreased, the adverse effects of excess zinc were exacerbated, whereas higher dietary levels of these elements were protective.

Effects of Excess Zinc in Humans

Acute Effects

Zinc in drinking water produces an undesirable taste that is noticeable at 15 mg/l and is marked at 40 mg/l. Food and beverages contaminated with zinc from galvanized containers produced vomiting and diarrhoea (Brown et al. 1964). Nausea, vomiting and fever occurred in a patient after kidney dialysis at home using water stored in a galvanized tank (Gallery et al. 1972). After several such episodes and rehospitalization, she had severe anaemia and elevated erythrocyte zinc. A 16-year-old male ingested 4 g metallic zinc in peanut butter and the following day consumed 8 g of zinc (Murphy 1970). He became very lethargic and had elevated zinc concentrations in whole blood and serum, but recovered after treatment with dimercaprol. A woman who consumed approximately 28 g zinc sulphate vomited, developed tachycardia and hyperglycaemia and died 5 d later from apparent haemorrhagic pancreatitis and renal damage (Cowan 1947).

Subchronic and Chronic Effects

Direct evidence for zinc antagonism of calcium has been observed in men. Spencer et al. (1982) reported that 140 mg zinc/d given as the sulphate decreased the amounts of calcium-47 (single dose) that appeared in the plasma and increased the excretion of calcium-47 in the faeces. These effects occurred when dietary calcium was low (230 mg/d) but not when it was normal (800 mg/d).

The most extensive evidence for zinc antagonism of another essential element is that for copper. Sandstead (1982) showed by regression analysis that the amount of copper required to achieve balance in men consuming mixed Western diets increased with increasing dietary zinc. Intake levels tested were 5, 10, 15 and 20 mg zinc/d. Sensitive effects of dietary zinc on copper were also reported by Festa et al. (1985). Nine men received a diet of mostly purified components that supplied 2.6 mg copper/d and zinc at 1.8, 4.0, 6.0, 8.0, 18.5 or 20.7 mg/d for 1- or 2-week periods in a 63-d study. An intake of 18.5 mg zinc/d (following a lower zinc intake) produced increased faecal copper, and apparent copper retention was decreased during the 2nd week, but there were no effects on plasma copper. A 27% decrease in apparent copper absorption was observed in elderly subjects fed 2.33 mg copper/d when zinc was increased from 7.8 to 23.3 mg/d (Burke et al. 1981). However, other studies with zinc and copper variations within ranges near requirement have not shown an antagonism of copper by zinc. Some of the zinc variations were small and there may have been unrecognized factors affecting the zinc–copper relationship.

Most other studies of zinc–copper antagonisms involve higher levels of zinc. Fischer et al. (1984) administered two doses daily of 25 mg zinc as the gluconate to 13 normal men and placebo capsules to a control group. After 6 weeks of such treatment, zinc had not affected plasma copper or ceruloplasmin activity; however, copper, zinc-superoxide dismutase activity in erythrocytes decreased and plasma zinc increased.

Large amounts of zinc (75–300 mg/d) over a 2-year period were used to treat a leg ulcer and sickle cell anaemia in an adult male (Prasad et al. 1978). He developed microcytosis, relative neutropenia and low plasma copper and ceruloplasmin. Administration of 1 mg copper/d for 11 d followed by 0.5 mg copper/d reversed these effects. Seven of 13 other sickle anaemia patients who had received zinc therapy for 4–24 weeks had a mean ceruloplasmin level in the "low normal" range. All values returned to normal with supplemental copper.

Severe anaemia has been observed in other patients receiving high levels of zinc. A man who had taken 450 mg zinc/d for 2 years for "prostate trouble" was easily fatigued and on exercising suffered from angina, claudication and dyspnoea (Patterson et al. 1985). His haemoglobin level was 5.1 g/d and the haematocrit was 15.6%. His plasma copper and ceruloplasmin were low and plasma zinc was elevated. He was transfused with 3 units of packed erythrocytes, he discontinued zinc supplements and by day 83 all clinical indices were normal.

Porter et al. (1977) successfully treated coeliac disease with zinc for 6-week periods. One woman who continued to take the 660 mg dose of zinc sulphate/d (presumably 150 mg zinc) for 2 months developed severe hypochromic anaemia, neutropenia and low serum iron and copper levels. Transfusion of 4 units of packed red blood cells and administration of 4 mg copper sulphate/d (presumably 1 mg copper) returned all values to normal by 4 weeks.

Severe anaemia occurred in patients exposed to zinc in water used for dialysis (Petrie and Row 1977). One teenage male had a haemoglobin level of 3 g/dl despite transfusions and his red cell survival time was 3 d. Elimination of excess zinc from the dialysis fluid corrected the anaemia in all cases. The involvement of copper was not investigated.

Evidence of copper deficiency due to excessive zinc treatment was reported by Hambidge et al. (1978) in an infant with acrodermatitis enteropathica. She had low serum copper, neutropenia and radiological evidence of a copper deficiency syndrome.

Changes in serum lipid patterns resembling those shown in copper deficiency have been associated with excess zinc supplement use in humans. Hooper et al. (1980) found that consumption of 80 mg zinc/d as the sulphate for 5 weeks by 12 healthy men decreased plasma high-density lipoprotein-cholesterol. There were no changes in total cholesterol, triglyceride and low-density lipoprotein-cholesterol. Similar results were reported by Chandra (1984), who administered 150 mg zinc as the sulphate twice daily to 11 healthy men for 6 weeks. Serum high-density lipoprotein decreased and low-density lipoprotein increased slightly.

Freeland-Graves et al. (1980) observed no consistent changes in plasma cholesterol or high-density lipoprotein-cholesterol in 32 healthy women given 15, 50 or 100 mg zinc/d as the acetate for 60 d. There was, however, a significant negative correlation between dietary copper and plasma cholesterol.

Chandra (1984) demonstrated that immune responses measured in vitro were also impaired in 11 adult male subjects who received 300 mg zinc/d. He observed reductions in the lymphocyte stimulation response to phytohaemagglutinin and in chemotaxis and phagocytosis of opsonized bacteria by polymorphonuclear leucocytes. It is not known what the in vivo significance of these data is for immune function.

Indian women during the third trimester of pregnancy were supplemented with 100 mg zinc sulphate three times per day (68–122 mg zinc/d, depending on water of hydration). In the first four subjects there were three premature births and one stillbirth (Kumar 1976), presumably from the effects of excess zinc. The study was discontinued. Such indications that a high intake of zinc may not be tolerated readily during the terminal stages of pregnancy are consistent with evidence of high perinatal mortality of the offspring of rats and sheep exposed to high zinc intakes during pregnancy.

Significance of Adverse Effects

The FAO/WHO provisional maximal tolerable intakes of zinc for humans are based on body weight; however, in relation to zinc requirement, they are most restrictive for the infant and least for the adult male (Table 23.1). The adult FAO/WHO value is above some of the levels that produced evidence of reduced copper status described above.

Most of the effects of excess zinc appear to be due to its antagonistic influence on the metabolism of copper, iron and/or calcium. Since intakes and status of each of these elements are low for many individuals, this would exacerbate the effect of a given level of excess zinc. Copper status can be further compromised by elevated intakes of ascorbic acid, fructose, sucrose and some antacids, and by low intakes of protein. The functions of copper in elastin and collagen formation suggest that low copper status may contribute to defective vasculature as well as to the atherogenic processes discussed above. The significance of excess zinc intake and low copper status in relation to ischaemic heart disease has been reviewed (Klevay 1984).

Table 23.1. Requirements and provisional maximum tolerable intakes of zinc for humans

Sex	Age (years)	Require-ment[a] (R) (mg/d)	Safety standard[b] (SS) (mg/d)	SS/R
Both	0.5–1	5	9	1.8
Both	7–10	10	28	2.8
Male	25–50	15	70	4.7

[a] National Research Council (1980b).
[b] Provisional maximum tolerable intakes from food and drinking water, 1 mg/kg body weight (FAO/WHO Expert Committee on Food Additives 1982).

Unresolved Problems

Animal studies are needed to establish the levels and the range of interactions between copper, iron, manganese or calcium, and an excess of zinc that influence human health. Dietary and other factors that affect status of the above elements need to be investigated. These include such substances as ascorbic acid, phytic acid, fibre, carbohydrates and drugs. The short-term effects established for excess zinc need to be investigated in long-term studies at lower levels of zinc intake.

Better indices of nutritional status of humans with respect to zinc, copper, manganese and calcium are of critical importance. Many individuals consume zinc supplements, but we lack adequate criteria for assessing either their beneficial or their harmful effects in population surveys.

Several of the effects observed in animals exposed to excessive intakes of zinc have not been investigated in humans or else there are limited data. Areas that need further study in humans include zinc effects on manganese and calcium, and effects on the immune system, bone and pregnancy, and on the risks of cardiovascular disease. Detailed study of individuals who have taken zinc supplements should be useful. It is also difficult to predict and understand problems in humans until the mechanisms of toxicity are more completely understood.

Acknowledgement. This chapter was written by M.R. Spivey Fox in her private capacity. No official endorsement by the Food and Drug Administration is intended or should be inferred.

References

Brown MA, Thom JV, Orth GL, Cova P, Juarez J (1964) Food poisoning involving zinc contamination. Arch Environ Health 8: 657–660
Burke DM, DeMicco FJ, Taper LJ, Ritchey SJ (1981) Copper and zinc utilization in elderly adults. J Gerontol 36: 558–563

Chandra RK (1984) Excess intake of zinc impairs immune responses. J Am Med Ass 252: 1443–1446

Cowan GAB (1947) Unusual case of poisoning by zinc sulphate. Br Med J I: 451–452

FAO/WHO Expert Committee on Food Additives (1982) Evaluation of certain food additives and contaminants. WHO Tech Rep Ser 683: 14–15, 32–33

Festa MD, Anderson HL, Dowdy RP, Ellersieck MR (1985) Effect of zinc intake on copper excretion and retention in men. Am J Clin Nutr 41: 285–292

Fischer PWF, Giroux A, L'Abbe MR (1984) Effect of zinc supplementation on copper status in adult man. Am J Clin Nutr 40: 743–746

Fox MRS, Jacobs RM (1986) Human nutrition and metal ion toxicity. In: Sigel H (ed) Metal ions in biological systems, vol 20, Concepts in metal ion toxicity. Marcel Dekker, New York Basle, pp 201–228

Freeland-Graves JH, Han W-H, Friedman BJ, Shorey RAL (1980) Effect of dietary Zn/Cu ratios on cholesterol and HDL-cholesterol levels in women. Nutr Rep Int 22: 285–293

Gallery EDM, Bloomfield J, Dixon SR (1972) Acute zinc toxicity in haemodialysis. Br Med J 4: 331–333

Hambidge KM, Walravens PA, Neldner KH, Daugherty NA (1978) Zinc, copper and fatty acids in acrodermatitis enteropathica. In: Kirchgessner M (ed) Trace element metabolism in man and animals, Tema-3. Institut für Ernährungsphysiologie Technische Universität München, Freising-Weihenstephan, pp 413–417

Hambidge KM, Casey CE, Krebs NF (1986) Zinc. In: Mertz W (ed) Trace elements in human and animal nutrition, vol 2. Academic Press, New York, pp 1–137

Hooper PL, Visconti L, Garry PJ, Johnson GE (1980) Zinc lowers high-density lipoprotein-cholesterol levels. J Am Med Ass 244: 1960–1961

Klevay LM (1984) The role of copper, zinc, and other chemical elements in ischemic heart disease. In: Rennert OW, Chan W-Y (eds) Metabolism of trace metals in man, vol 1. CRC Press, Boca Raton, pp 130–157

Kumar S (1976) Effect of zinc supplementation on rats during pregnancy. Nutr Rep Int 13: 33–36

Murphy JV (1970) Intoxication following ingestion of elemental zinc. J Am Med Ass 212: 2119–2120

National Research Council (1979) Zinc. University Park Press, Baltimore, pp 249–268

National Research Council (1980a) Mineral tolerance of domestic animals. National Academy of Sciences, Washington, pp 3–7, 553–577

National Research Council (1980b) Recommended dietary allowances. National Academy of Sciences, Washington

Patterson WP, Winkelmann M, Perry MC (1985) Zinc-induced copper deficiency: megamineral sideroblastic anemia. Ann Intern Med 103: 385–386

Petrie JJB, Row PG (1977) Dialysis anaemia caused by subacute zinc toxicity. Lancet I: 1178–1180

Porter KG, McMaster D, Elmes ME, Love AHG (1977) Anemia and low serum-copper during zinc therapy. Lancet II: 774

Prasad AS, Brewer GJ, Schoomaker EB, Rabbani P (1978) Hypocupremia induced by zinc therapy in adults. J Am Med Ass 240: 2166–2168

Sandstead HH (1982) Copper bioavailability and requirements. Am J Clin Nutr 35: 809–814

Spencer H, Kramer L, Norris C, Osis D (1982) Effect of the calcium intake on the inhibitory action of zinc on calcium absorption. Am J Clin Nutr 35: 816 (abstract)

Chapter 24

The Biological Significance of Zinc for Man: Problems and Prospects

C.F. Mills

This final chapter is not intended to provide a comprehensive overview of the topics discussed previously. Instead it will highlight those topics for which a better understanding is essential before the significance of zinc for human health and wellbeing can be fully assessed. It will also consider some troublesome controversies.

The Functional Roles of Zinc

As with other trace elements and micronutrients, discovery of the essentiality of zinc was followed by an extensive series of accounts of studies of purified enzymes for which either intrinsic zinc plays a functional role or extrinsic zinc modifies activity.

The list of more than 200 enzymes now known to contain zinc is impressive and, as indicated in Chap. 2, the range of their functions is extensive (Hambidge et al. 1986). However, the most notable feature of our newer biochemical understanding of the roles of zinc is not so much the variety of its functions as an essential catalytic component of such enzymes but rather its role as a structural constituent of enzymic and non-enzymic proteins and probably of polynucleotides. Zinc appears either to confer structural and metabolic stability or govern the tertiary architecture and thus the biological function of a wide range of macromolecules. The first suspicions of such an organizational, as well as a catalytic, role for zinc, emerged from investigations of the functions of zinc in alcohol dehydrogenase, within which enzyme it is possible to differentiate zinc associated with the functional, catalytically active centre from the zinc, less firmly bound, which stabilizes the polymeric structure of the enzyme. Such an organizational role is apparent for a growing range of other proteins (see Morisawa and Mohri 1972; Hesketh 1981; Williams 1984). Typical of the diversity of such roles is evidence that differences in zinc status of the donor tissue markedly influence the in vitro stability of α-mannosidase, and the more recent finding that zinc determines receptor-site structure of the enzyme transcriptase factor IIIA (as considered in Chaps. 2, 5 and 7). It will be surprising if evidence for such structural roles does not increase substantially in the near future.

The tacit neglect of the major proportion of body zinc associated with tissue proteins that relegates it virtually to the category of a trivial adventitious contaminant is inconsistent with the behaviour of the element at times when growth rates change or tissue integrity is prejudiced. Thus, it is well established that tissue catabolism provoked for example by protein/energy deficiency or calcium deficiency can be the stimulus to release of sufficient zinc to ameliorate or preclude development of pathological responses to a low zinc intake (Masters et al. 1986). Conversely, the anabolic response to nutritional rehabilitation or following tissue injury not only depletes the plasma pool of zinc but frequently provokes the first appearance of clinical signs of deficiency. The complexity of equilibria governing such relationships is evident from the initial response to zinc repletion of the severely deficient infant (as described for example by Golden and Golden 1981). The resumption of growth following zinc therapy can be accompanied, paradoxically, by a fall in circulating plasma zinc. Consistent and significant decreases in plasma zinc have also been noted recently when zinc-depleted mature subjects showing no clinical signs were first returned to diets normal in zinc content (C. Bosworth, personal communication). Such observations are incompatible with any concept that zinc is merely an exchangeable, mobile constituent of the increment of new tissue synthesized in response to an improved zinc status. More likely, zinc stimulates protein synthesis – perhaps indirectly – while also modifying tertiary protein structure in a way which promotes more complete exposure of zinc-binding ligands in association with which the element serves its structural role(s).

If zinc has such extensive structural roles rather than being present merely as a "pollutant" of the mass of body protein, it is to be expected that a decline in body zinc status might be accompanied by an acceleration of protein turnover. Many studies have been made of the effects of zinc deficiency on protein synthesis; of these, many have given equivocal results and few have been accompanied by corresponding studies of protein degradation rate. Thus, in the context of our suggestion that zinc may have a stabilizing role it is notable that among the very few studies of effects of zinc deficiency on protein turnover, one with rats (Giugliano and Millward 1987) and one very recently with a zinc-deficient infant (P.J. Aggett and T. Stack, personal communication) indicate that a decline in zinc status is accompanied by marked increases in protein turnover rate. Such effects require confirmation and thus their full significance is not yet known. However, it is clear that such a metabolic lesion might well account for the particular rapidity of responses to zinc depletion seen in young growing subjects (or tissues). It would also limit expression of growth potential.

Clinical Manifestation of Zinc Deficiency

The great diversity of the known functions of zinc, greater than for any other inorganic nutrient, certainly justifies its case for essentiality but may also account for the frequent difficulty in identifying specific pathognomonic features of deficiency among an abundance of other lesions.

Chapters 15–18 have described the clinical manifestations of zinc deficiency in experimental animals and human subjects. Chapter 19 has described many other clinical conditions for which there is clear evidence of deranged zinc metabolism or evidence that zinc has prophylactic or therapeutic value. The range and the nature of such conditions is substantial, probably reflecting both the multiplicity of metabolic roles of zinc and the fact that suboptimal supply becomes most clearly apparent when tissue growth or repair is normally rapid or has accelerated following tissue damage. This point has often escaped attention when the pathology of zinc deficiency in man is being considered. It is very clearly evident from work with all other mammalian species that zinc deficiency is a deficiency disorder most clearly manifest during growth, whether "growth" is defined in terms of the whole organism, an individual organ or a specific cell population. In the absence of the stimulus of tissue growth or repair, clinical signs of deficiency may often not be evident. For example, during many years work on zinc deficiency at the Rowett Institute it has become clear that the development of the skin lesions of zinc deficiency in experimental animals is provoked very strongly by minor topical surgery but is ameliorated by the transfer of experimental animals to "minimal disease" housing conditions with rigorous exclusion of ectoparasites which, previously, have caused minor skin damage. Such lesions are strongly exacerbated if topical damage occurs as a consequence of abrasion or skin infection.

One neglected aspect of pathological responses to zinc depletion is that such effects can either be general, for example influencing the growth of the entire subject, or can be highly localized at specific sites of tissue injury or repair. Both from studies on the efficacy of zinc in promoting the healing of leg and foot infections in farm animals (Demertzis and Mills 1973) and from investigations of its effectiveness in the treatment of chronic leg ulcers in human subjects (Husain 1969) it is evident that establishment of a zinc-responsive lesion is certainly not contingent upon the development of a generalized (whole-body) deficiency of zinc. It is difficult to identify the zinc-responsive populations in either of the above situations from plasma zinc measurements or estimates of zinc intake. Zinc responsiveness in either is certainly not contingent upon the presence of cutaneous lesions of zinc deficiency at sites other than those subject to chronic infection. The logical conclusion is that such chronic cutaneous infections with their enhanced demands for protein synthesis create conditions in which the entry of zinc into the local area of repair is insufficiently rapid to maintain tissue repair rates. Claims that such zinc-responsive conditions existed when plasma zinc data suggested that zinc status was normal or, at worst, only "marginal", were initially greeted with strong scepticism. Now we are beginning to appreciate that such situations reflect the inadequacy of existing biochemical criteria for anticipating when a local increase in zinc demand cannot be met by endogenous zinc.

Biochemical Criteria of Zinc Status

Detection of a suboptimal zinc status on the basis of clinical evidence is frequently difficult. Chapters 18 and 20 have indicated the frequently non-specific character of clinical signs of zinc deficiency. Chapter 19 indicated the wide variety of conditions for which beneficial effects from the administration of zinc have been claimed. Such diversity in pathological response complicates diagnosis. The situation is rendered more complex by limitations to the biochemical diagnostic criteria currently available. There is no doubt that changes in plasma zinc levels in experimental animals maintained under infection-free and regulated environmental circumstances while given constant diets, rapidly reflect any change in overall zinc status. Under the same conditions, hair zinc reflects longer term changes. However, as indicated above and in Chaps. 1 and 20, plasma zinc can be highly variable. It increases in response to tissue catabolism and declines when growth resumes after a dietary amino acid deficiency is corrected. It declines rapidly in response to infection or stress or, surprisingly, if growth is stimulated by a modest increase in zinc supply. The detection of zinc-responsive acrodermatitis enteropathica in some subjects in whom plasma zinc is normal (Garretts and Molokhia 1977) causes additional diagnostic difficulties. The latter exception apart, there can be little doubt that few cases of suboptimal zinc status would escape detection if repetitive blood analysis indicated low plasma zinc concentrations. Unfortunately such repetitive sampling is rarely feasible in population studies and thus the search for alternative, more specific, biochemical criteria is being urgently pursued.

Immunoassay of the protein metallothionein in blood constituents (Bremner and Morrison 1986) or urine may offer considerable diagnostic advantages over measurement of plasma zinc levels. The technique exploits the importance of zinc both as a primer and a constituent of zinc metallothionein. The presence of metallothionein in blood or tissues indicates that sufficient zinc has entered cells to permit metallothionein synthesis. In this respect, its presence is more informative than merely the presence of zinc in plasma. Nevertheless, it must be anticipated that metallothionein analysis may still have practical limitations as a "marker" of zinc status. Principal among these is the probability that no direct relationship exists between metallothionein concentrations and any functional defect; it is a "storage" metalloprotein with, at present, no known metabolic role and no immediate relevance to any fundamentally essential step in cell metabolism or function. Metallothionein analysis could well be supplanted as a diagnostic aid once a more functionally relevant and pathologically sensitive marker is discovered on the basis of growing knowledge of the functions of zinc in tissues. Despite this, monitoring of changes in zinc metallothionein status has interim value both for surveys of population zinc status and as a sensitive criterion less readily influenced than plasma zinc by concurrent infection.

Whatever the future may hold for the development of improved biochemical criteria of zinc status, unequivocal diagnosis of zinc-responsive disorders is always feasible by the expedient of increasing zinc supply. Tolerance of zinc is high and with the exception of suspicions as to its toxicity during the later stages of pregnancy, it is relatively innocuous. However, few, if any, of the authors contributing to this volume would subscribe to the view occasionally voiced in the "popular" medical press that "if a little doesn't improve things try more . . . and more"! If zinc status is sufficiently low to cause pathological changes, restitution of zinc supply by supplements providing up to twice the daily requirement for zinc is normally sufficient to produce rapid recovery of growth or repair of skin lesions even when deficiency has been chronic and severe. Furthermore, behavioural changes are rapidly evident in the acrodermatitis child treated with zinc, and defective cell-mediated immunocompetence induced by zinc deficiency is restored to normal shortly after initiating zinc repletion. Even more impressive, but perhaps of less relevance to man, is the elimination of inappetance in zinc-depleted rats within 2 h of commencing dietary repletion. The biochemical origins of this and other behavioural effects of zinc depletion (see Chaps. 11, 14 and 15) may not yet be explicable and, certainly, many are not solely and specifically influenced by zinc supply. Nevertheless, behavioural changes observed after modifying zinc intake can facilitate a rapid if provisional diagnosis.

Zinc Utilization and Zinc Requirements

Eventually it will be possible to assess the population significance for health of anomalies in zinc supply from estimation of the intake of absorbable dietary zinc and its relationship to zinc requirements and tolerance. Progress towards this objective is certainly being made. However, to those formulating dietary policies our present efforts frequently appear like the first faltering footsteps of a child learning to walk! The need for improved definition of such relationships is particularly urgent for zinc for which the pathological sequelae of deficiency are often non-specific and biochemical diagnostic criteria have substantial limitations.

In Chap. 21, Janet King and Judith Turnlund reviewed past work on the estimation of human zinc requirements and illustrated some of the achievements and difficulties in this field. Whatever experimental techniques are used to estimate requirements it is inevitable that the protocols selected must involve long periods on experimental diets, with monotonous experimental conditions and the continual need for rigorous quality control of sampling and analytical procedures. Ethical considerations limit the possibility of investigations with the subjects at greatest risk and therefore of greatest interest, namely, infants and pregnant females.

Although these difficulties are not unique to work with zinc, they have had a marked effect both on the nature and the validity of the estimates of zinc requirements and the dietary allowances suggested for zinc. With zinc, as with iron, the derivation and use of estimates of "requirements" has often been fraught with technical problems and semantic uncertainties.

It is abundantly clear from the studies with experimental animals and from those with human subjects described in Chaps. 1, 3 and 21 that the efficiency with which orally administered zinc is utilized is related, inversely, to the zinc status of the subject. Furthermore, maintenance of a normal or supranormal zinc status is accompanied by increased endogenous losses. Since, overall, the deficient or marginally deficient subject absorbs and retains zinc with greater efficiency than the zinc-replete subject, it is clear that most estimates of zinc requirement need qualification as to their applicability to subjects differing widely in zinc status.

Thus, the diagnostician or epidemiologist may need an estimate of requirement which defines the minimum supply of absorbed nutrient consistent with long-term maintenance of physiological normality and tissue integrity. The supply of absorbable zinc meeting this "functional" or "basal" requirement would merely be sufficient to maintain normal functions and freedom from pathological change, and to maintain the equilibria associated with the probably lower than "normal" tissue concentrations of zinc that nonetheless are consistent with health. By definition, it would not confer protection against future, if transient, zinc depletion since it offers no surplus for tissue storage.

Alternatively, the preferred form of statements of "requirement" for zinc may describe the supply of absorbable zinc which not only establishes and maintains freedom from the pathological effects of deficiency but also maintains a zinc status conferring an acceptable measure of protection against transient periods of low zinc intake.

It has become common practice to describe such estimates incorporating margins of safety as "allowances" rather than "estimates of requirement". However, the very techniques whereby the data for such estimates are obtained make it questionable whether such a distinction is either real or justifiable. Unless relationships between zinc intake and physiological normality are based upon overt or covert pathological criteria (as when defining the *functional* requirement), the alternatives are to use input/output relationships or factorial approaches to determine requirements. Unfortunately, both of these alternatives are influenced by changes in the equilibria which control the total pool of exchangeable body zinc. Thus any estimate of requirement based thereon has a validity restricted to subjects of similar zinc status.

Discussing the use and misuse of balance studies in this context, Mertz (1987) has emphasized that the typical balance study does not yield data suitable for determining *the* requirement of a mineral element except for situations in which requirement is defined as the intake needed to maintain the existing pool size. For the purposes of the present discussion, this conclusion does not detract from the value of balance data provided that the pool size of utilizable and exchangeable zinc is defined during the balance studies and the requirement estimate is regarded as applicable only in the context of subjects of similar zinc status. Many balance studies have lacked this vital dimension. These studies have certainly contributed to our appreciation of the effectiveness of homeostatic control of zinc metabolism but have not greatly improved our ability to define requirements needed to maintain body "reserves" of zinc.

There is no question as to the value of defining needs for zinc in terms of such a "*nominal storage requirement*". However, progress towards this objective is contingent upon agreement as to the desirable size and kinetic characteristics of the body zinc compartments which it is desirable to maintain in the "normal" healthy population. It will also be contingent upon a wider application of isotopic techniques used for determining the turnover rate and size of exchangeable zinc pools such as that typified by the work of Jackson et al. (1984). The term "nominal" will remain applicable until such time as we can identify with certainty the component of tissue zinc whose availability most closely reflects the maintenance of normal physiological, zinc-dependent processes.

Failure to distinguish between these "functional" and "storage" options in defining requirements for zinc is leading to controversies and discrepancies. Their quantitative differences are substantial and not only reflect the need to replace the higher endogenous losses associated with the maintenance of a higher-than-minimal zinc status, but also are influenced by the lower fractional efficiency of absorption associated with the increased zinc flux needed to meet this higher demand.

The fact that estimates of the need for zinc often differ so markedly reflects not the technical incompetence of investigators, but rather that the conditions under which such estimates have been derived were different. Thus, unless experimentation has been preceded by depletion, the data derived from balance experiments on normal subjects are appropriate only for the derivation of *nominal storage requirements*. However, since we often lack facilities to define the precise zinc status of such ostensibly "normal" subjects it is a fruitless exercise to argue why the derived estimates of requirement so often differ.

Chapters 17, 18 and 20 summarize the results of the very few balance studies on subjects depleted of zinc sufficiently to induce early covert pathological manifestations of deficiency. These are the few observations demonstrating remarkably low endogenous zinc losses and high absorptive efficiencies from which estimates of the *functional requirement* for zinc in human subjects can be derived. It is clear that ethical considerations will always preclude deliberate investigation of such functional requirements during infancy, adolescence and pregnancy. Apart from the opportunities for biochemical definition of pathological responses occurring in subjects suffering zinc depletion by accident or neglect, it is likely that many provisional estimates of the functional requirements for zinc may have to be derived by extrapolation from non-human species.

The errors that may be introduced by such interspecies extrapolation may be less serious than currently imagined, and appreciably less so than, for example, the major discrepancies in estimates of dietary requirements for zinc during pregnancy in the human female (see Chap. 12). Thus, Armstrong (1986) and Swanson and King (1987) have pointed to the fact that successful pregnancy, normal foetal development and successful lactation is frequently achieved by human females consuming less than half the estimated minimum daily requirement for zinc in pregnancy suggested by the WHO (1973) and many other agencies. The cause of this apparent anomaly may well be failure to apply relevant evidence of changes in zinc metabolism during pregnancy provided by studies with laboratory animals. These indicate that pregnancy is associated with increased efficiencies of zinc absorption, and possibly with a reduction of endogenous losses as foetal zinc demands increase. Zinc demands for lactation are met largely by corresponding changes in the efficiency of zinc utilization. Our gross overestimates of dietary zinc requirements for the pregnant and lactating woman probably reflect our reluctance to apply findings already evident from studies with rats, sheep and laboratory primates.

In Chap. 21 and in an additional review (Swanson and King 1987) it has been suggested that such adaptation may well occur in the pregnant human female but has escaped detection because of limitations to the experimental protocols adopted in studies of stable zinc isotope absorption during late pregnancy. Whatever the explanation, it is clear that a substantial revision of our estimates either of requirements for absorbed zinc or, more likely, of the efficiency with which zinc is absorbed and retained, is needed before predictions and practice can be reconciled.

In contrast to suggestions which arise from use of many current estimates of requirement, vast populations of pregnant women in the Western world are not zinc deficient; adolescent populations may well be at risk from zinc deficiency but not to the degree that misinterpretation of experimental data derived from balance studies would lead us to suggest. Prediction of risks requires previous investigation with subjects at risk. Many of our data from which requirements have been estimated have been derived from normal subjects not called upon to exercise "zinc economy" but encouraged to maintain their zinc status by eating diets unquestionably well supplied with zinc. Not surprisingly, the estimates of "requirement" so derived are of limited value to those assessing relationships of zinc supply to health and morbidity.

Prediction of zinc requirements by "factorial" techniques frequently involves the summation of terms describing the accretion of zinc in foetal tissues and adnexa or lost to the body through the output in milk, other secretions or squames. Surprisingly, the influence of such processes on endogenous losses of zinc is unknown and, almost by default, it is frequently assumed that the zinc required is used, uniquely, with 100% efficiency. Until the "inefficiency" component of such processes has been quantified, it may well be prudent to identify as a *fundamental requirement* the minimal absolute quantities of zinc that must be incorporated into these products of tissue growth or essential secretory activity.

While this volume is not the place for more extensive discussion of such terminological problems it is worth questioning whether any other branch of biological or physical science has been endowed so liberally with imprecise or potentially confusing definitions, the misuse of which has been so extensively discussed but so ineffectively restrained. It is a problem that has arisen in the context of defining requirements for a wide range of inorganic nutrients.

Zinc "Availability" from Foods

Sandström's chapter "Dietary Pattern and Zinc Supply" in this volume strikingly illustrates the frequent disparity between total zinc content of a meal and its content of potentially utilizable zinc. A disquieting feature of the data presented is the fact that, with few exceptions, our understanding of the variables influencing the efficiency with which dietary zinc can be utilized from foods or meals is insufficient to explain many observed differences or predict others.

As indicated in Chap. 4, there are clear indications that soluble phytate, if consumed with high levels of dietary calcium, is a powerful inhibitor of zinc absorption. However, as pointed out by Forbes and Erdman (1983), the potency of phytate as a zinc antagonist is influenced markedly by the physical nature of its association with food proteins. The significance of other processes such as the release of zinc from insoluble zinc–calcium phytate complexes by amino-acid-rich media such as intestinal contents has been considered by Wise (1983). From work with the relatively high-calcium diets of laboratory rodents, pigs and poultry there is universal agreement that calcium potentiates the phytate/zinc antagonism (Mills 1985). The extent to which the substantially lower calcium content of most human diets provokes a similar phytate/zinc antagonism has not been explored adequately. Nor have the significance for zinc absorption of the growing practice of dietary calcium fortification or exclusive consumption of phytate-rich but relatively low-calcium vegan diets. Overall, there is substantial scope for investigation of the processes responsible for the major differences in zinc utilization illustrated in Chap. 22. Until we understand such mechanisms it will remain difficult to predict the influence of changed patterns of food sources and food processing upon the potentially available zinc content of typical Western diets.

It is probable that the greater dietary dependence upon phytate-rich cereals in the developing countries may be a much more important cause for concern. The addition of calcareous supplements during cooking and the practice of geophagia of calcareous soils establish all the conditions required to potentiate the action of phytate as a zinc antagonist.

In such situations, as in the use of phytate-rich protein isolates in infant milk substitutes or as meat replacers, there is a need for vigilance to ensure that supplies of physiologically available zinc are maintained. At present, however, there are insufficient data from human studies to establish dietary guidelines for acceptable levels of phytate consumption.

It is important to appreciate that the data of Chap. 22 describing the absorption of zinc from individual meals are derived from studies with experimental subjects that were of adequate zinc status. For reasons considered elsewhere, we must anticipate that higher estimates of the fractional efficiency of zinc utilization might apply if the same foods were consumed by rapidly growing children or by pregnant or lactating subjects with high zinc demands. While this reservation must be considered when interpreting data on zinc bioavailability, it is open to question or, better, to investigation, whether this "adaptation" is possible in zinc-deficient subjects consuming phytate-rich diets. This doubt arises firstly from evidence that, in rats, intestinal alkaline phosphatase requiring zinc for its functional activity is identical to intestinal phytase, and secondly, from the evidence that rat intestinal phytase activity declines during zinc deficiency (Davies and Flett 1978). If these enzymes are also identical in man, the finding that intestinal alkaline phosphatase activity declines in some human subjects when zinc-depleted (Baer et al. 1985) suggests that such individuals may be particularly at risk if offered high-phytate diets.

Conclusion

The contributions to this volume have indicated both the very wide range of functions of zinc and the multitude of pathological changes possible if zinc deficiency develops. Despite a more extensive knowledge of the biochemical roles of zinc than for any other mineral element, we still lack a clear understanding of the metabolic origins of the pathological changes caused by zinc shortage. At present, very regrettably, the best diagnostic criterion of a suboptimal zinc status is to administer zinc and monitor the physiological effects of such action. Despite such diagnostic problems, there are clear indications of the importance of maintaining an adequate zinc status in subjects whose requirements are likely to be high.

Estimates of zinc requirements have been difficult to derive and failure to appreciate their past limitations has led to their misuse. Clearer definition of the objectives and validity of such estimates is now becoming possible and should do much to restrain those who interpret their data as showing zinc deficiency in a major portion of Western societies. Sceptical reactions to such attitudes have unfortunately led to neglect of some situations under which the maintenance of growth, health and wellbeing are limited by inadequacies in zinc supply, by inadvertent selection of dietary ingredients from which zinc is poorly available or by excessive zinc demands during infection or recovery from major tissue damage.

Our most pressing need is for improved biochemical criteria both for diagnostic purposes and to monitor dose/response relationships during studies of the functional requirements for zinc. In the interval before such ideal diagnostic procedures have been developed, much can be done by the concurrent monitoring of changes in plasma or serum of zinc and either alkaline phosphatase (Baer et al. 1985; Weismann and Høyer 1985) or plasma zinc-metallothionein (Bremner and Morrison 1986). Used with care and in association with our growing knowledge of situations which exacerbate the risks of zinc deficiency, such multiple criteria could eliminate many of our present uncertainties as to the significance of zinc for the health and development of man.

It is hoped that the critical reviews contained in this volume will have indicated both the importance of zinc as a micronutrient and the challenges it presents in our efforts to clarify understanding of the circumstances under which a deficiency or excess of zinc can constitute a threat to health.

Acknowledgement. I am indebted to Dr. C. Casey and Miss C. Bosworth for helpful discussions during the preparation of this chapter.

References

Armstrong J (1986) Trace element metabolism in human pregnancy. M Phil Thesis, University of Aberdeen

Baer MT, King JC, Tamura T et al. (1985) Nitrogen utilization, enzyme activity, glucose intolerance and leukocyte chemotaxis in human experimental zinc depletion. Am J Clin Nutr 41: 1220–1235

Bremner I, Morrison JN (1986) Assessment of zinc, copper and cadmium status in animals by assay of extracellular metallothionein. Acta Pharmacol Toxicol (Copenh) [Suppl 7] 59: 502–509

Davies NT, Flett AA (1978) The similarity between alkaline phosphatase [EC 3.1.3.1] and phytase [EC 3.1.3.8] activities in rat intestine and their importance in phytate-induced zinc deficiency. Br J Nutr 39: 307–316

Demertzis PN, Mills CF (1973) Oral zinc therapy in the control of infectious pododermatitis in young bulls. Vet Rec 93: 219–222

Forbes RM, Erdman JWQ (1983) Bioavailability of trace mineral elements. Annu Rev Nutr 3: 213–231

Garretts M, Molokhia M (1977) Acrodermatitis enteropathica without hypozincaemia. J Pediatr 91: 492–494

Giugliano R, Millward DJ (1987) The effects of severe zinc deficiency on protein turnover in muscle and thymus. Br J Nutr 57: 139–155

Golden MHN, Golden BE (1981) Effects of zinc supplementation on dietary intake, rate of weight gain and energy cost of tissue deposition in children recovering from severe malnutrition. Am J Clin Nutr 34: 900–908

Hambidge KM, Casey CE, Krebs NF (1986) Zinc. In: Mertz W (ed) Trace elements in human and animal nutrition, vol 2, 5th edn. Academic Press, New York, pp 1–137

Hesketh JE (1981) Impaired microtubule assembly in brain from zinc deficient pigs and rats. Int J Biochem 13: 921–926

Husain SL (1969) Oral zinc sulphate in the treatment of leg ulcers. Lancet I: 1069–1071

Jackson MJ, Jones DA, Edwards RHT, Swainbank IG, Coleman ML (1984) Zinc homeostasis in man: studies using a new stable isotope technique. Br J Nutr 51: 199–208

Masters DG, Keen CL, Lönnerdal B, Hurley LS (1986) Release of zinc from maternal tissues during zinc deficiency or simultaneous zinc and calcium deficiency in the pregnant rat. J Nutr 116: 2148–2154

Mertz W (1987) Use and misuse of balance studies. J Nutr 117: 1811–1813

Mills CF (1985) Dietary interactions involving the trace elements. Annu Rev Nutr 5: 173–193

Morisawa M, Mohri H (1972) Heavy metals and spermatozoa motility. Exp Cell Res 70: 311–316

Swanson CA, King JC (1987) Zinc and pregnancy outcome. Am J Clin Nutr 46: 763–771

Weismann K, Høyer H (1985) Serum alkaline phosphatase and serum zinc in the diagnosis and exclusion of zinc deficiency in man. Am J Clin Nutr 41: 1214–1219

WHO (1973) Trace elements in human nutrition. Report of a World Health Organization Expert Committee. Tech Rept Ser No 532, WHO, Geneva

Williams RJP (1984) Zinc: what is its role in biology? Endeavour (New Series) 8: 65–70

Wise A (1983) Dietary factors determining the biological activity of phytate. Nutr Abstr Rev Rev Clin Nutr 53: 791–806

Subject Index